International Series in Operations Research & Management Science

Volume 345

The book series **International Series in Operations Research and Management Science** encompasses the various areas of operations research and management science. Both theoretical and applied books are included. It describes current advances anywhere in the world that are at the cutting edge of the field. The series is aimed especially at researchers, advanced graduate students, and sophisticated practitioners.

The series features three types of books:

• Advanced expository books that extend and unify our understanding of particular areas.

• Research monographs that make substantial contributions to knowledge.

• Handbooks that define the new state of the art in particular areas. Each handbook will be edited by a leading authority in the area who will organize a team of experts on various aspects of the topic to write individual chapters. A handbook may emphasize expository surveys or completely new advances (either research or applications) or a combination of both.

The series emphasizes the following four areas:

Mathematical Programming : Including linear programming, integer programming, nonlinear programming, interior point methods, game theory, network optimization models, combinatorics, equilibrium programming, complementarity theory, multiobjective optimization, dynamic programming, stochastic programming, complexity theory, etc.

Applied Probability: Including queuing theory, simulation, renewal theory, Brownian motion and diffusion processes, decision analysis, Markov decision processes, reliability theory, forecasting, other stochastic processes motivated by applications, etc.

Production and Operations Management: Including inventory theory, production scheduling, capacity planning, facility location, supply chain management, distribution systems, materials requirements planning, just-in-time systems, flexible manufacturing systems, design of production lines, logistical planning, strategic issues, etc.

Applications of Operations Research and Management Science: Including telecommunications, health care, capital budgeting and finance, economics, marketing, public policy, military operations research, humanitarian relief and disaster mitigation, service operations, transportation systems, etc.

This book series is indexed in Scopus.

Louis Anthony Cox Jr.

AI-ML for Decision and Risk Analysis

Challenges and Opportunities for Normative Decision Theory

Springer

Louis Anthony Cox Jr.
Cox Associates and University of Colorado
Denver, CO, USA

ISSN 0884-8289 ISSN 2214-7934 (electronic)
International Series in Operations Research & Management Science
ISBN 978-3-031-32012-5 ISBN 978-3-031-32013-2 (eBook)
https://doi.org/10.1007/978-3-031-32013-2

This Springer imprint is published by the registered company Springer Nature Switzerland AG
The registered company address is: Gewerbestrasse 11, 6330 Cham, Switzerland

To

Ron and Pam for years of friendship and inspiration

Mike, Warner, and Karen for many fruitful and enjoyable collaborations

Preface

Artificial intelligence (AI) and machine learning (ML) are shedding new light on an old problem of vital practical importance: How can and should intelligent agents—either human or machine—learn to act effectively in uncertain environments, or when the consequences of different choices are uncertain? This book is about how people, machines, teams, organizations, communities, and societies can, do, and should make decisions under uncertainty. It is also about how they can apply insights and principles from AI-ML to learn to make such decisions better. Much of the inspiration for the book comes from a conviction that ideas that have proved successful in helping AIs to deal with realistically complex, uncertain, and rapidly evolving situations—whether winning a robot soccer competition, maintaining the formation and purposeful coordinated actions of a swarm of drones flying safely through urban spaces despite wind gusts and other hazards, or monitoring and adjusting the numerous linked processes in a refinery or chemical plant to increase productivity, energy efficiency, and safety—can also be applied to help improve human decision-making and AI-assisted organizational responses to challenges that arise in managing human businesses, conflicts, teamwork, and policy-making processes.

For well over half a century, traditional decision analysis has provided an elegant mathematical framework for prescribing what a single rational decision-maker should do when the consequences of different choices are uncertain. This framework, introduced in Part I of this book and critically examined in Part 2, includes normative decision theory axioms implying that choices should maximize subjective expected utility (SEU) (see Chap. 7). It includes generalizations to deal with uncertain probabilities ("ambiguity"), incompletely defined preferences and risk attitudes, and beliefs that are modified over time by conditioning (typically via Bayes' rule) on new information as it becomes available. Decision psychology, risk analysis, behavioral economics, and related fields introduced in Part 1 have also studied how real people (and animals) make decisions and how their decision-making departs from the normative prescriptions of decision analysis and SEU theory. For example, many decisions over-weight near-term gains or losses relative to somewhat arbitrary reference points instead of focusing on long-term outcomes and making current choices that will reduce predictable future regret.

Part 2 of this book reviews psychological and technical obstacles and fundamental theoretical challenges to traditional decision analysis. These range from practical constraints on the ability of real people to form coherent preferences and beliefs and to obtain and process information needed to make causally effective decisions (Chap. 6) to more foundational issues such as the non-existence of well-defined coherent beliefs and preferences for groups of people or AI agents (Chap. 5) and the algorithmic undecidability of many questions about risk in complex systems (Chap. 4). Part 3 turns to possible ways forward drawing on AI-ML. To the well-established normative principles and descriptive insights of decision analysis and decision and risk psychology, AI-ML has recently added a host of new ideas and techniques. Many of these ideas were originally developed and refined in the context of engineering applications such as enabling autonomous vehicles, drone swarms, industrial robots, AI controllers of industrial processes, AIs in video games, and a wide variety of other AIs to take appropriate actions—or, for advisory systems, to make appropriate decision recommendations to humans—to help achieve preferred outcomes, or at least make them more likely, even in realistically uncertain and rapidly changing environments. They include methods for acquiring and improving needed skills (e.g., using reinforcement learning, imitation learning, and other types of ML introduced in Part 1 and discussed further in Part 2) so that actions can be taken effectively if the need arises; allowing for the realistic possibility that intended and attempted actions will be interrupted or abandoned before they are successfully completed; recognizing that not all possible outcomes and future situations can reasonably be anticipated, which implies that being able to cope with novel situations and adjust plans and goals quickly is often important; and using causal artificial intelligence (CAI) algorithms and ML to explore new environments and learn about how actions affect outcome probabilities even if this causal knowledge is not initially available.

Part 4 illustrates several applications of these AI-ML and CAI concepts and methods to public health problems. Developing causally effective regulations, policies, and interventions to protect and promote human health and safety is often challenging because of uncertainties about how people will respond and how changes in exposures or behaviors will affect health outcomes. Part 4 argues that both well-developed AI-ML and CAI principles (Chaps. 10–12) and new opportunities created by large language models (Chap. 13) have great potential to improve the clarity, technical quality, credibility, and practical value of public health and epidemiological risk assessments and risk management decision recommendations.

This book is written for readers curious about how AI-ML addresses decision problems and about how these ideas and methods can be used to improve human decision-making. It addresses topics at the intersection of many specific disciplines that are concerned with various aspects of decision science, individual and organizational learning, planning, data analytics, statistical and causal inference, and optimization. To avoid presupposing any specific disciplinary background or expertise, we introduce key concepts and background in Part 1 and then discuss challenges, possible constructive solutions based on AI-ML, and some practical applications in Parts 2-4, respectively.

Applications of AI-ML concepts and methods to improve decision-making are exciting, useful, fascinating, and fun for researchers in many fields. This book seeks to convey what the excitement is about and to show how these advances can be applied to decision problems of vital importance to humans.

Denver, CO, USA Louis Anthony Cox Jr.

Acknowledgments

It is a pleasure to thank those who have inspired and improved the work reported in this book. Warner North, Mike Greenberg, and Karen Lowrie, to whom this book is dedicated, have encouraged, engaged with, and improved book reviews that I have written for the journal *Risk Analysis* over the past decade. The book reviews in Part 1 of this book have benefitted in many ways from their suggestions. I have greatly enjoyed our teamwork on book reviews and on many other projects.

Ron and Pam Cole of Boulder Learning, to whom this book is also dedicated, have inspired me for years with fascinating discussions of how we might use speech recognition, artificial intelligence, and related technologies to improve human flourishing. Ron first alerted me to the possibilities for deep learning and urged me to consider it carefully as a possible alternative to traditional risk analysis, operations research, and statistical methods of risk analysis. Much of the research in Parts 2 and 3 and the overall plan for the book on how AI-ML complements traditional decision analysis grew out of those discussions.

I thank Terje Aven and Roger Flage for constructive suggestions on a draft of Chap. 4 and for encouraging me to write up these ideas for publication. I thank Vicki Bier and Susan Dudley for valuable comments on an early draft of Chap. 8 that improved its framing, content, and exposition; George Maldonado for very thoughtful comments on early drafts of the material in Chaps. 11 and 12; Seth Guikema for an invitation to deliver a Plenary talk at the 2019 Society for Risk Analysis Annual Meeting that shaped much of the expository approach in Part 3; and Fred Glover, Gary Kochenberger, and Jason Turner of Entanglement for opportunities to present and discuss ideas on how to use causal artificial intelligence (CAI) to improve upon current computational methods for decision support and optimization. Discussions of the probability of causation with Ken Mundt and Bill Thompson and with Peggy Murray, Jacob Traverse, and other colleagues at the Center for Truth in Science helped stimulate the research on health risk assessment applications in Part 4. I have greatly valued the opportunity to create and teach courses in Decision Analysis and Artificial Intelligence for Business in the Business Analytics (BANA) program at the University of Colorado and the enthusiastic support of Deborah Kellogg and Scott Dawson for this work.

Material from the following articles has been used with the kind permission of their publishers:

- Cox LA (2023). Causal reasoning about epidemiological associations in conversational AI. *Global Epidemiology*. (In Press) (Chapter 13).
- Cox LA (2023). Re-assessing human mortality risks attributed to PM2.5. *Environmental Research*. Apr 15; 223:115311. (Chapter 10).
- Cox LA (2023). What is an exposure-response function? *Global Epidemiology*. (Under review). (Chapter 12)
- Cox LA (2023). Book Review of *The Alignment Problem: Machine Learning and Human Values* by Brian Christian. *Risk Analysis*. 43(2): 423-428.
- Cox LA (2022). Book Review of*Morality*. *Risk Analysis*. 42: 653-655. (Chapter 1)
- Cox LA (2021). Toward practical causal epidemiology. *Global Epidemiology*. (3) Nov. (Chapter 11)
- Cox LA (2021). Thinking about causation: a thought experiment with dominos. *Global Epidemiology*. (3) Nov. (Chapter 11)
- Cox LA (2021). Information structures for causally explainable decisions. *Entropy* 23(5), 601; (Chapter 9)
- Cox LA (2021). Teaching data analytics *Risk Analysis*. April; 41(4): 694-699. (Chapter 2)
- Cox LA Jr. (2020) Answerable and unanswerable questions in risk analysis. *Risk Analysis*. Nov;40(S1):2144-2177. (Chapter 4)
- Cox LA Jr. (2020). Thinking better: Six recent books on natural, artificial, and social intelligence. *Risk Analysis*. 40(6):1302-19. (Chapter 3)
- Cox LA, Jr. (2020). Book Review of On Grand Strategy. *Risk Analysis* 40(20): 435-438. (Chapter 2)
- Cox LA. Book Review of The Model Thinker: What You Need to Know to Make Data Work for You. *Risk Analysis* 39(12): 2786-2789. (Chapter 2)
- Cox, T. (2019). Muddling-through and deep learning. *Journal of Benefit-Cost Analysis, 10*(2), 226-250. doi:10.1017/bca.2019.17 (Chapter 8)
- Cox LA. Book review: Behaving Better –*Behave: The Biology of Humans at Our Best and Worst*by Robert M. Sapolsky, Penguin Press, 2017, and*12 Rules for Life: An Antidote to Chaos*by Jordan B. Peterson, Random House Canada, 2018. *Risk Analysis*. 2019 39(2):505-508 DOI: 10.1111/risa.13266 (Chapter 1)
- Cox T. Review of Misbehaving: The Making of Behavioral Economics by Richard Thaler. *Risk Analysis*. 2017 Sep; 37(9): 1796-1798. (Chapter 1)
- Cox T. Review of Algorithms to Live By: The Computer Science of Human Decisionsby Brian Christian and Tom Griffiths. *Risk Analysis*. 2017 June; 37(6): 1201-1207. (Chapter 2)
- Cox T. Review of Superforecasting: The Art and Science of Prediction by Philip E. Tetlock and Dan Gardner. (2015). New York: Broadway Books. *Risk Analysis*. 2017 Feb;37(2): 396-397. (Chapter 2)
- Cox LA Jr. Overcoming learning-aversion. *Risk Analysis*. 2015 Oct; 35(10) (Chapter 6)

- Cox LA Jr. Decision and risk psychology. *Risk Analysis*. 2013 Sep; 33(9): 1749-57. (Chapter 1)
- Cox LA Jr. Community resilience. *Risk Analysis*. 2012 Nov;32(11):1919-34. (Chapter 5)
- Cox LA Jr. Book Review. Poverty and Risk: A Review of Poor Economics: A Radical Rethinking of the Way to Fight Global Povertyby Abhijit V. Banerjee and Esther Durflo. *Risk Analysis*. 2012 June; 32(6):1104-1108. (Chapter 2)
- Cox LA Jr. Confronting deep uncertainties in risk analysis. *Risk Analysis* 2012 Oct;32(10):1607-29. (Chapter 7)

I thank the publishers, reviewers, and editors of these works.

Contents

Part I
Received Wisdom

Chapter 1
Rational Decision and Risk Analysis and Irrational Human Behavior

Introduction

Progress in decision science and risk analysis has started to profoundly affect popular understanding of how real people do and should make decisions when the consequences of alternative decisions are uncertain. Many well-written and insightful popular books on decision-making under risk and uncertainty have appeared in recent decades, including several bestsellers that explain and popularize key concepts and experimental findings about choice, risk, and decision psychology. Much of the older literature on cognitive heuristics and biases in judgment and decision-making under uncertainty, going back to the 1970s, was brilliantly explained and synthesized in Nobel Laureate Daniel Kahneman's popular book *Thinking, Fast and Slow* (2011). Steven Pinker's 2021 book *Rationality: What It Is, Why It Seems Scarce, Why It Matters* provides an accessible survey of normative principles and models for decision-making as well as insights from behavioral economics and decision psychology. This material is becoming increasingly well-known to members of the general public as well as to professional risk analysts and decision analysts. Concepts such as confirmation bias, the anchoring effect, the availability heuristic, and the Dunning-Kruger effect (those who are least knowledgeable about a topic are often most confident about their opinions) have become part of the lexicon of those interested in sound deliberation, argumentation, and decision-making. This chapter surveys many of the challenges to rational decision-making that must be confronted by successful decision-support systems.

In parallel to the development and popularization of decision science, advances in artificial intelligence (AI) and machine learning (ML) have been revolutionizing scientific and engineering understanding of how to make practical decisions under uncertainty in applications ranging from self-driving cars and self-piloting drones to medical diagnosis and safe, efficient control of industrial processes, supply chains, and transportation networks. A principal goal of this book is to consider how recent

L. A. Cox Jr., *AI-ML for Decision and Risk Analysis*, International Series in Operations Research & Management Science 345,
https://doi.org/10.1007/978-3-031-32013-2_1

computational approaches to decision and risk analysis in AI/ML can complement and extend classical decision analysis developed since the 1940s by mathematicians, statisticians, economists, psychologists, and operations researchers. Thus, one intended audience for this book is members of the operations research and management science (INFORMS) community interested in how AI/ML can contribute to decision analysis. In addition, the following chapters are meant to be useful for managers, risk analysts, decision makers, and policymakers involved in financial, health and safety, environmental, business, engineering, and security risk management. By focusing on potential contributions of AI-ML to decision and risk science that can significantly change and improve how we make (and learn from) important practical decisions, this book aims to inform a wide audience in these applied areas, as well as to provide a stimulating resource for students, researchers, and academics in data science and AI-ML.

Extending Classical Decision Analysis (DA) with AI/ML Ideas

Classical decision analysis (DA) roots prescriptive analytics in three core concepts: a set of alternatives ("choice set") that a decision-maker must choose among; a set of possible outcomes among which the decision-maker has preferences; and a model relating the choice from the choice set to probability distributions over outcomes. Chapter 7 discusses this classical DA framework further. The usual prescription for action, implied by various sets of normative axioms, is to choose an alternative from the choice set to maximize the expected utility of the outcome, where the (von Neumann-Morgenstern) utility function represents preferences and risk attitudes. If decisions are made sequentially over time, then the objective function is typically modified to allow discounting or more flexible time preferences for delayed outcomes. Optimization over a static choice set is then replaced by multistage optimization methods such as backward stochastic dynamic programming (Howard 1960).

This classical DA framework has proved enormously productive in economics, operations research, decision psychology, and artificial intelligence (AI). However, over the past decade, pragmatic needs for computationally practical intelligent decision-making by autonomous agents executing complex behaviors and responding to changing conditions in realistically uncertain and changing environments have forced DA used in many AI applications to expand in several useful new directions that take it well beyond the traditional conceptual framework. Crucial developments for creating a more realistic and applicable normative theory of decision-making under uncertainty include the following, which are discussed further in Chaps. 4 through 9, especially in Chaps. 4 and 9.

- *Static choice sets have been replaced by dynamic skill sets.* What an agent can choose to do in any situation is constrained by what it knows how to do. A substantial modern literature in artificial intelligence and machine learning

(AI/ML) and robotics deals with the deliberate acquisition of skills—hierarchies of learned sequences or subroutines for accomplishing desired higher-level tasks or goals in various situations. An agent may invest in acquiring and perfecting skills to expand its repertoire of available behaviors for responding to future situations (Moradi et al. 2010; Shu et al. 2017). This replaces the classical DA paradigm's fixed choice set with a more flexible and dynamic set of constraints on choices that reflect acquired skills.

- *Actions take time and are uncertain.* The mathematical formalism of DA treats the selection of an alternative from a choice set (e.g., an action, or, more generally, a *decision rule* mapping information to actions) as a primitive component of the theory, without modeling implementation details. In many applications, however, executing an action takes time. Successful completion is not guaranteed. Interruptions may occur before an attempted action is successfully completed and multiple low-level behaviors must be sustained over an interval to complete an intended higher-level task (e.g., moving toward a goal while maintaining balance and avoiding collisions, in the case of a robot navigating through pedestrian traffic on a sidewalk). As discussed in Chap. 4, DA models have been extended to handle such realistic models of actions using semi-Markov decision processes and behavior trees that model not only the uncertain duration of attempted actions and the possibility that they will not be completed before further decisions must be made, but also the need to execute multiple behaviors concurrently and to select next behaviors to undertake while some are already underway (de Pontes Pereira and Engel 2015; Colledanchise and Ogren 2020). These extensions greatly expand the primitive concept of "act" or "choice" in classical DA to allow for realistic delays and uncertainties in implementation.
- *Not all possible outcomes are known.* In realistically complex and uncertain ("open world") environments, the set of possible futures is not completely known. Even simple questions such as whether it is possible to take actions to achieve a goal state in a given amount of time with at least a specified probability may be unanswerable (technically, undecidable), in the sense that no algorithm exists that is guaranteed to compute the answer in a finite amount of time (see Chap. 4). This makes it impossible to fully optimize actions based on the utilities and probabilities of all their possible consequences. Instead, many modern AI systems use available partial knowledge of consequence probabilities for alternative actions to simulate sets of multiple possible futures following different current decisions. They seek the best (e.g., highest estimated expected utility) plans—current decisions followed by intended future decisions contingent on future events—that can be found with available knowledge and computational resources by the time a decision must be made (Shen et al. 2019). Sampling possible futures using a probabilistic causal model, together with search optimization heuristics (such as Monte Carlo Tree Search (MCTS), discussed in Chaps. 4, 8, and 9, which selects actions to evaluate further based on Bayesian estimates of their current conditional probabilities of being best) provides heuristic estimates of the relative values of alternative current decisions. These estimates allow practical recommendations for what to do next even when full

optimization is not practicable. Chapters 4, 8, and 9 discuss these innovations further.

- *Simulation-optimization is needed to anticipate and prepare for possible futures.* A rich variety of simulation-optimization methods support adaptive optimization of decision rules when reliable probabilistic causal models are available to simulate the probabilistic consequences of alternative decision rules but the set of possibilities is too large and complex to enumerate or describe explicitly (Eskandari et al. 2011; Juan et al. 2015). Many simulation-optimization methods selectively sample and expand (via simulation) future trajectories for decisions and outcomes to estimate value functions and heuristically optimize immediate decisions (Du et al. 2020; Piché et al. 2019). If no reliable causal models are known, however, different principles are needed, as discussed next.
- *Causal models linking actions to outcome probabilities are often unknown.* As discussed in Chaps. 4, 8, and 9, the challenge of decision-making under uncertainty when no trustworthy causal model is available to predict conditional probabilities of outcomes for different choices has been addressed via reinforcement learning (RL) algorithms that select decisions probabilistically (possibly subject to safe learning constraints to avoid catastrophic outcomes while learning) and adjust selection probabilities in response to observed rewards. RL algorithms provably lead to asymptotically optimal or near-optimal ("low-regret") policies for important classes of sequential decision optimization problems, including many Markov decision processes (MDPs) or partially observable MDPs (POMDPs) with initially unknown transition and reward probabilities and with discounted, average, or total reward criteria (Leike et al. 2016). In effect, RL enables AI decision systems to learn from experience enough about the causal relationship between decisions and conditional probabilities of outcomes (i.e., state transition and reward) to identify optimal or nearly optimal policies. For a wide range of stochastic control decision problems, these methods yield the same policies, via guided trial and error, as formal optimization (e.g., stochastic dynamic programming), but without requiring initial knowledge of the parameters of the controlled process *(ibid)*. Model ensemble techniques that maintain a population of plausible causal models, discarding those that prove to be inconsistent with observations and combining predictions from those that are consistent with available data, can be used both to characterize current model uncertainty and to recommend decisions that hedge against it (Lee et al. 2020).
- *The real world is non-stationary.* In applications from personalized medicine to algorithmic marketing to predictive maintenance to industrial process control, optimal policies can seldom be learned once and applied thereafter without change. Background conditions change and new conditions emerge. Practical decision-making systems, therefore, use adaptive updating and learning methods, such as RL, as well as probabilistic causal models to predict the effects of interventions and changes in the absence of extensive historical data (Kiciman and Sharma 2021).

These innovations have greatly expanded the capacity of traditional axiomatic normative decision theory to support computationally practicable decision-making under realistic conditions of risk, uncertainty, novelty, and change.

The following chapters explain and illustrate these developments. They attempt to show how AI/ML has not only benefitted from classical decision analysis concepts such as expected utility maximization (Colledanchise and Ogren 2020) but also has contributed to making normative decision theory more useful by forcing it to confront realistic complexities such as skill acquisition, uncertain and time-consuming implementation of intended actions, open-world uncertainties about what might happen next and what consequences actions can cause, and learning to cope effectively with uncertain and changing environments. The result is a more robust and implementable technology for AI/ML-assisted decision-making under uncertainty.

The plan of the book is as follows. There are four sections. The first, "Received Wisdom," consisting of Chaps. 1–3, emphasizes concepts, findings, and methods that have been relatively well established and explained in popular books for a wide audience of general readers. They review the insights from a variety of books related to decision science and risk analysis published since about 2010, emphasizing topics and insights that we believe may be of interest to most people concerned with rational decision-making as well as to professional decision analysts and risk analysts. The rest of this chapter reviews aspects of human decision and risk psychology and behavior, contrasting them with the prescriptions of normative decision theories and rational decision-making. It also touches on aspects of morality, altruism, trust, ethics, and behaviors related to risky decisions and choices by individuals living in societies. Chapter 2 reviews data analytics, machine learning, mathematical modeling, strategic thinking, and empirical (randomized control trial) approaches to improving individual, organizational, and social decisions. Chapter 3 introduces AI-ML principles and methods and relates them to work on human reasoning, learning, perception of risks, and decision-making.

Chapters 4–6 turn to fundamental mathematical, conceptual, and practical challenges for effective individual and collective risk management decision-making, whether by humans or AI agents. These chapters seek to synthesize and extend insights from relatively specialized technical literatures on AI, ML, operations research, neuroscience, behavioral economics, game theory, collective choice theory, decision and risk psychology, and organizational design. They focus on identifying fundamental challenges and obstacles to more effective decision-making by realistically limited, rather than idealized, people, organizations, and computational agents. This comprises Part 2 of the book, "Fundamental Challenges for Practical Decision Theory."

Part 3 ("Ways Forward," Chaps. 7–9) explores approaches to overcoming, avoiding, or working around these challenges. These chapters make heavy use of concepts from AI-ML introduced in Parts 1 and 2. Part 4 ("Applications"), consisting of Chaps. 10–13, provides some examples of applications in human health risk assessment and epidemiology, emphasizing applications of causal artificial intelligence (CAI) concepts and methods to better understand how exposures to pollutants or other agents affect probabilities of health responses.

As far as possible, the chapters are written so that they can be read independently of each other. Key concepts and examples are briefly reintroduced where necessary to make each chapter relatively self-contained. However, the content grows somewhat denser and more specialized as the book progresses, with later chapters using technical concepts from earlier ones. Thus, Chaps. 1–3 are intended to be broadly accessible. Chapters 4–9 require more work but are still intended to be at a level less challenging than many technical articles in operations research and management science journals. Chapters 10–13 will be of greatest interest to readers with a specialized interest in public health and epidemiological risk assessment.

Decision Analysis for Realistically Irrational People

Findings from behavioral economics experiments and brain imaging studies, especially, functional magnetic resonance imaging (fMRI) studies in people and other primates (see Chap. 3), and investigations of the interplay between emotions, attention, learning, and cognitive decision-making, are entering the main stream of popular science expositions and shedding new light on what it means to be human. How and why humans exhibit unreasonably effective cooperation and altruism is being illuminated by such experiments and by the speculations of evolutionary psychologists, leading to exciting progress in understanding what kinds of improvements may be possible in individual, group, organizational, and societal risk management. Both scientists and journalists have felt called to explain these developments to the reading public.

The next few sections review insights from the following seven popular books published around 2010.

- *How We Decide*, by Jonah Lehrer (2009)
- *Bozo Sapiens: Why to Err is Human*, by Michael and Ellen Kaplan (2009)
- *The Science of Fear*, by Daniel Gardner (2008)
- *Why We Make Mistakes: How We Look Without Seeing, Forget Things in Seconds, and Are All Pretty Sure We Are Way Above*, by Joseph Hallinan (2009)
- *Predictable Irrationality: The Hidden Forces that Shape Our Decisions*, by Dan Ariely (2008)
- *The Numbers Game: The Commonsense Guide to Understanding Numbers in the News, in Politics, and in Life*, by Michael Blastland and Andrew Dilnot (2009)
- *Adapt: Why Success Always Starts with Failure*, by Tim Harford (2012)

These books provide a foundation for discussion of the following four more recent books:

- *Misbehaving: The Making of Behavioral Economics,* by Richard Thaler (2016)
- *Behave: The Biology of Humans at Our Best and Worst*, by Robert Sapolsky (2017)
- *12 Rules For Life: An Antidote to Chaos*, by Jordan Peterson (2018)
- *Morality: Restoring the Common Good in Divided Times*, by Jonathan Sacks (2020).

All of these books are largely about how real people *do* make decisions under risk, uncertainty, time pressure, and peer pressure; and about how they *can* make such decisions better. These are the aspects reviewed here. They do not deal much with how idealized, perfectly rational people *should* make decisions—the core of traditional decision analysis—but instead emphasize practical improvements in risk management decisions, given the realities of the human mind.

Together, these books tell a fairly coherent story of great interest to risk analysts and decision analysts. Its main elements are as follows:

1. Decisions are shaped by quick emotional responses ("System 1" thinking) and by slower reasoning and deliberation ("System 2" thinking. Both are necessary for effective decision-making and risk management.
2. We all make predictable mistakes in risk perceptions, judgments and decisions.
3. Marketers, politicians, and interest groups can exploit these mistakes to increase sales or to heighten concern about issues and acceptance of proposed actions.
4. Members of groups influence each other in ways that can further degrade the quality of group decisions. (However, adopting deliberate disciplines can greatly improve group and individual decisions.)
5. In organizations and markets, people respond to incentives and measurements (rather than to the intentions behind them); this can cause well-intended policies for risk management to have unintended adverse consequences.
6. Situations, organizations, and institutions that encourage poor choices by individuals and groups can be identified and avoided or re-designed.
7. Moral psychology enables groups and communities to cooperate better and manage many risks more effectively than purely rational economic agents could.
8. We can learn how to make better risk management choices, both as individuals and in groups, organizations, and institutions.

The different books emphasize these major themes to different degrees. The following sections explain them further.

Two Decision Pathways: Emotions and Reasoning Guide Decisions

How We Decide explains that, to a very useful first approximation, choices among alternatives can be understood as resulting from the interplay of two different systems: a quick, intuitive, emotional response (called "System 1," or simply

"Gut" in *The Science of Fear*), followed (time permitting) by a slower, more reasoned comparison and evaluation of the alternatives ("System 2" or "Head"). *How We Decide* relates these two systems to brain processes. It uses real-world examples and experimental findings to give a lively, accessible introduction to findings from neuroeconomics and decision psychology.

Applications of these insights range from the trivial, such as which brand of strawberry jam to buy (a decision for which gut reaction typically leads to greater post-decision satisfaction than rational identification and weighing of various attributes), to major decisions such as which home to buy or which job offer to accept, to life-and-death decisions, such as whether to fire upon an unexpected blip on a radar screen that might or might not prove to be an attacking missile instead of a friendly jet. Correct predictions are reinforced (by dopamine, the neural currency of reward); errors and surprises recruit conscious awareness and learning circuits (in part by activating the anterior cingulate cortex (ACC), which broadcast signals from dopamine neurons across the cortex using high-speed electrical signals transmitted via specialized spindle cell neurons (only in humans and great apes).

As discussed further in Chap. 4, computer programs can mimic such *reinforcement learning (RL)* by continually updating expectations (expected values of the rewards from taking alternative acts in various recurring situations) based on data, and adjusting the probabilities of selecting actions in each situation to reduce the "error signal" difference between expected and received rewards. Such program have produced world-class backgammon-playing programs and successful programs for scheduling flights, controlling banks of elevators, and making numerous other decisions in complex, changing, and only partly predictable environments. *How We Decide* argues that much of our unconscious emotional processing implements highly effective RL algorithms for improving decisions in risky situations, not only enabling rapid stimulus-response reactions when there isn't time for deliberation, but also providing a useful gut feel for the best choice in many complex situations.

The Science of Fear and *Bozo Sapiens* offer the additional evolutionary psychology perspective that System 1 evolved to help us survive by quick, unconscious, instinctive reactions ("Run!"), while System 2 can help us improve on instinctive reactions when time permits and necessity requires. *The Science of Fear* explains that "Gut decides, Head reviews." It notes that gut reactions are often triggered by logically irrelevant cues (such as photographs of attractive people on loan applications), and that these initial reactions are then inadequately adjusted when reviewed by Head. Indeed, Head too often merely rationalizes what Gut has already decided, suggesting more or less plausible conscious rationales to justify unconscious emotional decisions. Similarly, most moral judgments and intuitions have deep emotional and instinctive roots, to which Head typically adds a patina of rational justification, but seldom a compelling logical deduction.

How We Decide documents some life-saving exceptions to the dominance of Gut over Head via case studies in which even strong instinctive reactions—run from fire, or head a slowing plane downward to pick up speed—were successfully overruled by System 2. On the down side, System 2 has its pathologies. Over-thinking can impair the effectiveness of many intuitive decisions and behaviors, cause professional performers to choke, and distract analysts from the relatively few things that are most essential, producing increased confidence but significantly worse performance on tasks ranging from selecting stock portfolios to predicting academic performance in college on the basis of high school data. Simple quantitative models typically greatly out-perform expert judgment on a variety of prediction and decision tasks, in part because they ignore irrelevant details.

System 1 also has its pathologies, which contribute to systematic mistakes in risk perceptions and decisions. *How We Decide, The Science of Fear,* and *Bozo Sapiens* all emphasize that, in the presence of random data, we are prone to see (and act or bet on) non-existent patterns, feeling confident that we have spotted winning or losing streaks in sports, game shows, gambling, and stock price movements. Our emotions, which are easily manipulated, can dramatically affect not only which choices seem best intuitively, but also our attempts to judge objectively the attributes that might guide reasoned choice. The *affect heuristic* studied by Paul Slovic and co-workers, as discussed clearly and insightfully in *The Science of Fear*, lets holistic emotional judgments color our perceptions of the values of different attributes of a risk (e.g., frequency, severity, expected value, unfair distribution, etc.) that are assessed for purposes of deliberative decision-making. Perceived values tend to be significantly positively correlated, even if there is no logical or statistical reason for them to be so.

Since gains are usually seen as good and losses as bad, with losses looming about twice as large as equally sized gains in our metal accounting, the affect heuristic can help to explain loss aversion, the endowment effect (valuing what we own more after acquiring it than before, since we dislike losing it more than we enjoy gaining it), and some of the striking framing effects made famous by Tversky and Kahneman. It also explains why news reports that some product, activity, or exposure has been "linked" to an adverse effect (even if the link is not specifically a causal one) can powerfully engage our emotions, concern and political activity to redress the situation—even if the quantitative magnitude of the adverse effect is never stated, or is vanishingly small.

Indeed, many news accounts do not mention the magnitudes of risks at all. Risk communication that does not provide any information about the *magnitude* of a risk—the only question that really matters to Head, although perhaps irrelevant to Gut—is common in media accounts of suspected environmental and toxicity risks. As stated in *The Numbers Game*, "Size is especially neglected in the special case when mixed with fear. Here, there is often no need even to claim any great magnitude; the simple existence of a danger is enough, in any quantity .... When the headline reveals toxicity, the wise reader, aware than most things are toxic at some dose, asks, 'In what proportions?' ... This does not mean all claims of a risk from toxicity can be ridiculed, simply that we should encourage the same test we apply elsewhere, the test of relevant human proportion. Keep asking the wonderful question, with yourself in mind: 'How big is it?'"

We All Make Predictable Mistakes

The now-standard heuristics-and-biases effects made famous by Tversky and Kahneman (anchoring, representativeness, availability, framing, loss aversion, endowment effect, certainty effect, etc.) are discussed and illustrated with entertaining and thought-provoking examples in *Bozo Sapiens*, *Why We Make Mistakes*, *How We Decide*, *Predictable Irrationality*, and *The Science of Fear*. The exposition and discussion in *The Science of Fear* is especially thoughtful and thorough. It presents many other striking examples of systematic errors in decision-making and risk judgment, with discussion of their presumed underlying mechanisms, including the following:

- In risk-rating experiments, time pressure and physical or mental stress (e.g., hunger, tiredness, distraction, unhappiness) increases the strength of the affect heuristic, e.g., increasing the perceived risk and reducing the perceived benefit of nuclear power relative to other risks. Emotional evaluations and intuitive reactions play a larger role in rushed and stressed choices than in more considered ones.
- Reading about a technology's benefits changes perceptions of its risk, even if the text says nothing about its risks (as the affect heuristic predicts);
- Reading a news story about a tragic death caused by fire increases the perceived risks of fatalities from leukemia or murder (since it increases the emotional salience of mortality risks, which shape perceived attributes such as likelihood or frequency);
- Beef described as "75% lean" gets higher ratings in taste tests than beef described as "25% fat" (since products perceived as "good" tend to look and taste better to consumers);
- Surgery described as giving a "68% chance of being alive a year after surgery" is more than twice as likely to be preferred to radiation treatment (44% vs. 18%) as the same surgery described as giving a "32% chance of dying" within a year after surgery (since the former "gain frame" triggers less aversion that the latter "loss frame.")
- Psychiatrists are about twice as likely to keep a patient confined if told that "20 out of every 100" similar patients will commit an act of violence after release than if told that "20%" of similar patients will commit an act of violence after release (because "20%" has less visceral impact than the more concrete and real-sounding "20 out of every 100 patients);
- Students express stronger support for a purchase of airport safety equipment that would save "85% of 150 lives" than for equipment that would save "150 lives" (since "85%" sounds high, hence "good," but "150 lives" lacks a context to cue Gut as to whether this should be viewed as a big (good) or a small (bad) number).

Such examples—and *The Science of Fear*, *Bozo Sapiens*, and *Why We Make Mistakes* present dozens more—illuminate consistent differences between how real people make judgments and decisions, and how one might expect judgments and decisions to be made by more dispassionate rational beings.

Why We Make Mistakes provides a relatively short, accessible, and enjoyable account of systematic errors in reasoning and judgments. It covers the usual bases: Systems 1 and 2, Tversky and Kahneman's heuristics and biases (describing and explaining dramatic shifts in preferences for risky alternatives depending on whether they are framed to emphasize potential gains or potential losses); insights from Prospect Theory (e.g., that preferences commonly exhibit risk aversion for gains and risk-seeking for losses, with potential losses being weighted about twice as heavily as potential gains); consequences of hyperbolic discounting (over-weighting of current rewards (and temptations) compared to delayed ones); and greater willingness to take risks with delayed results. The discussion avoids all technical jargon, including the terms "Prospect Theory" and "hyperbolic discounting," but is rich in fascinating examples drawn from marketing and advertising, psychological experiments, medicine, and accident risk analysis. *Why We Make Mistakes* also introduces additional types of human decision errors. These include "proofreading" mistakes, where our minds automatically perceive (and do not even notice exceptions) to what is expected from context. Limited attention spans and sharply limited ability to search vigilantly for rare outcomes can lead to fatal mistakes, such as a high prevalence (up to 90%) of missed lung tumors in X-rays of patients who subsequently are diagnosed with lung cancer. They also include flawed mental models for understanding how the world works; misunderstanding and mis-predictions of our own future preferences and evaluations of outcomes; and failures to learn effectively. Failures to learn arise from a host of well-documented foibles. We tend to misattribute blame and misdiagnose underlying causes; fail to understand delayed feedback; misinterpret and mis-remember events in ways that protect and flatter our own egos; rush to act instead of first reflecting adequately on what we have seen and what we should be learning from it; and over-weight anecdotes, especially if they are vivid. Even experts make these mistakes. Singly and collectively, we place unwarranted confidence in plausible-sounding narratives that have no objective predictive power.

As social animals, we also tend to over-emphasize the faults of humans, and to under-emphasize the potentially preventable design flaws of systems and situations where humans tend make mistakes. This limitation is compounded by *hindsight bias*, which encourages us to mistakenly believe that whatever actually happened was (or should have been) obvious before the fact; and by well-documented limitations in our cognitive capacities, which take many unnoticed short-cuts (e.g., selectively allocating attention, skimming for meaning based on context, misremembering our actions and their consequences in ways that tend to flatter us, but that impede learning, and so forth.) A major theme of *Why We Make Mistakes* is that what we notice is strongly driven by context and by (possibly logically irrelevant) cues. These shape what we pay attention to, what we miss, and our perceptions and interpretations of what we notice.

Marketers, Politicians, Journalists, and Others Exploit Our Systematic Mistakes

That we make systematic errors in decisions involving risky and/or delayed consequences creates opportunities for others to exploit our decision weaknesses. *The Science of Fear* discusses several chilling examples in which activists, politicians, corporations, regulatory agencies, and scientists jointly exploit weaknesses such as the affect heuristic (e.g., the public's generally negative perception of chemicals or other threats) to manipulate public fears and preferences to support their own political and business agendas. *The Numbers Game* examines the tactics of such manipulation more closely. It gives examples of reports and headlines that exploit the fuzziness of language to create sensational-sounding stories to mold public perceptions. (For example, what is counted and reported as "unemployment" or as "bullying" or as deaths or illnesses "attributed to" a specific cause, in reporting statistics on risks of these events, may reflect the agendas of those doing the counting.) Other types of manipulation exploit the innumeracy of System 1, which tends to perceive worrisome disease clusters even in random data, and to interpret headlines about exposure-related risks (e.g., reporting "links" between food or environmental hazards and adverse health effects) as undesirable and worth avoiding, without considering such niceties as the magnitudes of relevant exposures or responses, the existence of thresholds for adverse effects, or whether the reported "link" has been shown to be causal.

Credit card and mortgage companies, retailers, advertisers, and marketers have also become adept at exploiting imperfections in decision-making, inducing consumers to spend more and take larger personal financial risks than they otherwise would. They prey on predictable patterns such as our over-weighting of immediate compared to delayed consequences; willingness to be led on by false hopes when gains and losses occur unpredictably ("variable-schedule rewards"); men's perception of mortgage rates as being more desirable if they are printed on brochures featuring attractive women; or our proneness to anchor quantitative estimates of value on quantitative cues, even if they are logically irrelevant. (Thus, for example, grocery stores will sell significantly more of an item if they advertise the pricing as "4 for 2 dollars" than if they advertise "2 for a dollar," since the former anchors the purchase quantity at 4 instead of at 2.) *Why We Make Mistakes* gives many such examples.

Predictable Irrationality: The Hidden Forces that Shape Our Decisions develops this theme further, noting that consumer preferences often exhibit "arbitrary coherence." This means that how much we are willing to pay for something may be driven by anchoring on irrelevant cues (creating opportunities for manipulation by savvy retailers and marketers), but we then tend to evaluate other goods consistently around this initially arbitrary level, being willing to pay more or less for things that we clearly desire more or less than the first good, respectively. *Predictably Irrational* also adds to the catalogue of exploitable patterns (or "decision illusions") in market and non-market (e.g., ethical) decision-making. For example, it notes the disproportionate power of "free" add-ons (even if they are not very desirable) to induce consumers to make otherwise unattractive purchases or other trade-offs (the "zero price effect"). It describes experimental evidence that even highly intelligent people over-value keeping their options open, thereby achieving much lower rewards in sequences of risky choices than if they would commit to a strategy and follow it, allowing some options to close instead of sacrificing resources to keep them open.

Predictably Irrational has a lively, engaging, anecdotal style. A great strength is its reports of ingenious and fascinating experiments with different groups of people, from trick-or-treating children choosing between candy offers at Halloween to business executives making ethical decisions, to patients choosing treatments. Many of these experiments were designed and conducted by the author and colleagues. The results vividly illustrate the predictable irrationalities discussed in the book. Some are well-known, such as loss aversion and the endowment effect, which cause us to value and cling to what we already possess (perhaps including our existing preconceptions and ideological beliefs) far more than we do to similar goods that are not ours. Others are new, such as demonstrations that people will tell bigger lies about their performance on a simple math test if the reward for claimed performance is paid in tokens that can be exchanged for money, rather than being directly paid in money itself. Cheating and "soft theft" (to obtain symbolic goods or tokens that can later be converted to cash) seems less morally offensive than cheating or stealing to directly obtain money. This finding applies to many people, from students in experiments to insurance policy holders submitting claims to office workers submitting expense reports.

People Respond to Incentives and Influences in Groups, Organizations, and Markets

To reduce decision errors (and opportunities to exploit them), it is natural for organizations and governments to deliberately design and enforce decision rules and procedures intended to increase the quality of decisions, as measured by objective performance metrics. *The Numbers Game* and *Adapt* explain why this often backfires. Performance metrics necessarily measure only selected aspects of the full distribution of performance. Individuals in organizations, and entrepreneurs in markets, learn to respond to the incentives created by the rules and metrics, rather than to make decisions that truly optimize (e.g., maximize the expected utility of) the resulting distribution of performance. For example, hospitals that measure (and assign rewards or penalties based on) the fraction of admitted patients who are treated within a certain amount of time may soon find that the admissions process is distorted to favor patients who can meet this target. Average waiting times may increase, and health outcomes worsen, as patients who are likely to miss the target (or who have already done so) are consigned to extremely long waits in order to accommodate those more likely to meet the target. Similarly, efforts to reduce risks of inadequate education, such as the "No Child Left Behind" initiative in the U.S., may incent teachers and schools to focus on moving a relatively small subset of children from just below a given standard-based line to just above it, while serving neither the worst nor the best students well. Gaming the rules, or "hitting the target but missing the point," is a common problem in organizations that seek to improve risk management decisions by using metrics.

If we organizational decision rules and performance metrics distort incentives and undermine true optimization of decisions, then might enlightened command-control regulation, centralized planning, and strong, benevolent, top-down leadership provide a satisfactory alternative for overcoming the natural limitations and defects of individual decision-making? *Adapt: Why Success Always Begins with Failure* examines the performance of top-down risk management systems and decision hierarchies in important risk management situations, including military conflicts (e.g., in Iraq and Afghanistan), evolution and survival of businesses in competitive markets, government investments to stimulate socially beneficial research and development (R&D), efforts to reduce poverty and child mortality risks in developing countries, environmental regulations, and management of the risks of tightly-coupled, complex engineering systems (such as nuclear power plants) and tightly coupled financial systems. The main conclusion is that, in all of these applications, effective risk management decision-making requires experimentation, innovation, and improvisation by multiple groups; feedback (including candidly recognizing when an attempt has failed); selection and sharing of what works; and effective learning (e.g., via social and economic imitation, or instruction), so that successful innovations spread through the population of interacting decision-makers. Detailed local information and understanding (a "worm's eye view" of ground-level details), coupled with ability to improvise, experiment, and spread successes, are crucial to successful adaptive decision-making and risk-taking in such situations.

These features imply that top-down command-control and leadership are limited in what they can accomplish to solve problems or to mitigate developing risks, insofar as the local information needed to improvise good solutions is simply not available at high, remote levels of decision hierarchies. We should not have unrealistically high expectations of what can be accomplished to solve diverse or changing local problems by even benign and competent top-down leadership, well-intended rules, and well-designed decision hierarchies of dedicated and honest workers. Geographically distributed systems and threats, and emergent risks arising in complex adaptive social systems (such as financial markets or social networks of intelligent agents, where each individual's behavior is influenced by the behaviors of others), require distributed control and adaptation based on local information.

To this insight, *Bozo Sapiens* adds that, even if the detailed information needed to manage a complex social or engineering system is available to a centralized controller, via delayed feedback about the consequences of interventions, still top-down, rule-based control usually fails. Whether one is managing a simple thermostat-controlled refrigerator or a simulated socioeconomic-environmental system, obtaining desired outcomes from top-down management of such systems requires "thinking probabilistically." In experiments discussed in *Bozo Sapiens*, "Managers who used probabilistic terms like *sometimes, in general, often, a bit, specifically, to a degree, questionable, on the other hand*, and so on, were successful. Managers who preferred the absolutes—*always, never, without exception, certainly, neither, only, must*—were not. A surprise that: all the strong, sure terms that stud the leadership-course lexicon turn out to be a diagnostic mark for *failure*. ... [W]hat makes someone expert [is] openness, avid curiosity, an eye for variation, and a lust for understanding. These are probabilistic, not absolute, virtues."

Adapt acknowledges the power of incentives in organizations and markets, but also notes that incentives are strongest when actions generate prompt feedback. It holds that markets or organizations of many creative individuals will always figure out how to out-smart, or game, any set of rules, regulations, or designed incentives. Therefore, the author, economist Tim Harford, recommends solving complex problems (such as how to reduce carbon emissions) by creating incentives and effective feedback loops (e.g., via carbon taxes) and letting the ingenuity of populations of businesses, engineers, and entrepreneurs evolve creative solutions to the resulting optimization problems.

Moral Psychology and Norms Improve Cooperative Risk Management

The books reviewed here emphasize the importance of ethics—reasoning about the right course of action—and of moral and social norms in determining how well individuals, groups, organizations, and populations can trust each other and cooperate to manage risks. Most individuals and many communities and societies exhibit remarkably high propensities to help each other, to share good fortune with those in need, to trust and to be trustworthy in cooperating in risky environments, and to punish those who succumb to incentives to cheat at the group's expense, even if punishment is costly. These innate tendencies, amplified by culturally transmitted norms, facilitate collective risk management by communities living in risky environments.

Bozo Sapiens describes both the brain science and the possible evolutionary psychology of such cooperative risk management. Networks of mirror neurons allow us to learn from, understand, and empathize with the observed behaviors and imputed motives and feelings of others. They enable sympathy, altruism, and cooperation even in the presence of purely rational incentives to act selfishly to the detriment of others. Bozo Sapiens explains that humans, like wolves, hyenas, vampire bats, and other "social animals that depend on securing rare, high-value nourishment have the habit of then distributing food among not just the helpless children but the group as a whole. It's a behavior that acknowledges the central role of probability: hunting as a pack increases the overall chance of any one animal being successful, but it remains a matter of chance—today you, brother; tomorrow, me. Probability is what makes ethics both necessary and complicated." In various primate societies, and even especially among humans, social norms and moral instincts begin with such basics as sharing good fortune, helping each other through occasional hardships, returning favors, and punishing community members who flout these norms. Among humans and some other primates, social norms are flexible and culturally transmitted. They create a need to think about and participate in complex social dynamics, remembering past cooperative and uncooperative behaviors by individuals and alliances, tracking resulting reputations for being trustworthy or not, and assessing the probable responses of others in risky situations. Arguably, these needs may have exerted a powerful selection pressure for the evolution of abstract reasoning, cognition, and flexible mental modeling in human and other primate societies.

But moral judgments are also intensely emotional. We often respond quickly and instinctively to judge the deeds of others as right or wrong, fair or unfair, admirable or contemptible, expiated or unforgiven, and to mete out social approbation or censure accordingly. Such quick, intensely felt, judgments and swift feedback help to maintain social order. (As *How We Decide* points out in a chapter on "The Moral Mind," psychopathic evil is associated with dysfunctional emotions and absence of normal empathy with victims. *Bozo Sapiens* adds examples where normal emotions of affection, trust, and pleasure in altruism are blunted by grim experiences and cultures, leaving individuals and communities cruel, spiteful, and uncooperative with each other and with outsiders.) On the other hand, strong moral emotions and norms such as loyalty, courage, honor, duty, hatred of injustice, and love for one's family and in-group, can be subverted to fuel suicide bombings and terrorist attacks.

Our emotional moral reactions—such as elevation when we contemplate profoundly good, right and noble actions; disgust and outrage when we witness the reverse; and willingness or unwillingness to trust others in seeking and sharing gains from cooperative risk-taking or risk management—usually precede any explicit rational ethical analysis of why these deeds are judged as good or bad, respectively. Thus, the idea of Systems 1 and 2 applies also to moral judgments. *Bozo Sapiens* discusses examples (many due to moral psychologist Jonathan Haidt) in which quick, strong, and definite moral judgments cannot easily be rationalized by abstract ethical reasoning, leading to the phenomenon dubbed "moral dumbfounding" (strong moral intuitions that are hard to explain rationally).

Predictably Irrational examines the psychology of dishonesty and erosion of trust and trustworthiness in the legal and medical professions (e.g., among physicians who over-prescribe treatments or tests that benefit them financially) and in business, finance, and the work-place. Replacing tangible goods such as cash with more abstract ones, such as shares in ownership of financial instruments or portfolios, lulls our moral intuitions, making it far easier to commit outright theft or to indulge in highly questionable risk-taking with other people's assets without provoking uncomfortable reactions from our consciences. Switching from social norms to market norms (perhaps in response to inadvertent cues) can make people both less likely to seek needed help, and less likely to give it. Trust in other people or in organizations and institutions is easy to destroy and hard to restore; without it, societies can lose the benefits of cooperative risk-taking and reliance on business and political systems that would allow profitable joint investments and mutually beneficial transactions and agreements. These facts may be of particular interest in risk analysis, where careful rational and ethical analysis of perceived threats to life, justice, or other core values may not always harmonize with intuitive emotional reactions, such as instinctive outrage over an exposure that turns out to pose little or no quantitative risk of harm, or collective apathy about a threat that is quantitatively large, but that fails to provoke moral outrage.

We Can Learn to Do Better

The books we have discussed agree that we can learn how to makes Systems 1 and 2 work together more effectively, and how to design and manage organizations, workplaces, and engineered systems to better respect the realities of human decision and risk psychology. Doing so can lead to a safer, more productive, and more enjoyable world.

How We Decide concludes by celebrating the decades-long program of successful reductions in aviation risks, making air travel far safer than almost any other human activity. Part of this progress is due to the use of simple checklists to avoid routine errors. Part is due to a deliberate high-tech fusion of Systems 1 and 2: pilots are trained in flight simulators until the probable consequences of their actions in unusual circumstances (such as loss of power in navigation controls) no longer need to be reasoned through afresh by System 2 each time an emergency occurs, but become instinctive. Pilots now have a better feel (System 1) for what might have gone wrong and what to do about it. Dramatic improvements in safety were also ushered in by the introduction of the Cockpit Resource Management ("See it, say it, fix it") strategy for breaking down authoritarian decision hierarchies and improving the capacity of cockpit teams to give prompt feedback and suggest changes or improvisations if needed; the same tactics have proved highly successful in improving the safety records of surgical teams. *How We Decide* explains: "The reason CRM is so effective is that it encourages flight crews and surgical teams to think together. It deters certainty and stimulates debate. In this sense, CRM creates the ideal atmosphere for good decision-making, in which a diversity of opinions is openly shared. The evidence is looked at from multiple angles, and new alternatives are considered. … The safety of flight is a testament to the possibility of improvement."

Bozo Sapiens and *Adapt: Why Success Always Starts with Failure* both see great hope in re-designing decision processes and organizations to encourage, and make better use of, honest, effective feedback. *Bozo Sapiens* notes that management of poorly understood systems (e.g., complex, nonlinear socioeconomic or engineering systems, especially with delayed feedback) tends to be most successful when managers think probabilistically, maintain a humble awareness of what they don't know and an active enthusiasm for learning from new data, and take care to learn from experience between incremental adjustments. Likewise, *The Numbers Game* emphasizes the importance of cultivating a culture that respects data, so that feedback from the real world can be correctly understood.

Adroit use of System 2 methods such as statistics and modeling can help to assure better learning from data, with statistical associations, successive events, and coincident trends not being automatically construed as evidence of causation, and with potentially biased or manipulated beliefs and judgments being tested, corrected, and informed by empirical data. *Adapt* recommends decoupling and down-scaling complex systems to allow time and scope to learn from feedback, as well as encouraging healthy self-doubt, active learning by deliberate experimentation, and willingness to acknowledge and learn from ("make peace with") our mistakes. *Adapt* concludes that "There are three essential steps to using the principles of adapting in business and everyday life ... First, try new things, expecting that some will fail. Second, make failure survivable: create safe spaces for failure or move forward in small steps. ... And third, make sure you know when you have failed, or you will never learn."

These seven books represent a much larger literature. Other popular books such as Malcolm Gladwells' *Blink*, Sunstein and Thaler's *Nudge*, Poundstone's *Priceless*, and Kahnman's *Thinking Fast and Slow* further develop and illustrate many of these points, and an enormous academic and technical research literature supports these popular works. They reflect an emerging, increasingly widely shared, understanding of how real people judge and manage risks, and how they can do so better. The task of building an applied discipline that incorporates these insights into practical methods for improving individual, group, organizational and societal decision and risk management processes is still in its infancy. Enough has been learned, however, to make development of such a practical, psychologically grounded, prescriptive theory of risk management decision-making a realistic goal for today's decision scientists and risk analysts. The challenge of developing such a theory has been taken up by behavioral economists, as discussed next.

The Rise of Behavioral Economics

Pioneering behavioral economist Richard Thaler concludes his 2016 autobiographical book *Misbehaving: The Making of Behavioral Economics*, his personal retrospective on the founding of the now flourishing field of behavioral economics, with the following advice for individuals and organizations seeking to behave more effectively in an uncertain world: *observe, collect data,* and *speak up*. To observe means to notice what actually happens in the world around us, including how real people behave in response to the situations, incentives, and policies they encounter, without being blinded by theory-driven expectations and predictions. This entails collecting relevant data and paying attention to them even when—or especially when—they do not conform to what we expected or hoped to see. Such observations and data collection enable us to learn what works and what doesn't, rather than simply assuming that well-intended policies produce their intended consequences. Speaking up means calling attention to observed data and anomalies during decision-making. Thaler advocates evidence-based economics and rigorous empirical evaluation of the performance of policies. Observation, documentation, and use of relevant facts provide the foundation for both.

Thaler's career, entertainingly discussed in the book, began with observation and documentation of anomalies in how people actually behave and choose in situations where economic theory makes clear predictions about what rational economic agents (called "Econs" by Thaler) *should* do. Very often, these predictions do not match observations of what real people ("Humans" in Thaler's terminology) actually do. For example, basic microeconomic theory might confidently predict that a shortage of snow shovels during a blizzard would cause their prices to rise to ration the scarce supply and allocate shovels to those most willing to pay for them. But humans typically do not behave this way: stores that raise prices during an emergency are perceived as unfair or exploitative and risk losing business in the long run. Likewise, a thirsty person may be willing to pay more for the same cold beer if it comes from a fancy hotel than if it comes from a run-down shack, even though Econs would value the same product the same no matter where it comes from. In countless ways, including many that involve choice under uncertainty and risk, Humans behave differently from Econs. Policies that are designed to benefit Humans may therefore have to differ significantly from those that would benefit Econs.

The first part of *Misbehaving*, comprising three of its eight sections, discusses how Thaler and others came to notice, confirm with experiments, and document in seminal articles a list of anomalies, meaning discrepancies between the predictions of economic theory for Econs and the observed behaviors of Humans. These included staples of current behavioral economics such as the endowment effect, in which simply possessing something makes it more valuable to the owner. Mental accounting, in which gains, losses, and budgets are not pooled but are kept in separate mental accounts (e.g., for household expenses, entertainment, retirement savings, college fund, etc.) explains many observed discrepancies between theoretically prescribed and empirically observed behaviors. So do the tendencies of Humans to adapt their aspirations and choices in light of sunk costs (which rational Econs would rightly ignore) and to reveal dynamic inconsistencies in preferences and choices. What we plan to do and what we actually do when the time comes often differ, as when we yield to short-run temptations that frustrate previous good intentions and resolutions and incur predictable long-term regrets. These are phenomena with which many human dieters, gamblers, and credit card debtor are very familiar, but which no Econ would ever experience. Moreover, Humans care about perceived fairness—a topic that Thaler explored with Daniel Kahneman in the mid eighties, as documented in Section 4 of the book. Concerns for fairness affect consumer purchasing and investment decisions; how employers treat employees; how businesses treat customers; and how employees respond to employers and customers to sellers.

Thaler and his colleagues proposed that "supposedly irrelevant factors" from the standpoint of traditional economic theory—factors such as the endowment effect, loss aversion, status quo bias, and narrow framing of decisions and incentives (see Chap. 6); sunk costs and mental accounting; bounded will power, rationality, and self-interest; and perceived fairness of transactions—are important in the real world. They must be taken into account to understand important observed microeconomic, macroeconomic, and financial economic behaviors. However, to persuade many mainstream economists of the importance of behavioral economics took many years, in part because established economic theory and prominent academic economists taught that the usual textbook assumptions of rational optimization and equilibrium, while admittedly not perfect, were unlikely to be importantly wrong. Section 5 of the book, entitled "Engaging with the economics profession: 1986-1994," describes the struggles of Thaler and his colleagues to get behavioral economics taken seriously as part of mainstream economics, rather than being perceived as little more than a collection of psychological experimental results and field observations with doubtful relevance to most important economic phenomena. This long battle was ultimately substantially won, despite the initially dismissive attitudes of some prominent economists, including several Nobel Prize winners. Section 6, entitled "Finance: 1983–2003" discusses applications to financial economics, especially stock price movements and overreactions of the market to weak information. Today, practical applications of behavioral economics insights are found throughout much of mainstream economics. They include better understanding and prediction of stock price movements; of savings, consumption, and investment decisions under uncertainty and over time; and of stickiness of wages during recessions and fluctuations in the labor market.

The final two sections of *Misbehaving* recount Thaler's experiences at the University of Chicago since 1995; and his efforts, together with Cass Sunstein and other colleagues, to apply behavioral economics to help design "choice architectures" that make it easier for employees, consumers, executives, bureaucrats, and others to make good choices, as evaluated by themselves, without expert help. At the University of Chicago, Thaler and colleagues have applied behavioral economics insights to law-and-economics. For example, they point out that reluctance to accept offers perceived as unfair interferes with the smooth operation and predictive power of Coase's Theorem on the Pareto efficiency of voluntary contracting, independent of the initial allocation of property rights, even if the other conditions of well-defined property rights, perfect information, and negligible transaction costs are satisfied. They have also used data from high-stakes settings for choices made under uncertainty and risk, from game shows to football, to examine how risk-taking varies as gains and losses are experienced and how narrow framing and excessive or inconsistent discounting of potential future gains lead to sub-optimal decisions by individuals, teams, and organizations, even when the expected returns from better decision-making are large and real.

Moving from description to prescription, Thaler and Sunstein, in their famous 2008 book *Nudge*, proposed that changing defaults and other aspects of choice architectures (the designed setting in which choices are made) could change behaviors and reduce subsequent regrets by making certain choices easy. For example, making enrollment in employee retirement plans or organ donation programs the default from which employees or drivers could then choose to opt out, rather than making them choices that have to be opted into, could greatly increase the numbers of people who start to save adequately for retirement or who became organ donors. Applying other lessons and principles from behavioral economics, such as deferring escalation of savings as part of a multi-year employee retirement savings plan ("Save More Tomorrow"), or telling people that most others in their community make a certain choice or exhibit a certain behavior, further facilitates changes in behavior to reduce predictable regrets. These changes in behaviors are accomplished without coercion, consistent with an ideology that Thaler and Sunstein dubbed "liberal paternalism."

The last section of *Misbehaving* recounts experiences in the United Kingdom and the United States applying these ideas in attempts to make government more effective and efficient. Thaler notes that it is usually difficult to devise nudges and choice architectures that elicit large changes in behaviors, but even small changes in large populations may be worthwhile. Wider use of randomized control trials (RCTs) and field experiments in evidence-based evaluation can help organizations and policy-makers to discover what works in the real world of humans, rather than only deducing what should work in a theoretical world of Econs. Willingness to experiment and learn, to collect data, and to pay attention to them during decision-making and evaluation of past decisions and policies can potentially help to make economics and related analytic disciplines dramatically more useful for improving policies and outcomes. However, capturing these potential benefits requires organizational cultures and incentives that encourage sensible exploration, innovation, and risk-taking in order to discover how to produce the best outcomes.

Misbehaving is a fun book to read. It is packed with amusing and instructive anecdotes and accessible accounts of the discovery, empirical testing, and eventual acceptance of principles of behavioraleconomics. Thaler's accounts of his collaboration with Daniel Kahneman and of his arguments with various luminaries of the academic economics world add human interest and back stories for many results that are now widely taught and used but that were new and controversial only a few decades ago. *Misbehaving* would make a fine supplementary book for graduate or undergraduate courses in economics, policy analysis, risk analysis, or marketing science that have strong decision psychology components. It provides a lively, informal account of key ideas and findings that also make it entertaining light reading for those with an interest in decision science, policy analysis, and efforts to improve the realism and practical value of economics.

Beyond Behavioral Economics and Rational Choice: Behaving Better in a Risky World

Does good risk management decision-making require good people? To what extent do effective personal, organizational, and societal risk management depend on social virtues such as responsibility, trustworthiness, honesty, generosity, and altruism; or on classical personal virtues such as temperance, fortitude, and prudence? Two very different books—*Behave: The Biology of Humans at Our Best and Worst* by biology professor Robert Sapolsky and *12 Rules For Life: An Antidote to Chaos* by psychology professor, clinical psychologist, and YouTube phenomenon Jordan B. Peterson—discuss sources of virtuous and malevolent behaviors and how they relate to many aspects of risk perception, communication, and risk management.

Both books draw freely on evolutionary psychology and related fields, including developmental psychology, behavioral economics, and neuroscience. Both explain relevant scientific findings (and their limitations) clearly and accessibly, using them to comment on human nature; social structures and dominance hierarchies; health consequences of poverty, inequality, and stress; and how individuals and groups respond to risk, uncertainty, and temptation. Sapolsky's enlivens his accounts of human behaviors and their causes with expert comparisons of human behaviors to those of other primates. He presents clear, fascinating informal summaries of pertinent neuroscience, as well as neural and brain anatomy correlates of behaviors and behavioral dispositions. His book is popular science at its best: entertaining, witty, irreverent, and in many places magisterial. It is lengthy (717 pages of text and appendices plus an additional 50 pages of notes), but moves quickly.

Peterson's much shorter self-help book (370 pages of text plus 18 pages of references) touches more lightly on relevant science, but also presents vivid examples—often quite moving ones—from his clinical psychology practice and life experiences. These are interwoven with archetypal stories from mythology, the Judeo-Christian Bible (heavily represented), and other religious traditions, regarded as distillations of ages of human experience in coping with risk, uncertainty, and tragedy. Loss of what is loved prompts an often agonizing search for enduring meaning, and Peterson offers his reflections and advice as a fellow seeker with some specific ideas on where such meaning may be found.

Each book is described separately below. The review concludes with parting comments on the intended audiences and the best use of the material from both books for risk analysts.

Review of *Behave: The Biology of Humans at Our Best and Worst*

Behave focuses primarily on *why* people behave as they do. Its first 10 chapters provide a brilliant popular science exposition and synthesis of explanations for behaviors—a momentary kindness, perhaps, or an act of impatience or cruelty. These chapters develop explanations on expanding time scales, from one second before to millennia before the behavior itself. Thesy first cover the interplay of emotional and cognitive structures in the brain, corresponding roughly to Kahneman's System 1 and System 2. Other topics covered are (a) Rapid emotional responses and impulses from brain structures such as the amygdala, and (b) Impulse control by more cognitive systems such as those coordinated by the prefrontal cortex. Successive chapters discuss the often nuanced roles of neurotransmitters, such as how the dopamine reward system helps to mediate delayed gratification, motivation and goal-oriented behavior, evaluation of risk and uncertainty, risk-taking, and gambling addiction. On a time scale of hours to days, hormone levels can powerfully shape behavioral predispositions. Prominent examples include testosterone, which can boost status-seeking and foolish risk-taking behaviors under some conditions; and oxytocin, which can both increase prosocial and altruistic predispositions within a family or in-group, but also increase hostility toward strangers and out-group members. On time scales of decades and lifetimes, neural plasticity and genetics contribute to some behavioral predispositions, albeit in complex and highly conditional ways. Sapolsky rightly emphasizes the huge importance of gene-environment interactions. A key take home message is that genes could under some conditions strengthen proclivities toward aggression, risk taking, depression, or other aspects of personality yet only do so in response to certain environmental triggers, such as an abusive childhood. A recurring theme is that stress affects health, brain development, gene expression, hormone secretion, risk perceptions, and subsequent behaviors. These influences begin with prenatal exposure to maternal stress hormones.

Chapters 1–10 also repeatedly mention the roles of social dominance hierarchies and aggression in creating and relieving stress in people and other primates. On time scales of centuries to millennia and longer, biological and cultural evolution shape social attitudes and expectations. Kin selection, newer and more respectable versions of group selection, reciprocal altruism, altruistic punishment to enforce norms, and sexual selection all help to form and maintain socially expected and accepted norms of behaviors. These explanations on different time scales overlap: sensitivity to cultural cues, for example, may depend on development of the brain and on prior life experiences that have triggered or failed to trigger heightened expression of genes that contribute to this development. Chapter 6, entitled "Adolescence: Or, Dude, Where's My Frontal Cortex?" explores aspects of adolescence, including risk-taking and heightened fear of social rejection, from the perspective of brain science and delayed development of integrated control over emotional responses.

The last half of *Behave*, Chaps. 11–17, applies the explanatory perspectives from the first half to behavioral issues of conflict, cooperation, morality, religion, political ideology, empathy and altruistic behaviors, violence, forgiveness, the criminal justice system, and war and peace. Perhaps unsurprisingly, this part may seem less coherent, authoritative, and insightful than Chaps. 1–10, although it is still well worth reading. There are beautiful examples of how different levels and types of explanatory factors may intersect in shaping world views and political orientations (estimated from twin studies to have a heritability of about 50%); e.g., that "the 'risk-taking' version of the D4 dopamine receptor gene is associated with liberals—but only in people with lots of friends." This rich interplay of explanatory factors is appropriately caveated (it is complex, many results have not been reproduced and well validated, some findings appear to be very nuanced if not outright contradictory, and much remains to be discovered and elucidated), but Sapolsky nonetheless draws some fairly sweeping conclusions. Among the conclusions are that religion and moralizing deities are cultural inventions devised to increase in-group prosociality and also serve to foster violence against out-groups; that concepts such as free will, the soul, evil, and punishment of criminal behavior are (or should be) on their way out as science increasingly explains the true causes of choices and behaviors; that, as Aristotle recognized, moral behavior is better inculcated by habit and practice than by reason, sympathy, and will; and that lasting peace among peoples depends on recognizing the irrational importance of values considered sacred by others. Chapter 11 is entitled "Us versus Them," and a repeated theme in the remaining chapters is the importance for decent behavior of resisting in-group/out-group psychology and refusing to demonize, stereotype, or diminish others.

12 Rules for Life: An Antidote to Chaos

12 Rules for Life, like *Behave,* acknowledges at its outset the layered, highly evolved biological systems and cultural antecedents that shape our perceptions, reactions, embodied minds, and behaviors. As Peterson puts it,

> Everything you value is a product of unimaginably lengthy developmental processes, personal, cultural, and biological. You don't understand how what you want—and therefore what you see—is conditioned by the immense, abysmal, profound past. You simply don't understand how every neural circuit through which you peer at the world has been shaped (and painfully) by the ethical aims of millions of years of human ancestors and all of the life that was lived for billions of years before that.

This quote succinctly summarizes several key themes developed at length in *Behave*. However, although Peterson draws adeptly on scientific explanations for *why* we perceive, reason, feel, and respond to various challenges as we do—pointing out, for example, that dominance hierarchies and neurotransmitters mediating evaluation of experiences are highly conserved in the animal kingdom, from lobsters to primates—his main concern is not with explanation, but rather with *how* to live a good and meaningful life despite inevitable limitations, challenges, suffering, and tragedy. What choices should we make? What habits and behaviors should we cultivate, to flourish as well as possible in a world full of risk and losses? The answers developed in *12 Rules* amount to a philosophy and recommendations for living well in a risky world.

Peterson does not ponder how or whether free will is possible in principle, but dives directly into the problems of choosing how to respond to a world that is often uncertain, unfair, and harmful—as well as surprisingly wonderful. The "chaos" in the subtitle of his book refers to the hazards, uncertainties, and known and unknown risks that continually threaten to collapse the secure order in which we seek to dwell. Chaos may manifest itself through catastrophic accidents, the illness or death of a child, or the ruin of a marriage, a company, an economy, a country, or an ecosystem. Moral and social virtues—including generosity (e.g., sharing unexpected good fortune), temperance and prudence (e.g., saving some for later), courage in exploring the unknown, fortitude and faith in the face of adversity, confidence and competence as well as mercy in navigating dominance hierarchies—all emerge as biologically and socially evolved solutions to the problems of living and thriving together in a risky world.

One possible reaction to the reality of painful risk and loss in life is to reject, resent, decry, and perhaps seek to destroy the partial order we have won from chaos, blaming and despising lives and a world that seem so manifestly imperfect. This possibility is illustrated by the life trajectories of some of Peterson's childhood friends as well as by quotes from several famous philosophers and authors. Peterson argues that this path of rejection, if followed far, can lead to deepening anger, outrage, and despair. It culminates in violent behaviors such as school shootings, domestic terrorism, substance abuse, or suicide, and the decay of the psyches and souls of those choosing it. Peterson sees the root of much malign will and behavior as rejection of a reality that is often painful as well as often sweet.

Alternatively, one may choose to acknowledge and confront—and, where possible, prevent and reduce—the suffering and risks that threaten the habitable order that we and our ancestors have created. Doing so requires humility and grace to learn well from experience and from others (past and present) about what is true and how it can best be met. It requires noticing and articulating unexpected and possibly threatening realities by seeing and saying what is true with precision, even when—or especially when—it is unpleasant or scary. It requires cultivating habits and developing capabilities and competence to reduce chaos and confront dangers as they arise. Finally, says Peterson, continuing to wrest and maintain habitable order from chaos requires teaching our children how to give and receive help and knowledge gracefully, so that they can participate more effectively and joyfully in the shared enterprise of fixing what is broken in the world, bit by bit, and creating and expanding the predictable and habitable order in which we thrive.

A major theme that Peterson develops and repeats largely in the context of Judeo-Christian religion is that being willing to sacrifice our preconceptions and wishes (and, indeed, everything else) to learn from reality how best to set and pursue high aims in this chaotic world *can provide a path* toward a meaningful life. Success is not assured: risk and tragedy are real. Courage and faith are necessary to undertake and persevere in trying to make increasing parts of the world stable, reliable, safe, and pleasant and in diminishing preventable harms and losses, or mitigating and repairing those that have occurred. But the struggle to do so gives zest and meaning to life. It is part of human nature to join in that struggle.

The struggle is waged in part via vigilant, realistic, and precise perception and articulation of risks, seeing and saying what is true about threats and dangers. These are essential first steps. Candid assessment, discussion, and deliberation with others can then help to refine and calibrate individual perceptions and reactions, set aspirations, and devise constructive responses to the risks at hand. Such deliberation is most likely to produce useful aspirations and plans if it is informed by knowledge of the hard-won lessons of the past on various time scales, some of which may be embodied in successful cultures, norms, religions, archetypal myths and stories, and institutions. Finally, personal commitments to persistently aim high and follow through by the daily practice of making achievable incremental improvements in what is in front of us, and in our own capabilities to oppose accident, injury, loss, and chaos, as time and opportunity and ability permit, make the struggle against chaos a practical personal reality. In Judeo-Christian religious language, Peterson equates the path of angry rejection and revenge on an already injured and imperfect real world, together with deliberate unwillingness to confront and learn about reality in favor of enjoying the simpler and more pleasing products of our own intellects, with roads to sin, death, and hell on earth. He identifies personal commitment to embracing reality, however painful, and seeking to make what we can better, one broken piece at a time, with laboring to bring the kingdom of God to earth.

Comments on *Behave* and *12 Rules*

Behave and *12 Rules* complement each other well for readers who are interested in two such different works. Reading *Behave* first highlights exciting advances, as well as important remaining gaps, in scientific understanding of the panoply of biological moderators of individual, group, and societal values and behaviors. But, after over 700 erudite, friendly, entertaining pages, the reader may be left wondering what to do to improve one's own perceptions and behaviors, other than refusing to oversimplify others by viewing them through us-them filters. *12 Rules* offers direct advice on how to elevate and improve one's own aspirations, values, habits, goals, and behaviors. To do so, *12 Rules* unabashedly goes beyond science to draw on insights from the lives of people the author knows and from archetypal stories, myths, and religions, which the author sees as distilling lessons from vast ages of painfully acquired and valuable experiences and insights. Sapolsky writes mainly for the head, Peterson largely for the heart.

For risk analysts interested primarily in brain science and risk psychology, the first 10 chapters of *Behave* are outstanding. They would make terrific supplementary reading for an undergraduate course on risk psychology and the biology of risk perception. They could also serve in such a course as a principle source for introducing the biology of human health effects of stress. For further philosophical reflection and some pragmatic common-sense advice on how our approaches to managing risks and coping with chaos express our evolved and embodied human nature, and shape our place in the world, *12 Rules* makes quick, interesting, and sometimes inspiring reading. However, its repeated, often passionate and lyrical, and sometimes mystical challenges to respond constructively to how the world really is while laboring to make it closer to what we want it to be, will not appeal to everyone. Still, the two books together provide an excellent start for understanding why we behave as we do and how we might choose to behave more effectively and constructively in the face of the uncertainties, anxieties, risks, limitations, and losses, that are important parts of even the best-lived lives.

Review of *Morality: Restoring the Common Good in Divided Times*

Is there such a thing as the *common good* in a fragmented society whose members hold sharply divided world views and political ideologies? If so, how can the field of risk analysis promote it or fail to do so? The late Jonathan Sacks' final book, *Morality: Restoring the Common Good in Divided Times* (2020) offers a diagnosis and proposes a remedy for many of the interconnected public health, social, political, psychological, spiritual, and moral discontents afflicting the developed world today. Along the way, it examines aspects of risk perception and psychology, personal and community resilience in the face of tragedy, and roles and prerequisites for cooperative analytic-deliberative thought, trust, open intellectual and policy conflict and debate, and effective self-governance in current Western democratic societies. Like *Behave* and *12 Rules*, *Morality* engages with moral, ethical, and spiritual questions that go well beyond the usual technical issues and insights addressed in most decision analysis and risk analysis textbooks, but that perhaps occupy a more central position in the reflections of many people who wonder how we can and should live better together.

The book consists of an introduction followed by 23 chapters organized into five parts on The Solitary Self; Consequences: The Market and the State; Can We Still Reason Together?; Being Human; and The Way Forward. Part 1, The Solitary Self, documents many interlinked current ills and tragedies in the U.S., U.K., and other developed countries. These include elevated teen suicide rates, prevalent anxiety and depression, loneliness, alcoholism, drug use, social unrest and violence, shaming and vigilantism on social media, increasing economic inequality, the rise of resentment-based politics (populism on the right and identity politics on the left), intoleranceand assaults on free speech in academia, the decline of families and volunteer organizations, incivility in public discourse, untrustworthiness of post-truth in news, and erosion of trust in governments and institutions. Sacks proposes as a common cause for these many symptoms the cultural pivot from "We" to "I" that took place largely in the 1960s, albeit with philosophical roots that he identifies as running back through postmodernism, Freud, Nietzsche, Marx, and others to the Enlightenment and Reformation. He states that "Divisive policies, inequitable economics, the loss of openness in universities, and the growth of depression and drug abuse ... are the long-term consequences of the unprecedented experiment embarked on throughout the West a half-century ago: *the move from 'We' to 'I.'*" He sees the liberal revolution—especially the conviction that it is not the task of the law to enforce a shared morality, but only to assure that people do not directly harm each other in exercising their individual autonomy—together with the abdication of traditional moral responsibilities by individuals during the expansion of free-market ideology in the 1980s and more recent technological shifts toward smartphones, internet, and social media, as undermining bonds of shared moral community that traditionally promoted trust and trustworthiness, self-restraint, and capacity for responsible individual freedom in a social world. Dissolution of these bonds also weakened traditional socialization, social capital, and cooperation essential for individuals to flourish together in communities and to be happy.

Part 2 (Consequences: The Market and the State) posits that neither the market nor the state can substitute adequately for the essential human need to help each other through altruistic activities and participation in community responsibilities. It argues that duties to care for each other and to cope with how our choices and behaviors might adversely affect ourselves and others—whether our own health, or our families, friends, and communities—cannot be outsourced successfully to government agencies or market institutions. Not only does self-serving behavior by companies and bureaucrats undermine their trustworthiness and credibility in serving the common good, but also personal contact, commitment, and caring are essential parts of effective help and healing. Sacks states that free markets and a free society need morals if they are to protect and respect the dignity and inalienable natural rights of individuals; that human flourishing and resilience and long-run ("eudaemonic") happiness despite life's vicissitudes depend on virtue, close social and family connections, and doing good rather than just doing well; and that moral communities where people help each other in face-to-face relationships are a historically important part of the Anglo-American political model that cannot and should not be left to the state, market, or private individual spheres. He sees risk-analytic thinking as having limited practical value for addressing policy problems with deep uncertainty, noting that utilitarian analytic-deliberative principles for calculating desirable conduct under uncertainty by considering risks, costs, and benefits for all affected parties are seldom helpful practical guideswhen the consequences of our present actions for ourselves and for others are long delayed and difficult or impossible to foresee accurately and confidently. Instead, moral norms grounded in experience about how to live well together over time are usually more compelling to most people. For example, people may take precautions not to harm others or future generations more readily if they see it as "the right thing to do"rather than as merely the outcome of a risk-benefit-cost calculation.

Part 3 (Can We Still Reason Together?) poses a central question for risk analysts and others involved in analytic-deliberative decision-making: in an age of unreliable news, misinformation, and disinformation, "narrowcasting" that tailors news feeds to appeal to and reinforce the confirmation biases of their recipients, and "post-truth" rhetoric that often emphasizes ethos and pathos more than logos, how much room remains for collaborative pursuit of objective scientific truth through reasoned debate? Sacks argues that shared respect for truth as an end in itself, together with the patience and humility required to test the truth of hypotheses and beliefs against data, as in the traditional scientific method, provides a moral foundation for mutual respect, trust, and collaboration in truth-seeking that can cut across individual differences and resist and transcend ideological fashions, political power and news media manipulation, and the pull of individual interests and preferences.

Although Sacks acknowledges the formidable power of recent campus politics—of safe spaces, deplatforming and no-platforming, concerns about perceived microaggressions and power imbalances, and intolerance of intellectual diversity—to chill and suppress open intellectual inquiry and discussion, he concludes that hearing both sides of arguments, and accepting the resulting risks of offense and disagreement, is essential for any person, institution, or community to pursue justice, expansion of knowledge, and truth. He uses moving accounts of Holocaust survivors who flourished later in life to emphasize that resilience to tragedy and trauma is founded on a focus on building the future—deciding what to do next and doing it rather than being consumed by pain and resentment for past evils and injustices. Sacks argues that grievance-oriented politics and cultures of victimhood, tragedy, catastrophe, and oppression undermine the very habits of future-oriented choices and politics and cultures of hope and resilience that are needed to thrive in an uncertain and painful world.

Part 3 also suggests that a shared focus on building the future, shared respect for truth, and shared devotion to serving the common good insofar as we can discern it fosters a highly desirable form of civility in politics and argumentation—"dining with the opposition"—rooted in mutual respect and dignity and in awareness that the game is larger than the teams and that the moral bonds that unite us are far stronger than the differences in policy preferences that divide us. Conversely, such civil deliberation and debate disappear when each side views the other as an enemy, less than fully human, not part of a shared moral community and when each side views the purpose of argument and deliberation as being to win and to enforce one's preferred views, rather than being to get at the truth and to discover the best way forward together. Incivility and inability to respect and cooperate effectively with each other are promoted by narrowcasting of news playing to different communities of confirmation bias; disinhibition and public shaming in social media comments in place of judicial establishment of guilt; and fragmentation of polities into identity and interest groups that shout at each other rather than listening respectfully and reasoning productively together. Sacks suggests that the dissolution of respectful deliberation informed by shared truth-seeking analyses—arguably, crucial underpinnings of applied risk analysis in the analytic-deliberative mold—poses an existential threat to Western democracies.

Part 4 (Being Human) begins by contrasting different views of humanity, from materialist and scientism propositions that mankind is nothing but an accidental product of physical laws and that free will is an illusion to the Kantian and Biblical view that dignity and morality are inseparable parts of our nature as freely choosing, responsible moral agents. It contrasts science rooted in testable predictions based on causal generalizations from experience with freedom and decision-making in which people choose how to act and live over time and choose the aspirations and envisioned better futures that they will pursue. Sacks argues that our capacity to ask ourselves what is right before choosing what to do, our ability to develop a moral sense through interactions with significant others, and our freedom to live by its constraints—refusing to engage in some activities that we might like because doing so would harm ourselves or others—or to reject these constraints are essential parts of what makes us human. Free will and moral reflection and decision-making are perhaps not readily amenable to scientific and reductionist understanding, but this does not diminish their reality or importance. Our capacity for moral growth over time through challenges and adversity ("redemptive suffering") and the social development of individuals as cooperating members of moral communities rather than isolated strangers are also developed in Part 4 as essential aspects of humanity, leading to a discussion of different ethical cultures (the civic ethics of ancient Greece and Rome; duty ethics; honor ethics; and the ethics of love instantiated in Judaism and Christianity) and different moral personalities (tradition-directed, inner-directed, and other-directed). Part 4 closes with a discussion of the emergence of altruism from kin-selection to reciprocity, the game-theoretic and historically consequential challenges of establishing social capital and trust among unrelated strangers, and the roles of religion in solving these and other challenges (such as inter-generational altruism andresponsible stewardship of resources) by creating long-lasting moral communities.

Part 5 (The Way Forward) consists of two short but vital chapters ("Morality Matters" and "From 'I' to 'We'") that set forth two crucial clusters of ideas for managing risks and improving lives in a highly uncertain, sometimes dangerous, and rapidly changing world. The first is that capacity for *collective self-help* at the level of families, communities, and volunteer organizations is an essential distinguishing feature of strong civil societies. Sacks urges that it will become increasingly necessary to restore this capacity as the limits of the caring state become increasingly clear, especially, its intrinsically impersonal approach that separates "care" delivery from shared moral codes, obligations, and rebukes that friends or family can deliver but that the state cannot. The second is the concept of covenant: a shared commitment to take care of each other no matter what the future holds. Where contracts are created and sustained by the self-interests of those involved, covenants are sustained by the loyalty, fidelity, and morality of those involved, even when sacrifice is necessary and self-interest must be transcended. Sacks argues that covenants create moral communities with strong bonds of trust and social capital. He argues that history shows that it is both possible, and necessary for the survival of democracies and civil liberties, that a covenant dimension be restored to politics. He states that "The politics of covenant does not demean or ridicule opponents. It honors the process of reasoning together. It gives special concern to those who need most help, and special honor to those who give most help. . . . Politics [in the United States and other Western-style democracies] has regressed from covenant to contract: we pay our taxes, the government delivers services, and we search for a deal that is most advantageous to us. This is a diminished view of politics, which can work for a while, but which cannot hold together divided societies." By contrast, covenantal politics is about "We the people" bound by a shared moral commitment to care for each other at the local and personal level as well as in questions of national policy. Sacks recalls Lincoln's reflections on the Civil War, and his recasting it in moral terms that invoked the covenantal language and aspirations of the Declaration of Independence, as an example of restoring the covenant dimension to politics in sharply divided times. The book concludes that the concept of covenant has the power to transform the world. A brief epilogue suggests that rebuilding after the Covid-19 pandemic may provide an opportunity to start applying these principles now.

For risk analysts, *Morality* offers insights into both the strengths and the vulnerabilities of our discipline. On the one hand, it emphasizes throughout the importance of being able to reason together and the importance of shared respect for pursuit of objective truth even when ideologies differ. Risk analysis in the analytic-deliberative mode perhaps epitomizes the process of reasoning together. It is therefore vulnerable to ideological or academic intolerance that subordinates open investigation, disagreement, and debate as part of a shared search for truth to conformity and homage to ideological positions. On the other hand, the book also emphasizes throughout the difficulty or impossibility of predicting future events in complex systems with high accuracy and confidence. It refers repeatedly to the "fatal conceit," in Hayak's term, of supposing that we can predict better than our predecessors the likely long-term consequences of policies and social experiments such as the move during the 1960s toward a society governed by self-interest without conventional moral codes. How, then, can we best manage unforeseeable risks and unintended consequences of policies, including risk management-related policies and interventions such as the war on drugs and the war on poverty? Sacks' answer is that the solidarity and resilience of self-helping communities supported by covenantal politics rooted in strong individual and local morality is the best response to such radical uncertainty. Even when risks are not easily quantifiable and capacity to manage them is limited, commitment to care for each other directly at the local level no matter what comes provides the best chance to overcome divisions and to thrive together under uncertainty.

Morality is an engaging and beautifully written book. It is enlivened by many examples drawn from the author's own experiences and those of others he talked to over the course of a fascinating career as a public intellectual, Professor of Law, Ethics, and the Bible at King's College London; Chief Rabbi of the United Hebrew Congregations of the Commonwealth from 1991 to 2013; and peer in the House of Lords since 2009. The book would be appropriate for an undergraduate or graduate course in ethics for decision analysts, especially since it considers social aspects of moral decision-making and evaluation of outcomes on multiple time scales that are often not emphasized in axiomatic normative decision theories. *Morality* would also be useful for courses or seminars in risk analysis as background reading for discussions of the sources and nature of personal, community, and societal resilience in the face of tragedy and disaster. Its argument that moral guidance is often more effective than technical calculations in changing behaviors gives risk communication specialists and the wider community of professional risk analysts much that is worthwhile to think about.

References

Ariely D (2008) Predictably irrational: the hidden forces that shape our decisions. Harper-Collins Publishers, New York, NY

Blastland M, Dilnot A (2009) The numbers game: the commonsense guide to understanding numbers in the news, in politics, and in life. Penguin Group, New York, NY

Colledanchise M, Ogren P (2020) Behavior trees in robotics and AI: an introduction. https://arxiv.org/pdf/1709.00084.pdf

de Pontes Pereira R, Engel PM (2015) A framework for constrained and adaptive behavior-based agents. CoRR, abs/1506.02312. https://dblp.uni-trier.de/db/journals/corr/corr1506.html#PereiraE15

Du S, Hu W, Li Z, Shen R, Song Z, Wu J (2020) When is particle filtering efficient for POMDP sequential planning? https://arxiv.org/abs/2006.05975

Eskandari H, Mahmoodi E, Fallah H, Geiger CD (2011) Performance analysis of commercial simulation-based optimization packages: OptQuest and Witness optimizer. In: Jain S, Creasey RR, Himmelspach J, White KP, Fu M (eds) Proceedings of the 2011 winter simulation conference. https://www.informs-sim.org/wsc11papers/212.pdf

Gardner D (2008) The science of fear: how the culture of fear manipulates your brain. Penguin Group, New York, NY

Hallinan J (2009) Why we make mistakes: how we look without seeing, forget things in seconds, and are all pretty sure we are way above average. Broadway Books, New York, NY

Harford T (2012) Adapt: why success always starts with failure. New York, NY, Farra, Straus and Giroux

Howard RA (1960) Dynamic programming and Markov processes. The MIT Press. https://gwern.net/docs/statistics/decision/1960-howard-dynamicprogrammingmarkovprocesses.pdf. Last accessed 1-28-2023

Juan AA, Faulin J, Grasman SE, Rabe M, Figueira G (2015) A review of simheuristics: extending metaheuristics to deal with stochastic combinatorial optimization problems. Oper Res Perspect 2:62–72. ISSN 2214-7160, doi:https://doi.org/10.1016/j.orp.2015.03.001. (https://www.sciencedirect.com/science/article/pii/S221471601500007X)

Kahneman D (2011) Thinking, fast and slow. Farrar, Straus, and Giroux, New York, NY

Kaplan M, Kaplan E (2009) Bozo sapiens: why to Err is human. Bloomsbury Press, New York, NY

Kiciman E, Sharma A (2021) Causal reasoning: fundamentals and machine learning applications. https://causalinference.gitlab.io/causal-reasoning-book-chapter1/

Lee K, Laskin M, Srinivas A, Abbeel P (2020) SUNRISE: a simple unified framework for ensemble learning in deep reinforcement learning. https://arxiv.org/abs/2007.04938

Lehrer J (2009) How We Decide. Houghton Mifflin Harcourt Publishing Company, New York, NY

Leike LT, Orseau L, Hutter M (2016) Thompson Sampling is asymptotically optimal in general environments. arxiv, abs/1602.07905

Moradi P, Shiri ME, Entezari N (2010) Automatic skill acquisition in reinforcement learning agents using connection bridge centrality. In: Kim T, Vasilakos T, Sakurai K, Xiao Y, Zhao G, Ślęzak D (eds) Communication and networking. FGCN 2010. Communications in computer and information science, vol 120. Springer, Berlin, Heidelberg. https://doi.org/10.1007/978-3-642-17604-3_6

Peterson J (2018) 12 Rules for life: an antidote to chaos. Penguin Random House Canada, Toronto

Piché A, Thomas V, Ibrahim C, Bengio Y, Pal C (2019) Probabilistic planning with sequential Monte Carlo methods. ICLR

Pinker S (2021) Rationality: what it is, why it seems scarce, why it matters. Viking, an imprint of Penguin Random House LLC, New York, NY

Sacks J (2020) Morality: restoring the common good in divided times. New York, NY, Hachette Book Group

Sapolsky RM (2017) Behave: the biology of humans at our best and worst. Penguin Press, New York, NY

Shen W, Trevizan F, Toyer S, Thiebaux S, Xie L (2019) Guiding search with generalized policies for probabilistic planning. In: Proceedings of the twelfth international symposium on combinatorial search (SoCS 2019)

Shu T, Xiong C, Socher R (2017) Hierarchical and interpretable skill acquisition in multi-task reinforcement learning. https://arxiv.org/abs/1712.07294

Thaler RH (2016) Misbehaving: the making of behavioral economics. W.W. Norton & Company, Inc, New York, NY

Chapter 2
Data Analytics and Modeling for Improving Decisions

Introduction

This chapter turns to data science and analytics methods for improving decision-making and for using data and modeling to help overcome the psychological obstacles to accurate risk perception and belief formation discussed in Chap. 1. It continues Chap. 1's survey of recent literature, summarizing key ideas from the following five books:

- *Superforecasting: The Art and Science of Prediction*, by Philip Tetlock and Dan Gardner (2015)
- *The Art of Statistics*: How to Learn from Data, by David Spiegelhalter (2019)
- *The Model Thinker: What You Need to Know to Make Data Work for You*, by Scott Page (2018)
- *On Grand Strategy*, by John Lewis Gaddis (2018)
- *Poor Economics: A Radical Rethinking of the Way to Fight Global Poverty*, by Abhijit Banerjee and Esther Duflo (2011)

The first three of these books present technical approaches to data analysis, modeling, and analytic thinking that can inform System 2 deliberations and improve predictions and formation of more accurate beliefs about event probabilities. *On Grand Strategy* examines lessons from history about how (and how not) to respond to opportunities and change to form and pursue goals over time. This is a topic that has not yet been formalized and incorporated as part of traditional decision analysis, which takes preferences as given. It will be an important theme in Chap. 3 and later chapters in considering AI methods for forming and coordinating goals and plans on multiple time scales. Finally, *Poor Economics* addresses the extent to which data analysis and risk analysis principles can be applied to successfully alleviate human misery and promote human flourishing by breaking self-sustaining poverty cycles. This important work, which contributed to a 2019 Nobel Memorial Prize in

L. A. Cox Jr., *AI-ML for Decision and Risk Analysis*, International Series in Operations Research & Management Science 345,
https://doi.org/10.1007/978-3-031-32013-2_2

Economic Sciences for authors Banerjee and Duflo, shows the practical value of data analysis for discovering causal relationships between interventions and consequences that can inform successful policymaking.

Forming More Accurate Beliefs: *Superforecasting*

How well can people predict what will happen in politics, wars, current events, sports, financial markets, and other areas not easily described by well-validated probabilistic models? Are some people much better at forecasting than others? If so, is it possible to discover why, and to teach this skill to others? These questions are addressed in the fascinating 2015 book *Superforecasting: The Art and Science of Prediction*, by political scientist Philip Tetlock and journalist Dan Gardner, the author of *The Science of Fear* discussed in Chap. 1. The answers, in brief, are that a small minority of people are indeed much better forecasters than most of us; that habits of mind used to seek, interpret, and update information explain a large part of the difference; and that these skills can be taught and learned, although they can be perfected only by long practice. The book's approach throughout is empirical, informed by results from the Good Judgment Project, which studies the performance over time of multiple forecasters on a variety of prediction tasks with clearly defined time frames and unambiguous criteria for whether events occur in those time frames, and with forecasts delivered as numerical probabilities.

Superforecasting begins with a review of the treacherous psychology of intuitive forecasting and use of evidence. This background covers a wide range. It discusses overconfidence in initial judgments, which is often abetted both by confirmation bias in seeking and interpreting evidence, and also by hindsight bias that leads us to believe, after the fact, that whatever happened was predictable. Other limitations of intuitive prediction include Kahneman's "what you see is all there is" heuristic and well-documented proclivities to confabulate and explain in the absence of adequate data—and then anchor on the results, perhaps seeing pattern and meaning where there is none. The history of Archie Cochrane's difficulties in persuading his colleagues to adopt basic principles of evidence-based medicine and stories of the frustrating but frequent conjunction of ignorance and confidence among experts and leaders as well as laypeople illustrate the powerful grip and needlessly poor performance of intuitive judgment and forecasting by individuals, groups, and professions. Technical concepts of calibration, resolution, and Brier scores are explained clearly, accurately, and in pleasant prose for non-specialists, to clarify what is meant by a "good" probabilistic forecast. The bottom line, convincingly supported, is that intuitive and expert forecasts are usually not very good and are usually regarded as better than they are, both by those making them and by those receiving them. We all tend to be complacent about and invested in our judgments and the evidence on which they rest—except for the rare "superforecasters" whose forecasts are much better than most people's, time after time and across multiple domains.

Superforecasters remain open-minded, always regarding their current beliefs as hypotheses to be tested and improved by new information. They are eager to update their current judgments frequently and precisely, actively seeking and conditioning on new data and widely disparate sources of data and evidence that might disprove or correct their current estimates. They make fine-grained distinctions in their probability judgments, often adjusting by only one or a few percentage points in light of new evidence, which is a level of precision that most people cannot bring to their probability judgments. Midway through the book, the authors offer the following rough recipe for improving probability forecasts: (1) "Unpack" the question to which the forecast provides an answer into its components. (2) Distinguish as well as possible between what is known and unknown and scrutinize all assumptions. Implicit assumptions that go unquestioned have been the bane of many a confident prediction. (3) Consider other, similar cases and the statistics of their outcomes (taking what the authors call "the outside view") and then (4) Consider what is special or unique about this specific case in contradistinction others (the "inside view"). For example, in predicting sales for a new business, consider relevant statistics for similar businesses, thus avoiding base rate neglect, but also consider what is unique about this business. (5) Exploit what can be learned from the views of others, especially those with contrasting informed predictions, as well as from prediction markets and the wisdom of crowds. (6) Synthesize all of these different views into one (the multifaceted "dragonfly view," in the authors' term) and (7) Express a final judgment, conditioned on all this information, as precisely as possible using a fine-grained scale of probabilities. Skill in forecasting using this guidance can be built through informed practice and clear, prompt feedback, provided that there is a deliberate focus on tracking results and learning from mistakes.

A striking theme developed toward the end of the book is that teams can reliably outperform individuals under certain conditions—but only if those conditions are met. Many pitfalls impede the performance of groups and organizations in making usefully accurate predictions. They range from groupthink and social loafing to ego investment in publicly stated positions to use of ambiguous language ("weasel words"). Such language can give people a false sense of engaging in useful discussion and reaching well-informed consensus even if participants agreeing to the final judgment have very different understandings of what it actually means and agree only because these discrepancies have been obscured by vague language. Good group dynamics can boost organizational performance and resilience in the face of risk and uncertainties. These dynamics include precise questioning, constructive confrontation, intellectual humility and focus by leaders on clearly defining goals while appropriately delegating tactics to better-informed subordinates, and willingness to acknowledge key uncertainties, plan for surprises, and improvise as needed. Sharing information and predictions effectively in teams and vigorously questioning assumptions enables teams to reliably out-perform even the best individual superforecasters by about 23% on a wide range of prediction tasks.

Superforecasting presents many practical applications of its recommended methods. One is a demonstration that it is relatively cheap and easy to improve upon the risk estimates produced by the intelligence community, with its annual budget of about 50 billion dollars, by averaging the forecasts of about 100 ordinary people, giving extra weight to the top performers on practice problems, and then tweaking the average to make it more extreme, i.e., closer to 0% or 100%, since many people are biased toward the middle. The book ends with reflections on the potentially revolutionary impact of a commitment to evidence-based forecasting, evidence-based policy making, and deliberate learning from experience and intelligent trial-and-error how to improve predictions and policies over time. Asking well-designed questions can help trained superforecasters provide the probabilistic information that trained policy makers and risk managers in business, government, and the military need to act more effectively in an uncertain world.

Superforecasting has generally been favorably reviewed, and it is easy to see why. It is engagingly written. It presents important ideas from risk psychology and decision science, as well as recent findings from the ongoing Good Judgment Project (http://www.goodjudgment.com/), with clarity, and vigor, enlivened by compelling examples. Its lessons have the potential to make probabilistic predictions far more informative than is usual today, and decisions and policies based on them correspondingly more likely to succeed.

Learning About the World Through Data Analysis: *The Art of Statistics*

How should statistics be taught in the present age of data science, big data, artificial intelligence and machine learning (AI/ML), and sophisticated computational statistics algorithms and software? Does progress in these areas provide new foundations and algorithms for statistical inference that can enable us to learn more or learn better from data? To what extent should it render parts of the traditional statistics curriculum obsolete, or provide essential new content that students of applied statistics should understand to be effective in the modern world? Eminent statistician David Spiegelhalter addresses these issues with skill and insight for general readers in his new book *The Art of Statistics: How to Learn from Data* (Basic Books. New York, NY. 2019). The book's Introduction and 14 chapters explain how to use modern data analytics to learn about the world. It illustrates key concepts and ideas of modern statistics and AI/ML with a host of predominantly medical, survival data, and public health examples. The writing is cheerful, the pace brisk, and the content technically well-informed and gracefully expressed in plain English, augmented by simple tables, graphs, and classification and regression trees (CART trees).

The Introduction sets the stage by showing how plotting and inspecting data in various ways can reveal unexpected and important patterns (such as about when and how a doctor started killing his elderly patients), despite the necessary imperfections of surveys and measurements that attempt to reduce the real world to data for purposes of analysis, and the occurrence of random variability in both measurements and measured phenomena. It also sounds a call for a modern pedagogy that emphasizes that statistics, as part of data science, should be taught and practiced as far more than a bag of techniques backed by probability theory. It should also embrace machine learning algorithms for discovering and validating patterns in data, including big data (many rows) and rich data (many columns), while avoiding the types of data dredging, investigator biases and modeling choices, selective reporting, and over-interpretation that have led to widespread (if not universally accepted) concerns about high false discovery rates and the reproducibility or replication crisis in science. To increase data literacy and competence in data science, applied statistics should be taught to emphasize its problem-solving structure, such as the "problem-plan-data-analysis-conclusion and communication" (PPDAC) cycle of investigation, with special attention to fully acknowledging, assessing, and communicating the limitations of evidence and the trustworthiness of statistical claims. The rest of the book deliberately takes real-world problem solving, rather than theory or textbook examples, as the principle vehicle for introducing modern statistical ideas and methods.

Chapter 1, "Getting things in proportion: Categorical data and percentages" notes the importance, familiar to risk analysts, of expressing rates and proportions in ways that the audience understands, taking into account that how data is displayed can greatly affect perceptions of risk and other quantities. Data visualization can support or undermine accurate, understandable understanding and comparison of relative and absolute risks and survival rates, and even the choice of whether to present mortality or equivalent survival rates (negative or positive framing) can greatly affect emotional responses and perceived risks. It will be gratifying to risk analysis to see these aspects of the psychology of risk perception emphasized immediately as part of thinking about how best to communicate statistical information about health risks (North 2012). As real-world examples, Chap. 1 discusses mortality rates from pediatric heart surgery, cancer risks from bacon, and confusion in the press between absolute and relative risks and between relative risks and odds ratios (leading to an increase in absolute risk of reported muscle pains from 85% to 87% for non-statin users vs. statin users being misreported by a major newspaper as a "20 per cent" increase in risk!)

Chapter 2, "Summarizing and communicating numbers. Lots of numbers" continues the theme of communicating data through numbers and plots. It briefly (in a few lines each) introduces histograms, box-and-whisker plots, and skewness (and log-transformations of axes to reduce it) for summarizing empirical distributions; count and continuous variables; mean, median, and mode as summaries of location; interquartile range (IQR), outliers, and standard deviations as indicators of spread; differences of means, medians, and distributions; scatter plots, Pearson's correlation coefficient and Spearman's rank correlation coefficient for pairs of variables (taking as a motivating example the question of whether hospitals that do more surgeries tend to have better safety records); and line plots showing temporal trends. Chapter 2 concludes with advice for communicating statistical information effectively, including fighting "the temptation to be too sophisticated and clever, or put in too much detail" and using data visualization ("dataviz") and infographics ("infoviz") to tell and illustrate compelling data-driven stories and engage audiences in understanding the data while trying to inform decision-making without manipulating audiences or triggering inappropriate gut reactions.

Chapter 3, "Why are we looking at the data anyway? Populations and measurement" reviews the sometimes tenuous links between what is true in survey responses and sample data and what is true about the larger study population being sampled, or about the target population to which statistical inferences are to be applied. It discusses challenges of measurement reliability and validity, biases (e.g., from the wording of survey questions), internal and external validity of inductive inferences, extrapolation from a study population (e.g., mice) to a target population (e.g., people), and the usefulness of random sampling, emphasizing that a population can be thought of not only as a group of individuals, but also as providing a probability model for random samples drawn from it. The chapter briefly introduces use of a few population parameters to summarize entire frequency distributions (e.g., the mean and variance for a quantity having approximately a normal distribution in a population, such as birth weights of babies) and the use of sample statistics (e.g., sample means and variances) to estimate population parameters. It concludes by noting that, in the age of big data, even data on all of the individuals in a population (e.g., all customers, all patients, all voters, etc.) can be conceptualized as a sample from a "metaphorical population of events that could have occurred but didn't." This concept of counterfactual events leads into Chap. 4, "What causes what?" which discusses the distinction between evidence of statistical associations and evidence of causality in epidemiology and risk assessment. Randomized controlled trials (RCTs), systematic review and meta-analysis of multiple studies, A/B testing, and use of observational data from cross-sectional, cohort, and case-control studies to draw causal inferences are all discussed in this concise (25-page) chapter, along with important caveats about adjusting for confounding, latent ("lurking") variables, selection biases, and Simpson's Paradox, in which the direction of an association is reversed by conditioning on additional variables.

The next three chapters deal with predictive analytics. Chapter 5 ("Modelling relationships using regression") extends Chap. 4's discussion of causality to the context of dependent and independent variables, where the observed value of the dependent variable is modeled as the sum of the output from a deterministic model (a deterministic function of the independent variables, or inputs) plus a residual error that is often assumed to be normally distributed with zero mean and constant variance. Using motivating examples such as evaluating the effectiveness of installing red light cameras at intersections to reduce car crashes, Chap. 5 explains the phenomenon of regression to the mean (that cases selected for intervention because of unusually high risks or poor performance will tend to perform better on average later, whether or not the intervention has any causal effect). It emphasizes the distinction, often blurred in epidemiology and public health risk assessments, between interpreting the slope of a regression lines as predicting the *expected difference* in observed values of a dependent variable per unit of difference in observed values of an independent variable; and interpreting it as a causal relationship (e.g., the potency of a causal dose-response relationship), i.e., as predicting the *expected change* in the dependent variable caused by an intervention that changes an independent variable by one unit. Chapter 5 also notes that human psychology encourages attributing observed changes to human interventions, and mentions Judea Pearl's work on building causal regression models from observational data. The concluding sections of the chapter introduce multiple linear regression models with many explanatory variables, some of which may be categorical rather than numerical; nonlinear regression models; and regression models with proportions, counts, or survival times for their dependent variables.

Chapter 6 ("Algorithms, analytics, and prediction") turns to artificial intelligence and machine learning (AI/ML) algorithms that analyze past data to create decision rules for classifying new cases (via "supervised learning," i.e., training on known correctly classified cases); discovering patterns or clusters ("unsupervised learning"); and predicting future observations from past ones. The challenges of selecting informative, non-redundant measures with high predictive power ("feature engineering") and of fitting flexible nonparametric models to data using ML methods such as deep learning are briefly mentioned. Then the chapter introduces classification trees as a core ML method, illustrating their use by modeling survival probabilities for passengers on the *Titanic* having different attributes. The results motivate discussion of how the performance of predictive models can be assessed via measures such as sensitivity, specificity, areas under receiver operating characteristic (ROC) curves, confusion matrices, and calibration plots, for binary and categorical dependent variables; or by mean-squared-error (MSE) of predictions for continuous dependent variables, and by Brier scores for probabilities. The crucial ML concepts of model over-fitting, bias-variance tradeoff (between models with enough flexibility to describe training data very accurately by "training on the noise" and models with better generalizability and predictive power), and cross-validation for model tuning are introduced for classification trees and regression models. The use of regularization (penalizing for complexity and non-zero coefficients), boosting (over-training on misclassified cases), and interaction terms in linear and logistic regression are described in a few lines each. More complex techniques including random forest, support vector machines, neural networks and deep learning, and k-nearest neighbor classifiers are mentioned but not explained in detail. The chapter ends by discussing the risks of developing overly complex "black-box" ML predictive algorithms that might perform well in Kaggle competitions but that lack (a) robustness for use in a changing world; (b) stability as additional samples are taken; (c) ethical sensitivity to what variables (e.g., age, sex, or race) should and should not be used in making consequential decisions; and (d) transparency and interpretability for human users. It concludes, again citing Judea Pearl's work on causality in AI, that the lack of understanding of causality in AI/ML algorithms substantially restricts their practical use to predicting what is expected to be observed in the absence of interventions, rather than predicting the probable consequences of interventions—which may be of far more interest to risk analysts and policy makers. Therefore, the last sentence of the chapter states, "A basic humility when building algorithms is crucial."

Chapter 7 ("How sure can we be about what is going on? Estimates and intervals") describes resampling methods—specifically, the bootstrap—for constructing interval estimates by randomly resampling from the sample data many times (with replacement) to create many "bootstrap samples," each the same size as the original sample. Calculating the statistics of interest (e.g., the mean or median of a distribution, or the slope of a regression line) for each random sample creates a bootstrapped distribution for its value. This illustrates part of the proposed new pedagogy for teaching applied statistics. It focuses on answering useful questions, such as about the distribution of a quantity calculated from sample data, by computational methods before developing underlying theory. This allows natural questions about the accuracy of sample-based estimates of quantities to be addressed immediately via computation, even before the probability theory needed for a traditional development (via the Central Limit Theorem (CLT) and approximate normal distributions of sample means around the true mean) has been covered. The finding that bootstrapped sample means for variables in large datasets are approximately normally distributed around the population means, regardless of the shape of the original data distribution, emerges here as an empirical discovery (justified only later by the CLT) that allows confidence intervals and margins of error to be computed without using probability theory or making strong assumptions. Chapter 8 ("Probability—the language of uncertainty and variability" starts introducing formal probability theory, using probability trees and expected frequencies to ease exposition and calculations. It introduces the key ideas of random variables, independence, conditional probability, Poisson distributions—and their remarkable success in describing a wide range of phenomena, from daily homicide counts to goals scored in football matches—and probability models for randomness and interindividual variability in a population. Chapter 9 ("Putting probability and statistics together") begins with a warning that the content of the chapter is challenging but worthwhile, and then introduces the key idea of treating statistics calculated from sample data as random variables with their own distributions. Binomial distributions, expected values, standard errors, and tail areas of distributions are introduced in a few pages, along with control limits, funnel plots, and the Law of Large Numbers for describing natural variability in populations of different sizes. A section on the Central Limit Theorem discusses normally distributed sample means for large sample sizes and calculation of standard errors for sample means. This is followed by a discussion of aleatory and epistemic uncertainty; 95% prediction intervals for sample statistics when population parameters are known; and construction of 95% confidence intervals for unknown population parameters from observed values of sample statistics. Chapter 9 cautions that statistical inferences based on probability models, including margins of error, can be incorrect and misleading if the assumptions of the models do not hold, e.g., if sources of systematic error and bias have been overlooked.

Chapter 10 (Answering questions and claiming discoveries") addresses hypothesis-testing. It first explains the concept of a null hypothesis and how formal hypothesis-testing can both help to overcome the very human tendency to see patterns where they do not exist (apophenia), and also reduce rates of false discoveries. Similar to Chap. 7's introduction of the bootstrap to answer confidence interval questions without assuming any specific probability model, Chap. 10 presents permutation tests, in which a distribution of outcomes is simulated under the null hypothesis of independence between two variables, and used to quantify the P-value, i.e., the probability of getting a result at least as extreme as the observed one if the null hypothesis and modeling assumptions are correct. One-sided and two-sided tests, statistical significance, and Null Hypothesis Significance Testing (NHST) are explained first in this computational context, where no mathematical probability theory is required. Then the chapter shifts gears, describing in words how chi-squared goodness-of-fit tests and tests for independence, hypergeometric distributions, t-distributions, 95% confidence intervals, and normal approximations to them (based on means plus or minus 2 standard errors), are used in traditional hypothesis testing. It illustrates these ideas with application such as testing whether taking statins reduces heart attack risks. These technical essentials explained, a section on "The danger of carrying out many significance tests" warns practitioners against multiple testing biases and false-positive results. As a memorable example, it describes a brain imaging study that found that a subject shown photographs of humans expressing different emotions had highly statistically significant responses (at the $P < 0.001$ level) responses in 16 specific brain sites (out of over 8000 monitored)—even though the subject was in fact a dead fish! Bonferroni corrections and replication are recommended for controlling false discovery rates, and the Neyman-Pearson theory of hypothesis testing is explained, with its implications for choosing sample sizes large enough to assure desired levels for both size and power of a test (i.e., probabilities of avoiding both false positives and false negatives). The theory of sequential testing is described, including Wald's Sequential Probability Ratio Test (SPRT), along with an important caveat that such statistical tests can identify significant differences between observed and expected values in quality control, medicine, and other applications, but they do not reveal *why* the differences occur. Chapter 10 ends with a discussion of misuse of P-values; the importance of publication bias, other biases, and the reproducibility crisis in science (including Ioannidis's famous 2005 claim that "most published research findings are false"); the American Statistical Association's six principles for using and interpreting P-values; and the distinction between statistical significance and practical significance. This paves the way for a discussion of Bayesian alternatives to NHST in Chap. 11 ("Learning from experience the Bayesian way").

Chapter 11 introduces Bayes' Rule by using expected frequency trees as a simple, intuitive approach, and then presented in the form: final ("posterior") odds for a hypothesis = initial ("prior") odds x likelihood ratio, where the likelihood ratio is the probability of the evidence given one hypothesis divided by the probability of the evidence given the alternative hypothesis. Applications to drug testing and forensic science are discussed, and the difficulty of obtaining priors that are widely accepted is acknowledged. The chapter strongly recommends using sensitivity analyses to deal with uncertainty about priors. Chapter 11 also discusses hierarchical Bayesian modeling and the important modern technique of multi-level regression and post-stratification (MRP) for Bayesian pooling and smoothing of sparse, geographically distributed data that evolve over time and spatial locations, such as disease data or polling data. The chapter ends with a discussion of the ideological battle between frequentists and Bayesians and the controversy over use of Bayes factors ("the equivalent of likelihood ratios for scientific hypotheses") as a substitute for NHST. It concludes that such philosophical differences are less likely to be the sources of important problems and limitations in statistics than "inadequate design, biased data, inappropriate assumptions and, perhaps most important, poor scientific practice." These practical challenges are the subject of Chap. 12.

The last step in the PPDAC framework for applied statistics is communicating conclusions. Chapters 12–14 deal with this crucial step. Chapter 12 ("How things go wrong") details what often goes wrong in forming and communicating statistical conclusions in epidemiology, health risk analysis, and other applications. Aside from deliberate fraud, prevalent challenges include mistakes in analysis or interpretation (such as invalid study designs or misinterpreting "non-significant" as "no effect"); uncorrected multiple testing bias and resulting false discoveries; selective reporting, publication biases, and failures of peer review; P-hacking, blurring of exploratory and confirmatory studies, and HARKing (inventing Hypotheses After the Results are Known); filtering and distorting results through the media (which tends to sensationalizes results, often by reporting only relative and not absolute risks); and questionable research practices, in which investigator choices throughout the data analysis and interpretation lead to conclusions of questionable validity, contributing to the reproducibility crises in science (or, less dramatically, the unreliability of many published scientific findings). Chapter 13 (How we can do statistics better") recommends potential solutions to these ills. It recommends using data sharing and pre-registration of studies (e.g., via the Open Science Framework); clearly distinguishing between exploratory and confirmatory studies; improving communication of probabilities and statistical results to make them more accessible, intelligible, assessable, and useable; calling out poor practice (journal referees, especially, should do this more); checking the statistical distribution of reported p-values for evidence of p-hacking (e.g., a spike or reported p-values just under 0.05); checking for rigor of studies and their conclusions (e.g., via checks for internal validity, control groups, pre-registered protocols, and representative samples); and candid presentation of remaining uncertainties. The book concludes with a two-page Chap. 14 ("In conclusion") that lists ten rules for good statistical practice: Statistical methods should enable data to answer scientific questions; signals always come with noise; plan ahead, really ahead; worry about data quality; statistical analysis is more than a set of computations; keep it simple; provide assessments of variability; check assumptions; replicate when possible; and make analyses reproducible, allowing others to access data and code.

The Art of Statistics is not the only book to survey statistical ideas and methods for a general audience, or to warn of potential pitfalls in data analytics and communication and propose ways to avoid them. Other worthy contributions to this genre include David Salsburg's *The Lady Tasting Tea: How Statistics Revolutionized Science in the Twentieth Century* and several of David Hand's books, including the 2020 *Dark Data: Why What You Don't Know Matters*. However, *The Art of Statistics* is distinguished by its suggestions for a new pedagogy for teaching applied statistics. This approach emphasizes the PPDAC cycle for purposeful data-driven investigation and question-answering. It introduces modern computational methods (e.g., the bootstrap and permutation tests) early on, thus postponing the need for probability theory (not introduced in the book until Chap. 8) and statistical theory (Chaps. 9–11). Finally, the new approach integrates AI/ML techniques such as classification trees, support vector machines, predictive analytics, and causal artificial intelligence with traditional statistical methods. *The Art of Statistics*, although written for a general audience, is well suited for supplementary reading in a graduate course on data analytics. Its perspectives on real-world applications and on principles of good practice and communication of results, as well as its many insightful observations about the practical value and limitations of statistical methods and theory, make it a valuable complement to more detailed technical expositions that teach hands-on analytics skills, such as Rober Kabacoff's excellent *R in Action*.

A possible limitation of any book on statistical methods—or modern AI/ML and statistical methods—is that most of these techniques tend to use very simple underlying models of relationships among variables, such as postulating that observations arise from simple, stable underlying deterministic relationships (e.g., regression curves) plus some random noise. But real-world data often involve complex, shifting causal relationships, perhaps with quite different effects on different time scales, that cannot easily be captured by simple patterns readily discoverable by statistical or machine-learning methods. Other techniques, from systems dynamics modeling to systems analysis, simulation, decision theory, and adaptive optimal control, also have much of value to offer about "how to learn from data," as the subtitle of *The Art of Statistics* puts it. Scott Paige's book *The Model Thinker*, discussed next, provides an excellent complement to the statistics and AI/ML methods emphasized in *The Art of Statistics* by introducing this larger set of methods for coming to grips with analysis, prediction, and decision-making for complex, changing, and uncertain systems.

To risk analysts, *The Art of Statistics* is likely to be of interest not only for its survey of modern data analytics and its pedagogical innovations, but also for the many practical problems that the fields of risk analysis and applied data analytics have in common. How best to communicate risk information and remaining uncertainties to inform rational deliberative ("System 2") decision-making without distorting or manipulating emotional responses ("System 1") is a familiar challenge to students of risk perception and risk communication (North 2012). Problems of interpreting data; assessing internal and external validity of studies; drawing valid causal inferences from observational data; using AI/ML and applying causal artificial intelligence to improve learning from data about how to manage risks and about what interventions are effective; avoiding p-hacking and false discoveries; and transparently disclosing modeling assumptions and the sensitivity of conclusions to remaining uncertainties, are themes at the intersection of risk analysis, data analytics, and AI/ML. The book's constructive suggestions for addressing these challenges and for improving current practices in learning about probabilities and causal patterns from data and communicating the results to non-specialist audiences are valuable for risk analysts as well as for data scientists.

Its pleasant, accessible style and many vivid examples make *The Art of Statistics* fun to read as well as useful. Its suggestions for teaching applied statistic using modern (computational and AI/ML) methods deserve to be widely considered by teachers of statistics, AI/ML, and data analytics at all levels. They may help to create a next generation of data scientists and data analysts that engages more easily with data analysis and communication challenges without relying on restrictive assumptions and approximations that, not many decades ago, were essential for obtaining numerical answers. The potential value of teaching more flexible and realistic data modeling skills from the start seems enormous, and should be very helpful in making health risk assessment, epidemiology, and public health research, as well as other areas of applied statistics, more useful, more intelligible, and more trustworthy.

Using Models to Interpret Data: *The Model Thinker*

We have seen that analysts who wish to more accurately forecast probabilities of future events are admonished in recent works such as *Superforecasting* to adopt multiple perspectives and to integrate information from diverse, conflicting hypotheses, models, and sources of evidence, rather than relying on any single "best" model, hypothesis, or expert. Predictors are advised to be "foxes" (knowing many little things) rather than "hedgehogs" (knowing one big thing). They are encouraged to be skeptical, humble, and flexible in acknowledging uncertainty about how the world works; in constantly questioning what is assumed, or thought to be known; in remaining open-minded to new evidence and interpretations of old evidence; and in using many different perspectives and models to inform probabilistic estimates and predictions. Current model-dependent conclusions should always be regarded as provisional, recognizing that other models may eventually prove more accurate than those that currently seem best, and that averaging results from multiple models often out-performs all of them. This advice to take a multi-model approach appears to be sound and useful: those who follow it substantially out-perform those who don't, on a variety of predictive tasks across many different domains (Tetlock and Gardner 2015).

But how can busy analysts with limited time to spend upgrading their probability forecasting skills achieve the advantages of multi-model thinking relatively quickly? *The Model Thinker: What You Need to Know to Make Data Work for You* by Scott E. Page (Basic Books, 2018) provides a possible answer in the form of a whirlwind tour of some 25 analytics models and related core ideas that have proved powerful in modeling the behaviors of both human and natural systems. Based on a wildly popular Coursera MOOC course on "Model Thinking," the book compresses into 355 pages of brisk, highly readable text (augmented with 25 pages of notes and another 25 pages of references) some of the most useful ideas and techniques for applying mathematical and computational modeling to data to reason, explain (via empirically tested assumptions and explicit causal chains), design, communicate, act, predict, and explore new hypotheses and what-if scenarios while learning from experience. These steps (summarized in the mnemonic acronym REDCAPE) are intended to help modelers and those who are informed by them move from data to information, knowledge, and wisdom. Page defines knowledge as "understandings of correlative, causal, and logical relationships," typically expressed via statistical, mathematical, or computational models; and wisdom as "ability to identify and apply relevant knowledge" to inform choice of actions when the consequences of alternative choices are uncertain. To a risk analyst, such wisdom overlaps heavily with data-driven, risk-informed decision-making and risk management.

Overview of Contents

The book's 29 chapters develop the following major themes.

- *Multi-modeling advantages.* Chapters 1–3 provide the philosophy, motivation, and framework for using multiple models to interpret data and inform choices of actions and policies. Chapter 1 notes that "models make us smarter" by helping to overcome well-known psychological heuristics and biases that compromise the accuracy of human judgments under uncertainty (Kahneman 2011, reviewed by North 2012): "in head-to-head competitions between models and people, models win." Different models illuminate different causal factors and paths between choices and consequences, and averaging or otherwise combining the predictions from multiple diverse models usually out-performs any single model. Chapter 2 explains the REDCAPE framework and illustrates it with compelling examples and paradoxes. Chapter 3 succinctly summarizes key ideas and results from modern computational statistics and data science that explain why multiple models out-perform single models. It briefly describes methods to take advantage of this (including model ensembles, bootstrap aggregation (bagging) to prevent model over-fitting, and managing bias-variance tradeoffs). Each key method and technical result is explained in a page or two using non-technical language and boxes to highlight key results. Mathematical notation is used seldom (mainly in boxes that develop technical points), and key concepts are illustrated with simple, clear examples. This friendly, jargon-free expository style is used throughout the book to make key results accessible to general readers.
- *Modeling human decisions and behavior.* Chapter 4 introduces the challenges of modeling human behaviors. It discusses and contrasts models based on (a) rational (optimizing) action; (b) modifications to make the models more descriptive of real behaviors under risk, uncertainty, and delayed rewards (including prospect theory to model loss aversion and other biases, and hyperbolic discounting to model present bias); (c) simple fixed behavioral rules; and (d) adaptive behavioral rules, e.g., models of learning and imitation. These themes are developed in several subsequent chapters. Chapter 10 introduces network models for social (and other) interactions, including power law, small-world, and random networks; and Chap. 11 ("Broadcast, Diffusion, and Contagion" presents Bass diffusion models for adoption ("diffusion") of ideas or practices or technology innovations in a population of potential adopters and compares them to susceptible-infected-recovered (SIR) models of disease epidemics, thresholds, and tipping points. It also examines the spread of diseases (or memes or fashions or other contagious things) through networks, highlighting the roles of "superspreaders" at hubs in contact networks. Chapter 20, on "Spatial and Hedonic Choice," presents models of preferences based on value and utility functions (hedonic preference models) and proximity of outcomes to ideal points (spatial preference models). Other ways to model human behaviors over time, including agent-based models (ABMs), cooperative game theory models, systems dynamics simulation models, and adaptive optimization and learning models, are spread throughout the other chapters.

- *Probability and statistics models.* Chapters 5–8 cover material on probability distributions and statistics that is fundamental for students of risk analysis. Chapter 5, in just under 10 pages, introduces the normal distribution, the Central Limit Theorem (one page with a box for the theorem), hypothesis testing, Six Sigma methods, and the log-normal distribution, and mentions examples such as distributions of farm sizes, disease latency periods, and income inequality. It also provides valuable practical advice on the limitations of statistical modeling: fitting models to past data does not necessarily provide valid predictions of future input-output relationships following an intervention or policy change. Chapter 6 (13 pages) discusses power law and heavy-tailed (high risk) distributions. It explains how they arise (e.g., via preferential attachment in networks, or self-organized criticality mechanisms) and how they can be used to model entrepreneurial risks and rewards and a variety of catastrophic events, from earthquakes and fires to financial collapses. Chapter 7 (10 pages) introduces linear statistical models, including simple and multiple linear regression, correlation vs. causality, and machine learning classifiers (and forests of decision trees, which need not be linear). Chapter 8 gives a similarly compact introduction to nonlinear (concave or convex) models with applications to economic growth. This brisk pace is maintained throughout the book: each chapter is packed with useful information for readers with limited time to absorb the key points.
- *Dynamic and stochastic models.* Chapter 12 discusses how entropy and information theory can be used to model uncertainty, classify time series, and motivate choices of probability distributions for representing uncertainty. Chapters 13–18, respectively, introduce random walk models with applications to the stock market and to species extinction; path dependence, tipping points, volatility, and value at risk (VaR) risk measures; agent-based and local interaction models; methods for proving that certain dynamic processes will converge to equilibrium outcomes; Markov models; and systems dynamics models, which are widely used in ecology and policy analysis to understand changes over time in causal networks with feedback loops. Chapter 19 discusses how agent-based dynamic models (ABMs) with feedback among the behaviors of individuals can lead to thresholds for collective behaviors, e.g., for sparking riots or stock market crashes.
- *Game theory models.* The book introduces a great deal of traditional and modern game theory, including cooperative games and political power indices (Chap. 9), zero-sum and sequential games (Chap. 21), emergence of cooperation in iterated Prisoner's Dilemma (Chap. 22), collective action problems with applications to renewable resource extraction (Chap. 23), mechanism design theory for auctions and public project funding decisions (Chap. 24), signaling models (Chap. 25), and learning in games (Chap. 26).
- *Adaptive optimization.* The book concludes with chapters on learning and optimization. Chapter 26 considers reinforcement learning for individuals and social learning (via replicator dynamics) for multiple individuals and shows that each can solve adaptive optimization problems of learning to choose the best alternative. It then turns to learning in games and shows that it may be difficult to learn to cooperate effectively (since learning processes may lead to inefficient "risk

dominant" equilibrium behaviors). Chapter 27 presents a 7-page summary of selected aspects of multi-armed bandit problems, in which a person or organization must experiment with different choices (e.g., medical treatments, technology choices, advertising strategies, etc.) to learn about their reward distributions and identify the best (highest average reward) choice. Chapter 28 presents "rugged landscape" (i.e., multiple-optima) models for combinatorial optimization, reviewing arguments that interactions among components or attributes of complex choices can lead to higher-value optimal choices but make them harder to identify. Finally, Chap. 29 concludes by sketching three models—a multi-armed bandit model, a Markov model, and a systems dynamic model—to gain insight into the opioid crisis, followed by a more extensive set of models for explaining income inequality over time.

Comments on *The Model Thinker*

The Model Thinker delivers a powerful bang for the buck. It has an astonishingly high average number of useful technical ideas and insights per page, an admirable simplicity and clarity of exposition of most of its topics for non-specialists, and numerous valuable practical perspectives on the strengths and limitations of different types of models. Risk analysts should ideally be familiar with many of these modeling methods. For those who aren't, this book provides a quick, easy, insightful introduction and useful references to the primary literature. Given its extraordinarily wide scope and brevity of treatments, its depth on each topic is necessarily limited, but the author's strategy of presenting the few most important points about each topic will be very welcome to readers who need a well thought-out integrative overview of many intersecting areas of analytics and how they can be used together in practice to improve understanding of complex systems and data driven decisions. The book provides enough information about each model, together with interesting examples, to motivate readers to explore some of them in greater detail. The descriptions, notes, and references provide follow-up readings to facilitate such exploration.

There are a few typos (e.g., "combing" instead of "combining" in a box explaining the Gittins Index for solving multi-armed bandit problems) and a few minor errors that specialists will notice (e.g., referring to present bias as immediacy bias, which is quite a different phenomenon, in discussing hyperbolic discounting). More importantly, the exposition in a few places may not be clear to those who have not previously encountered this material. Thus, payoffs in game theory are introduced using the standard "bimatrix" game notation, in which pairs of numbers in each cell refer to Row's and Column's payoffs, but this notation is not carefully explained for readers encountering it for the first time. Some major topics are treated so briefly, using simple examples, that the main points and power of the underlying techniques may not come through. For example, a one-and-a-half page box on Markov decision processes (MDPs) (pp 199–200) presents an example of a student who chooses between studying and surfing the internet and who makes probabilistic transitions between bored and engaged states. The example concludes that "As seen in this example, framing a choice as a Markov decision process can produce better actions. By taking into account the consequences of an action on our state, we choose more wisely." But this treatment probably will not convey to an uninitiated reader the generality and power of the MDP framework, or its practical value (and availability of practical algorithms) for solving more interesting and larger-scale instances.

Although the book is generally excellent at integrating ideas from multiple modeling techniques and exposing interrelationships among them, it misses a few important relationships, such as the optimality of reinforcement learning algorithms for solving MDPs. In short, the virtues of brevity and simplicity carry some costs. Perhaps a second edition could clarify the few expositions that are not already clear and trace a few more key relationships among techniques without much increase in length.

The topics covered are loosely bound together by themes from complexity theory and emergent behaviors. While the collection of models is impressively varied and unquestionably useful for thinking about a variety of individual and public policy decisions under risk, uncertainty, complexity, and delayed rewards, adding a few more topics would make a second edition even more useful for risk analysts. The book introduces systems dynamics modeling but not discrete-event simulation modeling. Although it refers to optimization models for rational decision-making, it omits standard operations research optimization methods such as linear, nonlinear, and dynamic programming and it does not cover decision and risk analysis methods such as decision trees, influence diagrams, and Bayesian networks.

Two categories of readers may find *The Model Thinker* especially valuable. On the one hand, researchers (and students) engaged in analyzing and modeling important policy issues and decision problems under uncertainty, but lacking extensive training in operations research, mathematical modeling, and analytics, will find it a cornucopia of useful and accessible modeling ideas, perspectives, and advice on how to use multiple models to gain useful insights. On the other hand, managers in the public and private sectors who are curious about what modeling can accomplish and who want to get more robust understandings, insights, and recommendations from their analytics teams will glean a rapid appreciation of possible approaches. Analytics specialists are often frustrated by organizational data analysis and decision processes that rely on simplistic regression models, or that fail to model at all the causal links between alternative decisions and their probable consequences over time. This book can help both managers and analytics teams to do better by drawing on a rich ecosystem of overlapping modeling approaches. Risk analysis practitioners and analytics specialists will have to dig more deeply into the references provided and other literature and software to implement the modeling approaches described, but simply being aware of them is an immensely valuable first step.

Responding to Change Strategically: Setting Goals and Acting Under Uncertainty

How can leaders use their limited means, together with realistically imperfect information and control, to navigate uncertain and changing environments, increasing the odds of achieving desired goals successfully? How can they select goals wisely, balancing aspirations and ambitions against realistic constraints and uncertainties? How should preparation, training, and planning be interleaved with improvisation and opportunism to deal with threats and opportunities as they arise? These grand themes are tackled in Yale historian John Lewis Gaddis's *On Grand Strategy* (Penguin Press, 2018).

The book is organized around the "hedgehog and fox" framework popularized by Isaiah Berlin (Berlin 1953) and more recently in *Superforecasting*. Recall that hedgehogs are those who "know one big thing," such as a totalitarian ideology or a divine plan, and use it to organize their own assumptions and plans. Foxes, by contrast, know many little things. They are more humble and flexible in acknowledging uncertainty about the future and in being willing to adapt their world views and plans as circumstances and information change. *On Grand Strategy* contrasts the historical actions, guiding principles, and outcomes for pairs of leaders and thinkers from Xerxes, Pericles, and Sun Tzu to Stalin, Churchill, and Franklin D. Roosevelt, illuminating how well different approaches to goal-setting and acting under uncertainty have succeeded in practice. In the concluding chapter of the book, Gaddis comments that freedom to choose how to respond to circumstances, using principles as a guide or compass, characterized "the younger Pericles, Octavian Caesar, Machiavelli, Elizabeth I, the American Founders, Lincoln, Salisbury, and especially Roosevelt, all of whom had the humility to be unsure of what lay ahead, the flexibility to adjust to it, and the ingenuity to accept, perhaps even to leverage, inconsistencies." These leaders and thinkers—foxes, all—are contrasted with hedgehogs: "the older Pericles, Julius Caesar [in the years leading up to his assassination], Augustine, Philip II [of the Spanish Armada], George III, Napoleon, Wilson, and the twentieth-century totalitarians, all of whom knew with such certainty how the world worked" that they sought to defy or overcome, rather than to work within, real-world constraints. The stories of these and a few other political, military, and thought leaders (including Clausewitz, Tolstoy, and Berlin himself) are recounted in the remaining chapters, and are used to illustrate essential points about successful philosophies and strategies for making success more likely and for avoiding the hubris-induced failures that tend to confound over-confident and overly blinkered hedgehogs.

Key themes of the book include the following.

- Common sense allows most of us to navigate through uncertain, changing, and cluttered physical environments (e.g., crowded sidewalks or streets) without much effort, planning and re-planning automatically and effectively as we go to skirt hazards and take advantage of unexpected short cuts. Political, business, and military leaders do something similar on a larger scale, fitting ends to means, sketching and revising plans as circumstances change, and aligning actions and goals across time (sequencing actions to lead to desired ends), space (coordinating actions in different places), and scale (from high-level diplomacy to lower-level implementation).

- As leaders or experts rise in visibility and popularity, they develop reputations and their stated plans and actions are scrutinized and second-guessed. This may lead to a loss of flexibility and common sense among those in authority, and hence a tendency to commit to dumb courses of action.
- Grand strategy is intended to preserve common sense in fitting ends—potentially unlimited ambitions—to means (necessarily limited capabilities) and in enabling effective pursuit of goals as information and conditions change. It does so by teaching strategic thinkers to combine Kahneman's System 1 (fast, intuitive) and System 2 (slow, cognitive) thinking; to recognize when to be a hedgehog and when a fox; to draw on history and training to recognize how past cases and (often conflicting) principles and lessons bear on present situations, and to plan and improvise as needed; to grasp interconnections; to think in terms of sequences that achieve goals; to develop a sense of the whole that reveals the significance of the parts; to seize opportunities while retaining objectives; and to extract and articulate purposes, goals, and plans for achieving them from masses of details and unpredictable events.
- To a large extent, grand strategy is a teachable and learnable skill. Training can inform planning about what has succeeded or failed in the past, although the collision of plans and theory with reality (called "friction" by Clausewitz) will always require dynamic replanning and improvisation and opportunism as unpredictable events occur. The most successful plans are usually light sketches, tethering practice to principles, that show how sequences of future actions can be connected (if all goes well) to increase the odds of achieving desired ends, and how unpredictable opportunities can be fit into already partly planned action sequences to expedite attainment of goals. Such sketches can provide the coordination needed for effective delegation and distribution of authority and control while preserving purposeful action and balancing factions at multiple levels.
- Conversely, effective implementation of strategy also requires graceful recovery from surprises, adaptation to changing conditions, and resilience in accommodating the unexpected. Procedures, preparation (e.g., checklists), experience with prior cases, and proportionality in responding to unforeseen threats and opportunities can all help to cultivate resilience and healthy opportunism, in which ambitions expand with opportunities when and if they materialize.

On Grand Strategy illustrates these themes through studies of exceptional successes and failures in strategic leadership in government, politics, statecraft, and war. It tethers the preceding general principles to an array of historical case studies that vividly illustrate and extend them. The 10 chapters of the book are as follows:

- Chapter 1, "Crossing the Hellespont," discusses the disastrous attempt by Xerxes to conquer Greece in 480 BC, based on the reasoning that "If you were to take account of everything . . . you would never do anything. . . . Big things are won by big dangers." It also introduces the hedgehog-vs. fox theme that runs throughout the entire book.

- Chapter 2, "Long Walls," describes the increasingly insular and aggressive stance of Athens after it enclosed itself and the port of Piraeus by defensive walls in 457 BC. It contrasts the strategies of persuasion vs. confrontation—steering within the flow of events or against them—characteristic of the young and old Pericles, respectively, and reflects on how the destabilizing forces of disease, fear, illogic, ambition, deception combined with Pericles' decreasing flexibility and mercy and his loss of proportionality in responding to perceived threats to usher in "the Athenians' descent from an extraordinary to an ordinary culture."
- Chapter 3, "Teachers and Tethers," recounts principles connected to practices in Sun Tzu's *The Art of War* and traces the history of Octavian Caesar's (later Augustus) battles against Antony and others, and his gifts for appropriate delegation for execution of his strategies, forging causal chains from contingent (and often unexpected and unpredictable) events, connecting sequences of decisions and victories to achieve larger purposes, and seizing opportunities while retaining clear longer-term objectives. The chapter argues that Octavian intuited and applied many of Sun Tzu's principles, and that both Rome and China developed the robustness to survive even terrifyingly bad rulers by diversifying their sources of power and allowing a robust ecosystem of political and economic might to flourish.
- Chapter 4, "Souls and States," recounts St. Augustine's efforts in his *Confessions* to reconcile faith with reason as the Roman empire crumbled; his influential theory of just war; and his strategies for determining obligations to Caesar and to God while seeking to attain to everlasting life. It develops the principle of *proportionality* (in Gaddis's words, "the means employed must be appropriate to—or at least not corrupt—the end envisaged") and applies it to proportionate use of force to achieve its purposes without destroying what it means to defend. The second half of the chapter examines Machiavelli's uses of a similar principle of proportionality without prayer, recognizing that ideals cannot always be achieved, as set forth in *The Prince,* his carefully crafted advice to Lorenzo de' Medici as Machiavelli sought to regain favor and status.
- Chapter 5, "Princes as Pivots," contrasts Elizabeth I of England and Philip II of Spain in their philosophies on aligning souls with states and in Elizabeth's deft, light touch and penchant for "strategic mischief" with Philip's heavy-handed efforts to conduct just wars while succumbing to heuristics and biases such as loss aversion that destroyed proportionality in his responses to potential losses. The chapter traces pivotal moments in history, such as the unleashing of fire ships on the Spanish Armada by Elizabeth's lord admiral, to improvisations empowered by clarity on goals, awareness of opportunities (such as a favorable wind one night), and confidence in using delegated authority to take advantage of transient local conditions to pursue larger long-term ends.
- Chapter 6, "New Worlds," focuses on the American Revolution and the founding of the United States. It discusses *The Federalist Papers* ("the most enduring work

of grand strategy since Machiavelli's *The Prince*") as an effort to align incompatible aspirations with limited capabilities by preserving and applying common sense principles to enable a republic to grow into an empire in an unpredictable future without replacing its liberties with tyranny. Allowing factions to compete at all levels helped bend the arc of United States history away from another Nero and toward Lincoln.

- Chapter 7, "The Grandest Strategists," compares Tolstoy's and Clausewitz's similar reflections on war, the limitations of theory, the "friction" of real-world events and exigencies that degrade planned coordinated actions, the reality that "something or someone will sooner or later break, but you can't know how, where, or when," and the practical value of theory as training to help overcome such friction.
- Chapter 8, "The Greatest President," describes how Lincoln responded to unpredictable events and opportunities, matching ambitions to capabilities as both expanded, to show "the practicality, in politics, of a moral standard . . . an external frame of reference that *shapes* interests and actions, not . . . an internal one that only *reflects* them."
- Chapter 9, "Last Best Hope," begins with Victoria's Prime Minister Salisbury, who observed that "There is no such thing as a fixed policy because policy like all organic entities is always in the making," and traces the evolution of national and international policies, conflicts, and alliances among the United States, England, Germany, and Russia and the new Soviet Union during the Twentieth Century and its two world wars. It examines the leadership of Lenin, Wilson, Churchill, and especially Franklin Delano Roosevelt, in part through essays by Isaiah Berlin.
- Chapter 10, "Isaiah," concludes with a further discussion of Isaiah Berlin's reflections on the increasingly brutal Soviet Union, the seductive inevitabilities claimed by Marxism, positive and negative liberties, and the adaptations to incompatible aspirations and capabilities and unexpected contingencies that constitute history.

What can decision and risk analysts learn from these reflections on historical events? The book offers a wealth of valuable ideas for risk analysis, some of which challenge and extend standard risk analysis paradigms. One is that the unpredictability and uniqueness of crucial future events, from accidents of wind and weather to the untimely and unexpected incapacity or deaths of leaders, makes adroit navigation of uncertainties more dependent on clear objectives paired with ability to connect and build on events as they unfold rather than on clear prediction, analysis, and planning. Recognizing patterns and improvising creative ways to exploit or redirect them may be more useful than being able to perform fault tree or event tree analyses for dealing with the open-world uncertainties on which history often pivots. Likewise, preparing to respond to unpredictable and unique events ("unknown unknowns," "black swans," and the like) as well as to known challenges and to risks that can be assessed probabilistically, require a realistic understanding of one's own capabilities and their limitations, together with training and plans—easily modified and adapted—to make best use of them come what may. Such preparation and assessment build resilience and capacity for fruitful innovation under stress.

These and related ideas suggest a discipline of strategic management of uncertainty that focuses not on the usual risk analysis questions (e.g., what can go wrong, how likely is it to do so, and the consequences if it does), but on new questions such as what we aspire to achieve with our limited capabilities and knowledge; how best to deploy our resources moment-to-moment and across time, space, and scale to improve the odds of achieving these ends, given what we know; and how to prepare now to respond to unpredictable future events, both favorable and unfavorable, while preserving proportionality between effort and ends. Strategy emphasizes what we can do to make successful achievement of our goals more likely, rather than what we can do to make probabilistic losses less likely. Grand strategy also emphasizes choosing goals wisely, taking into account current capabilities and constraints as well as ambitions and proportional use of resources to pursue them. Perhaps the field of risk analysis will eventually expand to encompass many of the principles of strategy that the book discusses and illustrates, and to give them more formal quantitative expression. Until then, the perspectives on purposeful action under uncertainty offered by *On Grand Strategy* provide a valuable, thought-provoking complement to traditional risk analysis.

Using Data to Discover What Works in Disrupting Poverty

How, if at all, can risk analysis help us to better understand and address the interlinked challenges of global poverty—the self-perpetuating feedback loops, or vicious cycles, of low income, education, and productivity; poor nutrition and health (especially for infants and young children, and especially for girls, in many poor countries); woefully inadequate medical care and disease prevention; denied opportunities for girls and women; unavailable or prohibitively costly access to credit, insurance, and savings accounts; low rates of individual and societal savings and investment; inefficient and corrupt institutions; and slow or negative economic growth? These problems are notoriously recalcitrant. They appear at first to be largely the consequences of political and economic conditions that have little to do with the usual risk analysis concerns of risk perception, assessment, management, and communication. Yet, if the well-researched claims in a recent book on the economics of poverty—Poor Economics: A Radical Rethinking of the Way to Fight Global Poverty, by Nobel Laureates Abhijit Banerjee and Esther Duflo—are correct, then these traditional risk analysis disciplines may be crucial in figuring out how to replace vicious cycles with virtuous ones, helping to shift people and countries trapped in cycles of poverty onto a path toward greater (and more predictable) growth in prosperity, health, education, nutrition for children, and financial well-being (Banerjee and Duflo 2011).

Conceptual Framework: Uncertain Risks and Rewards and Poverty Traps

Figure 2.1 shows a unifying conceptual framework for problems that Banerjee and Duflo examine in different chapters—the connected problems of hunger, health, education, and family size in Part I ("Private Lives"), and the problems of unemployment and income risks, microfinance and lending to the poor, family savings and investment, entrepreneurship and low-income businesses, and corrupt and inefficient politics and policies in Part II ("Institutions"). In this diagram, the x-axis shows the level of a resource now, and the y-axis shows the resulting level of the same resource one period from now (e.g., next planting season, next year, next administration, next decade, next generation, etc., depending on the specific problem being discussed).

The dashed 45-degree line shows points where the level of resource does not change from one period to the next. The S-shaped curve summarizes the input-output relation (perhaps mediated by a complex, only partly understood, causal network of social, political, economic, cultural, and other variables) determining next period's level from the current level. Where it intersects the 45-degree line, the level of resource is in dynamic equilibrium. Where it lies above the 45-degree line, the level of resource is increasing, and where it lies below the 45-degree line, the level of resource is decreasing. If the model curve is truly S-shaped as shown in Fig. 2.1, then there is a stable "poverty equilibrium" (or "poverty trap") at A and a stable "wealth equilibrium" at C. To escape the poverty trap and shift from A to C, it is necessary to boost the level of resources past threshold B. Then, the system will fall into the basin of attraction for C, instead of A (i.e., escape the poverty trap), and spontaneously move rightward to C.

The poverty trap metaphor has motivated a great deal of well-intended international aid policy and practice, with the key idea being to give or raise enough resources to get past B. But, as Banerjee and Duflo emphasize, whether the world

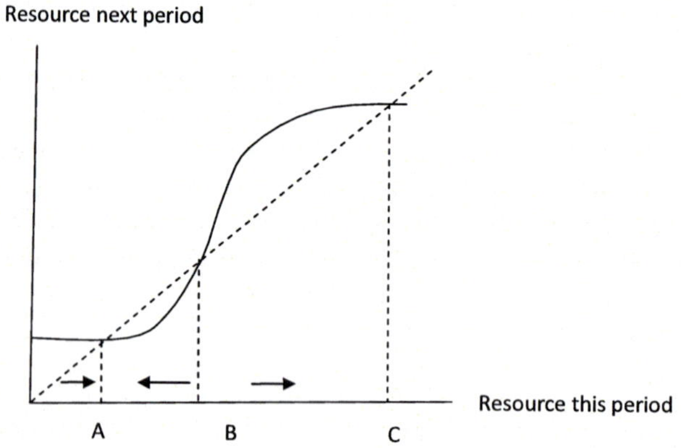

Fig. 2.1 Conceptual Framework: A Poverty Trap (A) and Wealthy Equilibrium (C)

really works as Fig. 2.1 suggests in particular locales is a crucial *empirical* question that cannot be settled by abstract logic, or by ideology and preconceptions, or by policy discussions carried out in ignorance of detailed local knowledge about how specific societies or communities or cultures actually work. Whether the S-shaped curve exists, or whether the causal relation between current and subsequent levels of resources has a different shape with very different implications (e.g., lying above the 45-degree line even at the origin), can only be determined via careful field investigation, ideally backed by randomized control trials (RCTs). The real and perceived shapes of the curve determine what interventions (if any) will be effective in different locations. Discovering what will work where, based on empirical evidence and RCTs, is the central task to which the book is devoted.

Extending and Applying the Framework: How Risk and Perceptions Strengthen Poverty Traps

Banerjee and Duflo add some important risk analysis insights to the familiar poverty trap model in Fig. 2.1, reflecting research and practical experience with the psychology of risk perception, motivation, and impatience, as well as the realities of constant uncertainty and informal risk management, in the lives of the poor. They point out—with supporting evidence from anecdotes and RCTs—that if sustained investment (whether of effort, savings, self-denial, or time) is required to move from A to B, and if the y axis in Fig. 2.1 only represents *expected* or average future resource levels, but unpredictable shocks and losses (e.g., due to sickness or drought) can sometimes deplete the stock of a resource (e.g., a farmer's crop, a worker's health, or a family's savings), then *the required investment may never be made.* The apprehension that current sacrifice and investment may not achieve the desired goal (surpassing B), or may do so only slowly, with an uncertain amount of effort and several false starts, is a powerful disincentive for even trying to achieve it. This is true even if the potential rewards (transition to C) are very high compared to the required investment.

Discouragement aside, the poor in many countries often cannot risk taking even very favorable gambles (e.g., risking a small investment to reap relatively high rewards with high probability) that might wipe them out. They end up spending relatively large amounts of time and effort seeking to manage routine risks, e.g., by diversifying their efforts across multiple plots of land in different locations, in agrarian communities; or across multiple activities or lines of low-yield businesses, in poor urban communities. This diversification forces them to forego the considerable economic advantages (on average) of specialization that greater security would make possible. Vulnerability to routine risks and setbacks, high opportunity costs and other costs of routine risk management, and low perceived return to investments that seek to promote a better future, all tighten the poverty trap.

Moreover, if the actual shape of the model curve in Fig. 2.1 is uncertain, then (possibly false) beliefs about its shape may powerfully affect incentives and behaviors. Misunderstanding or misperceiving the causal relation between current investments (e.g., in bed nets, or chlorine pills for drinking water, or vaccines, or education) and future rewards (e.g., reduced burden of malaria, or deadly water-borne illnesses, or childhood diseases and deaths, or improved earnings) can sap the will to make the investments, and induce self-fulfilling low expectations. For example, Banerjee and Duflo note that many poor people mistakenly believe that the returns to extra years of school follow an S-shaped curve, and hence believe that there is little potential gain from keeping their children in school unless they can be sent for a long time (e.g., through high school). (Empirically, the relation between lifetime earnings and years in school is approximately linear, rather than S-shaped; however, misperception of an S-shaped relation can generate self-fulfilling behaviors and expectations.) This misperception of the returns to schooling encourages poor families to under-value primary education for most children and to adopt strategies, such as selecting one child to educate, while allowing others to drop out early, that promote continued poverty from generation to generation.

Banerjee and Duflo apply the augmented poverty-trap framework to understand and interpret data on a host of important poverty-related problems. They examine the evidence for and against the reality of various types of poverty traps, and consider the institutional, social, psychological, and policy causes and implications of the relation between current and future resource levels, and between current savings and investments and future returns in improved nutrition, education, health, savings, earnings, security, and well-being. As one example of their insights and approach, an early chapter on nutrition notes that, in biological reality, most individuals in most locations do *not* have S-shaped curves for nutrition (e.g., consumption of calories per day), with consumption having to reach a certain threshold level (B in Fig. 2.1) before each extra calorie consumed today gives a worker enough extra strength to earn more than one extra calorie tomorrow. This logical possibility turns out not to be an important empirical reality in the countries for which they have collected data. Rather, the poorest and most under-nourished workers typically gain the most benefit per extra calorie, contrary to the poverty trap's S-shape. (Indeed, when income has increased in poor regions, consumption of calories per person per day typically has not increased; rather, people start buying more expensive, tastier food and other items, such as television sets. This does not deny that food and productivity affect each other. For example, Banerjee and Duflo cite studies showing that very poor workers typically buy more food when they are paid per-piece than when they are paid fixed wages, as the extra investment in food consumption boosts productivity, and hence payment under the per-piece system.) However, on a time scale with *generations*, rather than working days, as periods, nutrition poverty traps gain importance: parents who were malnourished in childhood typically earn less and are more likely to raise malnourished children, and these effects can be measured quantitatively: they are statistically and economically significant. Breaking this cycle requires providing at least a threshold level of nutrition (including essential micronutrients) for pregnant mothers, infants, and young children to enable their healthy development, effective participation in schools, and increased earnings in adulthood.

This example of nutrition illustrates the richness of Banerjee and Duflo's analyses and examples. They examine possible poverty traps, perceptions, incentives, behaviors, expectations, and outcomes using empirical data on multiple time scales (e.g., days to generations), for different poor populations (e.g., urban vs. agricultural) across different locations and countries. They deliberately avoid grand simplifications and generalizations, preferring to document real behaviors and responses—both desired and undesired—to attempted interventions, as revealed by RCTs and other studies. Their analysis integrates political and economic insights, in the best tradition of political economy, and traces the rich interactions over time among beliefs, investment behaviors, risk-taking or avoidance, and eventual outcomes in the linked areas of nutrition, health, family planning, education, earnings, savings, investment, employment opportunities, political freedom and opportunities, institutional integrity, and personal and economy-wide growth or decline in prosperity. Toward the end of the book, concepts from political science and sociology, such as the "Iron Law of Oligarchy" and the tendency of low-performing institutions to perpetuate themselves, are revisited from a helpful perspective that recognizes them as consequences of vicious—but potentially disruptable—cycles. The emphasis throughout is on discovering, through trial and error and rigorous evaluation (where possible), enriched by personal interviews, what measures best disrupt vicious cycles and alleviate poverty, and under what conditions.

Escaping Poverty Traps: Weakly Held Beliefs and Credible Communication

Poor Economics ends on an optimistic note. A vicious cycle can be broken at many points, and Banerjee and Duflo look for—and, in many cases, find—points where relatively simple and inexpensive changes in local rules, expectations, and beliefs generate relatively large benefits by disrupting the feedback loops that create or strengthen poverty traps. They observe that even weakly held beliefs (e.g., about the efficacy of modern medicine *vs.* traditional cures, or the effectiveness of bed nets in reducing malaria, or the possibility of teaching children from poor or low-caste families) can have large effects on behavior. A *weakly held belief is* one that can easily be overcome by new evidence, credibly communicated. It may be acted on by default, but can readily be replaced by a more accurate belief if credible new information that contradicts it becomes available. One of the encouraging conclusions that Banerjee and Duflo draw is that relatively small investments in information-sharing (e.g., conveyed through headlines exposing and tracking corruption, or via credible outreach and communication programs) and in changing the rules of local politics and cultures to allow old stereotyped expectations to be replaced by new ones (e.g., by observing that women can participate very effectively in village leadership), can lead to measurable improvements in multiple poverty-associated dimensions of life. Greater investment in roads and infrastructure, reduced theft of public monies, dramatically increased performance by teachers and nurses and civil servants, and higher rates of savings, employment, financial security, and investment in health are among the measurable benefits that can be achieved by such reform "from below".

Banerjee and Duflo emphasize that much real good can be accomplished through these relatively modest changes, working within existing imperfect institutions and despite very real impediments (such as corruption, opposition by vested interests, and limited time, interest, and "mental space" to think about changes among the poor who are struggling with present realities). Doing so requires no dramatic confrontations with strongly entrenched beliefs or political factions. Rather, the key is to focus on changing weakly held beliefs by credibly communicating relevant facts, and by sharing and illustrating new ways of thinking, behaving, and organizing. Simply correcting misperceptions, or sharing information (where no strongly held beliefs already exist) about the relation between various kinds of present precautions and investments and their likely future returns, can strongly inform and change incentives. They can help people muster the optimism and will required to risk investing now in a potential better future—especially when such investments can be coupled to accurate expectations about returns over time. Sometimes, in some locations, this is what is needed to break a vicious cycle and escape a poverty trap.

How Can Analysis Help Reduce Health Risks and Poverty?

Poor Economics emphasizes that financial and health risks and uncertainties are conspicuous in the daily lives of most poor people, exacting high costs in time, stress, and foregone opportunities, and sapping the optimism and faith in the future needed to initiate and sustain an upward cycle of savings and productive investments. Where richer countries and parts of societies have 401(k) plans, mandatory childhood vaccination programs, school lunch programs, piped water, modern sewage facilities, access to relatively inexpensive savings and credit plans, and countless "nudges" to be responsible and plan for the future, the poor must depend on their own initiative and resources to try to manage the many risk of everyday life. The inability of the poor to risk taking even highly favorable gambles (e.g., investments in disease prevention or education or household savings that would probably—but not certainly—yield very high returns in future health and earnings), combined with their poor information about true risk-return trade-offs, and weakly held beliefs that strongly affect behaviors, are key contributors to the vicious cycles that trap many countries, and part of societies, in poverty.

To combat these ills in a practical, near-term way, without waiting for major upheavals or unrealistic changes in institutions or ideologies, Banerjee and Duflo recommend reform of institutions and decision-making "from below," via a "quiet revolution" based on spontaneous changes in incentives, expectations, and behaviors in response to credible information and examples. Such reform requires key analytic and communication skills, to provide the following services:

- *Understand and explain the avoidable causes of risks and harms.* In many case studies, simply diagnosing and publicizing the preventable causes of present (and possible future) undesired events suffices to bring about change for the better. Banerjee and Duflo present examples of how describing and reporting causes and consequences has reduced certain types of risks, such as for premature road failures due to theft and diversion of high-quality construction materials; or from crop failure due to misuse and poor allocation of fertilizers.
- *Apply the psychology of risk perception and decision-making to facilitate better choices.* Banerjee and Duflo note that poverty is associated with more frequent stress (e.g., as indicated physiologically by cortisol levels) and poorer decisions, as defined by predictable future regret for actions taken or not taken now. Such "time inconsistency" of preferences (i.e., choosing now what we are sure to regret later), as well as misperceptions of the true risks and rewards for different choices, probably contribute to choices that increase many risks. Among these are risks of premature deaths (due not only to predictable failures of expensive but ineffective folk "treatments," but also to failures to use inexpensive preventive measures); childhood illnesses and fatalities (due to failures to use effective and cheap or free preventive measures, such as chlorine pills in drinking water, or bed nets against mosquitoes, or routine medical check-ups for children); financial hardships (due to lack of savings); and needlessly frequent crop failures or low crop yields (due to failure to use agricultural products and methods that reliably increase average yield). Banerjee and Duflo provide quantitative estimates of the effects of poor choices on health and earnings, and the effects of better choices on these and associated outcomes, such as years of schooling, height and weight gains for children, and average earnings in later life. They suggest that acknowledging the reality and importance of time-inconsistent preferences, and applying insights from the psychology of risk and choice, can help design improved incentives and systems that "nudge" participants to make choices that they are less likely to regret—from savings plans that make deposits easy and withdrawals less so, to payment schedules for fertilizers that are tied to receipt of revenues from crops.
- *Quantify trade-offs and uncertainties.* Discover and explain what returns (e.g., to investments in savings, crop productivity, education, or sanitation) are realistically possible to achieve with confidence, given the pragmatic constraints of existing resources, institutions, and constraints. Identify, quantify, and publicize the causal relation between actions now (e.g., investments in preventive measures to avoid diseases or boost expected crop yield) and probable consequences later.
- *Communicate credibly.* Tell people about how their choices affect probable outcomes, using credible, understandable, effective messages.

These are areas in which risk analysts excel. The core competencies of risk analysis include tools and techniques for hazard identification; quantitative risk assessment; clear description of decision and risk trade-offs and of causal relations between choices and their probable consequences; design of decision processes and institutions that improve risk management options and choices; and clear, credible, effective risk communication with a variety of audiences who need better information to act on. These are precisely the tools that Banerjee and Duflo suggest can be most effective in instigating "quiet revolutions" that change expectations and behaviors for the better.

Banerjee and Duflo conclude that there are probably no simple or general solutions to the tightly interconnected problems of poverty. What works best depends on detailed local conditions and may be different in different places and at different times. Yet, the tools of risk analysis appear to have great potential for application in development economics, to help figure out and communicate what information and changes in choices will best help individuals, communities, and societies break free of poverty traps. The need and opportunity for cross-fertilization between the disciplines of risk analysis and development economics is made apparent by *Poor Economics*. Its compassionate, insightful, and deeply informed accounts of the lives and struggles of the poor reminds us that risk analysis need not always deal with rare events and hard-to-measure outcomes, but can potentially contribute to measurably improving the lives of billions of people in the near term.

Poor Economics is engagingly written, accessible to risk analysis students and professionals at all levels, and well suited for use in both undergraduate and graduate courses dealing with risk, poverty, and development. It offers insights, data, and constructive suggestions for solving some of the world's most difficult and important challenges. Its humble, compassionate tone—focusing on trying to find out and report what works and what doesn't, and why—is refreshing in an area where intense policy debates have too often been driven more by ideological divisions and preconceptions than by data and experience. Many risk analysts will find the empirical approach familiar and welcome as a more productive way to solve problems.

References

Banerjee A, Duflo E (2011) Poor economics: a radical rethinking of the way to fight global poverty. Public Affairs Books, New York, NY

Berlin I (1953) The hedgehog and the fox: an essay on tolstoy's view of history. Weidenfeld & Nicolson, London

Gaddis JL (2018) On grand strategy. Penguin Press, New York

North DW (2012) Book review: thinking, fast and slow by Daniel Kahneman; Nudge: improving decisions about health, wealth, and happiness by Richard H. Thaler and Cass R. Sunstein; The better angels of our nature: why violence has declined by Steven Pinker. June. Risk Anal 32(7). https://doi.org/10.1111/j.1539-6924.2012.01821.x

Page SE (2018) The model thinker: what you need to know to make data work for you. Basic Books, New York, NY

Spiegelhalter D (2019) The art of statistics: how to learn from data. Basic Books, New York, NY

Tetlock PE, Gardner D (2015) Superforecasting: the art and science of prediction. Penguin Random House LLC, New York, NY

Chapter 3
Natural, Artificial, and Social Intelligence for Decision-Making

Introduction

Several recent books have offered new insights and summarized old ones about how to use data and experience to think more usefully about choices under uncertainty to reduce predictable regrets and to increase probabilities of achieving goals and preferred outcomes. This chapter first examines the complementary perspectives offered by six recent books in the overlapping fields of cognitive neuroscience, psychology of thinking and reasoning, artificial intelligence and deep learning, social science, and social statistics and data analysis. A seventh book on AI, machine learning, and human values completes the chapter. The first six books can be thought of as three pairs. Two address the biology and psychology of individual thinking and reasoning, including the fast, subconscious pattern recognition-driven reactions that constitute most of the information processing done by the brain, as well as the slower, deliberative reasoning beloved of many risk analysts. Both are in the wonderful series of "Very Short Introductions" published by Oxford University Press (OUP). They are

- *Cognitive Neuroscience: A Very Short Introduction* (Richard Passingham, OUP, 2016)
- *Thinking and Reasoning: A Very Short Introduction* (Jonathan Evans, OUP, 2017)

The next two books address the challenges of computational modeling of different types of thought. They introduce modern AI and machine learning methods that are used in automated risk management and control systems, among many other applications. These are

- *Deep Learning* (John Kelleher, MIT Press, 2019)
- *Artificial Intelligence: A Very Short Introduction* (Margaret Boden, OUP, 2018)

L. A. Cox Jr., *AI-ML for Decision and Risk Analysis*, International Series in Operations Research & Management Science 345,
https://doi.org/10.1007/978-3-031-32013-2_3

The last two books deal with the reality that people think together in groups and communities of knowledge. This has both negative consequences familiar to students of risk communication (e.g., misconceptions, delusions, and misperceptions of risks reinforced by like-minded others) and many positive ones (e.g., division of cognitive labor and astonishing collective accomplishments fueled by the combining expertise and skills that no single individual has mastered). These books are

- *The Knowledge Illusion: Why We Never Think Alone* (Steven Sloman and Philip Fernback, Riverhead Books, an imprint of Penguin Random House LLC, 2017)
- *Factfulness: Ten Reasons We're Wrong About the World—and Why Things Are Better than You Think* (Hans Rosling with Ola Rosling and Anna Rosling Ronnlund, Flatiron Books, 2018).

The seventh book is

- *The Alignment Problem: Machine Learning and Human Values* (Brian Christian, W.W. Norton & Company, 2020).

It provides a history, survey, and prospects for deep learning and related AI/ML methods that are currently revolutionizing our technological world and our capacity to automate decision-making under uncertainty in a wide variety of technology applications.

In the following sections, these book titles are abbreviated as *Cognitive Neuroscience*, *Thinking and Reasoning*, *Deep Learning*, *Artificial Intelligence, The Knowledge Illusion*, *Factfulness,* and *The Alignment Problem.* All of them are written for a general audience (although *Deep Learning* provides enough detail to give a genuine understanding of training and prediction algorithms for multilayer neural networks, and *Artificial Intelligence* will be appreciated most by those with some previous exposure to the field). All six books are enjoyable and interesting, and all shed light on aspects of risk psychology, including risk perception, communication, assessment, or management, including decision-making, planning, learning, and problem-solving. For busy risk analysts and policy-makers with time for only one of these books, we recommend *Factfulness* as most likely to change world views with striking data and advice on how to think better about key global trends and risks.

Biological Foundations of Thought: *Cognitive Neuroscience*

Cognitive Neuroscience: A Very Short Introduction surveys evidence about information processing in the brain, derived from brain imaging studies that compare differences in observed patterns of activation within the brain as subjects perform various mental tasks. Chapter 1 explains that imaging techniques such as positron emission tomography (PET) and functional magnetic resonance imaging (fMRI) help to visualize blood flow and oxygen consumption in different parts of the brain while subjects perform different tasks or report being in different mental states. By studying changes in these activation patterns across the brain in response to different kinds of thought, cognitive neuroscientists gain clues to how the parts of the brain share information and work together, shedding light on what the human mind can and cannot do and how it functions in health and when the brain is damaged. Chapter 2, "Perceiving," examines the parallel processing of sensory and control information via multiple diverging pathways leading to different interconnected brain areas with distinct localized functions. The hierarchical arrangement of natural neural networks into six main layers of neurons, with multiple lower-level neurons feeding into individual higher-level neurons, is briefly described, along with evidence about how this hierarchical arrangement enables object recognition by neural networks, irrespective of object orientation and angle of view, as high-level neurons work with the features identified (or extracted from sensory data) by specialized lower-level neurons. (This principle of hierarchical feature extraction is also important in the artificial layered neural networks discussed in *Deep Learning*.) Chapter 3, "Attending," examines the processes by which we scan the world, attending selectively to sense data based on the tasks we are engaged in (e.g., by matching stimuli to templates set up subconsciously by the prefrontal cortex based on expectations about the activities we are engaged in). Divided attention and distraction are explained in terms of top-down control over processing of input information streams to facilitate execution of the task at hand. Importantly for safety engineering, human decision-making in response to unexpected events takes place in a single bottleneck (in the ventral prefrontal cortex), where sensory input is transformed to motor output; this forces serial processing and consequent delays when multiple decisions are required simultaneously. For example, talking on a cell phone while driving, although seldom impairing reaction speeds and safety when all goes as expected, can lead to increased reaction time when a novel situation occurs, such as a pedestrian stepping out into traffic or a car in front slamming on its brakes.

Chapter 4, "Memory," examines the anatomically separate processes involved in different types of memory, distinguishing between *episodic memory* of past events, e.g., "flashbulb memories" (controlled largely by the hippocampus, which is activated in planning routes and in recalling experiences) and *semantic memory* of facts one has learned, such as people's names and what tools are used for (controlled largely by the perirhinal cortex in the temporal lobe). Memory allows learning from experience and by instruction. It improves ability to cope with uncertain environments by enabling a decision-maker to remember what has been tried before, and conditions under which it has worked or failed. This connection between memory, learning, and action is developed further in Chap. 6 on "Deciding."But first, Chap. 5, on "Reasoning" examines the brain anatomy of fluid reasoning measured in IQ tests. It traces the clustering of results in different subtests (and hence the existence of a measure of general intelligence, denoted by *g*) to common dependence on specific brain systems—especially the interconnection of the parietal cortex, which represents comparative relationships (e.g., larger or smaller, equal or unequal, etc.) with the dorsal prefrontal cortex, which provides executive control for coordinating multiple cognitive demands. Chapter 5 also presents evidence from brain lesion studies and imaging experiments showing that reasoning does not require language ("inner speech"), although being taught in language greatly increases our cumulative inheritance of knowledge and understanding.

For many *Risk Analysis* readers, Chap. 6, "Deciding," is likely to be of special interest. It begins by identifying anatomic structure at the top of the brain's information-processing structure: the prefrontal cortex (PFC). The PFC helps map information to actions. It has parts that receive inputs from all of the senses (the ventral PFC), as well as sensations of hunger, thirst, and temperature (orbital PFC). The dorsal PFC influences actions via direct connections with the premotor area. As a whole, the PFC generates situation-appropriate actions, compares alternative outcomes on a common scale, and evaluates alternative choices of actions in light of current needs and goals. It associates sensory inputs with actions, actions with outcomes, and outcomes with values (psychological rewards or losses), providing a biological basis for learning to make adaptive choices to meet changing needs. There is widespread activation in the dorsal PFC when one is learning a new task, such as a sequence of key presses on a keyboard or a sequence of turns to make in driving from one location to another, but as the task is learned, the correct sequence becomes habitual and attentive action involving the PFC is no longer required—a savings of limited cognitive resources that is valuable as long as the environment changes slowly enough so that habitual action sequences remain effective. Novelty forces engagement of the PFC to figure out what to do and to plan ahead. This typically involves *mental simulation*, in which possible consequences of different courses of action are imagined. In brain imaging studies, people with less vivid imaginations of future consequences, as indicated by activation in the ventromedial PFC (vmPFC) are more prone to exhibit present bias (favoring smaller immediate rewards over larger delayed rewards) and to take unfavorable gambles (overvaluing immediate rewards despite the danger of long-term losses). Damage to the vmPFC also undermines ability to imagine outcomes for others, decreasing empathy and inclination to behave morally toward others.

Chapter 7 ("Checking") highlights the distinction between deciding to do something and successfully doing it. This distinction forces us to be aware of our own intentions and to monitor our own behaviors to check whether intended actions or action sequences have been successfully executed. The chapter presents the notorious "Libet task," in which awareness of intention to make a voluntary movement lags the decision to make it, in the sense that the direction of motion can be predicted from brain scans before a subject is consciously aware of having made a decision. Monitoring our own intentions and performance and reengaging the PFC when we make mistakes on difficult tasks helps learn and perfect skills. Monitoring the intentions and movements of others (activating the rostral cingulate cortex and mirror neurons) helps learn social skills and to learn from others.

Chapter 8, "Acting," notes that the cerebellum is activated as skills are successfully learned and become automatic, being transferred from laborious, cognitively demanding trial-and-error efforts involving the PFC to precisely executed sequences of actions that require little or no conscious thought. Parts of the brain are also activated when novelty is detected or when actions do not have expected consequences (the cerebellum when the sensory consequences of actions are not as expected, and the anterior striatum when expected rewards for actions do not occur). Such *prediction errors*, or mismatches between expectations and observations, signal a possible need to rethink, updating our understanding and behaviors to cope with an uncertain and changing world. Chapter 9, "The Future," provides an appropriately humble assessment of the dramatic accomplishments and dramatic remaining challenges in cognitive neuroscience, emphasizing that "Now that we know so much about *where* there is activity in the brain while people perform cognitive tasks, the next step is to find out *how* that activity makes cognition possible. In other words, we need to understand mechanisms." The book ends, on pages 109 and 110 of its whirlwind tour, by stressing the need for biologically plausible computational models of how the brain works. It notes that recent advances in artificial intelligence and deep learning, such as the AlphaGo program, which uses planning and evaluation artificial neural networks to play Go with superhuman skill, make a start in such computational models. These themes are developed further in *Artificial Intelligence* and *Deep Learning*.

Like most of Oxford University Press's Very Short Introduction books, *Cognitive Neuroscience: A Very Short Introduction* is highly readable—an engaging, informal, and fun read that moves briskly through its topics. It provides 14 pages of references and 2 pages of further readings. Overall, the book succeeds admirably in delivering its promised very short (110 small pages) introduction to brain scanning results that are illuminating the major anatomic features and interconnections by which information is processed within the brain, making both subconscious and conscious thought possible. For risk analysts, key take-home lessons are that the human brain has evolved to allow both learning from experience (largely from feedback signals generated by prediction errors) and deliberative planning (via the PFC), with learned behaviors for functioning in a stationary environment rapidly becoming automatic (via the cerebellum), but with novelty and prediction errors triggering the PFC and deliberative thought when needed. There are clear parallels between this biological architecture and the Systems 1 and 2 (fast, intuitive vs. slow, deliberative thought) made famous in Kahneman's *Thinking, Fast and Slow* (reviewed by Warner North in the July 2012 issue of *Risk Analysis*).

Thinking and Reasoning

Jonathan Evans' *Thinking and Reasoning: A Very Short Introduction* explores the psychology, rather than the biology, of human thinking. "Thinking" is interpreted broadly as information processing in the brain, rather than as conscious thought only. The final chapter of the book, Chap. 7, contrasts what it calls Type 1 thinking (Kahneman's System 1) with Type 2 thinking (Kahneman's System 2). Type 1 thinking is typically fast, high capacity, parallel-processing, belief-based, nonconscious, biased, contextualized, automatic, associative, based on implicit learning and knowledge and experience-based decision-making, independent of cognitive ability and working memory, and autonomous. By contrast, Type

2 thinking is typically slower and more effortful, capacity-limited, serial processing, conscious, normative, abstract, controlled, rule-based, based on explicit learning and knowledge and on consequence-driven decision-making. It is positively correlated with cognitive ability and working memory, and involves mental simulation of outcomes of different courses of action. The author proposes that not only are there these two main systems for thinking ("dual process theory"), but that there are essentially two distinct minds involved in human thought. The older intuitive (System 1) mind uses experiential and associative learning to achieve goals by responding to the past and repeating what has worked before (as in animal learning and conditioning experiments). The more recently evolved, reflective mind draws on both System 1 and System 2. It pursues goals by reasoning and hypothetical thinking, trying to decide by imagining and reasoning about probable future consequences of alterative courses of action, i.e., by mental simulation. The old mind is associated with automatic, unconscious learning, analogous to training of neural networks in *Deep Learning*. The new mind is associated with imagination of possible futures, effortful planning, flexible intelligence, and deliberative reasoning (and hence with the PFC in *Cognitive Neuroscience* and with symbolic reasoning and planning methods in *Artificial Intelligence*). The field of risk analysis is quintessentially a new-mind development.

The first six chapters of *Thinking and Reasoning* prepare for this culminating discussion of the old and new minds. Page 1 explains that "Reasoning involves making suppositions and inferring their consequences" and that "Reasoning can help us to solve novel problems, to make one-off decisions, to develop mathematics, science, and engineering, to design our environments to suit ourselves," thus positioning reasoning as a distinctive type of future-oriented thinking. The rest of Chap. 1 introduces the field and history of the psychology of thinking and reasoning, noting that that the vast majority of thinking (information processing in the brain) is not reasoning, but rather automatic and subconscious processing. It discusses the high costs of cognitive errors and expert misjudgments (typically due to heuristics and biases, such as confirmation bias), e.g., in medical misdiagnosis. It distinguishes among deduction, induction, and abduction (reasoning to the most probable explanation) as forms of reasoning. It surveys early studies of the mind via introspection, psychoanalysis, behaviorism and conditioning experiments, and, more recently, cognitive psychology, which makes testable predictions based on models of information processing in the brain. Chapter 1 concludes that "The psychology of thinking deals primarily with novelty. How do we solve a problem, make a decision, or reason to a conclusion when we have never encountered a task of that kind before? Cognitive psychologists have been studying such questions intensively over the past 50 years or more and this book will summarize many of their findings."

Chapter 2 ("Problem Solving") examines the phenomenon of creating solutions to new problems through flashes of creative insight (studied by Gestalt psychologists) and by heuristics. Heuristics are used in both human and AI problem-solving. Forward search seeks to identify sequences of actions to take us from the current state to a desired goal state. Backward search starts from the goal state and recursively identifies subgoals until ones that can be attained from the current state (or from states that forward search reveals can be reached from the current state) have been identified. These techniques, which have been used in AI since Newell and Simon's General Problem Solver (GPS) program in the 1960s, are similar to event tree (or decision tree) and fault tree analysis in decision analysis and probabilistic risk analysis. Chapter 2 also summarizes findings on expertise and expert problem-solving. On the positive side, some experts can use well-trained intuitions to quickly spot useful patterns and focus on promising solutions, as well as using analogy and linking of ideas from different domains to solve problems creatively. On the negative side, the chapter examines the phenomenon of compelling but wrong intuitive solutions. These typically occur when Type 1 thinking is inappropriately engaged in place of Type 2 thinking (as when one answers the question "A bat and a ball together cost $1.10. If the bat costs one dollar more than the ball, how much does the ball cost?" with the intuitive but wrong answer of 10 cents). The extent to which people rely on intuition to answer questions, rather than reflecting and checking using reason (e.g., working out the correct answer of 5 cents, in this example), is a personality characteristic. Performance on problem-solving tasks depends not only on intelligence (correlated with IQ or general intelligence, *g*) but also on *propensity to reflect*, i.e., to engage in Type 2 thinking when needed, rather than trusting intuitive answers.

Chapter 3 ("Thinking Hypothetically") points out that experts often have much richer and more detailed mental models than non-experts, allowing then to more efficiently test a sequence of hypotheses to converge on the cause of an observed problem. (Use of mental models at varying levels of detail by people who specialize in different areas is discussed further in *The Knowledge Illusion.*) However, confirmation bias impairs intuition-guided hypothesis-testing and conclusions. For example, the Wason selection task experiment (1966) showed that, when asked which of 4 cards showing A, D, 3, and 7 must be turned over to determine whether it is true that "If there is an A on one side of the card, then there is a 3 on the other side of the card," most people correctly determine that the A must be turned over, but many also choose the 3 card (which is irrelevant for testing the hypothesis), and few choose the 7 card (which is logically necessary, as an A on its flip side would disconfirm the hypothesis). The author is an expert on the Wason selection task, and shares fascinating insights into variations that make it easier or more difficult for most people to solve correctly. The key insight remains that intuitive reasoning about how to test hypotheses is often logically flawed, in part because we do not readily seek disconfirming evidence and alternative hypotheses to explain observations once a favored hypothesis has been adopted. The remainder of Chap. 3 discusses similar biases in assessing probabilities of hypotheses and in causal and counterfactual reasoning. Chapter 4 ("Decision Making") notes that Type 1 information processes dominate most of our routine choices, but Type 2 reasoning and mental simulation are essential for deliberative decision-making in novel situations. Decision trees, expected utility theory, and decision analysis provide a normative theory for decision-making. More descriptively adequate theories (such as Prospect Theory) depart from such normative prescriptions due to heuristics and cognitive biases such as the certainty effect, loss aversion, overweighting of small probabilities, and framing effects. Human judgments, including those of expert groups, tend to be overconfident unless very accurate feedback on repeated judgments is available (e.g., for weather forecasters). Much of this material will likely be familiar to many readers of *Risk Analysis*.

Chapter 5 ("Reasoning") examines deductive (syllogistic) reasoning and belief bias, i.e., tendency to endorse believable conclusions as logically valid and to reject implausible conclusions as not logically valid, independent of their logical validity. That both abstract logical reasoning and probability judgments are often incorrect (violating logical consistency conditions and Bayes' Rule) leads to a discussion of a "new paradigm" for psychology of reasoning that examines how people use their existing beliefs as they reason with new evidence. The new paradigm is more concerned with how people reason than with the departures of their reasoning from normative theories such as logic, probability theory, and decision theory. This theme is continued in Chap. 6 ("Are we rational?"), which discusses various types and concepts of rationality: instrumental rationality (how should we behave to make our desired outcomes more likely?), epistemic rationality (what should we believe?), bounded rationality, normative (ideal) rationality, ecological rationality (behavior adapted to its environment, e.g., using "fast, frugal heuristics" that work well in specific environments although perhaps not in general), and evolutionary rationality (behavior suited to the survival of our genes). The chapter mentions the "great rationality debate" now unfolding in academic journals over the practical value of making choices in accord with normative theories, given the observation that people without special training such as that of risk analysts often do poorly on tests involving standard normative theories of logical, probabilistic, and decision-analytic reasoning (although subjects with higher IQ/*g* scores tend to perform better on such tests). Finally, Chap. 7 discusses how dual process theory (Type 1 and Type 2 thinking) can explain cognitive biases, and how certain cognitive styles—specifically, rational thinking disposition, as measured by a number of scales, which indicates inclination to check one's intuitive responses via reasoning—favor better performance on certain types of reasoning tasks. For example, in response to the question "If it takes 5 machines 5 min to make 5 widgets, how long would it take 100 machines to make 100 widgets?" many people find the intuitive answer of 100 min compelling and feel confident enough in the answer not to check it. People with high rational thinking disposition are more likely to engage System 2 and reason out the correct answer of 5 min.

For risk analysts, this very short introduction (115 small pages, a two-page appendix on Bayes' Rule, and 7 pages of references for the chapters and 3 pages of additional readings) presents a succinct, highly readable account of much of the psychology of reason, rationality, and normative thinking (as well as various departures from it) that are pivotal to our field. It confirms at the level of psychology key biological insights from *Cognitive Neuroscience*, such as the distinct (though linked) processes of fast automatic thinking and slow deliberative thinking, and the special role of the latter in dealing with novel and unexpected situations by imagining possible future consequences of actions and planning how to achieve desired outcomes.

The next two books deal, respectively, with computational modeling of fast, automatic responses to stimuli based on learning from experience, and with slower planning and deliberation about how to achieve desired goals.

Computational Models of Nondeliberative Thought: *Deep Learning* (MIT Press, 2019)

John D. Kelleher's *Deep Learning* is a delightful exposition for non-specialists of the multi-layered artificial neural network technology that has recently revolutionized face, image, handwriting, and speech recognition. This deep learning (DL) technology has enabled DeepMind's AlphaGo program and its successors to triumph over the world's champion Go players. It powers current state-of-the-art language translation and internet search apps and supports AI composition of text, music, and art (discussed further in *Artificial Intelligence*). It is improving countless automated decision and routine risk management systems in finance, business, and industry. DL is increasingly being incorporated into credit-granting, underwriting, claims processing, fraud detection, and cybersecurity decision algorithms. It is used to enhance safe operations of complex engineering systems, in autopilot and autonomous and unmanned vehicle navigation and control systems, robotics, and industrial automatic process control. DL algorithms have recently found successful applications in control of logistics and infrastructure networks, marketing and advertising, and medical diagnosis, and new applications are introduced weekly. *Deep Learning* spends little time on these applications, however, instead allocating most of its 250 brisk pages to describing how the underlying DL technology works. Assuming little or no technical background, the book walks readers through the details of how artificial neurons (simple mathematical functions), interconnected networks of such neurons, and multiple interconnected layers of such networks, are trained to solve the astonishing variety of classification and decision problems to which DL is being applied. It explains why DL can deal easily with incomplete, noise-corrupted, and ambiguous input data; how it learns hierarchies of increasingly abstract features and concepts for guiding robust decisions despite realistic uncertainties, gaps, and imperfections in input data; and how it can be used to generalize and abstract from specific cases to decide what to do when presented with novel inputs.

Chapter 1 formulates the general problem of how to learn a *decision function*—a decision rule that maps data to decisions (or input values to output values)—from training data consisting of many cases exemplifying the desired input-output behaviors. If a training set of correctly labeled or classified cases is provided (typically as a data frame, with each case represented as a row and each variable or feature used in describing cases corresponding to a column), then learning a decision function that correctly computes the desired output (e.g., the correct classification of a case) from the input features is the general problem of *supervised learning* within machine learning. Traditional linear regression modeling for continuous output variables, and logistic regression modeling for discrete outputs, can be viewed as simple examples of supervised learning with numerical inputs and outputs and with an assumed straight-line or logistic regression function relating the outputs to the inputs. However, DL also allows much richer inputs and outputs. For example, the input to a spam detector or machine translation DL system might be the text of an e-mail rather than a number. A digitized photograph could be the input to a face recognition or scene classification DL system on a smart phone; a sample of speech is the input to a speech recognition system; and DNA sequences and regions of the DNA molecule are the inputs and outputs of a DL gene prediction system. In automated process control and autonomous vehicles, inputs consist of data from sensors and outputs consist of control signals to actuators. These raw inputs are typically processed to create a set of feature values that can be used to represent a case for purposes of training and evaluating a classification, recognition, or control system. Traditionally, much time and effort in data analytics have been devoted to identifying useful predictive features derived from the raw inputs. (As a trivial example, the ratio of weight to height might prove a more useful predictor of health risks than either variable alone, or at least be a useful additional predictor.) DL automates such "feature engineering" by automatically discovering highly predictive featuresand how to combine them into higher-level features to improve output predictions; this is accomplished by the multiple layers in a DL system. Like regression models, neural network models can be "trained" (i.e., adjusted to provide good fits to the input-output patterns in the training data set) by an iterative numerical algorithm that compares desired to actual outputs for each set of input values in the training set, and that adjusts model parameters to reduce the difference between the desired and actual outputs. However, instead of adjusting only a few parameters (e.g., slopes and intercepts in a regression model), DL systems may adjust thousands or millions of parameters to obtain a close approximation to the desired decision function without assuming that it has any simple form.

Chapter 2 explains the key idea of *gradient descent algorithms* for solving simple regression problems numerically, by iteratively adjustingmodel parameters to reduce a measure of error such as the mean squared difference between current and desired outputs, averaged over all cases in the training set, i.e., the mean squared error (MSE). It makes the crucial point that gradient descent can also be used to fit much more complicated models, including models composed of layers of simpler models, with outputs from lower-level models serving as inputs to higher-level models. A simple example of a decision model in which a loan is granted to an applicant if and only if the credit score produced by a linear regression model exceeds an acceptable risk threshold is used for motivation and exposition. This simple decision model corresponds to a single artificial neuron. Its inputs are the values of the predictors in the regression model. The neuron outputs a value of 1 (accept application) if and only if a weighted sum of inputs exceeds the acceptance threshold; otherwise it outputs a value of 0 (reject application).

Chapter 3 generalizes this concept of artificial neurons as simple decision functions by allowing logistic or other nonlinear functions. A DL system typically contains several layers of neural networks, with each layer receiving inputs from the layer below it. The first, or lowest, layer is called the *input layer*; the last, or highest, layer is called the *output layer*; and those in between are called *hidden layers*. A DL network is one with multiple hidden layers; its "depth"reflects the number of hidden layers. Each layer consists of interconnected mathematical models of neurons, each of whichis a simple function that computes a weighted sum of its inputs and then applies a (usually nonlinear) *activation function* to the resulting total to determine the value of its output (called its "output activation"). For a threshold activation function, the possible output activation values are 0 or 1; for a logistic activation function, the activation values are numbers between 0 and 1. The *parameters* of a DL network (or of simpler neural networks with no more than one hidden layer) are the weights applied to each input to each neuron (i.e., to the activation values of the neurons that feed into it). Networks are trained by adjusting these weights to improve the fit between desired and actual input-output behaviors using either gradient descent or backpropagation algorithms, both of which are discussed in detail in Chap. 6. For computational purposes, the weights for a layer are conveniently represented as arrays of numbers, called weight matrices. The depth of a DL network is the number of layers with weight matrices. The weighted sums in successive layers can be calculated efficiently via matrix multiplication usingspecial hardware (graphics processing units (GPUs), originally developed for video games). Choices of how many layers to include, how many neurons to put in each layer, how to interconnect them, and which activation functions to use are currently more art than science, and the selection and adjustment of these "hyperparameters" can greatly affect the speed and quality of DL network training.

Principles of DL architecture and training were worked out from the 1940s until the 1990s. Chapter 5 briefly recounts this history, including various roadblocks and how they were overcome. Highlights include the development of the "Perceptron" in 1958 (essentially, a single artificial neuron with a threshold activation function and inputs from an array of 400 photocells); a proof in 1962 by Frank Rosenblatt, the developer of the Perceptron, that automatic iterative adjustments of weights to reduce output errors would eventually produce weights that correctly classified inputs in a binary (yes-no) classification task if such a set of weights existed; an extension from 2 to many output classes by Widrow and Hoff (1960) using a least mean squared error (LMS) algorithm, from which modern gradient descent algorithmsdeveloped; a proof by Minsky and Papert (1969) that single-layer models cannot learn some nonlinear decision functions, greatly limiting their expressive power; development of multilayer networks with nonlinear (sigmoid) activation functions for the neurons, which provably *can* learn nonlinear decision functions (the "universal approximation theorem"), overcoming Minsky and Papert's objection; development and popularization in the 1980s of the backpropagation algorithm for training multilayer networks by computing errors in the current output layer and using them to adjust weights in preceding layers based on their contributions to the errors (thus "backpropagating" the errors through the network to update the weights); and interpreting the weights as distributed representations of input features, with early layers representing low-level features of the inputs (e.g., edges and corners in a picture, or phonemes in a string of spoken words) and successive layers combining these features into higher-level, more abstract representations of the input (e.g., objects and scenes in a picture, or words and sentences in speech). Chapter 4 also introduces modern DL network architectures. These include *convolutional neural networks* (CNNs), which essentially scan images for features to automatically identify specific individuals, objects, or type of object (such as a cat, car, etc.) based on recognized features; *recurrent neural networks* (RNNs), which have a single hidden layer but are augmented with a memory buffer to allow processing of sequential inputs, such as speech or text strings, while treating each unit (e.g., a word) in the context of those that have preceded it; and *autoencoders*, which have a single hidden layer and few enough neurons in it so that simply training the system to "predict" (or reconstruct) the input from the activations in the hidden layer forces it to compress the input information efficientlyas the input is represented in the hidden layer, thereby identifying informative features and discarding superfluous information. Chapter 5 discusses the architectures for CNNs and RNNs in more detail, including the use of long short-term memory (LSTM) units to propagate the current activation patterns in a network over multiple time steps or units in a sequence, e.g., multiple words in a sentence being recognized or translated into another language. (Current activation patterns constitute "short term memory," in contrast to connection weights, which constitute "long term memory"). Chapter 6 discusses the details of the training algorithms (gradient descent and backpropagation) for DL networks.

The book concludes in Chap. 7 with a brief discussion of the future of DL, emphasizing the virtuous cycle of bigger data, faster hardware, and better algorithms. Bigger data comes from the internet of things (IoT) and increasing instrumentation, surveillance, and storage of data throughout technological society. Better hardware started with the use of graphics processing units (GPUs) a decade ago to support DL; today, specialized GPU hardware supports DL libraries such as TensorFlow and PyTorch. The energy consumption of DL hardware can be reduced by novel chip architectures (e.g., neuromorphic chips, in which the activities of artificial neurons are not synchronized by a centralized clock). Better algorithms include the recent (2014) advent of *generative adversarial networks* (GANs) used to produce "deep fake" images, text, and other outputs by pitting against each other two networks: a *generative network* that tries to produce samples indistinguishable from (i.e., drawn from the same distribution as) real data; and a *discriminative network* that tries to classify samples as real (i.e., drawn from the real data) or fake (i.e., drawn from the generated data). In the past 5 years, increasing attention has been paid to unsupervised learning, to avoid the need for large, costly training data sets of correctly classified ("labeled") cases. *Pretraining* and *transfer learning*, in which the lower layers in a DL system are pretrained on related tasks (e.g., general image recognition tasks) before the whole system is trained and optimized for a specific task (e.g., identifying cancerous lesions in a medical image processing application, or identifying enemy tanks in a military application) can greatly reduce training time and need for large training sets, which are currently limiting factors in DL technology. *Transformer models*, in which models dynamically select a subset of inputs to focus on in generating an output, are improving current state-of-the-art performance in machine translation, text generation, and other tasks. The book concludes by noting the privacy, legal, and technical challenges posed by black-box DL systems that make decision recommendations (e.g., on who should receive or be denied credit, be suspected of fraudulent or criminal behavior, be considered at risk of a disease, etc.) based on input-output patterns and algorithms that are distributed over layered networks and that cannot easily be explained. It discusses ongoing work on "explainable artificial intelligence" that uses techniques such as feature visualization, attribution of activation patterns and output decisions to specific inputs, and dimensionality reduction to interpret the flow of information processing within a DL system and to explain what drives its output classifications or decisions.

DL adds value to the risk analyst's toolkit in several ways. Most straightforwardly, it enables use of large amounts of high-dimensional data to automatically model (i.e., approximate) initially unknown, possibly nonlinear functions relating observed inputs to observed outputs. This use as a function approximator has many applications in routine risk analysis using big data. It provides an alternative to regression modeling, with the advantages that it automatically models nonlinearities and interactions among predictors, selects informative variables and features (i.e., combinations of variables), and uses available inputs despite missing data. These are technical conveniences for risk analysts striving to model the predictive relationship between multiple inputs (e.g., exposure concentration and duration, sex, age, socioeconomic variables, co-exposures and co-morbidities, etc.) and outputs (e.g., mortality and morbidity counts). Applications in medical, financial, credit-scoring, and security risk analysis, as well as in insurance underwriting and claims processing, industrial automatic control of engineering systems, and many other areas, attest to the practical value of DL in routine risk management applications. Conceptually, the use of successive network layers to discover and represent increasingly abstract (and, with autoencoders, succinct) summaries of the inputs for predicting outputs provides a way to learn from data a hierarchy of features and concepts for describing the world in terms that make it predictable, at least to the extent that outputs can be predicted from inputs. This principle of hierarchical knowledge representation and abstraction—using higher-level concepts and features built from lower-level ones, while discarding predictively irrelevant details—appears to be fundamental for generalizing from specific experiences and for quickly interpreting new situations and reacting, despite the complexity and constant changes in raw sense data about the world. Finally, DL can be seen as contributing to computational modeling of "System 1" thinking—quick, intuitive responses to changing stimuli that do not engage the slower, more deliberative and conscious cognitive thinking of System 2. DL networks can map inputs to outputs (e.g., perceived situations or stimuli to decisions or behaviors) very quickly based on past experience, as encoded in trained networks. But they cannot answer even simple questions that involve hypothetical, counterfactual, or causal reasoning about novel situations or about possible future consequences of current actions. For this, System 2 thinking is necessary. *Artificial Intelligence* addresses computational models of such thinking.

Computational Models of Deliberative Thought: *Artificial Intelligence: A Very Short Introduction* (Oxford University Press, 2018)

Artificial Intelligence begins by stating that AI seeks to give computers diverse information-processing capabilities—such as perception, association, prediction, planning, and control of actuators—corresponding to psychological skills in humans and animals that enable them to attain their goals in uncertain environments. Chapter 1 proposes that human minds, as well as many AI systems, can be understood as collections of interacting virtual machines carrying out information-processing tasks in parallel. This is consistent with the brain scan data in *Cognitive Neuroscience* and with cognitive psychology experiments in *Thinking and Reasoning*, as well as with layered neural network architectures for learning, prediction, and control in *Deep Learning*. However, *Artificial Intelligence* provides a broader view of computational models of intelligence. Chapter 1 distinguishes among five major types of AI based on different architectures for the virtual machines used to carry out information processing. Classical symbolic AI (such as Newell and Simon's General Problem Solver (GPS)), also called "good old fashioned AI" (GOFAI), provides algorithms for planning and reasoning about how to achieve goals using causal models or knowledge (e.g., represented by if-then rules, and-or graphs, and precedence constraints; or, more generally, by probabilistic causal relationships), as discussed in our review of Chap. 2 of *Thinking and Reasoning*. Artificial neural network ("connectionist") approaches allow desired input-output behaviors for pattern recognition and control to be learned directly from examples ("training"), as described in *Deep Learning*. Other computational models used in AI include cellular automata, consisting of arrays of simple finite-state machines (automata) each of which adjusts its own behavior in each time step based on the behaviors of its neighbors (via simple if-then rules); dynamical systems (e.g., simulation models), in which the values of variables adjust to each other over time based on differential equations; and evolutionary programming, in which populations of programs interact with each other over time. These five approaches, all of which have also been used in risk analysis, are introduced in Chap. 1 and explored further in subsequent chapters. Chapter 1 also traces the history of AI, including the recognition in the 1940s that statements of propositional logic could be represented by computations in neural networks or by Turing machines; the use of GOFAI heuristics and reasoning and planning algorithms in the 1950s and 1960s to prove theorems, solve problems, and play games such as checkers; the first studies of cellular automata by von Neumann; the use of cybernetics and feedback control systems to model certain aspects of goal-seeking and purposive behavior in living organisms; the schism between symbolic and connectionist approaches (due in part to the Minsky-Papert critique of single-layer neural networks described above for Chap. 5 of *Deep Learning*); and the current world of hybrid AI systems, with both symbolic and connectionist systems working together to recognize patterns, focus attention, and decide what to do.

Chapters 2–5 of the seven chapters in *Artificial Intelligence* elaborate the major themes from Chap. 1. Chapter 2 ("General intelligence as the holy grail") recognizes that, although AI has had phenomenal successes in certain narrowly defined applications, from image recognition to robotics, the goal of artificial general intelligence (AGI) remains a major unsolved challenge. AGI would integrate multiple AI capabilities to allow AI systems to acquire common-sense knowledge of the world and to act safely and effectively in it, even in novel situations, without extensive programming or training and restricted environments. The chapter surveys both heuristic search techniques for problem solving and goal-directed planning systems, including modern probabilistic planners for environments that are only partially observable, so that the system's causal model of the world may be incorrect or incomplete. Such planning systems generalize traditional methods of probabilistic decision analysis and risk analysis to allow for uncertainty about ability to implement chosen actions, as well as the possibility that attempted actions will be interrupted by unforeseen events before they are completed. Chapter 2 also considers techniques for making difficult planning and decision problems under uncertainty more tractable by adopting simplifying assumptions (such as specific probability distributions for uncertain quantities, or an assumption that observations are statistically independent and identically distributed). To describe causal knowledge about the world in which an AI system operates, various knowledge representation methods are used (e.g., if-then rules, natural language processing methods, symbolic logical formalisms, and default assumptions). However, none of these formalisms gives an AI system genuine understanding of what the represented knowledge means. This leaves them prone to respond to unanticipated situations in undesirable (unsafe or ineffective) ways. No set of if-then rules, for example, can anticipate and prepare to respond appropriately to all contingencies in open worlds where unforeseeable events occur. Current AI systems cannot easily cope with novel situations when not all possible consequences of all possible actions are known. This limitation, sometimes called the "open world" problem or the "frame problem,"remains a major challenge for developing more intelligent and trustworthy AI.

In some constrained and repetitive environments, it is possible to use machine learning (ML) to learn effective decision rules from experience. System 1 pattern recognition and learned stimulus-response reactions may prevail where reasoning would be of little use due to insufficient causal knowledge. Chapter 2 briefly introduces supervised learning (in less detail than *Deep Learning*), unsupervised learning, and reinforcement learning (RL). RL algorithms can eventually learn effective or even optimal decision rules (if-then rules mapping observations to action probabilities) in stationary random environments by trial and error, even if the probabilities of events and the causal relationships between actions and outcome probabilities are initially unknown. For example, hybrid DL and RL algorithms have learned to play some video arcade games with superhuman skill, starting with no knowledge of the games. Any future AGI system will probably include such hybrids of algorithms for low-level perception, learning, and control, as well as higher-level algorithms for goal-directed deliberative planning. However, no current AGI model comes close to human-level flexibility in dealing with multiple tasks and novel situations. Chapter 2 concludes that, in addition to the unsolved problems of endowing AIs with common sense and overcoming the frame problem, successful AGI systems will probably have to cover motivation and emotion as well.

Chapter 3 ("Language, creativity, emotion") examines natural language processing (NLP) and question-answering systems (including Apple's Siri, IBM's WATSON, and WolframAlpha) and the use of emotions, not only in communication (as when a companion robot is programmed to recognize or simulate emotional expressions), but also as heuristics for deciding which of various competing motives to attend to next. Chapter 4 ("Artificial neural networks")introduces key ideas that are covered in more depth in *Deep Learning*. It highlights the importance of hybrids of connectionist processing (parallel distributed processing (PDP), artificial neural network (ANN), and situated stimulus-response systems) and deliberative (symbolic, sequential, GOFAI) processing for responding flexibly to novel uncertainties while also reacting quickly to changing stimuli while planning what to do next. The chapter concludes that "In short, the virtual machines implemented in our brains are both sequential and parallel. Human intelligence requires subtle cooperation between them. And human-level AGI, if it's ever achieved, will do so too." Chapter 5 ("Robots and artificial life") considers cellular automata and self-organizing systems, evolutionary programming, and robots with hierarchies of behavioral patterns that respond appropriately to changing environments not by planning what to do, but by deploying applicable behaviors as conditions change. Chapter 6 ("But is it intelligence, really?") ponders the mysteries of consciousness, distinguishing between functional consciousness (e.g., what is a mind aware of or paying attention to at any moment) and phenomenal consciousness (awareness of sensations and qualia). It then turns to models of machine consciousness, including global workspace models that view the mind as a collection of specialized information-processing subsystems or modules broadcasting to each other and competing for attention and access to working memory, where they generate the stream of consciousness. Automatic skill components, long term memory, and adaptive specializations learned in coping with the environment are viewed as a large substrate of unconscious specialized processors or modules that generate a much smaller thread of conscious thought when they are strongly enough activated to gain access to working memory, the locus of awareness. None of this explains phenomenal consciousness, but it does lead to models of self-awareness and to theories of self as enduring computational structures that both generate and rationalize an agent's actions.

Finally, Chap. 7 ("The Singularity") provides a skeptical assessment of some of the more overblown predictions and concerns expressed about AI by various celebrities—such as the possibility of downloading human minds into computer hardware and software, or the possibility of machines evolving their own goals and enslaving or transforming mankind. It assesses other concerns as being more plausible. These include untrustworthy AI (due to programming errors, inability to solve the frame problem, lack of common sense, and lack of understanding of human goals); technological unemployment as AI and automation replace people in an increasing range of increasingly skilled jobs; risks from AI-enhanced cybersecurity and military attacks; and perhaps subversion of human dignity as robot nannies, caretakers, and companions with shallow simulations of human emotional responses increasingly interact with emotionally vulnerable people. The book closes by describing recent pioneering efforts to develop morally competent AI (e.g., equipped to decide whom to rescue first or when to withhold information) that will be safe, useful, robust, and beneficial, while avoiding potential hazards and performing transparently, predictably, and reliably. Some of these are goals that risk analysts have helped pursue for engineered systems for many decades. Current AI applications can often be viewed usefully as engineered systems, to which tools and concepts of risk analysis apply.

Artificial Intelligence is full of stimulating ideas. For risk analysts, these include the computational modeling of Type 2 deliberative reasoning in planning and problem solving (including planning under uncertainty), as well as the necessity of integrating Type 1 and Type 2 information processing in systems that sense, plan, and act effectively in open-world environments. Other key ideas include the roles of unconscious and distributed processing in computational models of intelligent behavior; reinforcement learning and deep learning for learning by trial and error how to behave in initially uncertain environments; and cellular automata for simulating the evolution of complex dynamic systems. The risks from AI systems themselves are also well explained, and will continue to be an important topic for risk assessment and risk management for AI systems that lack common sense knowledge, understanding of human goals and priorities, and guarantees of safe operation.

Communities of Knowledge

In *The Knowledge Illusion*, two cognitive psychologists take the reader "on a journey through the fields of psychology, computer science, robotics, evolutionary theory, political science, and education, all with the goal of illuminating how the human mind works and what it is for—and why the answers to these questions explain how human thinking can be so shallow and so powerful at the same time." (p. 5) The authors' main thesis, also articulated in the introduction (p. 5), is that "The mind is a flexible problem solver that evolved to extract only the most useful information to guide decisions in new situations. As a consequence, individuals store very little detailed information about the world in their heads. In that sense, people are like bees and society a beehive: Our intelligence resides not in individual brains but in the collective mind." The rest of the introduction and the remaining 13 chapters of the book expand on these points and their implications for how we understand, or misunderstand, scientific, technical, social, and policy issues, and how we might govern ourselves better.

The authors argue that deliberative thought evolved to guide more effective action in an uncertain world, and that humans are particularly well adapted to reason about causality and the effects of actions -- but only at a level that enables us to get things done. Our mental models of causal mechanisms are often shallow (at about the level of "pressing the gas pedal makes it go faster, pressing the brake pedal makes it go slower"), with the exception of a few experts who specialize in understanding the underlying details. Moreover, the important ability to generalize from past experiences to new situations requires understanding and remembering underlying regularities rather than superficial details. As a result, many people cannot correctly complete a sketch of a bicycle to show where the chain and pedals go and what the frame should look like; describe in detail how toilets or common household tools and devices work; or explain in detail how economic or other policies that they favor would cause the effects that their supporters ascribe to them. Yet, few people are fully aware of the extent of this lack of knowledge until they are challenged to offer clear, step-by-step explanations of things they think they understand pretty well. This is illustrated in Chap. 1 by results of experiments in which people rate their own understandings of everyday devices (e.g., zippers), then try to explain their operation carefully, and then re-rate their understanding. The authors suggest that part of the reason for political polarization in the United States at present is that neither politicians nor voters realize how little they understand about the complex causal networks and mechanisms underlying policy issues they care about. Instead, we tend to let groups with which we identify do much of our thinking for us, adopting the positions of those whom we trust and admire.

The *knowledge illusion*, as discussed in the introduction and Chapters 1 and 2, refers to the misperception that we know and understand much more than we actually do. It arises in part because we fail to distinguish between what we personally know and what othersin a community of knowledge know (e.g., friends, experts, or the internet). Knowing enough about something to make it work, and where to find out more details about it if needed, feels much like understanding it. In reality policy issues, as well as common household devices and more complex systems such as cars or radios, are often far more complex and less predictable than we typically appreciate. Unknown unknowns, unpredictability (e.g., for the future trajectory of chaotic systems beyond a short forecast horizon), and complexity of systems on multiple scales of resolution also limit understanding. A more realistic appraisal of the limits of our knowledge might inspire more humility and less hostility and polarization in policy debates. Chapter 3 argues that, although we have evolved to reason about causality (and to tell stories that illustrate and transmit knowledge of causal patterns, counterfactual reasoning about alternative behaviors, and general lessons or morals), such reasoning is far from perfect. Forward causal reasoning (mental simulation of the potential consequences of actions or events) is prone to ignore alternative causes when assessing probabilities of effects given causes. Backward causal reasoning (from effects to causes, as in diagnosis and explanation of observations) encourages fuller consideration of alternative causes, but is difficult for people to do well. Chapter 4 points to the value of deliberation among people for improving the intuitive causal models of individual participants. It discusses some of the same material as *Thinking andReasoning*, including Systems 1 and 2, reflective and less reflective personality types, and the distinction between intuitive and deliberative reasoning (illustrating them with the bat-and-ball and widgets problems also discussed in *Thinking and Reasoning*). However, *The Knowledge Illusion* stresses that effective deliberation depends on a community of knowledge in which participants refine and improve each other's mental models and individuals learn from others whom they respect or trust. Thus, thinking has a strong social and collective component. This is the key new insight developed in *The Knowledge Illusion.*

Chapter 5 consider several problems for GOFAI that are also discussed in *Artificial Intelligence*, including the frame problem—which it describes as understanding what will change and what won't when an action is taken, i.e. the causal consequences of actions—as well as the problems of common sense knowledge and the need to plan and re-plan and to adapt behaviors in real time as conditions change. It examines embodied intelligence and subsumption architectures in which, as the authors explain, "Sophisticated tasks get done not through exhaustive computation and planning but by engaging a hierarchy of actors [AI agents, as explained in Chap. 5 of *Artificial Intelligence*] in an organized way that, at the lowest level, are just responding directly to the environment." Much sophisticated computation that would be required in GOFAI can be avoided by using fast, frugal heuristics (e.g., to walk through a doorway successfully, make sure that both sides of the door frame are approaching you at the same speed), which have the effect of exporting potentially difficult thinking to easily verified properties of the body and the world. Chapter 6 reviews evidence that larger brains in humans evolved specifically to support living in communities (the "social brain" hypothesis). It comments that people are built to collaborate, easily sharing attention and intentionality and inferring each other's intentions, beliefs, and desires, thus facilitating effective teamwork and pursuit of shared goals. People also naturally divide up cognitive labor: even in couples, as well as in larger groups, each member tends to specialize in, and remember, some things that the other does not.

Chapter 7 points out that our habit of blurring the boundary between what we know personally and what those in our community of knowledge know (and we simply know how to access when needed) can create dangerous overconfidence. Patients who have spent a few minutes consulting WebMD may feel that they have the expertise to deny a physician's diagnosis or seek alternative treatments. People who have just looked up irrelevant information (such as "What is a stock share?") on the internet are willing to bet more on their own performance in an investment game, and end up earning less money, than subjects who have not had just had their confidence in their own presumed knowledge boosted by an internet search. The authors state that machines do not yet recognize human goals and intentions (although some "mixed autonomy" AI systems are designed to do just that, which the authors do not mention). They caution that our increasing reliance on technology and automation make us increasingly vulnerable to technological failures. They suggest that real superintelligence is likely to come from communities of people pooling their knowledge via techniques such as crowdsourcing, prediction markets, and distributed collaboration platforms.

Chapter 8, "Thinking about science," discusses the dangers from misperceptions of risk shared by large numbers of people, as when reasonable skepticism about science and technology hardens into antiscientific thinking. Antiscientific rhetoric and actions directed against climate change, genetic engineering and genetically modified organisms (GMOs), food irradiation, nuclear power, nanotechnology, and vaccination are cited as examples. A correlation is reported between basic scientific literacy (e.g., knowing whether electrons are smaller than atoms, or whether antibiotics kill viruses as well as bacteria, both of which are answered correctly by about half the US population) and acceptance of these technologies, with perceptions of fewer risks and greater benefits. However, attempts to change pervasive, strong antiscientific attitudes by outreach, education, and risk communication have proved ineffective. The authors trace the difficulty to the fact that beliefs are deeply intertwined with each other, with our identities, and with the communities to which we belong and the beliefs of others whom we love and trust. Culture overpowers cognition, rendering attempts at (re) education relatively powerless. False causal models—for example, the misconceptions that genetic modifications in GMOs might spread like germs, through ingestion or contact; or that food irradiation is like radioactivity and might "get stuck" in food and contaminate it—also contribute to reluctance to accept technologies that are not well understood.

Chapter 9 ("Thinking about politics") extends to the realms of policy and politics the finding that we often feel strongly about things we understand poorly, or not at all. A striking example is a 2012 Pew Research Center poll following a US Supreme Court decision upholding key provisions of the Affordable Care Act: 36% of respondents favored the ruling, 40% opposed it (only the remaining 24% expressed no opinion)—and yet only 55% responded correctly when Pew asked what the court had ruled! Thus, feelings of agreement or disagreement with politically charged decisions do not always depend on knowing what the decisions were. Preferences and opinions can be strong without being informed, especially when groupthink leads people to take cues from each other in adopting positions. In experiments, having subjects try to give detailed causal explanations for what consequences they expect, and why, from different policies (such as merit-based pay for teachers, increased retirement age for Social Security, or cap-and-trade programs for carbon dioxide emissions) significantly reduced their self-rated understanding of the issues, extremity of support for their favored positions, and polarization. However, causal explanation is seldom emphasized in policy debates and rhetoric. Rather, the author explain, politicians and interest groups frequently cast policies in terms of what moral psychologist Jonathan Haidt calls "sacred values"—that is, values that are held because they are perceived as being morally right, independent of the consequences they cause. For example, debates over healthcare are framed divisively, as a clash in values between universal human rights to decent care vs. rights to individual choices instead of government decisions about vital matters, rather than as a deliberative analysis of how best to get what nearly everyone agrees we all want: better health care, lower costs, and more people covered. Framing policy issues as value issues helps attract votes, polarize passions, and shut down constructive causal analysis and deliberative thinking. Chapter 9 concludes that effective political leaders should make appropriate use of experts, with deep causal understanding in relevant subject matter domains, who understand how different policies are likely to affect outcomes in the short and long runs. Careful causal analysis of consequences (System 2 thinking) may sometimes point to quite different policies than those favored by passionate preferences, feelings, and moral convictions (System 1 thinking, which is easily manipulated by framing and distorted by cognitive biases). A mature electorate should appreciate leaders who pay attention to relevant expertise in deciding what to do.

Chapter 10 goes further, proposing a measure of collective intelligence (how well do people work in groups?) to supplement the traditional measure of general intelligence, *g*, for individuals. Chapter 11 suggests that education can and should be changed to increase collective intelligence, acknowledging the cognitive division of labor and that most people need to trust experts for detailed and specialized knowledge. People must also know how to find trustworthy experts and how to exercise appropriate skepticism and critical thinking in interpreting claims made by the media and other sources. Chapter 12 ("Making smarter decisions") asks how we can help people make wiser choices, given realistic limitations on knowledge, attention, understanding, and tolerance for explanatory detail. These undermine intuitive understanding and decision-making in many areas, from personal financial decisions (e.g., most people don't fully appreciate the values of annuities or the nonlinearity of mortgage payments) to medicine (e.g., how do band-aids actually work in promoting wound healing, and under what conditions will they work more or less well)? The main solution that the authors advocate is the *Nudge* framework of Thaler and Sunstein (reviewed by Warner North in the July 2012 issue of *Risk Analysis*): by changing the choice environments within which individuals make their decisions, e.g., by changing default choices (what must be opted out of in order to make a different choice), governments and organizations can change probabilities of individual choices and behaviors to reduce predictable regrets. This idea can be applied to decisions made as part of a community of knowledge by reducing complexity of explanations (e.g., of consumer finance products), providing simple decision rules, and giving people relevant information just when they need it, as well as teaching people to check their own levels of understanding ("know what you don't know") so that they can seek more information if needed before making important decisions.

The final chapter of the book concludes by summarizing its three main themes. First, *ignorance of decision-relevant facts and causal mechanisms is prevalent.* Moreover, those who perform worst on tests of expert skills and knowledge typically overrate their own skills the most, reflecting ignorance about the extent of their own ignorance (the Dunning-Kruger effect). Second, *most of us have an illusion of knowledge* that copes with complexity by ignoring it, allowing us to think that we adequately understand things that we don't. Third, *knowledge resides in communities of knowledge*: no individual may have expertise in most areas, but most areas have experts. Effective leaders make good use of delegation to experts while remaining wary of those whose self-confidence stems from ignorance and illusion of knowledge rather than from dependable expertise and skills. The illusion of causal understanding encourages us to believe that that we understand enough to adopt and support firm policy positions and recommendations, even when we don't. Left unchecked, it encourages overconfidence and polarization, but it can often be dispelled by challenging people (including ourselves) to articulate detailed causal explanations for the predicted consequences of alternative policies. On the positive side, the knowledge illusion gives some people self-confidence to explore new territories, attempt new ventures, and take new risks, from trying to circumnavigate the globe to trying to reach the moon. As the authors state (p. 262), "In this book, we have pointed out how the illusion of understanding can lead to war, nuclear accidents, partisan gridlock, rejection of science, lack of fairness, and other misfortunes. But we have also shown that the illusion results from an incredible feature of the mind. The knowledge illusion is a result of living in a community of knowledge; it arises because we fail to distinguish what's in our own heads from what's in other people's heads . . . because, cognitively speaking, we're a team. . . . Those who live in a knowledge illusion are overconfident about how much they know. . . . Many great human achievements are underwritten by false belief in one's own understanding. In that sense, the illusion may have been necessary for the development of human civilization."

The Knowledge Illusion has a fun, breezy style that is noticeably less academic than the previous four books. It is less tightly edited, repeating key points about the limitations of individual knowledge and expertise and the importance (and potential pitfalls) of a community of knowledge in most of its chapters. At 265 pages of text, with an additional 16 pages of notes on sources and references, it is a quick read. For risk analysts, some of the most useful discussions are likely to be the ones on thinking about science and policy in Chaps. 8 and 9; the perils of groupthink and of beliefs formed largely by the beliefs of others (as in social amplification of risk and distortion of risk perceptions by social and news media by the internet); and the emphasis in early chapters on causal reasoning and System 2 thinking. A highly useful aspect of the book, shared by *Factfulness*, is presentation of questions and challenges that readers can work through to experience the strength of misconceptions and illusions before reading statistics on how very prevalent wrong answers are. Such exercises encourage humility and perhaps eagerness to learn, but also raise the hugely important point that ignorance and misconceptions of relevant facts and causal mechanisms, overconfidence in our own knowledge and policy convictions, and strong opinions about what to do based on our flawed mental models, tend to occur together. Those with the strongest opinions and clearest visions about what needs to be done are often those with the poorest understandings of the probable consequences of doing it. The *Knowledge Illusion* makes this case and encourages more humility, more self-appraisal, and more attempts to explain causal mechanisms in detail as antidotes to overconfidence and polarization. It advocates more and shrewder reliance on genuine experts to take advantage of the amazing potential for collective intelligence to solve problems better than any individual can.

Factfulness: Ten Reasons We're Wrong About the World—and Why Things Are Better Than You Think

Factfulness differs from the preceding five books by focusing on how to think more clearly and usefully not just in general, or with the help of algorithms, but specifically about the big trends and problems affecting human wellbeing at present. It picks up where *The Knowledge Illusion* leaves off, acknowledging that survey questions, many of which are listed in the introduction, consistently reveal that most people have dramatic misconceptions about key aspects of how the world works. (Example questions: "In the last 20 years, the proportion of the world population living in extreme poverty has ... (A) almost doubled; (B) remained more or less the same; (C) almost halved". Or, "There are 2 billion children in the world today, aged 0 to 15 years old. How many children will there be in the year 2100, according to the United Nations? (A) 4 billion; (B) 3 billion; (C) 2 billion.") Most respondents, in multiple countries and venues, answer most such questions incorrectly; many are confident about their wrong answers. For example, only 25% of Norwegians and only 5% of United States respondents correctly identify that the proportion of the world population living in extreme poverty has almost halved in the past 20 years (although admittedly the definition of extreme poverty is a low bar to clear). Most respondents, from teachers to business and political elites, to more general audiences in multiple countries, do not realize that the United Nations projects no further growth in the global number of children. Misconceptions and unawareness of key facts about global trends leave leaders and voters poorly prepared to engage in well-informed policy deliberations and decisions about how best to improve our joint, interdependent futures. Rather than repeating *The Knowledge Illusion*'s call for greater reliance on valid expertise and collective intelligence, *Factfulness* seeks to diagnose and repair key widespread misconceptions, which the authors call "mega misconceptions," by presenting relevant data andproposed principles of data-driven analysis, opinion formation, andrules for clear thinking based on data—the "factfulness" referred to in the title.

Factfulness is highly idiosyncratic and deeply engaging. It is a wonderful resource for risk analysts aspiring to use facts, data and analysis to help improve health and lives, from reducing extreme poverty to reducing risks of war and pandemics. Hans Rosling was a talented story teller and a compassionate, widely-traveled physician and public health expert. He enlivens important technical concepts conveyed in his TED Talks and, now, via this book with tales from his own experiences and adventures that vividly illustrate them. The dedication of the book signals this approach: "To the brave, barefoot woman, whose name I don't know but whose rational arguments saved me from being sliced by a mob of angry men with machetes." (The whole story is told in the last chapter, where it illustrates unforgettably how "factful" System 2 thinking can calm and redirect System 1 passions acting on preexisting misconceptions and flawed mental models about health risks.)

The Introduction notes that ignorance and misconceptions about aspects of the world—from the extent of poverty, to population growth rates worldwide and in different regions, to availability of basic primary health care—are persistent as well as widespread. Important false beliefs are hard to correct. Rosling et al. trace the tenacity of misperceptions to an "overdramatic" world view: we tend to perceive things as simpler, more extreme, and hence more dramatic than they really are. As they explain, "Think about the world. War, violence, natural disasters, man-made disasters, corruption. Things are bad, and it feels like they are getting worse, right? The rich are getting richer and the poor are getting poorer; and the number of poor just keeps increasing; and we'll soon run out of resources unless we do something drastic. At least that's the picture that most Westerners see in the media and carry around in their heads. I call it the overdramatic worldview. It's stressful and misleading. In fact, the vast majority of the world's population lives somewhere in the middle of the income scale. Perhaps they are not what we think of as middle class, but they are not living in extreme poverty. Their girls go to school, their children get vaccinated, they live in two-child families, and they want to go abroad on vacation, not as refugees. Step-by-step, year-by-year, the world is improving. Not on every single measure every single year, but as a rule. Though the world faces huge challenges, we have made tremendous progress. This is the fact-based worldview." The rest of the book presents data to back up these claims and recommends ways of thinking that make the data less difficult to accept when they conflict—as they often do, for many people—with strongly held prior beliefs.

The most salient themes of the book's 11 chapters are as follows. Chapter 1 ("The Gap Instinct" notes that histograms and clusters of variables show that the old dichotomous division of the world into haves and have-nots (or developed and developing, or we and they) is too simple. Empirically, many aspects of daily life—from child mortality rates to how people cook and eat to the transportation they use (walking, bicycle, motorcycle, car)—are much better described by four levels of income, with boundaries at \$1/day, \$4/day, \$16/day, and \$32/day for levels 1–4, than by two levels. Most of the world's population is in the two middle levels. Dividing the world into polar extremes, focusing on distinct average values rather than on overlapping distributions, and compressing the qualitatively very different lives of people in each of the distinct levels 1, 2, and 3 into a single "non-4" category obscures profound improvements in quality of life for most people in most countries in recent decades. Similarly, Chap. 2, "The Negativity Instinct," quantifies progress in recent decades on many dimensions, including indicators of health and longevity, education and literacy, girls in school, access to clean water, pollution control, adequate nutrition, and even guitars-per-capita (a proposed indicator of culture and freedom). It contrasts these improvements with the pessimistic views often conveyed by selective reporting, lobbying, and activism, which emphasize the many serious problems that remain to be addressed. News media and activists rely on drama to grab attention, often presenting stories of conflicts and stark dichotomies when reality is less dramatic. It is realistic to recognize that much in the world remains bad and in urgent need of improvement, while also recognizing that most indicators are much better now than 20 years ago. A more balanced ("factful") view can help to allocate concern and effort where they will be most effective.

Chapter 3, "The Straight Line Instinct," emphasizes the fact that many trends and relationships are not linear. For example, dental health is better at income levels 1 (where people cannot afford sweets) and 4 (where they can afford dentists) than at levels 2 and 3. Traffic death risks and childhood risks of death by drowning also peak between levels 1 and 4. World population growth has already leveled off in the younger age groups as increasing wealth has reduced average children per woman. Chapter 4, "The Fear Instinct," may be especially interesting for many readers of *Risk Analysis*, as it touches on several aspects of risk perception and risk communication. It notes that evolution has primed us to fear—and hence to pay attention to stories about—physical harm, captivity, and contamination by invisible substances that can harm us. Journalists reporting on natural disasters, wars and conflicts, terrorism, and health effects associated with or attributed to chemicals are likely to find attentive audiences. Lobbyists and activists, as well as journalists, prey on System 1 instincts and reactions, presenting scary information that grabs attention but yields a distorted view of the world. Chapter 4 emphasizes what most risk analysts know: fear and danger are different, and "The risk something poses to you depends not how scared it makes you feel, but on a combination of two things. How dangerous is it? And how much are you exposed to it?" The authors urge putting fear aside and getting calm before making decisions. Chapter 5 ("The Size Instinct") teaches the principles of putting rare events in context by looking at rates-per-capita and by focusing on what fault tree analysts call dominant contributors—the relatively few drivers that usually account for most of the variation in totals or outcomes, as in the 80–20 rule. It notes principles of numeracy, such as that "Single numbers on their own are misleading and should make you suspicious. Always look for comparisons. Ideally, divide by something" to get per-capita risks or other meaningful comparisons. Rosling et al. estimate that, in news coverage in early 2009, "Each swine flu death received 82,000 times more attention than equally tragic death from TB." Putting risks in the context of per-capita risk comparisons can help counteract the distortions induced by selective reporting that emphasizes the new, the scary, and the dramatic.

Chapter 6 ("The Generalization Instinct") warns against the dangers of using broad categories, stereotypes, and misleading generalizations to oversimplify complex reality by treating heterogeneous groups as homogeneous. For example, the observation that unconscious soldiers placed on their backs on stretchers had higher mortality rates than if they were placed on their fronts led to a movement in the 1960s to place sleeping infants on their tummies. But sleeping babies are not unconscious soldiers, and by the mid-1980s, it was realized that this false generalization was contributing to sudden infant deaths, rather than preventing them. Similarly, Chap. 7 ("The Destiny Instinct") warns against the human habit (akin to the "fundamental attribution error" in psychology) of believing that innate characteristics determine the destinies of people, countries, or religions. Chapter 8 ("The Single Perspective Instinct") warns against looking for single causes and single solutions. It emphasizes the importance of reflective thinking in practice: "Being always in favor or always against any particular idea makes you blind to information that doesn't fit your perspective. This is usually a bad approach if you like to understand reality. Instead, constantly test your favorite ideas for weaknesses. Be humble about the extent of your expertise. Be curious about new information that doesn't fit, and information from other fields." Chapter 9 ("The Blame Instinct") highlights the importance of looking beyond simple single-cause explanations for accidents or disasters (e.g., looking beyond a sleepy pilot to diagnose and fix the system that put a sleepy pilot at the controls). Chapter 10 ("The Urgency Instinct"), discusses how activists, sales people, politicians, and others use fear and urgency to shut down System 2 thinking and directly manipulate System 1 reactions. Of great relevance to many *Risk Analysis* readers, it highlights five global risks that we should worry about: global pandemic, financial collapse, world war, climate change, and extreme poverty. In advising against being rushed or panicked in managing these crucial risks, the chapter advises that "These risks need to be approached with cool heads and robust, independent data. These risks require global collaboration and global resourcing. These risks should be approached through baby steps and constant evaluation, not through drastic actions. . . . Ask how the idea [for risk management] has been tested. Step-by-step practical improvements, and evaluation of their impact, are less dramatic but usually more effective."

Finally, Chap. 11, "Factfulness in Practice," discusses how understanding the four income levels, recent global trends, the importance of humility and curiosity (and of willingness to recognize when old beliefs should be revised based on new data), and awareness of how news media and others use our dramatic instincts and how to defend against a distorted world view using "factfulness" thinking, can be used in education, business, journalism, organizations, communities, and citizenship. The book concludes that "When we have a fact-based worldview, we can see that the world is not as bad as it seems—and we can see what we have to do to keep making it better." The call to respect facts and evidence, remain open to new data and understanding, deliberately engage in System 2 thinking when making risk management decisions, be humble about one's existing understanding and expertise and curious to expand them, and refuse to be rushed or scared or distracted into adopting simplistic patterns of reasoning and problem diagnosis, are key components of the thinking of many top risk analysts, as reflected in biographical profiles in *Risk Analysis* over the past decade.

Aligning AI-ML and Human Values: *The Alignment Problem*

Decision and risk analysis seeks to provide useful frameworks for thinking about how to manage both known risks and uncertain risks, taking into account that knowledge of consequence probabilities for different choices, as well as current perceptions of risks and beliefs about their causes and likely consequences, are usually limited and may turn out to be inaccurate. Modern artificial intelligence (AI) and machine learning (ML) wrestle with many of the same challenges in guiding the decisions of robots and autonomous agents and teams of such agents operating under uncertainty or under novel conditions. The field of AI/ML has raised questions that have not been much addressed in risk analysis, yet that might be of great interest to many risk analysts. Among these are the following.

1. Is it possible to design risk-scoring systems that are both equitable and accurate, meaning that they yield well-calibrated risk predictions while giving all participants equal (preferably small) probabilities of false positives and also equal probabilities of false negatives?
2. What role, if any, should *curiosity* play in deciding what to try doing next in new, uncertain, and hazardous environments?
3. Which is preferable for an AI agent that manages risks on behalf of humans:
   (a) Do exactly what it is *instructed* to do;
   (b) Do what it infers that its users (probably) *want* it to do, even if they have not articulated it perfectly;
   (c) Do what it judges is *best* for them (e.g., what it deems they *should* want it to do, or what it predicts they *will* want it to have done in hindsight), even if that is not what they want now.

These are some of the questions explored in Brian Christian's thought-provoking and readable new book *The Alignment Problem*. Similar questions can be asked for human risk analysts seeking to identify what is best to do when making or recommending risk management decisions and policies on behalf of others. *The Alignment Problem* does an outstanding job of explaining insights and progress from recent technical AI/ML literature for a general audience. For risk analysts, it provides both a fascinating exploration of foundational issues about how data analysis and algorithms can best be used to serve human needs and goals and also a perceptive examination of how they can fail to do so.

The book consists of a Prologue and Introduction followed by nine chapters organized into three parts (titled Prophecy, Agency, and Normativity, each consisting of three chapters) and a Conclusion. All are worth reading. The three-page Prologue describes the seminal work and famous 1943 paper of McCulloch and Pitts introducing artificial neural networks and hints that finding out just what "mechanical brains" built from simplified logical models of neurons could do would soon become an exciting field.

The Introduction explains that "This is a book about machine learning and human values; about systems that learn from data without being explicitly programmed, and about how exactly—and *what* exactly—we are trying to teach them." It presents several examples of applications in which AI/M systems fail to perform as desired or intended. It begins in 2013 with the introduction of Google's open-source "word2vec" which uses modern neural networks to encode words as vectors that can be added and subtracted. This leads to both insightful equations of word arithmetic (e.g., Paris—France + Italy = Rome) and more problematic ones (e.g., shopkeeper—man + woman = housewife) that reflect biases built into our language that we do not necessarily want our machine assistants to inherit. Other challenges include video game-playing programs that learn to optimize the reward functions specified by their designers while failing to exhibit the behaviors those rewards were meant to elicit; image-recognition software that performs better for racial groups that were well represented in the data used to train it than for other groups; and crime risk assessment software programs for supporting parole decisions that turn out to have different error rates for blacks and whites.

Chapter 1, "Representation," traces the early history of neural nets, starting with the 1958 perceptron, a single artificial "neuron" or logic gate that outputs a value of 1 if and only if a weighted sum of inputs is above a threshold and outputs 0 otherwise, where 0 and 1 are typically interpreted as two classes. The chapter describes the "stochastic gradient descent" algorithm (which Christian explains in eight simple lines of plain English) for automatically adjusting the weights to reduce the classification errors that the system makes on a training set for which the correct outputs (i.e., classifications) are already known. This adjustment process can be considered a form of "supervised learning" in which the examples in the training set are used to adjust weights for the inputs until the system classifies cases in the training set accurately. The final set of weights can be interpreted as implicitly representing the knowledge needed to predict the output class from the inputs. Although the perceptron can only classify some patterns accurately (e.g., "left" vs. "right" for which side of a card a shape appears on, but not whether the number of dots on a card is odd or even), networks with multiple layers of perceptron-like artificial neurons arranged so that the outputs from the neurons at one level are inputs to neurons at the next layer can learn any desired input-output function exemplified in a training set of examples. Such multi-layer ("deep") artificial neural networks can be trained to map input signals to output decisions. Applications range from classifying images (is this tumor malignant, are those pictures of enemy tanks, does this photo show a cat?) to deciding what to do next in response to what is sensed (should this mortgage application or job application be accepted? Should an autonomous vehicle steer left or right or go straight to stay on the road?)

Variations of deep neural nets can also be used to detect anomalies and novelty (is this input *not* mapped with high confidence to any of the output classes learned about in the training set?) and to win a variety of games and control a wide variety of industrial processes while avoiding dangerous or uncertain conditions that make achieving desired outcomes too unpredictable for safe operation. These dramatic accomplishments are achieved via deep learning and other "supervised learning" algorithms that iteratively adjust weights to reduce errors in classification, prediction, and control rules for the cases in a "training set" of examples for which correct or desired responses are known for a variety of input conditions. The rules so learned typically perform very well as long as the new cases or situations to which they are applied are statistically similar to those in the training set. But they may perform poorly when applied to cases different from those in the training set. A face-detection or face-recognition system trained only on white male faces may perform poorly if applied to black female faces. No amount of sophistication in the training algorithms and representation of decision rules can overcome limitations and biases created by training data that are non-representative of the cases to which the decision rules are applied. In a world where useful software propagates quickly, biases and limitations in training sets may get locked into classification systems that are then widely deployed. This is perhaps especially true for natural language processing (NLP) systems that represent word meanings as vectors (e.g., Google's "word2vec" algorithm for embedding words into vector spaces based on how frequently they appear near other words). Such systems use deep learning to extract the most predictively useful representations of words as vectors. They enable numerous useful applications, from sentence-completion software to AI for question-answering and retrieval of relevant information from the web on smartphones. But the resulting systems are trained on past language use. Therefore, they reflect the biases, assumptions, and historical conditions built into past usage. This can lead to false, obsolete, or undesirable inferences in AI NLP systems, such as about whether a doctor is more likely to be male or female, or about whether a job applicant is likely to succeed at a company based on inferred superficial (e.g., demographic) similarities to past hires. Moreover, the distances between words (represented as points in a word2vec-type vector space embedding) turn out to correspond quite closely to human reaction times in tasks that require pairing words, such as implicit bias or implicit association tests: pairs of words that are more distant take longer to pair. Studying how word embeddings shift over time can help to identify social trends that are reflected in language use, including changing perceptions of identity groups and of hazards such as pandemics, climate change, or terrorism risks.

Chapter 2 ("Fairness") looks at how predictive risk-scoring algorithms have been used to inform decisions about which inmates should be classified as safe to parole or to release early based on predicted risks to society. It cites the dramatic swings in public and media perceptions of such systems, such as the *New York Times* veering from urging wider acceptance of risk assessment tools in parole in 2014 because "they have been proved to work" to writing in 2016 about "a backlash against using data to foretell defendants' futures" because, in the words of an ACLU Director, it is "kind of rushing into the world of tomorrow with big-data risk assessment." A potent catalyst for the backlash was a May 2016 article by ProPublica entitled "Machine Bias: There's software used across the country to p[redict future criminals. And it's biased against blacks." The chapter then moves into a fascinating discussion of recent mathematical research on exactly what algorithmic "fairness" and "bias" mean, and of the discovery that the standards that ProPublica advocated—essentially, that an acceptable risk assessment tool should not only be well-calibrated, rendering statistically accurate predictions of reoffense rates, but should also have the same misclassification rates (i.e., false-positive and false-negative rates) for different groups with different base rates of reoffense—are mathematically impossible. Theorems on the "impossibility of fairness" shine new analytic light on trade-offs and on what "fairness" and "bias" can and cannot mean. Chapter 2 also discusses the important distinction between *predicting* and *preventing* crime: being able to predict who is most likely to have an undesired outcome under the conditions for which data have been collected does not necessarily reveal how outcome probabilities would change under new policies or conditions. Yet this—the province of causal artificial intelligence rather than predictive machine learning—is typically what reformers and policymakers most want to know.

Chapter 3 ("Transparency") discusses the challenge of creating safe, trustworthy AI advisory systems that provide clear reasons for recommended decisions or courses of action. It starts with the cautionary tale of a neural net-based system that learned from data a predictive rule stating that asthmatic patients have *lower* risks of developing pneumonia than other patients. The pattern was correct, but the system did not model the causes for it: asthmatic patients have lower risks precisely because they would have much higher risks if they were treated the same as other patients, and they are therefore given more intensive care to prevent the onset of pneumonia. A system that simply classifies asthmatic patients as low risk and therefore allocates limited care resources elsewhere would be disastrous for asthmatic patients. This reinforces the larger methodological point that risk models that accurately predict risks under current conditions do not necessarily offer insight into how risks would change under new conditions or following interventions intended to improve the current system. Much of Chap. 3 is therefore devoted to "explainable AI" (XAI) that seeks to explain the basis for algorithmic decision recommendations, such as why a borrower is turned down for a loan or why a course of treatment for a patient is recommended. It reviews the striking finding from decades of research that even simple linear models (applying equal weights to several causally relevant factors) typically make more accurate risk predictions than expert judgments or more complex statistical models. Human expertise in the form of knowing what variables to look at—what is likely to be causally relevant for predicting an outcome—together with simple objective quantitative models for combining information from these variables typically greatly out-performs human expert judgment alone. Other recent developments, such as multitask learning—that is, using models to predict multiple causally related outcomes simultaneously instead of just one (such as disease, hospitalization costs and duration, and mortality risks instead of just mortality risk)—have not only improved predictive accuracy compared to predicting a single dependent variable at a time but have also allowed greater visibility into the features that allow accurate predictions. Studying the relative times that an ML model spends processing different features to make its predictions ("saliency" analysis) helps to identify which features it treats as most informative for purposes of prediction. This has led to unexpected and useful discoveries, such as that age and sex can be identified with astonishing accuracy from retinal scans.

Part II of the book ("Agency") begins with Chap. 4 on reinforcement learning (RL) in animals, people, and machines. All three can learn from experience by noticing when actions are followed by desirable or undesirable results and trying to infer decision rules that specify what actions to take in different situations to make preferred outcomes more likely. If results are long delayed and only learned after long sequences of actions, then the "credit-assignment" problem arises of attributing causation of outcomes to specific choices that preceded them. Since the 1950s, when IBM unveiled a checkers-playing program that learned from experience how to improve its game by adjusting its parameters based on wins and losses, machine learning researchers have learned how to use *prediction errors*—the differences between predicted and experienced future rewards following an action (e.g., moving to a state with a higher expected value, assuming optimal decision-making ever after that transition)—to simultaneously adjust action-selection probabilities and estimates of the expected rewards from taking each possible action in each possible state until no further improvements can be made. The resulting reinforcement learning (RL) algorithms appear to reflect the biology of learning how to act effectively in initially uncertain environments. In such biological learning, the neurotransmitter dopamine acts as a signal of prediction error that guides learning in the brains of a wide variety of species. In environments where the causal rules linking actions to probabilities of outcomes remain fixed, such as games ranging from checkers or backgammon to video games, RL has produced impressive levels of mastery in machines, including many examples of super-human skill. Chapter 4 concludes by discussing research linking RL, dopamine, exploration, and happiness suggesting that happiness comes less from satisfaction that things have gone well, or even from anticipation that things are about to go well, than from being pleasantly surprised that things are going better than expected. From this standpoint, "complete mastery of any domain seems necessarily correlated with boredom" in humans and animals. Risk, exploration, and surprise are key requirements for their flourishing. Subsequent chapters explore the follow-up questions of how to determine what is valued (i.e., whether surprises are evaluated as pleasant or unpleasant and when events are evaluated as going better than expected) and how to structure rewards to elicit desired behaviors in machines as well as in animals or people.

Chapter 5 ("Shaping") examines how to train animals or machines to exhibit desired complex stimulus-response behaviors by rewarding successive approximations of the desired behaviors. It emphasizes the importance of creating both a good curriculum and appropriate incentives, i.e., designing rewards that lead a reward-maximizing learner to master a sequence of progressively more difficult and more accurate approximations of desired complex behaviors. These principles are illustrated by examples that range from Skinner's seminal work with animals and behaviorism in the 1950s, to DeepMind's use of automated curriculum design to train AlphaGo and more recent world champion-level Go-playing programs, to child psychology and implications for parenting. Children in families, adults in organizations, and AIs endowed with RL algorithms are all adept at gaming the systems in which they are placed to maximize their rewards, often discovering loopholes and ways to exploit rules and incentives that were not intended or desired by those who created them. Design principles discovered in ML research, such as (a) rewarding states rather than actions (e.g., rewarding achievement of a goal state rather than behaviors that we hope might lead to it); (b) paying as much attention to movement away from goals as movement toward them; and (c) distinguishing between what is desired and what is rewarded (since rewards shape behaviors in ways that are not necessarily simply related to what is desired or intended) may be useful for improving the performance of learning individuals and organizations as well as the performance of learning AIs. The chapter ends with discussions of the interaction between evolution and learning in which evolutionary pressures shape what we value and count as positive rewards; and gamification, in which well-designed curricula and incentives are used to make acquiring real-world skills and knowledge as compelling—or even addictive—as playing well-designed video games.

Chapter 6 ("Curiosity") opens with a discussion of the integration of deep learning (discussed in Chap. 1) with reinforcement learning (discussed in Chapters 4 and 5) to automate first the construction of higher-level features relevant for gameplay from raw pixel-level data in dozens of Atari video games, and then the process of learning to play and win the games. The resulting "deep RL" technology pioneered by DeepMind in 2015 learned to play most of the video games on which it was tested with super-human skill (often more than 10 times more skillful than human expert game players). For a small minority of games in which no rewards or feedback (other than the death of the game character) occurred until far into the game, however, deep RL could not learn to play: feedback was too sparse for RL to gain traction. It turned out that what was needed for AIs to succeed in these high-risk, low-feedback environments was something parallel to human and animal curiosity and *intrinsic motivation*: the desire to explore new environments and master new skills not for any reward, but from curiosity and love of novelty and surprise. The recognition that curiosity and novelty-seeking were important for both infants and AIs to learn about how to act effectively in new situations led ML researchers to the further insight that the "newness" of situations could be defined and measured by the unpredictability of what would be observed next (e.g., using estimated inverse log probabilities) and used to reward novelty-seeking. Novelty-seeking together with surprise-seeking—experimenting further when actions produce unexpected results until predictions are corrected and the initially surprising anomalies are no longer surprising—provides a computationally practicable version of machine curiosity. Amazingly, such curiosity-driven exploration performs well in video games and other tasks *even when scores or rewards are not revealed*. That is, learning to predict the effects of actions in a new environment is often a successful heuristic for guiding behavior even when extrinsic rewards and incentives are removed. The chapter closes with reflections on human gambling addiction, boredom, and the downside of novelty-seeking, such as novelty-seeking AIs that abandon useful tasks to surf through TV channels when given the opportunity to do so.

The final third of the book, "Normativity," consists of three chapters on Imitation, Inference, and Uncertainty. Chapter 7, "Imitation," discusses imitation learning. Compared to learning by trial and error (including RL) and learning via explicit instructions (including being programmed, for machines), imitation learning—i.e., learning by imitating with increasing accuracy the successful behaviors and skills that others have already mastered—has the distinct advantages of efficiency; safety of imitating known successful behaviors instead of risking exploration of potentially disastrous ones; and the possibility of learning skills that cannot easily be described, but that are easier to show than to tell. Whether the task is learning to walk without falling or ride a bicycle or drive an autonomous vehicle safely in traffic under changing conditions or control a complex industrial facility safely and efficiently, imitation learning can help AIs (as well as infants, children, and new employees) learn from people who already have the needed experience and skills to accomplish these tasks. But such learning is vulnerable to the fact that experts seldom make mistakes so crucial lessons about how to recover quickly from errors are unlikely to be learned by imitation of successful behaviors. Shared control, in which the learner is allowed to make decisions and try out partially-acquired skills and a human expert can override to correct mistakes provides dramatic improvements in safe imitation learning including error-recovery skills. When a master has skills that a novice lacks, however, imitation learning may be impracticable: the learner simply cannot imitate the master's behaviors. For an imperfect agent, the question of what constitutes "optimal" behavior deserves close consideration: should the value of reaching a state be defined as the value earned by acting optimally from that point forward, or as the value earned by acting as well as the imperfect agent can from that point forward? These two concepts (referred to in ML as off-policy vs. on-policy methods, respectively) can yield quite different decision recommendations. For example, a self-driving car trained with on-policy methods might stay away from a cliff edge even if driving quickly and without error along the cliff's edge would in principle be a slightly more efficient route.

Imitation learning raises the challenge of how an AI can learn to outperform the experts from which it learns. An architecture reminiscent of the dual-process "thinking, fast and slow" in people has proved successful in creating AI/ML systems such as DeepMind's AlphaGo Zero that taught itself to play world champion-level Go in three days without any examples of human games or instructions, guidance, or advice from human experts. The key idea in this approach is to have a system repeatedly play against itself and learn how to reliably imitate its own most successful strategies. To do so, a "fast thinking" component (implemented as a "value network" that estimates the value, i.e., probability of a win, for each position, together with a "policy network" that estimates the probability of selecting each possible move after further evaluation) is paired with a "slow thinking" component (implemented using a Monte Carlo Tree Search decision-optimization heuristic algorithm) that simulates possible future plays for the most promising-looking possible next moves to help decide what to do next. Each component improves the performance of the other over time. Beyond the context of board games—for example, in urban planning or transportation system design—this approach may yield super-human decision-making and design skills that reflect the values of users but the search and optimization capabilities of machines as people and AIs work together to create options and select among them.

Chapter 8, "Inference," deals primarily with inferring the goals, beliefs, and intentions of others from their observed behaviors and then using these inferences to help them overcome obstacles and achieve their inferred goals. Infants as young as 18 months engage in collaborative behaviors requiring such sophisticated cognition. "Inverse reinforcement learning" (IRL) algorithms endow AI with similar capacities for inferring goals and values (modeled as reward functions) from observed behaviors, including inferring goals (e.g., for the safe operation of drones) even from imperfect human attempts to achieve them. IRL has the advantage that goals are often much simpler to infer and describe than the complex actions and plans that might be undertaken in trying to reach them. An AI that infers human goals as explanations for their observed behaviors can use this understanding and its own skills to help achieve the inferred goals. In collaborations between AIs and humans, the AIs may need to learn about human goals as the two cooperate and interact to complete goal-directed tasks.

For risky applications ranging from diagnosing patients to recommending actions for increasing the safety and efficiency of industrial processes, it is essential that an AI's predictions, classifications, and recommendations be accompanied by indications of confidence. Chapter 9, "Uncertainty," introduces techniques for training multiple ML models on available data and then using the extent of disagreement among members of this *model ensemble* to estimate the uncertainty in its predictions and recommendations. Autonomous vehicles guided by such methods automatically slow down and drive more cautiously when they encounter unfamiliar conditions and the multiple models make highly variable predictions. The chapter discusses the challenges of developing safe AI, noting that the precautionary principle may fail to deliver useful recommendations when the possibility of harm is unavoidable. AI safety research has discovered that keeping options open—taking only actions that will allow several other goals (even randomly generated ones) to be pursued in the future—is often a valuable heuristic for avoiding premature commitment to actions that will be regretted in hindsight. The chapter also touches on effective altruism and on philosophical issues such as how to behave (or how machines should be designed to behave) when there is "moral uncertainty" about what is the right thing to do and when the interests of potential far-future generations are considered in present decision-making.

The book's final chapter, "Conclusion," discusses lessons and themes from the previous chapters. It notes that "Research on bias, fairness, transparency, and the myriad dimensions of safety now forms a substantial portion of all the work presented at major AI and machine-learning conferences." Reflecting on these themes, the chapter reminds us that no decision system, human or machine, can satisfy various proposed criteria for "fairness" that may seem intuitive and desirable; that "humans place greater trust in transparent models even when these models are wrong and ought *not* to be trusted;" and that RL and other ML methods inevitably make modeling assumptions (such as that selecting actions does not reshape our goals and values) that may prove to be erroneous. The modeling assumptions and data reflected in predictive and prescriptive models may also become outdated even while the models based on them continue to be used. However, human-machine cooperation and collaboration can ameliorate these limitations as people and AIs learn enough to work together safely and productively to achieve human goals with super-human efficiency and effectiveness. The book concludes with 64 pages of notes and a 50-page bibliography providing sources and technical literature references for the preceding chapters.

For risk analysts, a useful aspect of *The Alignment Problem* is its focus on clearly explaining technical challenges and possible solutions (or, in some cases, the mathematical impossibility of solutions) for creating fair, transparent, trustworthy data-driven prediction and decision-support models aligned with human values despite realistic limitations in available data and knowledge. The challenges and possibilities for developing and using trustworthy AI/ML algorithms are similar in many ways to those of developing and applying trustworthy risk analyses. (Indeed, the sentence from the concluding chapter quoted above could well be rewritten as "Research on bias, fairness, transparency, and the myriad dimensions of safety now form a substantial portion of all the work presented at risk analysis conferences.") AI/ML suggests some possible ways forward that have been little discussed in risk analysis to date. Deep learning teaches that accuracy of risk perceptions and risk assessment in multi-layer artificial neural networks, as measured by average prediction or misclassification errors, depends on extracting a hierarchy of predictively relevant higher-level features from low-level (e.g., sensor) input data, as well as on learning mappings from abstract features to risk predictions that minimize prediction error (e.g., using stochastic gradient descent). Such feature extraction is also widely used, though less commonly discussed, in risk modeling and risk assessment. An important part of the art of successful risk assessment is understanding and using the relevant features of a situation to predict risks.

AI/ML algorithms in applications such as classifying a tumor as benign or malignant, or a transaction as legitimate or fraudulent, or estimating the probability of heart attack for a patient or of default for a borrower with stated levels of confidence, use model ensemble techniques to gauge uncertainty about their own best predictions and to identify anomalous and novel situations for which confident predictions cannot be made. Such uncertainty characterization is also a key part of good practice in quantitative risk assessment. Likewise, the roles of curiosity-driven exploration and intrinsic motivation, trial-and-error (reinforcement) learning, and shaping of incentives to align behaviors with goals in AI/ML also have parallels in human and animal psychology and individual and organizational risk management. The possibility of using imitation learning and inverse reinforcement learning to infer goals and value trade-offs that are hard to articulate and teach explicitly suggests a fascinating constructive approach for dealing with the inexpressible and ineffable in risk analysis—a topic not often emphasized in past discussions of risk communication, but perhaps timely to consider in a world where clearly stated and defended, widely accepted values for use in risk analysis often seem increasingly hard to find, and yet collective decisions about threats to life, health, and wellbeing must still be made. Finally, the questions of how AI/ML agents can best serve human interests—e.g., by doing what they are told, or what they infer is intended, or what they predict will be most beneficial, whether or not it is what is asked for—are analogous to questions that arise in risk governance.

The insights that *The Alignment Problem* offers into how AI/ML systems are being designed to tackle these challenging questions may prove useful in thinking about how to improve human risk analyses. Both risk analysis and AI/ML must confront similar challenges in using realistically imperfect data and knowledge to make, explain, and defend predictions and risk management recommendations on behalf of people who may not care about the underlying technical details, but who want trustworthy predictions and recommendations with rationales that can be clearly explained if desired. *The Alignment Problem* shows that recent and ongoing advances in AI/ML are likely to be part of the solution to these challenges, as well as increasing the urgency of finding pragmatic solutions that society can apply and accept.

The Alignment Problem will appeal to readers who want to understand the main ideas of how AI/ML algorithms work and the challenges that they must overcome in more detail than is covered in other recent popular books. An easier introduction to the field is Polson and Scott's 2018 book *AIQ: How Artificial Intelligence Works and How We Can Harness Its Power for a Better World*. Henry Kissinger et al.'s 2021 book *The Age of AI and Our Human Future* discusses at a less technical level implications of current and emerging AI/ML technologies (including the revolutionary GPT-3 language model and its successors) and their implications for AI governance and the future co-evolution of AI and humanity. *The Alignment Problem* is distinguished from these and other recent books by the clarity and depth of its exposition of technical topics. It gives readers a real understanding of key research issues and insights in dealing with uncertainty and biases in AI/ML at a level not usually found in popular books. This is a triumph of exposition, making accessible to general readers key ideas and breakthroughs in AI/ML that are transforming our technological world. The author interviewed many of the innovators at the forefront of recent advances in AI/ML and took careful notes. This extensive research has paid off in clear plain-English explanations of intellectually and technically exciting topics that are usually discussed only in the technical literature. The exposition is intended for novices—there are no equations or mathematical symbols—but it successfully conveys the challenge and progress that make the area enthralling to participants, explaining both why obstacles are hard to overcome and how various ingenious ideas have been developed to overcome them. *The Alignment Problem* is an affordable, accessible introduction to how modern AI/ML deals with prediction and decision-making under uncertainty. It would make an ideal complement to technical textbooks for an undergraduate or graduate course on AI/ML methods in decision and risk analysis. Long after such a course is over, students are likely to remember the research challenges and applications, the people who tackled them, and the research ideas that inspired them as explained in *The Alignment Problem*.

All seven of the books reviewed in this chapter are pleasant, engaging, quick reads. They teach valuable lessons about how deliberative and subconscious thinking work together—and how to make them work together better—in understanding our uncertain and changing world better so that we can act more effectively in it. Any or all of them would be fine supplementary readings for general undergraduate or graduate courses in risk analysis, with *Thinking and Reasoning*, *The Knowledge Illusion*, and *Factfulness* being relevant for risk perception and risk communication; and *Cognitive Neuroscience*, *Artificial Intelligence*, *Deep Learning*, and *The Alignment Problem* being suitable for learning about biological and computational principles underlying System 1 and System 2 and modern automated risk management systems. *Factfulness* and *The Knowledge Illusion* will be of greatest interest to practitioners. *Factfulness* is especially rich in insights and advice that cut across risk perception, risk communication, risk assessment, and risk management. Taken together, these books provide an excellent introduction to the kinds of thinking used in risk analysis, and to current ideas about how people and machines can think more effectively.

Conclusions

The field of risk analysis has flourished in part by clarifying how System 1 and System 2 thinking—gut and head—reinforce each other, as well as how and when they diverge; and by studying how to use both to make wiser decisions under uncertainty and time pressure. While much research and practice in risk analysis will almost certainly continue to emphasize improved use of rational analysis, prediction, planning, and deliberative decision-making to improve outcomes and to better prepare for contingencies, there is also growing appreciation of the emotional, intuitive, social, and collective nature of risk management decision-making, as well as of risk perceptions and risk communication. To this mix of System 1 and Systems 2 thinking in current risk analysis, advances in artificial intelligence and machine learning add a third component, which might be dubbed algorithmic analysis or *machine thinking*. This includes automated, ongoing data analysis and interpretation combined with goal-directed reasoning and planning under uncertainty to help notice, identify, describe, respond to, and recover from a variety of risks that threaten human interests. Without exaggerating the current capabilities or understating the current limitations and pitfalls of current AI, it is perhaps not premature to predict that machine thinking is likely to play an increasingly influential role in risk analysis in the years ahead, and that AI-assisted risk management teams will probably be able to out-perform today's risk management teams by taking fuller advantage of a wide range of knowledge and thinking styles, not constrained by human psychology or biology, but attuned to human needs and goals.

References

Boden M (2018) Artificial intelligence: a very short introduction. Oxford University Press, Oxford

Christian B (2020) The alignment problem: machine learning and human values. W.W. Norton & Company, New York, NY

Evans J (2017) Thinking and reasoning: a very short introduction. Oxford University Press, Oxford

Kelleher JD (2019) Deep learning. MIT Press, Cambridge, MA

Kissinger HA, Schmidt E, Huttenlocher D (2021) The age of AI: and our human future. Little Brown & Company, New York, NY

Passingham RE (2016) Cognitive neuroscience: a very short introduction. Oxford University Press, Oxford

Polosn N, Scott J (2018) AIQ: how artificial intelligence works and how we can harness its power for a better world. St. Martin's Griffin, New York, NY

Rosling H, Rosling O, Ronnlund AR (2018) Factfulness: ten reasons we're wrong about the world – and why things are better than you think. Flatiron Books, New York, NY

Sloman S, Fernback P (2017) The knowledge illusion: why we never think alone. Riverhead Books, an imprint of Penguin Random House LLC, New York, NY

Part II
Fundamental Challenges for Practical Decision Theory

Chapter 4
Answerable and Unanswerable Questions in Decision and Risk Analysis

Introduction: Risk Analysis Questions

Risk analysis, including the risk management problem of deciding what to do to make preferred outcomes more probable, has historically addressed questions such as the following:

- *Risk assessment:* What can go wrong? How likely is it to happen? If it does happen, what are the consequences? (Kaplan and Garrick 1981; Aven 2020). More generally, how long is a system likely to operate as intended, or at least without catastrophic failure? How might it eventually fail, and how probable are different failure modes over time? How do the answers depend on design and operating decisions and on features of the environment in which the system operates?
- *Risk management:* What policies, plans, and decisions maximize expected utility or expected reward, or minimize expected loss or regret? (Raiffa 1968; DeGroot 2004) How should policies and plans be implemented, i.e., who should do what when, with what resources?
- *Risk management process evaluation and learning:* When a selected risk management policy or plan is applied to a system, how well does it perform? How large are the remaining risks, and how frequent and sizable are the losses that might still occur? For how long can the managed system be relied on to operate safely? What are the probability distributions for average reward (or loss) per unit time and for total reward or loss over the system's lifetime, or over some planning horizon? When and how should the current policy or its implementation be revised in light of experience?

In this context, a risk management *policy* is a decision rule that maps available information, including observations, to actions (i.e., to choices or decisions). A *plan* is a sequence of actions (sometimes called a course of action) for pursuing a goal or

L. A. Cox Jr., *AI-ML for Decision and Risk Analysis*, International Series in Operations Research & Management Science 345,
https://doi.org/10.1007/978-3-031-32013-2_4

optimizing an objective function. In an uncertain world, actions, plans, and sub-plans may take uncertain amounts of time to complete, and some may prove impossible to complete at all; moreover, plans may be partly contingent on future events, as when it becomes possible to attempt some actions in a plan only when others have been completed. *Open world* uncertainty arises when a planner does not currently know everything needed to perform a task, and must therefore plan to discover what it needs to know (if possible) as it goes, recognizing that some future events may be unforeseeable and that other agents with their own plans and actions may unexpectedly arrive or depart while a plan is being executed (Hanheide et al. 2017). For brevity and uniformity, we include plans as special types of policies by viewing them as mapping current information to a choice of which action to undertake next, recognizing that it may be only the first in an intended sequence (or, if contingency plans are made explicit, in a tree of possible future actions contingent on future events). The case of a team of multiple agents undertaking multiple tasks simultaneously is discussed later.

Policies are also called strategies, decision rules, or control rules. Actions may include gathering or communicating costly information, as well as intervening to change the design or operation of a system. Decision-analytic value-of-information (VoI) calculations address the question of when and whether it is worth gathering more information, e.g., by performing experiments or surveying respondents, to improve understanding of the probable consequences of different choices before intervening (DeGroot 2004). An example of a policy, in this sense, is a feedback control law for controlling a managed engineering system (e.g., an electric power grid or a chemical plant or a passenger plane) based on its observed behavior and on desired goals for its performance. A wildlife or fishery management policy might specify when to increase sampling, or to suspend fishing or hunting, and for how long, based on observed catch statistics. A criminal justice policy might specify penalties for observed crimes. Credit-granting and underwriting decisions typically implement policies based on actuarial statistics and on observed features of individuals. Hospitals use managed care policies to decide when to release patients. Systems administrators implement access control policies to reduce cybersecurity risks. Macroeconomic policies advise when and how much to adjust interest rates and money supply based on economic performance indicators. To the extent that such decisions under uncertainty are made by applying algorithms to data to decide what to do next, they implement policies in the sense used here. Likewise, multi-period plans for putting a person on the moon, or building a working fusion reactor, or treating a cancer patient to maximize length and quality of remaining life, would constitute policies in the sense used here, insofar as they specify what to do next given information available so far.

In these and countless other applications, the fundamental questions of risk analysis are how to identify, assess, reduce, characterize, and evaluate risks of harm or loss in managed systems. More generally, both desired and undesired outcomes—potential gains and losses—should be considered in formulating policies (or plans or decisions) with uncertain consequences. Table 4.1 summarizes the fundamental questions. They can be asked about deterministic systems as well as

Table 4.1 Fundamental Questions of Risk Analysis

1. *Hazard identification.* What can go wrong? What might happen to cause loss or harm?
2. *Risk assessment.* How likely are adverse events to occur, when, and with what consequences? More generally, what are the probabilities of different outcomes (e.g., gains or losses of different sizes) over time? How do the answers depend on available information, knowledge, and assumptions? How much and how soon might they change in light of future information?
3. *Risk management and policy optimization.* How can risks best be reduced or managed? What risk management decisions, plans, and policies should be used, e.g., to maximize expected utility or to minimize expected loss? More generally, what should a decision-maker do next?
4. *Implementation and coordination of policies.* How should risk management policies be implemented by teams or organizations, i.e., who should do what when?
5. *Policy evaluation and communication.* How should the anticipated performance of polices be quantified and communicated? What guarantees can be given that they will keep risk of failure or loss acceptably small, probability of success or gain acceptably high, and times and costs to recover from accidents or disruptions acceptably low?
6. *Learning, adaptation, and updating.* When and how should policies be changed to increase expected utility in response to remaining uncertainties or to changes in conditions or information?
7. *Characterizing remaining risk and uncertainties.* Even the best available risk management policy often does not eliminate probabilities of failure, loss, or catastrophe. Returning to steps 1 and 2, what hazards and risks remain, how large are they, and how sure are the answers? What is the value of collecting additional information (VoI) to improve risk management policies?

probabilistic ones, but arise so often in risk analysis that we refer to them as risk analysis questions.

Risk analysis, decision analysis, and policy analysis develop and apply causal models to answer these questions. Such models typically take as inputs possible decisions and probabilities of uncertain events. As outputs, they predict probabilities of different outcomes. Dynamic decision models allow sequences of decisions, uncertain events, and outcomes to unfold over time. In traditional normative decision analysis, outcomes are evaluated via a utility function, and the causal model is used to calculate the expected utilities of different decisions and to recommend decisions that maximize expected utility (Raiffa 1968).

Risk analysis questions can be answered simply in very simple situations that can be modeled well by small decision trees, decision tables, or influence diagrams, as discussed next. They can be answered with more effort by applying well-developed risk models and computational techniques for more complicated systems. These models and methods are reviewed next. The remainder of the paper then discusses the striking fact that *no techniques are able to answer these risk analysis questions correctly for some systems*, and suggests some alternative approaches for managing risks in such cases. That is, no algorithm can be guaranteed to provide correct answers, or even useful approximate answers, to these questions for many types of systems with realistic complexity and uncertainty (see Table 4.2), even though well-developed techniques are available to answer them for a wide variety of important and useful special cases. (For the impatient reader, types of systems for which the questions cannot be answered include general dynamical systems with failure states to be avoided, Turing machine models of computation, agent-based models, a

Table 4.2 Summary of main results on answerable and unanswerable risk analysis questions

Risk analysis causal models and policy optimization methods	What can and cannot be answered?
1. Small decision tables, decision trees, fault trees, event trees, bowtie diagrams, Bayesian networks (BNs), Influence Diagrams (IDs), Dynamic Bayesian Networks (DBNs)	All risk analysis questions can be answered in principle (e.g., via backward dynamic programming in small decision trees). Large problems can be computationally complex, but useful approximate answers can often be obtained using sampling methods.
2. Response surface models and methods (RSM)	Optimal polices (input settings) can be discovered via iterative exploration and adjustment if the response surface is sufficiently smooth, globally convex (unique optimum), and static (does not change during exploration).
3. Markov Decision Processes (MDPs)	Can find optimal policies via stochastic dynamic programming (e.g., value iteration and policy iteration algorithms) or linear programming. Decidability of reachability questions is unknown for Markov chains with more than 5 states.
4. MDPs with initially unknown parameters	Can eventually find optimal policy via reinforcement learning when regularity conditions hold that assure convergence.
5. Partially observable MDPs (POMDPs) and semi-Markov Decision Processes (SMDPs)	Approximately optimal policies can be found via dynamic programming and sampling (MCTS-type) algorithms for discounted rewards. Other risk analysis questions are undecidable in general, although they can be answered for some special cases of practical interest.
6. Simulation-optimization (SO) for continuous, discrete, and hybrid (mixed discrete and continuous) simulation models.	Risk assessment questions based on reachability, and risk management questions based on controllability, are undecidable for many nonlinear systems, and for some linear systems with constraints on values of controlled inputs.
7. Control of deterministic and stochastic systems via synthesis of adaptive, robust, or optimal control laws implemented by centralized, decentralized, distributed, or hierarchical control	Synthesis of controllers and guarantees of performance based on reachability and controllability are undecidable for many nonlinear systems. They can be answered for special cases, such as linear dynamical systems with simple control sets (in discrete time or continuous time).
8. Agent based models (ABMs) and cellular automata (CA); multi-agent systems (MAS)	Risk assessment questions based on reachability of a target configuration from an initial configuration are undecidable for many ABM and CA models with an unrestricted number of agents (e.g., spatialized Prisoner's Dilemma). They are decidable in some MAS models but not others, depending on what

(continued)

Table 4.2 (continued)

Risk analysis causal models and policy optimization methods	What can and cannot be answered?
	agents observe and how they can communicate.
9. Game theory models	Existence of winning strategies and of Nash equilibria in pure strategies are undecidable in many games. Winning strategies may be uncomputable even if they exist.

variety of 2-person games, and partially observable Markov decision processes: these are all sufficiently complex so that no algorithm can be guaranteed to yield correct answers to risk analysis questions. As a practical application, for many controlled probabilistic dynamic systems with even a few states, there is no way to determine whether a plan exists that achieves a desired goal state while avoiding undesirable states with at least a certain probability.) That fundamental risk analysis questions are inherently unanswerable for many systems of practical interest suggests a need to rethink the kinds of questions that risk analysts seek to answer in helping decision-makers thrive in a world of realistic complexities and uncertainties. The last part of this chapter reviews constructive methods from artificial intelligence (AI), machine learning, and related fields to help meet this need. The rapidly expanding scope of practical applications may soon include large-scale and complex policy and risk management applications that have not been adequately supported by earlier decision and risk analysis methods.

Some Models and Methods for Answering Risk Analysis Questions

This section quickly reviews some popular analytics models and methods for answering risk analysis questions. The goal is to survey essential concepts and causal models used in risk analysis in a relatively accessible way, providing a quick inventory of key ideas and methods useful in applied risk analysis in a variety of fields. Precise definitions, mathematical formulations, and technical details are relegated to the references. These concepts, models, and methods are used to examine how well risk analysis questions can be answered in settings with realistic complexities and uncertainties (Table 4.2) and to explain how modern AI methods meet practical needs to learn, plan, and act effectively in such settings.

The Simplest Causal Models: Decision Tables and Decision Trees

Risk analysis questions can be answered well for small, well-understood decision problems. If the possible sequences following an initial decision are known and there are not many of them, they may be diagrammed explicitly as a *decision tree*, with choice nodes and chance nodes representing decisions and random variables, respectively, and their branches representing decisions and uncertainty resolutions, respectively. Alternatively, very small problems may be represented via a *decision table*—that is, a table with a row for each possible choice (or policy, in a dynamic setting, specifying what to do at each decision node in a corresponding decision tree); a column for each possible state of the world (showing how uncertainties are resolved at each chance node in the corresponding decision tree, and possibly including "other" for states not explicitly envisioned and described); a consequence and its expected utility for each cell (i.e., each action-state pair); and known probabilities for the states. In a small decision table it is straightforward to calculate the expected utility for each row (interpreted as a choice, decision, action, course of action, etc.) and to identify the one with the greatest expected utility (Raiffa 1968). The state probabilities and the consequences of the action-state pairs for the chosen action yield the probability distribution for consequences when this choice is made. Likewise, small decision trees, showing possible sequences of decisions and resolutions of chance events (represented by branches out of choice and chance nodes, respectively) leading to terminal nodes at which utilities are assessed are easy to solve for the best choice at each decision node (by backward induction, a simple form of stochastic dynamic programming). This best policy, together with the conditional probabilities of the different possible outcomes (branches) at chance nodes, determines the probabilities of reaching each terminal node ("leaf") of the tree, and hence the probability distribution of rewards and the expected utility of the optimal policy. Modifications of decision trees to represent Bayesian inference from observations and to support VoI calculations are well understood (*ibid*). Recent advances in computational methods for large trees now allow even very large decision trees and game trees (i.e., trees with different players making choices at different decision nodes) to be searched efficiently for good policies. *Monte Carlo Tree Search* (*MCTS*) algorithms answer the question of what to do next at an initial decision node by generating and evaluating samples from relevant portions of the tree of possible choices and consequences (Fu 2017). MCTS, in conjunction with other techniques, has proved to be effective for games such as Go, in which the rules of the game are well enough understood to allow the tree of possibilities from the current position to be generated and evaluated; and for policy optimization for a variety of robot control tasks (Ma et al. 2019).

Fault Trees, Event Trees, Bayesian Networks (BNs) and Influence Diagrams (IDs)

Complete, explicit evaluation of decision trees is restricted to relatively small decision problems. Somewhat larger decision problems can be represented with the help of *Bayesian networks* (BNs) showing probabilistic dependencies among random variables. These are networks of random variables without directed cycles ("directed acyclic graph" (DAG) models), with the conditional probability distribution of each variable depending on the values of the variables that point into it, and with the marginal probability distributions of input nodes (i.e., nodes with only outward-pointing arrows) specified (Koller and Friedman 2009). BNs can be used to compactly represent the fault trees, event trees, and bow-tie diagrams of classical reliability theory and probabilistic risk assessment (PRA) (Khakzad et al. 2011). *Fault trees* are logic trees showing how an undesirable "top event" such as a catastrophic failure could occur via conjunctions ("cut sets") of lower-level probabilistic events, such as accumulation of unrepaired component failures (Ruijters and Stoelinga 2014). *Event trees* show what consequences might ensue, i.e., possible event sequences (scenarios) branching out from an initiating event; these are decision trees with only chance nodes. A *bow tie diagram* shows the fault tree for a top event, together with the event tree of possible event sequences that may follow its occurrence. BNs generalize these classical techniques: fault trees and event trees can be mapped automatically to BNs, and BNs can also easily handle probabilistic dependencies and Bayesian inferences that cannot easily be modeled by fault trees (Khakzad et al. 2011). Even without decision and value nodes, BNs have proved useful in wide variety of risk analysis applications. For example, if occurrence of a catastrophic accident or system failure is postulated, a BN model of the system can be used to identify the most probable explanation, revealing relatively probable failure pathways so that they can be protected against. *Dynamic Bayesian networks* (DBNs), in which the probability distributions of variables in one time period may depend on the values of variables in earlier periods, but not in later periods, have been widely used in risk analysis applications in the past decade, e.g., to diagnose illnesses and assess medical risks, predict risks of disease progression (Zandonà et al. 2019) and model risks of cascading failures and domino effects, in which initial accident and failure events propagate to cause later ones, in chemical plants, industrial fires, and other complex systems (Khakzad et al. 2017).

Augmenting a BN with decision nodes and a value or utility node yields an *influence diagram* (ID) model of a decision problem: the goal is to choose values of the decision nodes to maximize the expected value of the utility node. IDs and decision trees are equivalent, in that either can be automatically converted to the other without loss of information, but IDs are often much smaller (Koller and Friedman 2009). *Bayesian inference algorithms* for calculating the probability distributions of some variables given the observed or assumed values of others in a BN or ID model, as well as *most-probable-explanation* (MPE) algorithms for finding the values of some unobserved variables that best explain, i.e., maximize the probability of, specified observed or assumed values for other variables, are now widely available for BN software (Kwisthout 2011). They can be repurposed to provide *decision optimization algorithms* for ID models (by maximizing expected utility instead of finding an MPE). Special-purpose algorithms and solver software for deriving optimal policies in IDs have been extensively developed for over 30 years (Koller and Friedman 2009).

Modern decision and risk analysis software, including BN and ID software, deals effectively with decision problems having up to dozens or hundreds of decision (choice) nodes and random variable (chance) nodes, in applications ranging from engineering trouble shooting and fault diagnosis to medical decision support to environmental and public policy applications. The worst-case computational complexity of calculating Bayesian inferences and MPEs in BNs (and IDs) and of finding optimal policies in IDs increases rapidly with the number of variables. [Technically, these are NP-complete problems (Cooper 1990; Mauá et al. 2013).] Fortunately, this does not prevent the rapid solution of many problems of practical size and interest for which the required knowledge and data are available. Input requirements typically include the qualitative structure (the "topology" or connection pattern) of the causal decision tree or ID network relating choice, chance, and value nodes, as well as quantitative information about the marginal and conditional probabilities of values for chance nodes (representing random variables, which may be viewed as components of the random state); and perhaps deterministic formulas or functions relating the values of variables, including utility functions for value nodes.

Markov Decision Processes (MDPs) and Reinforcement Learning (RL)

In many real-world risk-analyses, the relevant states, their probabilities, and the causal relationships among states, choices, and consequence probabilities are initially unknown or highly uncertain. The DAG structure of a BN or ID network can sometimes be learned from relevant data, if sufficient data are available and the data-generating process remains stationary. Although the structure learning task is computationally demanding and scales poorly with the number of variables for large data sets (it is both NP-hard and NP-complete), current structure-learning algorithms have nonetheless proved tractable and useful in many applications, e.g., in systems biology and epidemiology (Scutari et al. 2019). Alternatively, instead of learning an ID model from data and using it to optimize decisions, it is often possible to experiment directly with alternative policies and to modify them based on experience to improve their performance. This has proved spectacularly successful for *reinforcement learning* (RL) algorithms that iteratively adjust probabilistic policies (i.e., policies that specify the probability of selecting each feasible action when in each state) in Markov decision processes (MDPs) (Sutton and Barto 1998). In an MDP, a system makes stochastic (i.e., random) transitions among states over time and generates immediate rewards as it does so, whenever an action is taken in a state. Both the transition intensities between states and the probability distributions for immediate rewards may depend on the decision or action taken in a state. For example, the system might be a reliability system with many components that can fail and be replaced repeatedly over time; the state of the system at any time is then the set of states (e.g., working or failed) each of its components, and transitions among states occur as components fail or are replaced. RL algorithms provide simple formulas for iteratively adjusting action-selection probabilities (and, in many versions, estimating the expected long-run value of taking each feasible action in the current state and choosing optimally thereafter) based on the differences between predicted and received rewards. These iterations provably converge to optimal policies (and the estimated value functions converge to the optimized value function solving the Bellman equation for stochastic dynamic programming) for many MDPs of practical interest. Successive adjustments increase the probabilities of taking value-maximizing actions in each state, where "value" may be defined as discounted or average reward generated by the MDP.

Extremely usefully, the RL approach works even if the conditional probability distributions for rewards and transition rates given the state and action are initially unknown (or are slowly changing): successive adjustments allow optimal policies and value functions to be learned, even as the causal relationship between states, actions, and consequences (i.e., the value of rewards received)s revealed by experience. RL algorithms have intuitive interpretations, e.g., as "actor-critic" methods in which one agent (the "actor") decides which action to take in each state (such as the action maximizing the currently estimated value function of the action and state) and another agent (the "critic") evaluates the performance of the choices, comparing predicted to received rewards and providing this feedback to the actor (Sutton and Barto 1998). In *deep RL (DRL) algorithms*, one or both of these components are implemented using artificial neural networks; these may include convolutional neural networks (CNNs) for extracting hierarchies of informative features and descriptions from data (Francois-Lavet et al. 2018). For example, the estimated value function may be approximated using a neural network. RL and DRL algorithms for MDPs have powered successful machine learning and AI applications such as AphaGo Zero for playing Go, control of energy consumption in data centers, on-line advertising and content recommendation systems, various types of robot training, and rational drug design in medicinal chemistry (Zhou et al. 2019). Modified versions of RL algorithms for risk analysis have been developed to avoid catastrophic accidents while learning ("safe learning") and to maximize risk-sensitive objective functions such as worst-case or risk-penalized performance (García and Fernández 2015). However, RL algorithms typically take massive amounts of training data from a stationary (or nearly stationary) MDP environment and engage in extensive, slow, and error-prone trial-and-error learning before converging to optimal policies; hence they are usually not suitable for mission-critical applications or for rapid adaptations to new situations or conditions.

Simulation-Optimization for Continuous, Discrete-Event, and Hybrid Simulation Models

RL algorithms trained on computer-simulated MDPs can meet the need for massive training data without incurring the cost of real-world mistakes. Such algorithms have been very effective in machine learning and AI programs that learn to play games and in enabling robots to learn how to perform complex tasks. They can be viewed as special cases of a general *simulation-optimization* (SO) paradigm in which a simulation model is used to describe or sample the modeled causal relationship (which may be probabilistic) among decision variables and an objective function, and an optimization loop is used to seek combinations of values for controllable inputs to optimize the objective function (Amaran et al. 2016; Fu 2015; Sörensen and Glover 2013). In dynamic SO models, typical objectives include maximizing average or risk-adjusted return per unit time or total return over a planning horizon. Optimization is typically subject to constraints on feasible input values, and perhaps also to constraints on acceptable probability distributions for outcomes. SO methods have been extensively developed for both deterministic and stochastic optimization problems with discrete (combinatorial), continuous, or mixed (both discrete and continuous) decision variables. These methods search the set of feasible input combinations using sophisticated mathematical programming techniques and metaheuristics such as Tabu Search (Juan et al. 2015; Sörensen and Glover 2013). Dynamic simulation models used in SO include the following:

- *Continuous simulation models,* also called system dynamics models. These are typically represented by systems of differential and algebraic equations (DAEs), with ordinary differential equations (ODEs) or partial differential equations (PDEs) describing the flows of substances over time among compartments or locations in an ecosystem, organism, economy, or other dynamical system with continuous flows.
- *Discrete-event simulation (DES) models* describe stochastic (random) transitions of individual entities and processes among states over time, e.g., resources and patients in a health care delivery system; components in a complex reliability system such as an electric power grid or a nuclear power plant; cars in a road network; shipments in a supply chain or network; or calls in a call center or telecommunications network (Riley 2013; Raska and Ulrych 2014; Chen et al. 2019). MDPs are examples of discrete stochastic transition models.
- *Hybrid systems simulation models* include both discrete transitions and continuous flows. For example, operation of a chemical manufacturing plant might involve continuous flows with parameters that change abruptly when components fail or when controlling computers issue commands to change the production process (Avraam et al. 1998).

Simulation-optimization methods for hybrid systems represent the current state-of-the-art for managing and controlling risks in many industrial applications (Lennartson et al. 2015).

Response Surface Methodology

Even when simulation is not possible, perhaps because not enough is known about a managed system or its environment to simulate them realistically, many of the ideas and methods used in SO can be applied directly to the real world. For example, one can still adjust controllable variables in light of experience in an effort to improve the performance of the system over time. This is most likely to be tractable and valuable if the initially unknown causal relationship between controllable inputs and objective function values (called a "response surface") is smooth and remains relatively stable (i.e., it remains stationary or changes only slowly compared to the time needed to discover effective policies). *Response surface methodology* (RSM) iteratively applies design of experiments (DOE) to perturb the current levels of controllable inputs and estimate how the mean values (or probability distributions) of one or more performance metrics change in response, keeping changes that lead to preferred outcomes (Myers et al. 2016). RSM can be thought of as climbing a hill (the response surface) with an initially unknown shape by using designed experiments to set values of controllable inputs to new levels, estimating the local shape of the response surface from observed responses—usually approximated by a quadratic statistical model—and then using this estimated shape to adjust the controllable inputs to levels that increase the predicted yield or reward from the managed process. Repeating this process of experimentation and adjustment can optimize the reward if the response surface is sufficiently smooth and has a single maximum. Other hill-climbing methods for initially unknown response surfaces include *stochastic approximation* algorithms that iteratively adjust controllable inputs in the direction of estimated steepest ascent (i.e., estimated gradient or slope of the surface at the current values of the controllable inputs). These have been combined with RSM methods to speed convergence to the optimum in applications such as designing cost-effective systems or structures to perform reliably in random environments (Marti 1997).

Adaptive and Robust Control

Unlike response surfaces, which ideally represent initially unknown but stable causal relationships between user-controlled inputs and mean responses, many dynamic systems shift their internal states in response to user-selected inputs, as when exposing a person or animal to a stimulus sensitizes it to future exposures, or if stressing a system during testing changes its future performance. In effect, such systems have memory, captured in changes in their internal states, which mediate how outputs (and further state changes) depend on inputs. Experimenting with such a system may change its future input-output behaviors by changing its internal state. Risks in systems with unknown or highly uncertain response characteristics (i.e., input-output and state transition dynamics), or with response characteristics that change substantially while the system is being controlled (e.g., due to occurrence of internal faults or component malfunctions or failures, or to regime changes in the environment in which the system is operating) can be managed using *automatic control* techniques designed to cope with uncertainty and change. These include the following.

- *Adaptive control* methods use observed input and output histories to update policies (e.g., by updating estimates of the system dynamics and current state from observed input-output data and synthesizing new or modified feedback control loops).
- *Robust control m*ethods seek to design policies that work well across a broad range of uncertain and/or changing conditions (Annaswamy 2014; Ioannou and Sun 1995).
- *Risk-sensitive* optimal control methods modify the objective function to reflect risk aversion or penalize risk (Chow et al. 2015; Runolfsson 2000; Miller and Yang 2017).

Hybrids of these control techniques are often used in practice. Automatic control methods for systems operating in uncertain environments are implemented by sophisticated numerical control software In applications ranging from robotics and drone control to control of medical devices, manufacturing processes, or autopilots on ships, planes, or cars. The control software implements policies by mapping data from sensors to adjustments in actuators. To assess and reduce risks of hardware failures in automatic control, the techniques of probabilistic risk assessment and risk management we have already discussed can be applied, e.g., reliability engineering methods using fault tree analysis, Bayesian networks, MDP models of component and system failures, maintenance, and repair, and so forth. Both hardware and software failure risks are also addressed by a mix of fault-tolerant designs and formal verification techniques, such as model checking, which applies temporal logic to a model of the hardware or software to examine whether any feasible sequence of transitions in response to inputs can lead to a failed state, e.g., due to the simultaneous conjunction of conditions that jointly suffice to cause a software crash or hardware failure (Grobelna et al. 2014).

Distributed and Hierarchical Control of Uncertain and Non-stationary Systems

In practice, both human decisions and automatic control decisions about how to manage risks in complex systems are often distributed among multiple decision-makers or controllers (generically referred to as "agents"), each with its own information (e.g., from local sensor data) and opportunities for action. Ordering and inventory decisions in supply chains and networks; management decisions in bureaucratic organizations and multi-division firms; and control of drone swarms, teams of autonomous robots, flows in traffic networks or chemical processing plants, power generation and storage in electric grids, packet routing in data networks, and operations in manufacturing plants and production processes, all require coordinating the decisions and activities of multiple agents. This raises questions about how to design communication protocols, delegation and role-assignment rules, and collective choice rules for the decision makers in a team, organization, or distributed control network.

Hierarchical control algorithms (including adaptive and robust hierarchical control algorithms for systems with uncertain dynamics) typically feature low-level feedback-control regulators making many quick control decisions, subordinated to mid-level supervisory controllers that make less frequent decisions about how to set or adjust goals for the low-level regulators (e.g., what levels of their controlled variables to seek to achieve and maintain) (Leigh 1992). These supervisory control decisions are usually based on optimization calculations made with a model of the controlled system. Still higher-level controllers update estimates of model parameters. Top-level controllers may monitor performance and occasionally adapt the model itself to better describe observations as conditions change. In large hierarchical organizations and distributed control systems, higher-level controllers may communicate with peers to share information and coordinate their decisions. They may also receive information from their own subordinates and supervisors to assist in updating goals and instructions. Rather than developing hierarchies of automatic controllers based on explicit models and optimization of the controlled system and of communication and decision protocols, contemporary *hierarchical multi-agent reinforcement learning* (HMARL) algorithms apply RL principles (and extensions such as abstract descriptions of control actions or policies on different time scales) to automate learning of effective communication and decision protocols for coordinating the policies, decisions and behaviors of the "agents" (e.g., lower-level controllers) whose joint activities affect outcomes (Nguyen et al. 2020; Ossenkopf et al. 2019).

Decentralized Multi-agent Control: POMDP, decPOMD, SMDP, and POSMDP Models

At the opposite end of the decision-and-control spectrum from a single top-down centralized controller is *decentralized control*, in which autonomous agents make their own observations and decisions about how best to achieve a shared goal or maximize the team's expected utility. In the absence of explicit communication, agents must infer each other's intentions, goals, and plans from their observed behaviors: these "hidden" variables cannot be directly observed. The basic concepts of MDPs for a single decision-maker—probabilities of transitions among states and probability distributions for immediate rewards being influenced by the decisions made in each state—must be extended to allow for hidden variables, including information about the current state that may not be accurately observed by all agents. *Partially observable MDPs (POMDPs)* provide such an extension. They allow decision-makers to make observations that depend (via conditional probabilities of observations given states) on the underlying state of a Markov decision process, but that do not necessarily fully reveal it. For example, a physician or a trouble-shooter for complex equipment may see symptoms (observations) and draw Bayesian inferences about probabilities of underlying states (diseases or system states causing the observed symptoms), without observing the states themselves. In the decentralized control context, agents on the same team can observe each other's actions and draw Bayesian inferences about their underlying goals and intentions (Yin et al. 2016), as well as about the uncertain state of the environment in which they act, leading to *decentralized POMDP* (decPOMPD) models of multi-agent coordination and decision-making under uncertainty (Bernstein et al. 2002; Oliehoek and Amato 2016).

A further extension of MDPs is to allow the time required for agents to execute actions to be random variables, rather than assuming that actions occur once per period. MDPs extended with probability distributions for the time to transition from one state to the next, where the probability distribution may depend on the action taken, are called *semi-Markov* decision processes (SMDPs). They are widely used to model single-agent planning and sequential decision-making in random environments. Algorithms developed for inference and learning in probabilistic graphical models such as dynamic Bayesian networks can be modified and applied to optimize polices for SMDPs (Hoffman and de Freitas 2012; Yin et al. 2016). Multi-agent planning and coordination tasks (e.g., multi-robot cooperation in robot soccer or in search-and-rescue operations, or in engagements in military and security applications) typically require asynchronous decisions by the agents, as well as frequent updating of inferences about each other's goals and plans, and hence revisions in each agent's own goals and plans (Jiao et al. 2017). They are thus naturally modeled by combining the partial observability of POMDPs with the random timing of SMDPs. The resulting POSMDP models have been applied to complex cooperative tasks, such as multi-robot package delivery under uncertainty (Omidshafiei et al. 2017). More generally, *multi-agent systems* (MAS) allow multiple autonomous agents to communicate via specified protocols to coordinate in cooperatively deciding what actions each will take in order to jointly cause state transitions that achieve shared goals (Dorri et al. 2018).

Agent-Based Models (ABMs) and Cellular Automata (CA)

Decision-makers need not be on the same team. They may have their own private incentives and goals (or utility functions), and take actions that attempt to promote their own individual good rather than, or in addition to, that of a larger team. This is typically the case in economic models of hierarchical organizations or business relationships (e.g., principal-agent models); in financial models of investor behaviors; in marketing models of consumer choices; and in engineering models of autonomous vehicles sharing a road network, or individuals in a crowd fleeing a dangerous area in an emergency (Salze et al. 2014). *Agent-based models* s (ABMs), in which each agent adjusts its own behavior in response to the observed behaviors of others with which it interacts (e.g., its physical or social network neighbors) allow simulation of emergent behaviors at the population level (Page 2018). These collective behaviors arise as individual agents interact with and respond to each other. An important special case consists of *cellular automata* models (CAs). CAs are simple spatial ABMs in which each agent occupies a cell in a grid and all agents update their choices or behaviors (selected from a small set of possible alternatives) synchronously, once per period, based only on the most recent behaviors of its immediate neighbors. ABMs and CAs can also be used to model the spatiotemporal dynamics of epidemics, rumors, memes, risk perceptions, forest fires, and many other phenomena involving interacting agents influenced by the behaviors of neighbors.

Game Theory Models and Adversarial Risks

In some important risk analysis settings, individual decision-makers ("agents") are partitioned into separate teams that actively compete with or oppose each other. This is the case in adversarial risk analysis models of physical and cyber attacks and defenses; in robot soccer; in some multiplayer video games; and in military conflicts (Bier and Azaiez 2009; Das et al. 2015; Pangallo et al. 2019). In traditional game theory terminology, decision-making entities are called "players" and their policies are called "strategies." Thus, a player might be an individual agent or a team, organization, army, or nation, depending on the context. Decisions of interacting agents or teams are often studied using *game theory models* of strategic interactions, with each player seeking to use a policy (i.e., strategy) that delivers the maximum possible payoff (reward, expected utility) for itself, given the choices of others. If all players are using strategies such that no player can increase its own payoff by unilaterally changing its strategy, then their strategies constitute a Nash equilibrium. Out of Nash equilibrium, agents with incentive and ability to change their strategies may do so, creating a dynamical system of co-evolving strategies. Game theory models of conflict among rational agents are used in terrorism risk analysis, cyber security, competitive marketing and advertising, and military strategy. Game theory also models the emergence and maintenance of cooperation among agents over time and the formation of coalitions and negotiation and stability of agreements with imperfect monitoring and enforcement (Berthon et al. 2017).

Undecidability: Not All Risk Analysis Questions Can Be Answered in All Causal Models

The foregoing concepts, models, and methods have been applied in many fields to answer risk analysis questions about managed systems. They are used in probabilistic risk assessment (PRA) for engineering systems and in public and occupational health, safety, and environmental risk analysis. They provide some of the best-developed and most widely used frameworks for policy optimization, and risk management decision-making under uncertainty for individuals and for multiple agents, whether human or robotic. But how well can risk analysis questions be answered in such models? This section collects a number of negative results, essentially stating that risk analysis questions cannot be answered by any general computational procedures (algorithms) for most of the classes of models we have reviewed, except in special cases (which may still be of great interest and practical value). These limitations are implied by results originally developed in mathematical logic, computer science, operations research, automatic control, computational game theory, and related fields. Most have the following flavor: *For any dynamic system that is sufficiently complex, questions about the system's possible or probable long-term behavior cannot be answered by any algorithm.* Here, "sufficiently complex" means complex enough to carry out general computations (technically, to simulate a universal Turing machine), and an "algorithm" can be thought of as a computer program, written in any modern computer language, which can be run on input data. (To keep the exposition relatively accessible, we again settle for informal summaries of key points, leaving precise technical concepts and definitions—such as for effective computability, decidability, and Turing machines—as well as mathematical formulations and proofs, for the references.) To the extent that "risk analysis" comprises a set of computational techniques, i.e., algorithms, that are applied to data about systems to compute answers to risk questions about their possible behaviors, the results that follow suggest that risk analysis is impossible for some interesting systems. It is both possible and valuable for simple and well-understood systems, but not for large classes of more complex systems.

These negative results follow from two main sets of ideas. The first are fundamental theorems of computer science, such as that nontrivial properties of behaviors of programs are undecidable in general, i.e., uncomputable by any algorithm (Rice's Theorem) (Prasad 1991; da Costa and Doria 2014); indeed, even the probabilities of key behaviors, such as the probability that a randomly constructed program will eventually halt, are uncomputable (Chaitin 1975). The second is that many classes of models of systems used in risk analysis are complex enough to simulate computation. These include POMDP, ABM, hybrid (discrete and continuous) simulation, optimization, and game theory models. For these and other classes of models, limitations on what can be decided by computation imply limitations on the risk analysis questions that can be answered about the behaviors of the modeled systems by applying algorithms to data. These principles imply some startling specific restrictions on the possibility of computing answers to risk analysis questions.

Undecidable Questions in Hazard Identification and Probabilistic Risk Assessment

Many risk assessment questions about what can go wrong in a system can be viewed as questions of *reachability*: can some undesirable state of a system be reached from its current state via a sequence of feasible transitions, e.g., by components failing and remaining unrepaired until the failures (and perhaps other events) jointly cause a catastrophic failure of the system? If so, what sequences of inputs (i.e., events in a passively observed setting, or events and decisions in a managed system) could cause the undesired state(s) to be reached? From this perspective, qualitative *hazard identification* addresses these questions by determining what undesired states, if any, can be reached, and how. Quantitative *risk assessment* then address how probable it is that transitions will lead to an undesired state (e.g., system failure), and how soon this could occur. Finally, *risk management* seeks interventions and control policies to prevent or delay reaching undesired states.

Hazard identification, construed as identifying whether and how specified undesired states can be reached starting from the current state, is easy in small decision trees: paths that that lead to the undesired states show what can go wrong and how (i.e., what sequences of events and decisions could cause the undesired states to be reached). However, it is impossible for many larger systems of practical interest that have too many possible sequences to enumerate, including many physical, engineering, biological, mathematical, computational, and economic systems. In these systems, *reachability is undecidable*—no algorithm can solve (i.e., be guaranteed to return correct answers to) the general problem of determining whether a specified set of states is reachable from a specified starting state. Hence, qualitative hazard identification based on reachability, as well as quantitative risk assessment based on probability of reaching undesired states, are not possible in general for these systems (i.e., there is no effective computational procedure for determining whether there is positive probability of moving from an initial state to an undesired future state, or how large the probability is), although these risk analysis questions can sometimes be answered for special cases if possible initial states and allowed changes over time are restricted by additional constraints. Classes of systems for which reachability is undecidable arise in all of the following contexts (and many others):

- *Computer science.* In theoretical computer science, Rice's theorem states that any nontrivial property of the set of strings (language) recognized by a Turing machine is undecidable (where the terms "nontrivial," "property," "language," "recognized,""Turing machine," and "undecidable" all have specific technical meanings) (da Costa and Doria 2014). This fundamental result implies the undecidability of many practical questions, including the halting problem (will execution of a program eventually end?), automatic program verification (roughly, will a program always perform as required by its specifications?), automated debugging (e.g., will a given buffer ever overflow?), malware and virus detection (roughly, will this code eventually do something harmful?) and other questions about the behaviors of programs (or of equivalent logic circuits). Many of these questions, such as whether a program will stop, can be viewed as reachability questions. Of course, all of these questions can be answered for some programs, e.g., one that consists of a single command to stop, but none of them can be answered automatically (via an algorithm) for all programs that can be written in a modern programming language.

- *Engineering.* In a hybrid dynamic system, continuous flows occur among compartments at rates that may change when the contents of the compartments reach certain levels. For example, a compartment may fill at a constant rate until it is full, and then stop filling. Such hybrid systems occur in many areas of applied science and engineering, including chemical engineering, biological networks, robotics, and avionics. In general, reachability of one state from another in such a hybrid system is undecidable, and whether the system will remain in a pre-specified set of "safe" states cannot be verified. However, for systems that can be simulated accurately, such reachability questions are semidecidable, meaning that simulation can show whether a trajectory leads from an initial state to an unsafe state over the simulated time interval, but not whether it would do so later (Asarin et al. 2012).
- *Cell biology, pharmacology, toxicology.* Suppose that molecules move between localized membrane-bound compartments within cells; that the membranes are selectively permeable to different types of molecules; that compartments can contain sub-compartments (e.g., mitochondria within cell nuclei); and that some compartments can merge or divide to form fewer or more compartments, respectively. Then the question of whether specified types of molecules can reach specified target compartments is undecidable in general, although certain restrictions (e.g., on merging) can make it decidable (Delzanno and Zavattarob 2012).
- *Cellular automata (CAs).* For CAs with an agent occupying each cell of a grid and only 2 possible choices for each agent, four qualitative classes of spatial aggregate behaviors are observed: (i) all agents end up in the same state; (ii) stable structures or periodic patterns emerge and persist; (iii) non-repeating, chaotic behaviors occur without reaching stable final patterns; and (iv) complex patterns and structures emerge and propagate locally through the grid (Wolfram 1983). Which of these aggregate behaviors will eventually be reached in the long run, and whether specific patterns will ever be generated, are undecidable questions for CAs in general, although many results are known for specific CAs and initial configurations of agent behaviors (Wolfram 1985).
- *Cybersecurity:* Given an access control policy and an initial assignment of rights to users in a secure IT system or network, can a user can gain access to resources in violation of the safety or security specification that the access control policy was meant to enforce? (Roughly, can the system be hacked?) This can be viewed as a reachability question somewhat analogous to the cell biology one: can a user with certain access privileges (analogous to a molecule with the ability to cross certain types of membranes) reach (i.e., gain access to) a target asset or resource (e.g., confidential data) that was supposed to be protected? This question is

undecidable—there is no algorithm that can answer it in general—although restricted access control policies and rights allocations can be designed that allow some safety specifications to be achieved (Kleiner and Newcomb 2007).

- *Physics (classical mechanics).* Suppose that asteroids in a swarm interact by mutual gravitational attraction. Is there a non-zero probability that one or more of them will eventually break free from the rest? Is there a positive probability that the swarm will eventually collapse, i.e., that the volume of space in which the asteroids move will shrink until they collide? Such questions cannot be answered in general (i.e., for unrestricted choices of the masses and initial momenta of the asteroids) by any algorithm or simulation program (Parker 2005), although of course they can be answered in special cases (e.g., if one asteroid is in stable orbit around another). For some deterministic physical systems, future trajectories cannot be computed even if their initial conditions (e.g., particle positions and momenta) are known precisely. Even a single particle moving deterministically in a smooth potential may exhibit unpredictable motion (its future trajectory cannot be computed) such that even the probabilities that it will enter and remain in different sets of possible values (basins of attraction) cannot be computed, and whether its motion is chaotic cannot be determined (Moore 1990).
- *Reliability and probability (Markov chains).* Given a finite Markov chain (e.g., a model of stochastic failure and repair in a reliability system) with a specified initial state and target state, and given a rational number p, is it possible to go from the initial to the target state with probability greater than p in a finite number of steps? Although this question is decidable for very small finite Markov chains (with at most 5 states), its decidability is unknown for Markov chains with more than 5 states, despite decades of research on equivalent problems (Akshay et al. 2015). The special case of $p = 0$ is the question of whether the target state can be reached in a finite number of transitions from the initial state. Although the theoretical (un)decidability of these basic reachability questions for Markov chains has not been proved, no algorithm is known for answering them.
- *Economics and game theory.* There are many undecidable questions in economics and game theory (Kao et al. 2012). In "spatialized Prisoner's Dilemma," each agent in in each time period must decide which of two actions to take, often called "cooperate" and "defect" (e.g., whether to burn its own leaves in the autumn, thereby polluting the local air, vs. abstain from burning them). Each agent in each period chooses the action with the higher payoff, given the most recent choices of its neighbors; it is never advantageous to be a lone cooperator surrounded by defectors. If this game is played on a grid (e.g., a piece of graph paper with no boundaries and an agent in each cell), then it is a special CA model. Many questions about what spatial patterns of choices can arise from a starting configuration are undecidable (Grim 1997). Even whether a strategy of "cooperate" will spread from a finite initial configuration or will eventually become extinct is undecidable, unless the set of initial configurations is restricted to special cases. (It is decidable for some initial configurations, of course, if simulations of finite length show that one of the choices becomes extinct.)

As illustrated by these examples, undecidability arises in many systems that are complex enough so that the set of possible sequences or trajectories of future states starting from an arbitrary initial state cannot be enumerated, e.g., because an infinite number of trajectories, or trajectories of infinite length, can follow the initial state. Undecidability of reachability is usually proved by contradiction, by showing that, if reachability were decidable in general in one of these systems, then it could be used to solve the halting problem for Turing machines (contradicting an implication of Rice's Theorem). Systems in classical mechanics (and other areas of physics), hybrid continuous-and-discrete dynamical systems, general continuous dynamical systems, many types of biological networks, and agent-based models such as that in spatialized prisoner's dilemma have all been shown to be capable of simulating universal Turning machines (roughly, by treating the initial configuration of the system as input data and interpreting its passage through subsequent states as computations), and hence their long-term behavior must be undecidable in general.

Although the undecidability of reachability implies that qualitative hazard identification and quantitative risk assessment questions are unanswerable in general for these systems, they can be answered in many important special cases. For example, in a coherent structure reliability system with a finite number of components that can fail at random times, but that are not repaired (as in a space craft), reachability of an undesirable state, such as a system failure, is clearly decidable, since all possibilities can be enumerated in principle. For such systems, fault tree analysis and discrete-event simulation provide constructive answers about how specific failure paths (cut sets) can lead to system failure and how likely they are to do so during the life of a mission or system. Thus, the lesson from the foregoing examples is not that probabilistic risk assessment (PRA) cannot be carried out at all, but that it can only be carried out for special cases—although these may still be of great practical interest and value. Clarifying what restrictions are needed to make reachability decidable in systems with infinite sets of possible trajectories (e.g., coherent reliability systems with repairs allowed) is the subject of much current research in each of the above areas.

Undecidable Questions in Risk Management

One way to restrict the transitions or state trajectories in a discrete or continuous (or hybrid) system, respectively, is to *control* the system by deliberately selecting inputs (i.e., making decisions or interventions) over time in an effort to avoid undesirable state and to achieve desired performance goals, such as maximizing or minimizing a specified objective function. The system's behavior will then be driven by a combination of decisions and random (uncontrolled) events. Risk analysis questions for controlled systems continue to include reachability—is there a set of input decisions that can guarantee (or make probable) achieving some goals while avoiding undesired states, despite the random inputs?—as well as risk management decisions about what decisions or policies (generating input decisions from observations) to use and how to implement them when more than one agent is involved in control (Table 4.1). Again, undecidability results imply that there is no effective computational procedure for answering these questions for many systems of practical interest. Some important examples follow.

Control of Deterministic Dynamic Systems

Before turning to the probabilistic systems of greatest interest to many risk analysts, it is worth considering the extent to which control questions can be answered in the absence of aleatory uncertainty. Even for deterministic systems, control can pose challenges. Whether an appropriately chosen sequence of inputs can drive a deterministic dynamic system from a specified initial state to a specified final state—the *point-to-point reachability problem* for a controlled system—depends on what inputs are allowed and how the dynamic system responds to them. This problem is undecidable even for some of the simplest and best-studied classes of deterministic dynamic systems (e.g., discrete-time linear time invariant (LTI) systems) if the feasible inputs are constrained (e.g., if the set of allowed inputs is non-convex, consisting of a disjoint union of a finite number of convex polytopes) (Sousa-Pinto 2017). It is also undecidable for many simple nonlinear systems e.g., with piecewise linear dynamics that are linear, in each of multiple regions; or with saturated linear dynamics, where the dynamics are linear up to a maximum possible response rate and flat above it (Fijalkow et al. 1997; Sousa-Pinto 2017). However, it *is* decidable for subsets of LTI systems with certain stability properties if the feasible inputs allow limited movement in any direction (more precisely, if they form a bounded convex polytope around the origin (Fijalkow et al. 1997). Detailed study shows that whether an arbitrary initial state can be driven to the origin is undecidable in piecewise linear systems with states having 22 or more dimensions, or, more generally, more than $21/(n-1)$ dimensions, where n is the number of different regions or "pieces"; $n > 1$ for any piecewise linear system (Blondel and Tsitsiklis 2000). Undecidability can create epistemic uncertainty about how to control even deterministic systems, by making it impossible for the controller to discover whether or how inputs can be chosen to achieve desired states.

The difficulties of control for deterministic systems extend to distributed control. Suppose that a team of multiple agents in a network cooperate in trying to maintain specified input-output behaviors for a jointly controlled system. That is, they seek to implement local policies that jointly guarantee that a specified global policy will be implemented for controlling the system's behavior. Local policies map observations (possibly including messages received from neighbors in the network as well as from the environment) to local actions, and the system responds to these actions and to inputs from the environment. Several models of such systems have been studied in which the challenge is to create (synthesize) local policies for reacting to local inputs to achieve a specified global input-output behavior for the system as a whole. Whether this distributed control task can be accomplished is undecidable in many models (Pnueli and Rosner 1990), even for small systems with as few as two agents (Janin 2007).

Risk Management of Uncertain Systems Modeled by POMDPs

Risk analysts usually care about decision-making and control of systems that respond to random events as well as to the controller's choices. Uncertainties about both the current state of the managed system and uncontrolled inputs from the environment are important in applications ranging from system reliability and mission-critical hardware to managed care of patients to prudent management of forests, fisheries, farms, and factories. Partially observable Markov decision processes (POMDPs) model both types of uncertainty, by allowing epistemic uncertainty about states, as well as aleatory uncertainty due to stochastic transitions among states at rates that are affected by the choices of a decision-maker or controller. Risk management questions for POMDPs include the following:

1. Given a finite set of states, a subset of which are identified as *goal states* (representing successful execution of a plan or completion of a mission); a probability distribution describing uncertainty about the initial state; and a finite set of possible actions that affect transition probabilities between states, is there is a sequence of actions (i.e., a plan) that will leave the system in a goal state with probability greater than p? *That is, can actions be chosen to make success probability acceptably high?* (Here and later, p is any user-specified rational number between 0 and 1, and "acceptably high" just means exceeding the user-specified threshold.)
2. Conversely, is there a sequence of actions that guarantees that a set of designated unacceptable states (perhaps representing catastrophic failure) is avoided with probability greater than p? In other words, *can the system be managed to keep the risk of failure (i.e., probability of ever entering an unacceptable state) acceptably small?*
3. If each action taken in a state generates a known reward, is there a sequence of actions that guarantees that the expected discounted value of rewards over an infinite planning horizon exceeds a specified threshold level? In other words, *can the system be managed to make the net present value of rewards acceptably high?*
4. Is there a sequence of actions that guarantees that the average reward per unit time generated by the managed process exceeds a specified threshold level? In other words, *can the system be managed to make average (expected or mean) reward per period acceptably high?*
5. Given any proposed sequence of actions, is there a different sequence that yields a higher average reward per unit time? In other words, *can an optimal policy be identified for maximizing average reward per unit time?*

All five of these questions are undecidable in general for POMDPs (Madani et al. 2003), although they can be decided for some important special cases (Chatterjee et al. 2016a, b). The third problem—determining where a policy can guarantee an acceptably high discounted reward over an infinite horizon—can be solved to an arbitrarily close approximation for many systems via stochastic dynamic programming algorithms, although such solutions may be time-consuming to produce (Chatterjee et al. 2016a; Zhang and Zhang 2001). Running optimization algorithms typically produces a sequence of increasingly good polices and value estimates, and if these eventually exceed the specified target threshold, then question (3) can be answered in the affirmative. For the other problems, however, finding even approximate answers (within a stated additive or multiplicative factor of the user-specified thresholds for acceptable solutions) is also undecidable (Madani et al. 2003; Chatterjee et al. 2016b).

The undecidability of risk management questions for POMDPs arises in part because they allow the state of a system to be uncertain. By contrast, in MDPs, where the decision-maker observes the state of the system before making decision, optimal policies for maximizing discounted reward, average reward per unit time, or total reward can be found to any desired degree of precision by several numerical algorithms [including linear programming; value iteration and policy iteration algorithms from stochastic dynamic programming (Papadimitriou and Tsitsiklis 1987; Puterman 1990); or by reinforcement learning, if model parameters are initially unknown and certain regularity conditions hold (especially ergodicity, i.e., any state can eventually be reached from any other with positive probability) (Majeed and Hutter 2018)]. POMDPs are sufficiently expressive to be used as models of *open world* situations in which an agent is initially uncertain about what objects exist in its environment and how its different actions might affect them. The agent may encounter and experiment with new objects over time, thus adding to its knowledge of possible states of the environment in which it acts (Srivastava et al. 2014). Moreover, POMDPs can model uncertainty and ambiguity in sensor data—is that blip outside the unambiguous range on a radar screen evidence of an enemy launch, or perhaps of several close together, or might it just be noise or a reflection from something else?—and uncertainty about actions and states (was that last course of vaccines effective in eradicating the disease, or did it miss a few carriers?) The price of this flexibility is undecidability of the above risk management questions in general (e.g., for POMDPs with integer-valued rewards), although they are still decidable in principle (although often computationally complex in practice) in some important special cases (e.g., for POMDPs with finite horizons or with positive rewards) (Chatterjee et al. 2016b).

Monitoring and Probabilistic Fault Diagnosis in Partially Observable Systems: Timeliness vs. Accuracy

Risk assessment methods are applied not only prospectively, to predict what *can* go wrong, but also retrospectively to diagnose what probably *did* go wrong when a system fails or exhibits unexpected behavior. If the states of a system cannot be directly observed, but must be inferred from the system's observed behavior, then the question arises of whether the occurrence of faults can be correctly inferred from observations, avoiding both false-positive diagnoses and false-negative failures to detect faults that have occurred. If this requirement is formalized as detecting faults correctly with probability 1 within a certain time window of their occurrence, while keeping the probability of false positives acceptably small (i.e., smaller than some user-specified acceptable level), then the question is undecidable for a general class of partially observable systems (Bertrand et al. 2016a, b). Thus, it may be impossible to design monitoring systems that reliably detect and report faults without an unacceptable rate of false positives if the delay between occurrence and report is required to meet certain guarantees, although this design goal can be achieved if the reporting time is left unconstrained (and if the diagnoser is allowed unlimited memory) (*ibid*). This illustrates a basic trade-off between the timeliness and the accuracy of the risk information that can be guaranteed.

Guaranteeing Timely Resolution of Tasks with Uncertain Completion Times

A different type of uncertainty about timeliness guarantees arises when an agent must decide in what order to undertake multiple tasks with deadlines, with the time required to complete each task being uncertain (lying in some interval). If new tasks may arise that preempt the one currently being worked on, depending on precisely when earlier-arriving tasks are finished, then the general problem of determining whether all tasks can be finished before their deadlines is undecidable (Fersman et al. 2007). (For intuitive motivation, think of an agency allocating its resources to investigate and remediate problems of different priorities, where new problems that preempt older ones may arise depending on whether and when older ones have been resolved.) In this case, uncertainty about what tasks will arise when is sufficient to make the risk of failing to meet deadlines on one or more tasks undecidable.

Multi-agent Team Control

Undecidability results for POMDPs extend immediately from the single-agent case to the team context of decentralized POMDPs and POSMDPs, in which multiple agents take time-consuming actions to pursue a shared goal (e.g., to win a contest against another team) or to maximize a joint team objective function using their own local observations and actions (Omidshafiei et al. 2017). In principle, a team of decentralized agents cannot do better than a single centralized planner that has access to all of their information, and that develops plans telling each agent what to do. If such a higher-level controller coordinates the agent's activities by assigning them various time-consuming tasks, informed by their estimates of time requirements and success probabilities for the tasks that they might be assigned, then, even though there is no way to develop an optimal plan (i.e., sequence of task assignments for agents) in general, a practical approach is to search for the best (highest-expected-utility) plan (i.e., assignment of tasks to agents) that can be discovered with a given computational budget. This satisficing approach for coordinating team activities and managing risks has yielded promising results in real-world planning and team coordination under uncertainty, such as having a team of robots (quadcopters and trucks) retrieve and deliver packages from base locations to delivery locations, despite multiple uncertainties (e.g., in wind, actuators, and sensors), obstacles, constraints on allowed paths, and needs for coordinated action among agents in tasks such as joint pickup of large packages (Omidshafiei et al. 2017).

The ability to solve such planning and multi-agent control problems under realistic real-world uncertainty highlights the fact that undecidability of optimal policies need not prevent useful (although presumably not optimal) plans and policies from being devised—at least as long as feasible solutions can be generated and improved fairly easily to obtain good (or even, with enough iterations, approximately optimal) solutions. This is the case for POMDPs with discounted reward criteria, as well as for certain extensions of POMDPs to robust optimization settings, where model parameters are known only to lie within a specified uncertainty set, and worst-case expected cumulative reward is to be maximized (Osogami 2015; Rasouli and Saghafian 2018). Although computational complexity remains a formidable challenge for large POMDPs, state-of-the-art solvers use a combination of ideas (including random sampling of search trees, discretization of value functions and updates ("point-based" value iteration), together with dynamic programming and linear programming techniques) to provide solutions whose quality improves with available computational budget (Smith and Simmons 2005; Shani et al. 2013). These techniques are often surprisingly effective in practice (Hsu et al. 2007; Zhang et al. 2016). Such POMDP solvers are useful for a variety of single-agent and multi-agent planning tasks, even when optimality cannot be guaranteed. Today, they provide solutions to POMDP problems in artificial intelligence and robotics planning with thousands of states instead, of being restricted to around a dozen states, as was the case in 2000 (Shani et al. 2013). Deciding how individual agents or teams of agents should plan and act under uncertainty can be solved in many practical settings with the help of POMDP solvers, even if performance guarantees are precluded by undecidability.

In formal models of multi-agent systems (MAS), the general question of determining whether the cooperating agents can meet desired performance specifications (e.g., reaching a desired goal state in a finite amount of time without first passing through any of a set of specified unsafe states) is undecidable if the agents have imperfect or incomplete information about the state of the system and communicate with each other via private channels. However, it is decidable if either the state or all actions taken by agents are public, i.e., observed by all agents (possibly after a finite number of rounds of non-public communication during which collusion among subsets of agents may take place) (Belardinelli et al. 2018). Thus, there is a close connection between the information available to agents and the decidability of safety and reachability questions in MAS models (Berthon et al. 2017).

Risk Management with Intelligent Adversaries: Game Theory Models

The goals, beliefs, intentions, and actions of other agents, including adversaries and attackers in cyber security, terrorism, and warfare, create many uncertainties for rational agents making choices under uncertainty. Game theory is a rich source of important unanswerable questions about risk analysis and optimal decisions, including the following:

- Does a player in a deterministic game have a winning strategy? This question is undecidable for 2-person zero-sum games in which players take turns moving a pointer on a grid, with one trying to reach the origin and the other trying to avoid it, and with each player choosing from a set of legal moves on each turn (Niskanen et al. 2016). (If this is reinterpreted as a game played by a single player against "Nature," which selects moves at random, it illustrates that the reachability of the origin from a starting point via appropriate choice of control inputs is undecidable.) Whether a player has a winning strategy is also undecidable for some real multi-player non-cooperative games (e.g., the popular card-trading game *Magic: The Gathering*) (Churchill et al. 2019). Restricting games in various ways (e.g., to assure that they end in finite time with probability 1, or that player strategies use only finite amounts of memory) can sometimes restore decidability of the existence of winning strategies, making exploration of the exact boundary between decidable and undecidable questions for games an exciting area for current mathematical research (Auger and Teytaud 2012; Berthon et al. 2017).
- Does a Nash equilibrium exist in pure strategies (i.e., if players do not use randomization to choose actions, but make deterministic choices)? (Sofronidis 2004).
- Does a pure-strategy equilibrium exist where one player wins almost surely in a multi-player stochastic game (a generalization of a Markov decision process in which state transition probabilities and payoffs depend on the actions of multiple players)? This is undecidable in games with more than 4 players, even if each player has only a finite set of possible strategies (Das et al. 2015).
- If a player has a winning strategy, how can it be computed? (Rabin 1957)

For some games, existence of a winning strategy can be proved by non-constructive means (e.g., topological fixed-point arguments), and yet the player who in theory can win is unable to do so in practice because there is no effective way to find or construct, i.e., compute, the winning strategy (Rabin 1957), or even an approximation to it (Auger and Teytaud 2012). For many other games, teams, and MAS, reachability questions and performance guarantees about what can be done, as well as strategy questions about how to do it (e.g., how to compute a distributed winning strategy), although they are undecidable in general, become decidable if information available to the players (agents) has a hierarchical structure, with more-informed players having all the information that less-informed players do, and possibly additional information as well (Berthon et al. 2017).

Table 4.2 summarizes the main results about which risk analysis questions can be answered in different classes of causal models. A rough summary is that all of the risk analysis questions in Table 4.1 can be answered in small finite models; none can be answered in sufficiently general dynamic models, even if they are deterministic; and some questions (e.g., about what to do next) but not others (e.g., about performance guarantees, such as achieving a target return or avoiding a set of undesired states with at least a certain probability) can be answered in intermediate cases.

Learning Causal Models from Data

To apply the decision and risk analysis models and methods in the left column of Table 4.2, a managed system or situation must be understood well enough to be able to simulate or predict its responses or outcome probabilities for different policies. Even a small decision tree requires enumerating possible sequences of choices and chance events and their outcomes, together with their conditional probabilities. Although reinforcement learning (RL) algorithms allow gradual learning of the relationship between actions in states and resulting reward and transition probabilities in ergodic MDPs [and some POMDPs and other systems (Majeed and Hutter 2018)], they still require the set of possible actions in each state be known, the state to be accurately observed, and the reward from taking an action in a state be promptly revealed. MDP models also make no distinction between choices and the actions that implement them: in an MDP, the possibility of error in implementing chosen actions (analogous to the "trembling hand" sometimes posited in game theory) is disregarded.

In practice, however, accurate causal models relating actions to conditional probabilities of outcomes may initially not be available. One expedient then is to try to learn the information needed to optimize policies by trial and error, provided that the relevant causal relationships remain stable for long enough to be learned [and that no catastrophic events occur during learning—the topic of safe reinforcement learning (García and Fernández 2015)]. This is, in effect, what RL accomplishes in ergodic MDPs, and what response surface methods and stochastic approximation do for static statistical input-output relationships. A great deal of recent progress has been made in algorithms for learning from observational data the causal relationships for causal Bayesian networks (CBNs) and influence diagrams

(IDs) (Goudet et al. 2018; Jabbari et al. 2017). These generalize response surface models for a single response by allowing the settings of multiple controllable inputs to change the conditional probability distributions of multiple other variables, including those of multiple outputs (or response) variables. Observations that are informative about the conditional probability distributions for outputs of interest can then be taken into account in deciding how to set the controllable inputs to obtain the most desirable (e.g., expected utility-maximizing) joint probability distribution of response variables. A CBN probabilistic graph model shows dependence relationships between variables by directing an arrow from *X* into *Y* if changing the value of the variable at its tail causes changes in the aleatory conditional probability distribution of the variable at its head. Quantitatively, the dependence of a variable on its parents (the variables that point into it) is expressed via a conditional probability table (CPT) or model specifying the aleatory conditional probabilities (or conditional probability density functions, for continuous variables) of its possible values as functions of the values of its parents. In well-specified causal models, a variable's CPT is the same no matter the setting of other variables and policy interventions in which it occurs; this *invariant causal prediction* property, along with the facts that effects depend on their direct causes and that information flows from causes to their effects over time, have been used to develop causal discovery algorithms for learning CBNs from data (Pfister et al. 2019). Algorithms are now well developed for using observational data both to infer the qualitative structure of a CBN, showing which variables depend on which others, and to estimate CPTs quantifying these dependencies; recent advances also allow for the possibility of unmeasured ("latent" or "hidden") causes and multi-period changes in the distributions of variables (Blondel et al. 2017; Goudet et al. 2018; Jabbari et al. 2017). Conditions under which CBN models can be uniquely identified from observations and generalized to predict effects of interventions under new conditions have also been elucidated (Blondel et al. 2017). Moreover, a rich theory has been developed within the CBN framework for determining what questions can (and cannot) be answered about effects of interventions and about effects of counterfactual conditions (e.g., how many more people would have died had it not been for an intervention or policy change?) (Shpitser and Tchetgen 2016). Thus, the challenges of learning causal network models from data and answering risk analysis questions within the CBN framework has been addressed by a substantial body of theory and computational techniques.

A limitation of CBNs is that they do not model time or dynamics in much detail, although dynamic Bayesian networks (DBNs) do allow information flows and probabilistic dependencies between variables within a period and from variables in one period or time slice to variables in its successor. Moreover, causal discovery algorithms for CBNs (and more general causal graphs, e.g., with bidirectional arrows or cycles) do not necessarily work well when applied to data from continuous-time dynamical systems (Aalen et al. 2016). Although they have proved highly useful for identifying and quantifying the causal roles of confounders, modifiers, and mediators in epidemiological risk assessment (Pearl 2009) and social science applications, where population responses to changes in causal factors may unfold on a time scale of years to decades, the basic conceptual framework of CBNs is that changing a cause changes the probability distribution of responses, and that enough time passes for this altered distribution to manifest itself in observed data. This is similar both to the response surface model (RSM) framework, where changes in input settings lead to changes in mean values of response variables but this causal relationship remains stable while being investigated; and also to the comparative statics framework widely used in economics and econometrics, where an intervention shifts a system from one equilibrium configuration to another. (The main difference is that CBNs describe responses of multiple dependent variables by conditional probability distributions rather than deterministically or by conditional mean values in a regression model.) This is a very different context from continually evolving dynamical systems.

If a system's state changes while it is being studied or controlled, then the fundamental question of which variables depend on which others (and how) cannot necessarily be answered based on observations, implying that a CBN description of the system may not be possible. Indeed, even if a program that accurately simulates a dynamic system is available, the question of which variables depend on others is undecidable in general: there is no algorithm for mapping programs (capable of simulating dynamic systems) to corresponding causal graph models, showing which variables depend on which others in the simulation program (for at least some values of the other variables), as any such algorithm might never halt (Icard 2017). This implies that there is no general algorithm for learning a causal graph description of dependency relations among variables from the data generated by such a program, or by the real-world data-generating process that the program simulates. Thus, while the possibilities for answering risk analysis questions summarized in the right column of Table 4.2 apply when the causal models in the left-hand column are given, they may not be answerable from observations alone. This may happen for any of the following reasons:

- Questions about causal effects of interventions or counterfactual questions about what would have happened (or would happen in the future) if different policies were used may not be uniquely quantifiable from the causal model and available observations; this is the problem of *identifiability of causal effects* of interest from data, given a model. It has been studied in depth for causal graph models such as CBNs (Shpitser and Tchetgen 2016).
- The causal models needed to answer risk analysis questions may be unknown and cannot necessarily be learned from data. This is the problem of *identifiability of a causal model* from data. It, too, has been studied in detail, and algorithms have been developed for identifying equivalence classes of CBNs that satisfy the same constraints implied by the data (e.g., conditional independence constraints) (Shpitser and Tchetgen 2016).
- No causal models capable of answering the questions exist (i.e., computationally effective models and algorithms that are guaranteed to return correct answers to causal questions in finite time may not exist), due to undecidability of causal relationships and inferences (Icard 2017). This is a relatively recent topic of investigation.

As usual, the undecidability of causal inference questions for some systems of interest does not imply that they cannot be decided for other interesting and useful special cases. Causal models can be learned from data generated by classes of POMDPs and policies for which all finite action-observation sequences have stable probabilities that can be reliably estimated from empirical data (i.e., learned from experience). Within this class, recurrent neural networks (RNNs) trained to predict future observations from past ones can be used to construct estimated causal states, representing the smallest set of clusters of histories (more precisely, the coarsest partition of histories into equivalence classes) that best predict responses to actions, as measured by mutual information (Zhang et al. 2019). Thus, in many environments that are both stationary enough and non-lethal enough for continued observations and learning to take place, current machine-learning techniques can eventually learn the causal relationships among actions, observations, and consequences (probabilities of transitions between the hidden states, and state-dependent rewards or costs) needed to guide effective action.

Responding to the Intractability of Risk Analysis

The undecidability results summarized in the right column of Table 4.2 imply that not all of the risk analysis questions in Table 4.1 can be answered for all of the causal risk models on the left side of Table 4.2. Indeed, questions about what might go wrong, whether and how adverse outcomes can be prevented, whether and how goals can be achieved and expected utility maximized, what guarantees can be placed on expected rewards or on loss probabilities, and whether and for how long systems can be operated safely (avoiding undesired states) in the face of random events or adversarial actions are unanswerable (undecidable) in general for many dynamic systems, although all can be answered for small decision trees and other sufficiently simple and static risk models. This suggests a rough division of risky situations into *tractable* ones, which can be modeled effectively (i.e., by algorithms that produce correct answers to risk analysis questions in finite time) for which decisions can be optimized; and *intractable* ones, for which effective modeling and decision optimization are not possible. A less crude division would acknowledge the many shades of intermediate complexity and tractability, including problems for which various risk analysis questions can be answered at least approximately with sufficient computational effort, but cannot be answered quickly and easily. Much worthwhile research is devoted to improving computational techniques so that more systems can be effectively modeled and managed, and recent breakthroughs such as deep learning and MCTS have rapidly expanded the size and scope of decision problems for which risk analysis and decision optimization are now tractable (Shani et al. 2013). However, the main point of Table 4.2 is that there are many systems and situations of practical interest that no amount of ingenuity can render tractable. It is still necessary to manage risks in these cases, even though there are hard limits (set by decidability) and practical limits (set by computational complexity and current computational technology) on how well their behaviors and responses can be predicted and controlled. To make progress in risk management in such difficult cases, it is possible to either restrict goals for risk management; to design systems for which risk management goals can be accomplished; or to apply risk analysis methods only to systems and environments that can be successfully analyzed. Alternatively, accepting that traditional risk analysis questions cannot be answered in some important situations, a different constructive response is to seek new bases for managing risk and making decisions in such cases by turning to questions that can be answered—for example, questions about which behaviors are feasible and appear usual, appropriate, or promising. These possible responses are discussed next.

Risk Analysis for Restricted Systems: Complexity-Tractability Trade-Offs

One constructive response to tractability constraints is to live within them by designing, using, and analyzing systems for which risk questions can be answered. Restricting the world (e.g., the systems, situations, or environments) for which risk analysis is undertaken; the set of policies considered for managing risks and maximizing a measure of return (e.g., limited-memory controllers); or the goals and specifications set for the performance of a risk management policy (e.g., maximizing expected discounted reward without insisting on bounds on failure probabilities) can allow risk analysis questions to be answered in many useful special cases, even if they are undecidable for more general settings. Similarly, if computational complexity, rather than decidability, is the limiting factor for finding or creating effective risk management policies or plans, then further restricting the systems, environments, policies, and objectives considered may restore tractability. For example, it is often possible to design restricted policies (such as limited-memory controllers for POMDPs) for which behaviors can be guaranteed to conform with desired specifications within a limited set of allowed specifications (Jansen et al. 2019). Likewise, there are many restricted but useful classes of discrete, continuous, and hybrid dynamic systems for which risk analysis questions can be answered, including questions of reachability, controllability, and decision optimization that are undecidable for less restricted systems (Lafferriere et al. 1999; Bertrand et al. 2016a, b). The price of such tractability is a loss in generality in the problems that can be modeled, and perhaps a loss of performance for the policies discovered, compared to what is possible (but less tractable) with fewer restrictions.

Much of applied risk analysis deals with the design and operation of systems to reliably achieve specifications by appropriately restricting the systems, operating environments, operating policies, and performance specifications and metrics considered in answering risk analysis questions. Recent advances in design and control techniques have allowed great progress in relaxing the restrictions that must be imposed to obtain answers for many realistically nonlinear and uncertain dynamic systems. For example, incorporating deep learning networks into control systems has recently rendered important practical decision problems for complex and nonlinear systems under uncertainty more computationally tractable, from managing logistics (Yang et al. 2018b) to landing space-craft and drones safely and smoothly under novel conditions (Sánchez-Sánchez and Izzo 2018) to optimizing building energy use management (Chen et al. 2019). Thus, successful risk analysis involves making adroit trade-offs among the following:

- *Systems:* Range of systems and situations to which risk analysis is applied (e.g., allowing linear vs. nonlinear input-output relationships; slowly changing vs. rapidly changing dynamics; and intelligent agents vs. random variables as sources of uncertainty);
- *Environments:* Breadth of environments and conditions in which the managed systems are assumed to operate for purposes of risk analysis;
- *Policies:* Flexibility in the risk management policies considered (e.g., what constraints are placed on possible actions and how much history is remembered and used in reacting to current observations?);
- *Risk models:* Realism and generality of the causal models used to represent systems and their environments and to analyze how policies affect risks (e.g., decision trees or influence diagrams vs. estimated response surfaces in RSM or value functions in MDPs vs. hybrid dynamic system simulation models);
- *Objectives:* Aspirations and goals for the risk analysis (e.g., maximizing expected discounted reward vs. maximizing expected discounted reward subject to specified constraints on probabilities of loss and probabilities of successful achievement of stated goals).

The main lesson from Table 4.2 is that trade-offs among these aspects of risk management are inescapable: if too much generality is allowed in systems dynamics, environments, and policies, then realistic risk models may be difficult or impossible to formulate and validate and risk analysis questions become undecidable.

Design of Resilient Systems for More Tractable Risk Management

To make risk management more tractable, increasing attention recently has been devoted to the design and cultivation of *resilient* sociotechnical systems and infrastructures. Ideally, these systems tolerate a wide range of changes and unexpected events in the operating environment (robustness); allow reconfiguration of internal resources to meet new needs as the environment changes (flexibility); fail gracefully when they fail at all, e.g., by losing functionality in small increments rather than via catastrophic failure, and only in response to multiple independent failures (fault tolerance); and recover quickly and efficiently following a loss of functionality (resilience), preferably to a fitter state than previously (antifragility) (Pagani et al. 2019; Martinetti et al. 2019). Such features allow a system to absorb and resist disruptive changes, minimize damage when they fail, and recover some or all functionality quickly following a disruption, while increasing capacity to withstand other disruptions in future. To our knowledge, there has been little formal analysis of whether the design of such systems is decidable and computationally tractable for environments with different types of uncertainties, but the fact that several versions of the comparative relationship of "faster than" between semi-Markov decision processes are undecidable in general (although they can be approximated in some cases) (Pedersen et al. 2020) suggests that determining which of two designs is more resilient, in the sense of leading to faster recovery of specified levels of functionality, may not be trivial.

Open-World vs. Closed-World Risks

In the relatively tractable models in rows 1-4 of Table 4.2, unanticipated and unforeseeable events do not occur. In small decision trees and equivalent models (e.g., influence diagrams) and in Markov decision processes, all possible sequences of future events are known, and probabilities can be assigned to each of them. Even in very large trees, for which it is impracticable to explicitly display all possible sequences and their probabilities, the probabilities of disjoint outcomes sum to 100% in principle, although in practice sampling techniques such as MCTS may be required to make calculation of recommended strategies computationally tractable. Likewise, in ergodic MDPs with infinite horizons and discounted rewards, although the set of possible futures is infinite (since states are visited infinitely often), all possible states and sequences of states are known in advance and probabilities of all possible outcomes again sum to 100%. In these tractable models, the "rules of the game" (i.e., the causal models) are understood in advance in enough detail to generate possible future scenarios and to assess their probabilities. Probabilities of different outcomes when different policies are followed can then be calculated (or estimated via sampling approaches similar to MCTS) and used to answer risk analysis questions. A sophisticated technical literature on probabilistic planning for reaching goal states under uncertainty via uses MDPs to generate a limited number of possible future scenarios starting from the initial state. It plans for these sampled scenarios (e.g., by using a deterministic planning algorithm for each of them, or for the most likely scenario(s) only) and combines the results into a final plan. This plan is executed until a state is encountered that it did not anticipate. Whenever an unplanned-for state is encountered, the scenario-sampling and planning process is repeated (dynamic "replanning"). Such algorithms, often referred to as *hindsight optimization* algorithms because they generate plans that make sense in hindsight if the envisioned scenarios come true, have proved surprisingly effective in competitions for assessing planning under uncertainty (Pineda and Zilberstein 2017; Yoon et al. 2010). They have also been modified to apply to POMDPs (Olsen and Bryce

2011). They provide practical heuristics for deciding what to do when full analysis is impracticable by considering a plausible, but not exhaustive, set of future scenarios.

By contrast, in the less tractable models in the rest of Table 4.2, not all possible futures can necessarily be foreseen, and plausible future scenarios cannot always be computed. Whether a system will enter a certain set of states may be undecidable, and relevant probabilities, expected values, and optimal policies cannot necessarily be calculated (or even approximated) for some systems, including many POMDPs (Lusena et al. 2001). Even for planning and control problems with finite horizons, for which all possibilities can be enumerated in principle, and hence optimal polices and their expected values can in principle be computed, the computational complexity of actually finding them (or close approximations to them) is prohibitive in practice for many realistically sized POMDPs (*ibid*, Table 4.1). Common sources of unpredictability, non-stationarity, and novelty in these less tractable settings include the behaviors of others (distributed control, MAS, ABM, and game theory models); computational complexity and undecidability of the future states of complex systems (e.g., hybrid control systems, CAs); and unobserved components of the current state of the world (POMDP models). These are common elements in *open-world risks*, meaning risks that arise from a decision maker's limited information, knowledge and understanding of how the world works. In open worlds, a decision-maker may be ignorant both about what exists (e.g., what dangers, foes, or prizes might exist where, and what actions might lead to them) and might happen next (e.g., an encounter with a new type of player or agent, or a transition to a state not previously encountered or imagined). Open-world novelty is quite different from the statistical novelty of unprecedentedly large catastrophic events governed by heavy-tailed distributions (e.g., unexpectedly large earthquakes, forest fires, landslides, bank runs, riots, financial crises, disruptive innovations, epidemics, or cascading failures in electric power grids or other network infrastructures). For such extreme events, the main source of novelty is that the past does not typically contain much (or perhaps any) experience with very rare, very large events that might occur. However, the causal mechanisms of larger and smaller events are usually the same. By contrast, the main source of open-world novelty is that what can happen is unknown until it occurs.

Open-world novelty was described as follows in a program announcement from the Defense Advanced Research Projects Agency (DARPA):

> Current artificial intelligence (AI) systems excel at tasks defined by rigid rules—such as mastering the board games Go and chess with proficiency surpassing world-class human players. However, AI systems aren't very good at adapting to constantly changing conditions commonly faced by troops in the real world—from reacting to an adversary's surprise actions, to fluctuating weather, to operating in unfamiliar terrain. For AI systems to effectively partner with humans across a spectrum of military applications, intelligent machines need to graduate from closed-world problem solving within confined boundaries to open-world challenges characterized by fluid and novel situations. The Science of Artificial Intelligence and Learning for Open-world Novelty (SAIL-ON) program intends to research and develop the underlying scientific principles, general engineering techniques, and algorithms needed to create AI systems that act appropriately and effectively in novel situations that occur in open worlds. The program's goals are to develop scientific principles to

quantify and characterize novelty in open-world domains, create AI systems that react to novelty in those domains, and demonstrate and evaluate these systems in a selected DoD domain. (www.darpa.mil/program/science-of-artificial-intelligence-and-learning-for-open-world-novelty).

To a useful first approximation, the relatively tractable models and methods in rows 1-4 of Table 4.2 are most useful for closed-world risk analysis, in which the causal rules linking choices to consequence probabilities are either well understood or can be learned in time to be useful. The relatively intractable models in the rest of Table 4.2 are more useful for describing open-world risks, but such risks cannot always be predicted or quantified. A great deal of useful risk analysis is accomplished by focusing on situations that that are well-approximated by closed-world modeling assumptions. However, extending risk management to open-world settings requires different principles and methods when closed-world models do not adequately represent the uncertainties that must be dealt with.

Artificial Intelligence (AI) Methods for Coping with Open-World Novelty and Risk

Useful principles and techniques for open-world risk management have started to emerge, in part from methods developed to guide autonomous vehicles, robots, drone swarms, and other embodied agents and multi-agent systems; and in part from AI methods developed for NPCs (non-player characters) in open-world on-line games. Even basic competence in navigating uncertain and changing real open-world environments (e.g., crossing rough terrain or a busy street to reach a destination) requires noticing and reacting quickly and appropriately to unforeseen changes and dangers, such as avoiding moving obstacles or compensating for unexpectedly soft ground. Developing, executing, and modifying plans to achieve multiple goals despite uncertainties about what can be accomplished, how long it might take, and what else might occur meanwhile, requires additional skills. Many of these skills are cognitive and strategic, such as recognizing and seizing opportunities and dynamically re-planning as conditions change and uncertainties are resolved, while maintaining purpose and reacting effectively to threats, delays, and interruptions.

Behavior Trees (BTs) Enable Quick Responses to Unexpected Events While Maintaining Multiple Goals

To endow robots and NPCs with such capabilities, a popular option is to control their behaviors with *behavior trees* (BTs) (Colledanchise and Ögren 2017; Martens et al. 2018). BTs are control architectures that allow hierarchies (i.e., trees) of tasks, each of which may be completed by completing lower-level tasks. (We use the generic word "task" for all nodes in a BT.) Tasks may be time-consuming. Whether they can be successfully completed, and, if so, how quickly, may remain uncertain until they are tried and either succeed or fail. The leaves of the BT represent primitive (i.e., not further decomposed) actions or behaviors that the agent can attempt to execute; these may include acquiring information (checking whether a condition holds) as well as executing actions. Higher-level tasks (i.e., internal nodes of the BT) control the flow of execution via logic gates And, or, k-out-of-n, and customized logic gates are standard. (They are called sequence, fallback, parallel, and decorator nodes, respectively, in BT terminology.) Unlike similar logic gates in fault trees, the children of these gates are attempted in a particular order (left-to-right for sequence and fallback nodes, simultaneously for parallel nodes, and customized for decorator nodes), and they can return a value of "running" as well as values of "success" or "failure". The tree structure of a BT controls both the order in which tasks are attempted (with left-to-right as the default for sequence and fallback nodes) and also switching among tasks over time as previously attempted tasks are completed successfully, fail, or are left running (neither succeeded nor failed). Tasks that have not yet succeeded can be left running (e.g., a robot may continue moving toward a destination without yet having arrived) while other tasks are worked on. Thus, undecidability resulting from potentially non-terminating algorithms need not render a BT-guided agent inactive or indecisive as long as other tasks can be undertaken.

BTs allow sequential composition of simple behaviors (sub-trees) to form more complex behaviors while preserving properties such as robustness (i.e., reaching goal states from a large set of initial states), safety (i.e., avoiding states that can harm the agent), and efficiency (i.e., reaching goal states quickly enough to be useful, or at least terminating in finite time) (Colledanchise and Ögren 2017; Sprague and Ögren 2018). Low-level behaviors (e.g., maintaining balance and avoiding collisions while walking) are subsumed into higher-level behaviors (e.g., stalking, attacking, fleeing, fighting or hiding from enemies in an open-world game; or exploring for new opportunities to perform useful tasks, for a real-world robot). Such "subsumption architectures," which control the behaviors of several embodied AIs, including commercial robots such as iRobots' Roomba vacuuming robot, enable smooth coordination of multiple goal-directed behaviors (e.g., wandering and cleaning while avoiding collisions and staying adequately charged) without top-down planning and optimization.

Several additions and refinements have increased the practical value of BTs and related techniques based on more general graph models. One allows next tasks to be selected based on their estimated expected utilities, as assessed using user-specified or machine-learned utility scoring functions (Merrill 2019). These are usually heuristic scoring functions. They allow very flexible, adaptive behaviors and priority-setting based on the estimated needs and urgencies of the moment rather than on pre-specified orderings. Competing priorities with different urgencies can be resolved by comparing estimated utilities for taking different actions next. More generally, several techniques—subsumption architectures, BTs, and probabilistic planners and hierarchical and distributed planning and control algorithms (discussed below)—allow decisions and plans to be made quickly enough to guide actions in real time, despite open-world uncertainties and unanticipated events, without becoming paralyzed by undecidable questions or by algorithms or tasks with excessively long execution times. By switching among multiple tasks and updating progress on each active task in each time step, BTs enable actions (i.e., work on tasks) to proceed while some attempted tasks are still running, without yet having returned success or failure. BTs with expected utility scoring, as well as any-time probabilistic planners, allow heuristic estimates of the relative values of alternative next actions whenever a decision must be made, even if there is not time for additional deliberation, planning, and optimization. These tactics allow planning and decision systems to keep up with the frequent changes and occurrence of unpredictable events typical of many open-world environments.

A second advance is to use machine learning techniques to evolve high-performing BTs for environments that can be simulated (Zhang et al. 2018; Colledanchise et al. 2019; Banerjee 2018). Variants include safe learning algorithms that avoid potentially harmful states during training, e.g., by restricting controls to those that avoid disallowed states (Sprague and Ögren 2018; for related work on safe navigation of traffic circles by autonomous vehicles, see Konda et al. 2019). Safe learning of BTs or other controllers is especially valuable for robots that must learn by interacting with the real world. A third advance applies BTs and BT-learning to teams or swarms of agents (Neupane and Goodrich 2019). Related work on composing sequences of behaviors for multiple agents while allowing needed communication among them enables teams of agents to share information, organize themselves, and coordinate to execute complex tasks such as searching urban environments to locate and rescue victims who need help (Pierpaoli et al. 2019). Finally, combining machine learning techniques such as reinforcement learning (RL) (Banerjee 2018; Dey and Child 2013), evolutionary programming, and deep learning (Zhang et al. 2018) with BTs enables agents to add to their capacities and skills over time by learning to perform new tasks, to perform old ones more efficiently, and to adapt their behaviors to new situations by modifying and adding to existing BTs or by learning new ones (Colledanchise et al. 2019).

Integrated Machine Learning and Probabilistic Planning for Open Worlds

A BT-controlled agent uses current conditions to decide what to do next. In this sense, its behaviors are reactive: they are driven by the pre-stored BT and by the current state of the world (as revealed through the agent's sensors, possibly including communications with other agents) rather than by envisioning and pursuing new desirable possibilities as novel opportunities arise. Low-level control, such as for maintaining balance and avoiding collisions while a robot moves to a destination, can often be delegated to low-level controllers that map sense data to actions moment by moment, with no need for higher-level symbolic thinking and planning.

By contrast, higher-level planning requires a different set of cognitive skills, including causal reasoning, consideration of different counterfactual scenarios and their outcome probabilities, and synthesis (rather than just selection) of action sequences to cause desired future states (Ghallab et al. 2016). Causal reasoning and planning, in turn, requires manipulating symbolic representations of causal models specifying constraints, such as what conditions must hold before a course of action can be attempted, and how completing various tasks would change the state of the world—or, in probabilistic causal models such as MDPs and POMDPs, how actions affect the conditional probabilities of different states and rewards.

Algorithms for automated probabilistic planning are now well developed (Kolobov et al. 2012). Many of the most successful ones incorporate Monte Carlo Tree Search (MCTS) principles to sample multiple possible futures and search for high-reward plans to inform current choices. These yield "any-time" decision algorithms in which initial plans are formed very quickly via limited search, and then improved by additional search while time remains, while still being able to return a decision whenever needed (Kolobov et al. 2012). Competitions have compared the performance of different probabilistic planning algorithms on a variety of MDP and POMDP benchmarking problems [typically represented as dynamic Bayesian networks and influence diagrams using the Relational Dynamic Influence Diagram Language (Sanner 2010)] in problem domains including manufacturing with randomly varying prices of goods; control of planetary rovers as they explore, take pictures, and recharge; eradication of invasive species from ecosystems; and deployment of a limited number of rangers to defend wildlife preserves against intelligent poachers (https://ipc2018-probabilistic.bitbucket.io/). Such probabilistic planners are increasingly integrated into standard software packages for robots and other autonomous agents, e.g., by including some of the most competitive ones in free software for the Robot Operating System (ROS) (Canal et al. 2019). Multi-agent planning (MAP) algorithms extend planning algorithms to enable teams of agents (e.g., drone swarms) to cooperate in planning joint actions to achieve shared goals (Shvo et al. 2018; Torreño et al. 2017).

Integrating planning, doing, and learning—that is, combining symbolic causal reasoning and planning with reinforcement learning and low-level control—to achieve goals or high rewards under uncertainty (e.g., in SMDP and POSMDP models) is a topic of much recent research (e.g., Ames et al. 2018; Ghallab et al. 2016; Illanes et al. 2019; James 2018; Konidaris et al. 2014; Yang et al. 2018a). A key insight from this work is that an agent can expand its capabilities not only by mastering new behaviors, but also by improving its understanding of how its actions affect the world, and hence of how and when it can deploy its possible behaviors to cause desired states (James et al. 2019). In human terms, both adding to *skills* (i.e., learned procedures or action sequences for completing tasks) and adding to declarative causal *knowledge* (specifically, understanding the conditions under which actions cause desired changes) can improve an agent's performance in uncertain environments, as measured by the speed and certainty with which it achieves goals or high rewards. In planning algorithms, causal understanding is typically represented by abstract symbolic causal models that map low-level sense data and action sequences to higher-level concept symbols such as states, tasks, and effects of tasks on states (Ghallab et al. 2016). (Abstract, temporally extended actions or behaviors such as "Fetch a needed tool" are often called *options* in this literature, but we will continue to refer to them as tasks, using BT terminology.) These concepts are abstract in the sense that many detailed low-level descriptions are mapped to the same high-level conceptual descriptions. Abstract concepts can, in turn, be used to develop higher-level concepts, creating a concept hierarchy. Planning can take place at a level of abstraction sufficient to identify causally effective sequences of actions without having to address all of the lower-level implementation details. The abstractions that are most useful for planning depend on an agent's current repertoire of possible behaviors. Concept symbols thus provide ways to partition the world into simplified representations useful for causal reasoning, planning, and decision-making based on the anticipated effects of actions.

Several recent AI programs have demonstrated that data gained by experience as an agent interacts with the world can be used both to learn relevant abstractions (i.e., symbolic terms for describing the world and the agent's actions within it and their effects) and also to develop effective plans and decision rules, expressed using those abstractions (Illanes et al. 2019; James et al. 2019). Using relevant abstract concepts learned from experience to guide reinforcement learning (RL) has been shown to help agents to achieve goals (or high rewards, in POSMDP and SMDPs) far more efficiently than RL alone in open-world settings with randomly generated objects and obstacles (Illanes and McIlraith 2019; James et al. 2019). Moreover, expressing plans and decision rules abstractly allows them to be generalized and applied to new situations with causal structures similar to those of previously encountered cases. Similarly, many statistical and machine learning methods have been developed to extract predictive features from data (Nguyen and Holmes 2019). For example, deep learning autoencoders compress input information used to predict outputs to identify essential information and discard the rest. Stacking successive autoencoders generates a hierarchy of features at increasingly high levels of abstraction, with each new layer composing more abstract features from those in the previous layer. Representing observed data in terms of informative features (often accomplished via statistical algorithms for "projecting" the original data onto a lower-dimensional space) abstracts from superfluous details and noise, yielding relatively parsimonious ("reduced dimensionality") descriptions (Nguyen and Holmes 2019). These more abstract descriptions enable subsequent planning and decision-making algorithms to focus on the most relevant information for predicting outcomes of actions. They also can help reinforcement learning algorithms quickly learn relevant reward functions from examples and feedback on successful (high-reward) and unsuccessful (low-reward) behaviors (Daniel et al. 2015) and can facilitate detection of significant changes, anomalies, and novelty in an agent's environment or in the behavior of a controlled system when observations fall outside the range predicted based on previous observations during normal operations.

Anomaly Detection Helps Focus Attention When Needed

Detecting unexpected events or changes that could not reasonably have been predicted can alert an intelligent agent—whether a robot, a drone swarm controller, a human planner, or a planning or regulatory organization—that old expectations and assumptions need to be revisited, and that perhaps new plans responding to the new information will be needed. Anomaly-, novelty-, and change-detection algorithms are now well developed in statistics and machine learning (Chalapathy and Chawla 2019). Many of them use deep learning (e.g., autoencoder) or statistical methods to compare reduced-dimensionality descriptions of observed and predicted behaviors in order to identify unexpected observations, i.e., anomalies (Aminikhanghahi and Cook 2017 for time series data; Chalapathy and Chawla 2019; Chandola et al. 2009; Nolle et al. 2018). Such anomalies can indicate a change in the underlying causal mechanisms generating observations and used in previous planning, and hence a need for increased attention and new causal reasoning in deciding what to do next. Despite some limited applications to business processes (Nolle et al. 2018), to date, anomaly detection has been used mainly in rather tactical applications, such as automatically identifying (in human terms, noticing) illegal traffic flows in video images (e.g., a car coming the wrong way on a one-way street, or a pedestrian stepping into traffic); cyberattacks and fraudulent activity in financial networks; pathologies in patients; and system failures or loss of process control in industrial applications (Chalapathy and Chawla 2019). In these and other applications, detecting anomalies can trigger a switch from automatic low-level control to higher-level symbolic causal reasoning, planning, and intervention, possibly in conjunction with warnings to human operators that something significant has changed and that previous ongoing low-level automatic control may no longer be adequate.

The potential to apply anomaly detection more strategically—perhaps to notice unexpected changes in the performance of public policies and regulations, or of business or military operations compared to what was predicted—is currently limited by lack of relevant data and causal models capable of making strong predictions about expected observations. As big data, predictive analytics, and causal analysis methods continue to advance and become increasingly applied to such large-scale strategy issues, it is likely that anomaly detection algorithms (probably in conjunction with most-probable-explanation (MPE) algorithms and other causal inference and diagnosis techniques previously discussed) will also become more widely and routinely used to detect deviations from planning assumptions. Anomaly detection can help achieve goals more quickly and reliably under uncertainty by calling attention and triggering timely review and revision of plans, policies, and their underlying assumptions when conditions change unexpectedly or results depart significantly from expectations.

Summary: AI Capabilities for Dealing with Open-World Risks and Uncertainties

In summary, a mix of the following components provides a pragmatic approach to decision-making under open-world uncertainty:

1. *Quick heuristic procedures* for selecting the most valuable-seeming choices if decisions must be made without time for further analyses. For example, BTs with expected utility scoring heuristics allow multiple tasks and goals to be pursued simultaneously while continually adjusting priorities and updating planned next actions in reaction to current information. Such heuristics can provide any-time decision recommendations, enabling an agent to select actions whenever needed, while leaving open the possibility that further analysis and deliberation might uncover better choices when and if time permits.
2. *Slower deliberative planning procedures.* These use causal knowledge and symbolic (often, causal model-based) reasoning to synthesize high-level plans to achieve goals or rewards. The resulting plans then guide lower-level control over detailed moment-to-moment actions.
3. *Lower-level learning and control procedures.* These are largely automated procedures, e.g., feedback control and reinforcement learning algorithms tuned by deep learning, that enable agents to execute behaviors; learn, use, and improve skills needed to implement plans in detail; and detect and recover from failures (Hammond et al. 2019).
4. *Automatic procedures for learning from experience* the informative features, hierarchies of abstract concepts, and causal knowledge needed to create effective plans. This knowledge can also be used to speed further learning and to help identify new goals that can be reached as conceptual and causal knowledge, as well as skills in applying them, increase.
5. *Automatic procedures for detecting novelty and change.* When time permits, further causal reasoning and deliberative planning (or re-planning) can be triggered when anomalous or novel situations are encountered or significant changes are detected in the causal relationship between actions and consequence probabilities.
6. *Social skills for multi-agent planning (MAP) and learning.* These include capabilities for communicating, collaborating (e.g., agreeing on allocation of tasks), and formulating a joint plan to accomplish shared goals (Shvo et al. 2018; Torreño et al. 2017). Other social skills include recognizing the goals and surmising the beliefs, desires, and intentions of others if they are not explicitly communicated; and learning from others by imitating successful behaviors and by asking for advice, demonstrations, and feedback when in doubt (Daniel et al. 2015). These skills are useful not only in MAP, but also in speeding reinforcement learning (Daniel et al. 2015; Singh et al. 2019) and in mixed-initiative collaborations between AIs and humans, such as when a human operator uses a drone swarm to conduct search-and-rescue operations under the operator's

general guidance, but with the drones making detailed local decisions autonomously (Bevacqua et al. 2015).

This mix of capabilities is reminiscent of aspects of human decision psychology, including a rough division of labor between "System 1" (quick, intuitive evaluations of alternatives) and "System 2" (slower, cognitive, rational, symbolic reasoning) for decision-making (Kahneman 2011). They enable agents to decide what to do next, even in the presence of open-world uncertainties, without becoming paralyzed by undecidable questions or by algorithms or tasks with excessively long execution times.

These adaptive capabilities are most useful in settings where causal laws remain stable for long enough for agents to learn them and use them to select actions that make desired outcomes more probable. Such stability, or invariance across different settings, times, and policies, is often taken as a defining characteristic of causal laws, including probabilistic ones, in natural and social systems (Pfister et al. 2019), and conditions under which it is possible for a single agent to learn optimal or near-optimal (low-regret) policies from experience have been elucidated (Rakhlin et al. 2010). However, when intelligent learning and strategic behaviors are important, the behaviors of multiple interacting agents in games, MAS, and CA systems may not be learnable. Successive responses that agents make to each other's behaviors may not converge (Pangallo et al. 2019), or convergence may be undecidable (Grim 1997). Even within teams of cooperating agents and in cooperative games, the existence of distributed plans or strategies for achieving goals may be undecidable when the agents have only partial information, unless special information conditions hold, such as if all members of the same coalitions have the same information, or if all agents have limited memories (Berthon et al. 2017).

In short, current AI methods can equip agents to plan and act purposefully to achieve goals in the presence of open-world uncertainties, even with realistically limited observations, causal knowledge, computational capacities and time for decision-making. But risk analysis questions about the probability of achieving stated goals or performance specifications can only be answered in special cases. In many cases, especially when multiple agents interact repeatedly, the only way to find out how well plans and strategies will perform and what will happen next is to keep playing. In such cases, traditional risk analysis methods focused on anticipating and acting to prevent or reduce potential losses must be complemented by capacities to respond to and recover from unanticipated and unpredictable events when they occur.

Discussion and Conclusions: Thriving with a Mix of Answerable and Unanswerable Questions

To learn, plan, and act effectively to cause desired outcomes despite uncertainties and unanticipated events, successful techniques do not ignore fundamental undecidability results (Table 4.2) and computational complexity results indicating that questions cannot always be answered, or cannot be answered quickly enough to be useful. Rather, they focus on a subset of questions that can be answered and that are needed to guide effective action: what to do next; how to sequence successive tasks to increase the probability of achieving goals or high rewards; how to share control and learn from others efficiently (including asking for and giving help and transferring knowledge and experience); how to detect and recover from failures and re-plan when necessary; and how to improve relevant causal and conceptual knowledge, skills, plans and policies when possible.

These questions can be answered completely for simple settings, such as those that are well modeled by small decision trees. They can be answered in many important risk analysis applications where the range of systems, environments, and policies considered is limited by assumption or by design. They can be answered practically, although not necessarily optimally, for more complex models suitable for describing open-world uncertainties (Table 4.2). The contemporary AI paradigm of (a) quickly generating plans that guide actions; and then (b) gradually improving them if time permits; and (c) updating them quickly as new information arrives, is able to sustain purposeful action and generate appropriate responses to an uncertain and changing world in many current applications in robotics and open-world games. It is being used in a host of real-world applications and demonstration projects, from automated package delivery by drones and robots to planning of military and space exploration missions. Cognitive skills that enable autonomous agents to generate and attempt different plans (both individual and collective) and to learn from their own and each other's successes and failures have proved powerful for mastering new domains, in the sense of learning to take causally effective actions that make preferred outcomes more probable. However, none of these useful advances can guarantee success, or even avoidance of catastrophic failure, if open-world uncertainties are large enough. Too much generality in the systems, environments, and policies considered brings undecidability or unmanageable computational complexity in the causal models for these applications [e.g., POMDPs, decPOMDPs, DBNs, MAS and their generalizations (Table 4.2)]. Fortunately, the size and generality of causal models that can be solved (at least to a useful approximation) in practice to yield high-performing decisions are being rapidly increased by the development of new algorithms, new insights into the precise restrictions and information conditions required for tractability, and new combinations of principles such as MCTS, deep learning, and distributed planning and control.

Many of these advances reflect a recognition that, in an open world, an agent or team can usually only control its own behaviors in response to conditions and events (which are often only partially observed), rather than controlling the conditions and events themselves. This rather stoical outlook emphasizes managing unpredictable risks by developing and maintaining capabilities to behave as effectively as possible when unanticipated events or conditions occur, even if they cannot be predicted, prevented, or avoided. Surviving and flourishing under open-world uncertainty may draw on capacities to anticipate, avoid, withstand, and recover quickly from harm to the extent possible; but it also requires acquiring information, causal knowledge, and skills needed to pursue valuable goals and high rewards while reacting to changes and uncertainties in the environment as they arise. Emphasis on building capacities to respond effectively to events as needed while maintaining and pursuing long-term goals is consistent with much recent work on reducing losses from natural disasters and other threats by fostering resilience in human communities and organizations (Aven 2019). It complements both traditional decision analysis paradigms that seek to predict the conditional probabilities of possible outcomes from alternative choices in order to identify choices maximizing expected utility (Raiffa 1968); and also traditional risk analysis paradigms that seek to anticipate and mitigate deviations from plans caused by uncertainties. The AI perspective accepts that probabilities of some events cannot be computed from facts and data, at least in time to be useful, and perhaps at all. [An example is that halting probabilities for certain randomly generated Turing machines and initial configurations, are well-defined but uncomputable (Chaitin 1975).] Rather than following subjective expected utility (SEU) theory by using subjective probabilities in such cases, behavior trees and other AI techniques allow decision-making to proceed without waiting for the results of long-running or impossible computations, e.g., by treating "running" as a possible value for an attempted task and branching to new tasks when some old ones have not yet been completed, i.e., have not yet returned either success or failure. The success of plans in AI can be evaluated by expected utility in simple cases where probabilities and utilities are known or can be estimated from available knowledge and data (including preference information on value tradeoffs and risk attitudes. More generally, however, a goal of many AI systems is to learn to act effectively in open-world environments, accepting that relevant causal models and rewards or payoff functions are initially unknown and must be learned from experience via trial and error, and that unanticipated events and outcomes may occur that make it difficult or impossible to compute informative expected utilities for different decision rules *a priori*.

Traditional decision and risk analysis questions about how to avoid unacceptable risks and losses and quantify remaining risks and uncertainties are often unanswerable in open-world settings. Instead, focusing on capabilities and resilience highlights answerable questions about whether and how the current best plans for achieving goals and making preferred outcomes more likely can be improved—e.g., by investing in additional thought, preparation, information, conceptual and causal knowledge, and execution skills. Ability to generate, improve, and replace current plans and policies as needed can help to cope with open-world uncertainties that are not well modeled in traditional decision and risk analysis or in more recent predictive and prescriptive analytics. It seems plausible that, in the near future, AI techniques supporting such flexible and adaptive planning and responses to unforeseen events will add valuable new tools for open-world risk management to the analytics toolkits used today to think systematically about how best to formulate effective policies and make decisions under uncertainty.

References

Aalen OO, Røysland K, Gran JM, Kouyos R, Lange T (2016) Can we believe the DAGs? A comment on the relationship between causal DAGs and mechanisms. Stat Methods Med Res 25(5):2294–2314

Akshay S, Antonopoulos T, Ouaknine J, Worrel J (2015) Reachability problems for Markov chains. Inf Process Lett 115(2):155–158. https://doi.org/10.1016/j.ipl.2014.08.013

Amaran S, Sahinidis NV, Sharda B, Bury S (2016) Simulation optimization: a review of algorithms and applications. Annals of Operations Research 240(1):351–380

Ames B, Thackston A, Konidaris G (2018) Learning symbolic representations for planning with parameterized skills. 2018 IEEE/RSJ International conference on intelligent robots and systems (IROS), 526–533

Aminikhanghahi S, Cook DJ (2017) A survey of methods for time series change point detection. Knowl Inf Syst 51(2):339–367. https://doi.org/10.1007/s10115-016-0987

Annaswamy AM (2014) Robust adaptive control. In: Baillieul J, Samad T (eds) Encyclopedia of systems and control. Springer, London. https://doi.org/10.1007/978-1-4471-5102-9_118-1

Asarin E, Mysore VP, Pnueli A, Schneider G (2012) Low dimensional hybrid systems – decidable, undecidable, don't know. Inform Comput 211:138–159

Auger D, Teytaud O (2012) The frontier of decidability in partially observable recursive games. Int J Found Comput Sci. World Scientific Publishing, Special Issue on "Frontier between Decidability and Undecidability" 23(7): 1439–1450. ffhal-00710073f

Aven T (2019) The call for a shift from risk to resilience: what does it mean? Risk Anal 39(6): 1196–1203. https://doi.org/10.1111/risa.13247

Aven T (2020) Three influential risk foundation papers from the 80s and 90s: Are they still state-of-the-art? Reliab Eng Syst Saf 193:106680. https://doi.org/10.1016/j.ress.2019.106680

Avraam MP, Shah N, Pantelides CC (1998) Modelling and optimisation of general hybrid systems in the continuous time domain. Comput Chem Eng 22(Suppl 1):S221–S228. https://doi.org/10.1016/S0098-1354(98)00058-1

Banerjee B (2018) Autonomous acquisition of behavior trees for robot control. 2018 IEEE/RSJ International conference on intelligent robots and systems (IROS), Madrid, 2018, pp 3460–3467. doi: https://doi.org/10.1109/IROS.2018.8594083k

Belardinelli F, Lomuscio A, Murano A, Rubin S (2018) Decidable verification of multi-agent systems with bounded private actions. In Proceedings of the 17th international conference on autonomous agents and multiagent systems (AAMAS '18). International Foundation for Autonomous Agents and Multiagent Systems, Richland, SC, 1865–1867

Bernstein DS, Givan R, Immerman N, Zilberstein S (2002) The complexity of decentralized control of Markov decision processes. Math Oper Res 27(4):819–840. https://doi.org/10.1287/moor.27.4.819.297

Berthon R, Maubert B, Murano A (2017) Decidability results for ATL with imperfect information and perfect recall. In S Das, E Durfee, K Larson, M Winikoff (eds) Proceedings of the 16th international conference on autonomous agents and multiagent systems (AAMAS 2017), May 8–12, 2017, Sao Paulo, Brazil. http://www.ifaamas.org/Proceedings/aamas2017/pdfs/p1250.pdf

Bertrand N, Bouyer P, Brihaye T, Carlier P (2016a) Analysing decisive stochastic processes. 43rd International Colloquium on Automata, Languages, and Programming (ICALP 2016), 2016, Rome, Italy. pp. 101:1–101:14, ff10.4230/LIPIcs.ICALP.2016.101ff

Bertrand N, Haddad S, Lefaucheux E (2016b) Accurate approximate diagnosability of stochastic systems. In: Dediu AH, Janoušek J, Martín-Vide C, Truthe B (eds) Language and automata theory and applications. LATA 2016. Lecture Notes in Computer Science, vol 9618. Springer, Cham

Bevacqua G, Cacace J, Finzi A, Lippiello V. (2015) Mixed-initiative planning and execution for multiple drones in search and rescue missions. In: Proceedings of the twenty-fifth international conference on international conference on automated planning and scheduling (ICAPS'15). AAAI Press, pp 315–323

Bier VM, Azaiez MN (2009) Game theoretic risk analysis of security threats. Springer, New York

Blondel VD, Tsitsiklis JN (2000) A survey of computational complexity results in systems and control. Automatica:1249–1274

Blondel G, Arias M, Gavaldà R (2017) Identifiability and transportability in dynamic causal networks. Int J Data Sci Anal 3:131–147. https://doi.org/10.1007/s41060-016-0028-8

Canal G, Cashmore M, Krivić S, Alenyà G, Magazzeni D, Torras C (2019) Probabilistic planning for robotics with ROSPlan. In: Althoefer K, Konstantinova J, Zhang K (eds) Towards autonomous robotic systems. TAROS 2019. Lecture Notes in Computer Science, vol 11649. Springer, Cham, pp 236–250

Chaitin GJ (1975) A theory of program size formally identical to information theory. J Assoc Comput Mach 22:329–340

Chalapathy R, Chawla S (2019) Deep learning for anomaly detection: a survey. ArXiv, abs/1901.03407

Chandola V, Banerjee A, Kumar V (2009) Anomaly detection: a survey. ACM Comput Surv 41, 3, Article 15 (July 2009), 58 pages. doi:https://doi.org/10.1145/1541880.1541882

Chatterjee K, Chmelík M, Tracol M (2016a) What is decidable about partially observable Markov decision processes with ω-regular objectives. J Comput Syst Sci 82(5):878–911

Chatterjee K, Chmelík M, Gupta R, Kanodia A (2016b) Optimal cost almost-sure reachability in POMDPs. Proceedings of the twenty-ninth AAAI conference on artificial intelligence. Artificial Intelligence, vol 234, Issue C May 2016

Chen Y, Shi Y, Zhang B. (2019) Optimal control via neural networks: a convex approach. International conference on learning representations (ICLR). https://arxiv.org/abs/1805.11835

Chow Y, Tamar A, Mannor S, Pavone M (2015) Risk-sensitive and robust decision-making: a CVaR optimization approach. Proceeding NIPS'15 Proceedings of the 28th international conference on neural information processing systems – vol 1, pp 1522–1530. Montreal – December 07–12, 2015. MIT Press, Cambridge, MA

Churchill A, Biderman S, Herrick A (2019) Magic: the gathering is turing complete. https://arxiv.org/abs/1904.09828

Colledanchise M, Ögren P (2017) How behavior trees modularize hybrid control systems and generalize sequential behavior compositions, the subsumption architecture, and decision trees. IEEE Trans Robot 33(2):372–389. https://doi.org/10.1109/TRO.2016.2633567

Colledanchise M, Parasuraman R, Ögren P (2019) Learning of behavior trees for autonomous agents. IEEE Trans Games 11(2):183–189. https://doi.org/10.1109/TG.2018.2816806k

Cooper GF (1990) The computational complexity of probabilistic inference using Bayesian belief networks. Artif Intell 42(2–3):393–405

da Costa NCA, Doria FA (2014). On an extension of rice's theorem and its applications in mathematical economics: dedicated to the memory of Professor Saul Fuks (1929–2012). In Horowitz S, Koppl R (eds) Entangled political economy (Advances in Austrian economics, vol 18), Emerald Group Publishing Limited, pp 237–257. doi:https://doi.org/10.1108/S1529-213420140000018011

Daniel C, Kroemer O, Viering M et al (2015) Active reward learning with a novel acquisition function. Auton Robot 39:389–405. https://doi.org/10.1007/s10514-015-9454-z

Das A, Krishna SN, Manasa L, Trivedi A, Wojtczak D (2015) On pure nash equilibria in stochastic games. In: Jain R, Jain S, Stephan F (eds) Theory and applications of models of computation. TAMC 2015. Lecture Notes in Computer Science, vol 9076. Springer, Cham

DeGroot MH (2004) Optimal statistical decisions (Wiley Classics Library edition). Wiley, Hoboken, NJ

Delzanno G, Zavattarob G (2012) Reachability problems in BioAmbients. Theor Comput Sci 431(4):56–74. https://doi.org/10.1016/j.tcs.2011.12.056

Dey R, Child C (2013) QL-BT: enhancing behaviour tree design and implementation with Q-learning. In 2013 IEEE Conference on computational intelligence in games (CIG), pp 1–8

Dorri A, Kanhere SS, Jurdak R (2018) Multi-agent systems: a survey. IEEE Access 6:28573–28593. https://doi.org/10.1109/ACCESS.2018.2831228

Fersman E, Krcal Pettersson P, Yi W (2007) Task automata: Schedulability, decidability and undecidability. Inf Comput 205(8):1149–1172. https://doi.org/10.1016/j.ic.2007.01.009

Fijalkow N, Ouaknine J, Pouly A, Sousa-Pinto J, Worrell J (1997) On the decidability of reachability in linear time-invariant systems. In Proceedings of ACM woodstock conference (WOODSTOCK'97). ACM, New York, NY, 11 pages. doi:10.475/123_4

Francois-Lavet V, Henderson P, Islam R, Bellemare MG, Pineau J (2018) An introduction to deep reinforcement learning. Found Trends Mach Learn 11(3–4):219–354. https://doi.org/10.1561/2200000071

Fu MC (ed) (2015) Handbook on simulation optimization. Springer, New York

Fu MC (2017) Markov decision processes, AlphaGo, and Monte Carlo tree search: back to the future, Chapter 4. In: Batta R, Peng J (eds) Tutorials in operations research. INFORMS, Catonsville, MD, pp 68–88

García J, Fernández F (2015) A comprehensive survey on safe reinforcement learning. J Mach Learn Res 16(1):1437–1480. http://www.jmlr.org/papers/volume16/garcia15a/garcia15a.pdf

Ghallab M, Nau D, Traverso P (2016) Automated planning and acting. Cambridge University Press

Goudet O, Kalainathan D, Caillou P, Guyon I, Lopez-Paz D, Sebag M (2018) Learning functional causal models with generative neural networks. In: Escalante H et al (eds) Explainable and interpretable models in computer vision and machine learning. The Springer series on challenges in machine learning. Springer, Cham, pp 39–80

Grim P (1997) The undecidability of the spatialized prisoner's dilemma. Theor Decis 42:53–80. https://doi.org/10.1023/A:1004959623042

Grobelna I, Grobelny M, Adamski M (2014) Model Checking of UML activity diagrams in logic controllers design. Proceedings of the ninth international conference on dependability and complex systems DepCoS-RELCOMEX, Advances in intelligent systems and computing, vol 286, Springer International Publishing, pp 233–242

Hammond JC, Biswas J, Guha A (2019) Automatic failure recovery for end-user programs on service mobile robots. arXiv Preprint arXiv:1909.02778

Hanheide M, Göbelbecker M, Horn GS, Pronobis A, Sjöö K, Aydemir A, Jensfelt P, Gretton C, Dearden R, Janicek M, Zender H, Kruijff GJ, Hawes N, Wyatt JL (2017) Robot task planning and explanation in open and uncertain worlds. Artif Intell 247:119–150

Hoffman M, de Freitas N (2012) Inference strategies for solving semi-markov decision processes. Decision theory models for applications in artificial intelligence: concepts and solutions. IGI Global, pp 82–96. doi:https://doi.org/10.4018/978-1-60960-165-2.ch005

Hsu D, Lee WS, Rong N (2007) What makes some POMDP problems easy to approximate. In: Proceedings of advances in neural information processing systems (NIPS), pp 689–696

Icard T (2017) From programs to causal models. Proceedings of the 21st Amsterdam colloquium. https://web.stanford.edu/~icard/ac2017.pdf

Illanes L, McIlraith SA (2019) Generalized planning via abstraction: arbitrary numbers of objects. Thirty-third AAAI conference on artificial intelligence, pp 7610–7618

Illanes L, Yan X, Toro Icarte R, McIlraith SA (2019) Symbolic planning and model-free reinforcement learning: training taskable agents. 4th Multidisciplinary conference on reinforcement learning and decision making. www.cs.toronto.edu/~lillanes/papers/IllanesYTM-rldm2019-symbolic.pdf

Ioannou PA, Sun J (1995) Robust adaptive control. Prentice-Hall, Upper Saddle River, NJ. ISBN:0-13-439100-4

Jabbari F, Ramsey J, Spirtes P, Cooper G (2017) Discovery of causal models that contain latent variables through Bayesian scoring of independence constraints. Mach Learn Knowl Discov Databases 2017:142–157. https://doi.org/10.1007/978-3-319-71246-8_9

James S (2018) Learning portable symbolic representations. In: Proceedings of the 27th international joint conference on artificial intelligence (IJCAI'18). AAAI Press, pp 5765–5766

James S, Rosman B, Konidaris G. (2019) Learning portable representations for high-level planning. https://arxiv.org/abs/1905.12006

Janin D (2007) On the (high) undecidability of distributed synthesis problems. In: Proceedings of SOFSEM 2007: theory and practice of computer science, vol 4362 of LNCS, pp 320–329. Springer. https://hal.archives-ouvertes.fr/hal-00306387/document

Jansen N, Junges S, Katoen J, Quatmann T, Becker B, Wimmer R, Winterer L (2019) Correct-by-construction policies for POMDPs. In: Proceedings of the Fifth international workshop on symbolic-numeric methods for reasoning about CPS and IoT (SNR '19). ACM, New York, NY, pp 6–8. https://doi.org/10.1145/3313149.3313366

Jiao P, Xu K, Yue SWei X, Sun L (2017) A decentralized partially observable Markov decision model with action duration for goal recognition in real time strategy games. Discrete dynamics in nature and society, vol 2017, Article ID 4580206, 15 pages. doi:https://doi.org/10.1155/2017/4580206

Juan AA, Faulin J, Grasman SE, Rabe M, Figueirae G (2015) A review of simheuristics: extending metaheuristics to deal with stochastic combinatorial optimization problems. Oper Res Perspect 2:62–72. https://doi.org/10.1016/j.orp.2015.03.001

Kahneman D (2011) Thinking, fast and slow. Farrar, Straus and Giroux, New York

Kao Y-F, Ragupathy V, Vela Velupillai K, Zambelli S (2012) Noncomputability, unpredictability, undecidability, and unsolvability in economic and finance theories. Complexity 18(1):51–55

Kaplan S, Garrick BJ (1981) On the quantitative definition of risk. Risk Anal 1:11–27

Khakzad N, Khan F, Amyotte P (2011) Safety analysis in process facilities: comparison of fault tree and Bayesian network. J Reliab Eng Syst Saf 96:925–932

Khakzad N, Landucci G, Reniers G (2017) Application of dynamic Bayesian network to performance assessment of fire protection systems during domino effects. Reliab Eng Syst Saf 167: 232–247. https://doi.org/10.1016/j.ress.2017.06.004

Kleiner E, Newcomb T (2007) On the decidability of the safety problem for access control policies. Electron Notes Theor Comput Sci 185:107–120

Koller D, Friedman N (2009) Probabilistic graphical models - principles and techniques. MIT Press, Cambridge

Kolobov A, Mausam M, Weld DS (2012) LRTDP versus UCT for online probabilistic planning. AAAI'12: Proceedings of the twenty-sixth AAAI conference on artificial intelligence. Toronto, ON. Sheraton Centre Toronto, July 22–26, 2012, pp 1786–1792. https://www.aaai.org/ocs/index.php/AAAI/AAAI12/paper/view/4961. Last accessed 9-20-2020

Konda R, Squires E, Pierpaoli P, Egerstedt M, Coogan S. (2019) Provably-safe autonomous navigation of traffic circles. 2019 IEEE Conference on control technology and applications (CCTA), pp 876–881. https://ieeexplore.ieee.org/abstract/document/8920597

Konidaris G, Kaelbling LP, Lozano-Perez T (2014) Constructing symbolic representations for high-level planning. In: Proceedings of the twenty-eighth AAAI conference on artificial intelligence (AAAI'14). AAAI Press, 1932–1940. https://cs.brown.edu/~gdk/pubs/orig_sym_aaai.pdf

Kwisthout J (2011) Most probable explanations in Bayesian networks: complexity and tractability. Int J Approx Reason 52(9):1452–1469. https://doi.org/10.1016/j.ijar.2011.08.003

Lafferriere G, Pappas GJ, Yovine S (1999) A new class of decidable hybrid systems. In: Vaandrager FW, van Schuppen JH (eds) Proceedings of the second international workshop on hybrid systems: computation and control (HSCC '99). Springer, London, pp 137–151

Leigh JR (1992) Applied digital control: theory, design and implemenation (2nd ed). Prentice Hall International (UK) Ltd, London. Republished by Dover books (2006)

Lennartson B, Wigström O, Riazi S, Bengtsson K (2015) Modeling and optimization of hybrid systems. IFAC-Papers On Line 48(27):351–357. https://doi.org/10.1016/j.ifacol.2015.11.199

Lusena C, Goldsmith J, Mundhenk M (2001) Nonapproximability results for partially observable Markov decision processes. J Artif Intell Res 14(1):83–103

Ma X, Driggs-Campbell K, Zhang Z, Kochenderfer NJ (2019). Monte Carlo tree search for policy optimization. IJCAI'19 Proceedings of the 28th international joint conference on artificial intelligence, pp 3116–3122 Macao, – August 10–16, 2019. AAAI Press

Madani O, Hanks S, Condon A (2003) On the undecidability of probabilistic planning and related stochastic optimization problems. Artif Intell 147(1–2):5–34

Majeed SJ, Hutter M (2018) On Q-learning convergence for non-Markov decision processes. In: Lang J (ed) Proceedings of the 27th international joint conference on artificial intelligence (IJCAI'18). AAAI Press, pp 2546–2552

Martens C, Butler E, Osborn JC (2018) A resourceful reframing of behavior trees. ArXiv, abs/1803.09099

Marti K (1997) Solving stochastic structural optimization problems by RSM-Based stochastic approximation methods - gradient estimation in case of intermediate variables. Math Methods Oper Res 46:409–434. https://doi.org/10.1007/BF01194863

Martinetti A, Chatzimichailidou MM, Maida L, van Dongen L (2019) Safety I-II, resilience. Int J Occup Saf Ergon 25(1):66–75. https://doi.org/10.1080/10803548.2018.1444724

Mauá DD, de Campos CP, Zaffalon M (2013) On the complexity of solving polytree-shaped limited memory influence diagrams with binary variables. Artif Intell 205:30–38. https://doi.org/10.1016/j.artint.2013.10.002

Merrill B (2019) Building utility decisions into your existing behavior tree. In: Rabin S (ed) Game AI Pro 360: guide to architecture. CRC Press, pp 127–136

Miller CW, Yang I (2017) Optimal control of conditional value-at-risk in continuous time. SIAM J Control Optim 55(2):856–884

Moore C (1990) Unpredictability and undecidability in dynamical systems. Phys Rev Lett 64(20): 2354–2357

Myers RH, Montgomery DC, Anderson-Cook CM (2016) Response surface methodology: process and product optimization using designed experiments, 4th edn. Wiley

Neupane A, Goodrich M (2019) Learning swarm behaviors using grammatical evolution and behavior trees. Proceedings of the twenty-eighth international joint conference on artificial intelligence (IJCAI-19)

Nguyen LH, Holmes S (2019) Ten quick tips for effective dimensionality reduction. PLoS Comput Biol 15(6):e1006907. https://doi.org/10.1371/journal.pcbi.1006907

Nguyen TT, Nguyen ND, Nahavandi S (2020) Deep reinforcement learning for multi-agent systems: a review of challenges, solutions and applications. IEEE Trans Cybern 50(9): 3826–3839. https://doi.org/10.1109/TCYB.2020.2977374

Niskanen R, Potapov I, Reichert J (2016) Undecidability of two-dimensional robot games. In: Faliszewski P, Muscholl A, Niedermeier R (eds) 41st International symposium on mathematical foundations of computer science (MFCS 2016), Article No. 73, pp. 73:1–73:13. https://pdfs.semanticscholar.org/02be/2448e3430e2bf69b40d4b0ab9eb057b38c8c.pdf

Nolle T, Luettgen S, Seeliger A et al (2018) Analyzing business process anomalies using autoencoders. Mach Learn 107:1875–1893. https://doi.org/10.1007/s10994-018-5702-8

Oliehoek FA, Amato C (2016) A concise introduction to decentralized POMDPs (PDF). SpringerBriefs Intell Syst doi:https://doi.org/10.1007/978-3-319-28929-8. ISBN 978-3-319-28927-4

Olsen A, Bryce D (2011) POND-hindsight: applying hindsight optimization to POMDPs. https://pdfs.semanticscholar.org/c88a/ae1aa57c768e1597ae05455e0a37c458ba73.pdf

Omidshafiei S, Agha-Mohammadi A-A, Amato C, Liu S-Y, How JP, Vian J (2017) Decentralized control of multi-robot partially observable Markov decision processes using belief space macro-actions. Int J Robot Res 36(2):231–258. https://doi.org/10.1177/0278364917692864

Osogami T (2015) Robust partially observable Markov decision process. In Bach F, Blei D (eds) Proceedings of the 32nd international conference on international conference on machine learning - volume 37 (ICML'15), vol 37. JMLR.org, pp 106–115

Ossenkopf M, Jorgensen M, Geihs K (2019) When does communication learning need hierarchical multi-agent deep reinforcement learning? Cybern Syst 50(8):672–692. https://doi.org/10.1080/01969722.2019.1677335

Pagani A, Mosquera G, Alturki A, Johnson S, Jarvis S, Wilson A, Guo W, Varga L (2019) Resilience or robustness: identifying topological vulnerabilities in rail networks. R Soc Open Sci 6(2):181301. https://doi.org/10.1098/rsos.181301

Page SE (2018) The model thinker: what you need to know to make data work for you. Basic Books, New York, NY. https://arxiv.org/abs/1812.11794

Pangallo M, Heinrich T, Farmer JD (2019) Best reply structure and equilibrium convergence in generic games. Sci Adv 5(2):eaat1328. https://doi.org/10.1126/sciadv.aat1328

Papadimitriou CH, Tsitsiklis JN (1987) The complexity of Markov decision processes. Math Oper Res 12(3):441–450

Parker MW (2005). Undecidable long-term behavior in classical physics: Foundations, results, and interpretation. Ph.D. Dissertation, University of Chicago

Pearl J (2009) Causal inference in statistics: an overview. Stat Surv 3(2009):96–146. https://doi.org/10.1214/09-SS057i

Pedersen MR, Bacci G, Larsen KG (2020) A faster-than relation for semi-Markov decision processes. Electron Proc Theor Comput Sci 312(2020):29–42. arXiv:1810.11243v2

Pfister N, Bühlmann P, Peters J (2019) Invariant causal prediction for sequential data. J Am Stat Assoc 114(527):1264–1276. https://doi.org/10.1080/01621459.2018.1491403

Pierpaoli P, Li A, Srinivasan M, Cai X, Coogan S, Egerstedt M (2019) A sequential composition framework for coordinating multi-robot behaviors. arXiv preprint arXiv:1907.07718

Pineda L, Zilberstein S (2017) Generalizing the role of determinization in probabilistic planning. https://arxiv.org/pdf/1705.07381.pdf

Pnueli A, Rosner R (1990). Distributed reactive systems are hard to synthesize. In: Proceedings of FOCS, pp 746–757. IEEE Computer Society

Prasad K (1991) Computability and randomness of Nash equilibrium in infinite games. J Math Econ 20(5):429–442. https://doi.org/10.1016/0304-4068(91)90001

Puterman ML (1990) Markov decision processes. In: Heyman DP, Sobel MJ (eds) Handbooks in operations research and management science, vol 2. North-Holland, Elsevier Science Publishers, New York, NY, pp 331–434

Rabin MO (1957) Effective computability of winning strategies. In: Dresher M, Tucker AW, Wolfe P (eds) Annals of mathematics studies, No. 39: contributions to the theory of games, vol III. Princeton University Press, Princeton, NJ, pp 147–157

Raiffa H (1968) Decision analysis: introductory lectures on choices under uncertainty. Addison-Wesley, Reading,MA

Rakhlin A, Sridharan K, Tewari A (2010) Online learning: random averages, combinatorial parameters, and learnability. In Proceedings of the 23rd international conference on neural information processing systems - volume 2 (NIPS'10). Curran Associates Inc., Red Hook, NY

Raska P, Ulrych Z (2014) Testing optimization methods on discrete event simulation models and testing functions. Procedia Eng 69:768–777. https://www.sciencedirect.com/science/article/pii/S1877705814002999

Rasouli M, Saghafian S (2018) Robust partially observable Markov decision processes. HKS Working Paper No. RWP18-027. Available at SSRN: https://ssrn.com/abstract=3195310 or doi:https://doi.org/10.2139/ssrn.3195310

Riley L (2013) Discrete-event simulation optimization: a review of past approaches and propositions for future direction. SCSC '13 Proceedings of the summer computer simulation conference, Article No. 47. Toronto, ON – July 07–10, 2013. Society for Modeling and Simulation International. Society for Modeling & Simulation International Vista, CA ISBN: 978-1-62748-276-9

Ruijters EJJ, Stoelinga MIA (2014) Fault tree analysis: a survey of the state-of-the-art in modeling, analysis and tools. (CTIT Technical Report Series; No. TR-CTIT-14-14). Enschede: Centre for Telematics and Information Technology (CTIT)

Runolfsson T (2000) Risk-sensitive control of stochastic hybrid systems on infinite time horizon. Math Probl Eng 5(6):459–478. https://doi.org/10.1155/S1024123X99001192

Salze P, Beck E, Douvinet J, Amalric M, Bonnet E, Daudé E, Duraffour F, Sheeren D (2014) TOXI-CITY: an agent-based model for exploring the effects of risk awareness and spatial configuration on the survival rate in the case of industrial accidents. Cybergeo: European Journal of Geography, Systèmes, Modélisation, Géostatistiques, document 692. http://journals.openedition.org/cybergeo/26522; doi: https://doi.org/10.4000/cybergeo.26522

Sánchez-Sánchez C, Izzo D (2018) Real-time optimal control via Deep Neural Networks: study on landing problems. J Guid Control Dyn 41(5):1122–1135

Sanner S (2010) Relational dynamic influence diagram language (RDDL): language description. http://users.cecs.anu.edu.au/~ssanner/IPPC_2011/RDDL.pdf.

Scutari M, Vitolo C, Tucker A (2019) Learning Bayesian networks from big data with greedy search: computational complexity and efficient implementationLearning Bayesian networks from big data with greedy search: computational complexity and efficient implementation. Stat Comput 29:1095. https://doi.org/10.1007/s11222-019-09857-1

Shani G, Pineau J, Kaplow R (2013) A survey of point-based POMDP solvers. Auton Agent Multi-Agent Syst 27(1):1–51. https://doi.org/10.1007/s10458-012-9200-2

Shpitser I, Tchetgen ET (2016) Causal inference with a graphical hierarchy of interventions. Ann Stat 44(6):2433–2466. https://doi.org/10.1214/15-AOS1411

Shvo M, Sohrabi S, McIlraith SA (2018) An AI planning-based approach to the multi-agent plan recognition problem. In: Bagheri E, Cheung J (eds) Advances in artificial intelligence. Canadian AI 2018. Lecture Notes in Computer Science, vol 10832. Springer, Cham

Singh A, Yang L, Hartikainen K, Finn C, Levine S (2019) End-to-end robotic reinforcement learning without reward engineering. Electrical Engineering and Computer Sciences University of California at Berkeley Technical Report No. UCB/EECS-2019-40. https://arxiv.org/pdf/1904.07854.pdf

Smith T, Simmons R (2005) Point-based POMDP algorithms: improved analysis and implementation. Proceeding UAI'05 Proceedings of the twenty-first conference on uncertainty in artificial intelligence, pp 542–549, Edinburgh – July 26-29, 2005 AUAI Press Arlington, VA, ISBN:0-9749039-1-4. https://arxiv.org/ftp/arxiv/papers/1207/1207.1412.pdf

Sofronidis NE (2004) Undecidability of the existence of pure Nash equilibria. Econ Theory 23(2): 423–428. https://doi.org/10.1007/s00199-003-0394-z

Sörensen K, Glover FW (2013) Metaheuristics. In: Gass SI, Fu MC (eds) Encyclopedia of operations research and management science. Springer, New York, NY, pp 960–970

Sousa-Pinto JM (2017) Decidability boundaries in linear dynamical systems (PhD thesis). University of Oxford, Oxford

Sprague CI, Ögren P (2018) Adding neural network controllers to behavior trees without destroying performance guarantees. *ArXiv, abs/1809.10283.*

Srivastava S, Russell S, Ruan P, Cheng X (2014) First-order open-universe POMDPs. UAI'14: Proceedings of the thirtieth conference on uncertainty in artificial intelligence. July 2014, pp 742–751. Morgan Kaufmann Publishers Inc. 340 Pine Street, Sixth Floor San Francisco, CA. https://people.eecs.berkeley.edu/~russell/papers/uai14-oupomdp.pdf, Last accessed 9-15-20.

Sutton RS, Barto AG (1998) Introduction to reinforcement learning. MIT Press/Bradford Books, Cambridge, MA

Torreño A, Onaindia E, Komenda A, Štolba M (2017) Cooperative multi-agent planning: a survey. ACM Comput Surv 50(6): Article 84 (Nov 2017), 32 pages. Doi:https://doi.org/10.1145/3128584.

Wolfram S (1983) Statistical mechanics of cellular automata. Rev Mod Phys 55(3):601–644. https://doi.org/10.1103/RevModPhys.55.601

Wolfram S (1985) Undecidability and intractability in theoretical physics. Phys Rev Lett 54(8): 735–738

Yang F, Lyu D, Liu B, Gustafson S (2018a) PEORL: integrating symbolic planning and hierarchical reinforcement learning for robust decision-making. In Proceedings of the 27th international joint conference on artificial intelligence (IJCAI'18). AAAI Press, pp 4860–4866

Yang F, Jin T, Liu T, Sun X, Zhang J (2018b) Boosting dynamic programming with neural networks for solving NP-hard problems. Proc Mach Learn Res 95:726–739. http://proceedings.mlr.press/v95/yang18a/yang18a.pdf

Yin Q, Yue Q, Zha Y, Jiao P (2016) A semi-Markov decision model for recognizing the destination of a maneuvering agent in real time strategy games. Math Probl Eng 2016 |Article ID 1907971 | 12 pages | doi:https://doi.org/10.1155/2016/1907971.

Yoon S, Ruml W, Benton J, Do MB (2010) ARTICLE Improving determinization in hindsight for online probabilistic planning. ICAPS'10: Proceedings of the twentieth international conference on international conference on automated planning and scheduling. AAAI Press, pp 209–216

Zandonà A, Vasta R, Chiò A, Di Camillo B (2019) A dynamic bayesian network model for the simulation of amyotrophic lateral sclerosis progression. BMC Bioinform 20(Suppl 4):118. https://doi.org/10.1186/s12859-019-2692-x

Zhang NL, Zhang W (2001) Speeding up the convergence of value iteration in partially observable Markov decision processes. J Artif Intell Res 14:29–51. https://arxiv.org/pdf/1106.0251.pdf

Zhang Z, Fu Q, Zhang X et al (2016) Reasoning and predicting POMDP planning complexity via covering numbers. Front Comput Sci 10:726–740. https://doi.org/10.1007/s11704-015-5038-5

Zhang Q, Yao J, Yin Q, Zha Y (2018) Learning behavior trees for autonomous agents with hybrid constraints evolution. Appl Sci 2018(8):1077

Zhang A, Lipton ZC, Pineda L, Azizzadenesheli K, Anandkumar A, Itti L, Pineau J, Furlanello T (2019) Learning causal state representations of partially observable environments arXiv preprint arXiv:1906.10437

Zhou Z, Kearnes S, Li L et al (2019) Optimization of molecules via deep reinforcement learning. Sci Rep 9:10752. https://doi.org/10.1038/s41598-019-47,148-x

Chapter 5
Decision Theory Challenges for Catastrophic Risks and Community Resilience

Introduction

Extreme and catastrophic events are notoriously challenging to learn from, prepare for, and protect against. They are rare and unfamiliar—the bigger the loss, the less frequent and familiar catastrophes of that magnitude tend to be. This makes them hard to envision and plan for adequately in our daily lives, and perhaps a natural area in which to consider applying AI-ML methods to assist human planning and decision-making (Guikema 2020). This chapter focuses on some of the challenges to normative decision theory that any decision support system for catastrophes and disasters must confront, especially that disasters are often inherently unpredictable, in that past data does not enable credible early warnings of the approximate time, place, or magnitude of the next occurrence. This unpredictability arises even under ideal conditions, with unrestricted access to all past data and computational power and modeling expertise to analyze it, largely because causes cannot always be discerned in advance. Seemingly trivial events sometimes precipitate large consequences, such as massive avalanches, forest fires, power blackouts, stock market slides, epidemics, or wars, even though they usually do not (Mandelbrot 1964). Several examples are discussed in this chapter.

Incentives are seldom structured to facilitate managing rare (and possibly hypothetical) catastrophes. Investing scarce resources to prepare for extreme events that seldom or never occur in a community's lifetime can place those who take such precautions at a competitive political or business disadvantage compared to those who do not. Moreover, when disasters strike, compassion moves us to rush help to victims, regardless of blame or calculations about whether optimal precautionary investments were made. This creates a degree of moral hazard, in which potential victims expect that others will help when and if needed, even if expensive precautions and mitigation measures were not purchased in advance.

L. A. Cox Jr., *AI-ML for Decision and Risk Analysis*, International Series in Operations Research & Management Science 345,
https://doi.org/10.1007/978-3-031-32013-2_5

Conversely, after the fact, the recent horror of a catastrophe can stimulate passionate and expensive (but not necessarily effective) attempts to prevent similar occurrences in future. These efforts last while the memory remains vivid, but seldom with a dispassionate eye toward calculating optimal risk reductions achieved for resources spent. In general, societies often spend too little in prevention or mitigation, and too much in reaction, to rare disasters, as judged in hindsight. Experience and incentives do little to encourage preparing well for disasters before they occur, and, indeed, disaster relief has been consuming a steadily growing share of national resources for several decades, as expensive relief efforts continue to dominate much less expensive preparation and avoidance activities (Michel-Kerjan and Slovic 2010).

These behavioral generalities are matched by challenges to the normative theory of how people (or AI-assisted teams) ideally *should* assess and manage rare but potentially catastrophic risks. One of the simplest and most useful formulations of small decision analysis problems is the *normal form*, which associates with each choice of an act a from a set A of feasible acts, and with each state s from a set S of possible states, a consequence $c(a, s)$ in a set C of possible consequences. If preferences for consequences are represented by a von Neumann-Morgenstern utility function and beliefs are represented by subjective probabilities for states (or probability measures for events, i.e., subsets of states), then consistency with various normative axioms implies that one should prefer acts that maximize expected utility (EU) to acts that do not (Luce and Raiffa 1957).

The following sections argue that this traditional decision-analytic conceptual framework, although highly useful for many purposes, is not ideally suited to analyzing and improving risk management decisions for catastrophic events. Informative probabilities for states cannot necessarily be assessed; realistically parsimonious descriptions of acts and consequences may mislead; coherent aggregation of individual beliefs may be impossible; and coherent preferences for acts do not necessarily exist, especially when social or moral norms link the behaviors and preferences of different individuals (Elster 2007). Thus, risk management decision-making for catastrophic risks needs different foundations to complement those of traditional decision analysis and EU theory.

A possible source of new foundations for catastrophe risk management comes from observations and models of how *communities* make decisions about when and how to prepare for disasters, and how they recover (or fail to do so) after disasters happen. Treating communities of interacting agents, rather than individuals, as the units of risk management decision-making suggests new primitives for normative decision theory, such as cooperation, coordination, organization, responsibility, trust and trustworthiness of individuals and institutions within a community, rather than the primitives (e.g., individual preferences, beliefs, and risk attitudes) emphasized in normative models of rational individual decision-making (NRC 2006).

The chapter is organized as follows. The following sections review a series of challenging issues for traditional risk assessment and decision-analytic risk management frameworks. Each issue is briefly described, and then illustrated with simple examples (some original, and some based on relevant social science, economics, and decision science literature) intended to clarify how it arises, using only a minimum amount of technical background. These examples illustrate how developing a normative theory of community-based decision-making for catastrophe risk management raises new modeling and prediction issues not encountered in single-decision-maker decision theory. We then reviews insights from an alternative framework, community based disaster risk management. Normative theories for community-based risk management focus on social and group-level variables (e.g., extent of shared expectations and behavioral norms for responding to crises; capacities to communicate, coordinate, and take local action effectively when needed) that have no counterparts in individual decision theory, but that appear promising for better understanding and prescribing how to more effectively prepare for, and respond to, catastrophic risks.

Challenges of Rare Catastrophic Events to Traditional Analytical Methods

Rare and catastrophic events pose challenges to traditional decision and risk analysis due to their unpredictability; the difficulty of adequately describing, envisioning, or evaluating their consequences; and the difficulty of organizing (or defining) coherent and effective responses among the many affected individuals. Some aspects of these challenges are explained and illustrated next.

Unpredictability of Catastrophes in Physical, Biological, and Social Systems

Axiomatic subjective expected utility (SEU) theory establishes conditions under which a decision maker with coherent preferences should act as if his or her beliefs are represented by probabilities for states, which then enter into the calculation of expected utilities for acts. An important part of applied decision analysis deals with techniques for eliciting, calibrating, and de-biasing such subjective probabilities for events that affect the consequences of decisions. However, probabilities for some events, if they are calculated from any finite amount of past data, may have little value (Howard 1966) for predicting some future events, as the future events are statistically independent of any finite amount of past data. This unpredictability is especially relevant for rare and catastrophic events.

An insight from complex systems theory is that *seemingly identical causes can have vastly different effects*, so that the approximate effects of some causes cannot be predicted. For example, the future effects of a specific initiating event or set of initial conditions cannot always be predicted to within one (or more) orders of magnitude, even from perfect knowledge of how a system operates and from extensive observational data. The system's future behavior can be *inherently unpredictable*, meaning that what the system actually does cannot be predicted any better with past data than without it (i.e., the mutual information between predicted and true outcomes is zero), no matter how the available information is used to make predictions.

Example: Self-Organizing Criticality Makes the Size and Timing of System Responses Unpredictable

Theoretical and, to a lesser extent, empirical studies of avalanches and landslides, wildfires, earthquakes, financial crashes, wars, epidemics, electric power blackouts, and species extinctions suggest that local interactions among components in many such complex systems (e.g., among grains in a pile, slip surfaces in an earthquake, individuals in biological populations, or traders in a market) can lead the system as a whole to a state of "self-organized criticality" (SOC) (Bak et al. 1988; Buchanan 2001). In such a state, an additional small stimulus (e.g., dropping a single additional grain of rice onto a pile) can trigger a response (e.g., an avalanche) of unpredictable magnitude. More precisely, simple models of such systems predict scale-invariant frequency distributions of response sizes (i.e., power law, or Pareto-Levy "heavy-tailed" distributions). This implies that an initiating event (e.g., a single falling grain) can cause a response (an avalanche) of any size, across many orders of magnitude. Such systems have no typical size scale for the response caused by an initiating event, and no typical time scale for the time between extremely large responses, which might be called "catastrophes." Empirically, the frequency distributions of sizes for many catastrophic events—including intense rain falls, fire damage, war casualties, electric grid blackouts, and unexpected losses in insurance and financial markets—have often been found to follow such heavy-tailed distributions (e.g., Mandelbrot 1964). For these distributions, past means and standard deviations do not provide useful guides to what to expect for future losses, since losses much larger than any previously observed ones will continue to occur, and sample means and sample variances of the empirical loss distribution will continue to increase, as experience accumulates. Adverse events with magnitudes many sample standard deviations greater than the previous historical sample mean occur far more often than a normal distribution would predict. These occurrences are large enough to substantially increase the historical mean of the severity distribution when they occur.

Example: Poisson Arrival of Rare Catastrophic Events

Abaimov et al. (2007) characterize the statistical distribution of times between large earthquakes (e.g., those of size 7 or larger on a Richter scale) in "slider-block" models of earthquake activity and energy dissipation. Times between large slip events are exponentially distributed (provided that the systems are not so stiff, i.e., coupling among plates is not so strong, that system-wide events occur). Thus, arrivals of large earthquakes follow a Poisson process, with average waiting times between occurrences that can be long compared to the times for communities to rebuild and adapt to (temporarily) earthquake-free conditions.

The exponential distribution of times between catastrophic events holds in many such models of "self-organizing criticality" (SOC) systems, including models of large earthquakes, forest fires, landslides, and other catastrophic events. The exponential distribution of inter-occurrence times implies that the time until the next one cannot be predicted better from detailed analysis of past data than from simply knowing the average time between occurrences. This is due to the "memoryless" property of the exponential distribution, which guarantees that the time until the next large event is statistically independent of all past data, including the time since the previous one (Abaimov et al. 2007; Solow 2005). In this setting, an expert's deep knowledge and ability to model the causal mechanisms of slip processes and their statistics contributes no more value-of-information to support early warnings or decisions about costly precautionary investments than the novice's simple average time between occurrences.

By contrast, smaller and more frequent events often follow different statistical laws (e.g., with approximately Weibull distributions between occurrences). This makes prediction, planning, and risk management much easier. For example, the sum of many small, independent losses over an interval (such as a year) typically follows a normal distribution. This makes it relatively easy to estimate the reserves needed to compensate for such cumulative losses, at any desired confidence level. As another example, if the failure rate of components in a large batch (e.g., for bolts in a bridge) increases with age, then an optimal risk management strategy (minimizing the sum of replacement, deterioration, and failure costs over a time horizon, or per unit time) is often very simple: wait until the components reach a certain age, and then replace them all (Sheu et al. 2011). Other optimized screening, inspection, and intervention scheduling policies for managing risks are routinely used in medicine (e.g., age-specific cancer screening tests), and in reliability and industrial engineering. Such simple and effective time-based risk assessment and risk management tactics are unavailable for rare and catastrophic events in SOC systems with exponentially distributed inter-occurrence times, since the passage of time in these systems provides no information about when the next catastrophe is likely to occur.

Example: Unpredictability in Deterministic Physical and Ecological Models

Before considering probabilistic systems and catastrophes on a human scale, it may be useful to recall that even many deterministic natural systems, with well-understood dynamics and initial conditions, also generate unpredictable outcomes.

In classical mechanics, for example, three or more masses orbiting each other according to Newton's law of gravitation can produce complex orbital trajectories for which it is impossible to predict whether any of them will eventually escape from the gravitational pull of the rest (Aguirre et al. 2001). (Such unpredictability might be of direct interest to catastrophe risk analysis if a possible consequence of escape is eventual collision with Earth, but the point of this example is simply to stress the unpredictability of even fairly simple natural systems.) The reason is that the eventual escape or non-escape of the masses can depend on their *exact* initial conditions (positions and velocities), and exact initial conditions can never be known from any real (finite-precision) measurements. In many other simple mechanical systems, too, any of several different discrete outcomes can occur, starting from any neighborhood (no matter how small) of any point. Improving the precision of measurement of the initial conditions does not improve ability to predict outcomes. Outcome probabilities, conditioned on any finite-precision measurement of initial conditons, remain unchanged by more precise measurements (e.g., Camargo et al. 2010). Similar mathematics describe catastrophic species loss (extinction) in ecosystems (Vandermeer 2004).

Example: Deterministic Chaos Limits Possible Forecast Horizons

Consider an epidemic that spreads in a population of N initially susceptible individuals according to the following deterministic logistic difference equation for an SIS (susceptible-infected-susceptible) disease process (Sae-jie et al. 2010):

$$I(t+1) = k^{*}I(t)^{*}[N–I(t)]$$

Here, $I(t)$ denotes the number of infecteds at the start of week t, N is the population size, and k reflects the infectivity of the epidemic. Assuming that each person is either infected or not at any time, so that $I(t + 1)$ must have an integer value, this continuous model can be rounded to give the following discrete model:

$$I(t+1) = round(k^{*}I(t)^{*}[N - I(t)]),$$

where *round* denotes the function that rounds numerical values to the closest integer. (This changes the dynamics from chaotic to ultimately periodic, for all initial conditions.) Suppose that a surveillance program is able to accurately estimate the fraction of infecteds in the population at the start of any week to three significant digits (better than most real-world surveillance programs). The initial infected fraction at the time that risk management of the epidemic begins is found to be 0.101. Both k and N are perfectly known: $N = 100{,}000$ and $k = 0.00004$. If the epidemic can be reduced to fewer than 0.01% of the population (i.e., 10 cases)

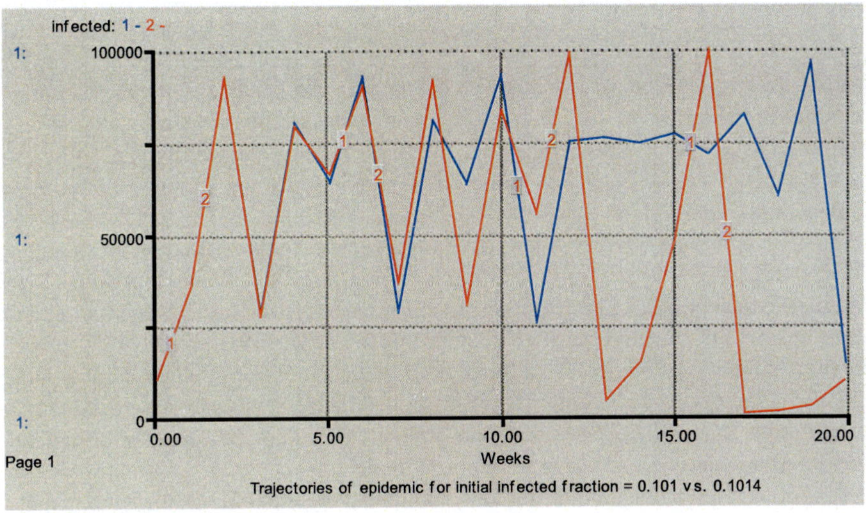

Fig. 5.1 Amplification of Rounding Error Limits Useful Prediction to <15 Weeks

infected at any one time, then it can be eradicated. A question of great interest for planning the epidemic risk management program is: *How many weeks (approximately) will the epidemic last,* before it becomes small enough to be eradicated? The question is especially important if an epidemic that lasts too many consecutive weeks becomes catastrophic, e.g., by exhausting bed spaces in hospitals, stockpiles of medical supplies, or other limited resources. In this case, quantifying the probability of a catastrophe amounts to quantifying the probability that the epidemic will last more than a certain amount of time that the response system can handle.

The information available in this example is almost perfect—complete and accurate to several significant digits. Yet, the answer to the prediction question depends on the information that is lost in the rounding error (beyond the three significant digits available) for the initial infected fraction. For example, one can easily verify with a calculator that the answer is 16 weeks when the initial infected fraction is 0.10117, but 366 weeks when the initial infected fraction is 0.10118. These very different answers correspond to the same measured initial infected fraction, to four decimal places. Thus, even with extremely accurate and precise (to four decimal places) knowledge of how this epidemic spreads, it is impossible to specify how long it will last, to within the correct order of magnitude.

Figure 5.1 shows why not. Two trajectories that start at almost the same point (initial infected fraction of 0.1010 vs. 0.1014, in this example) begin to diverge widely by about 10 weeks, as the dynamics of the system amplify the small differences (in the fourth decimal place and beyond) exponentially, until they are of the same order of magnitude as the phenomenon being studied, i.e., total number of infected cases. The same thing would occur slightly later if even more precise measurements of initial conditions were available (e.g., after 11 weeks for the trajectories that start from initial infected fractions of 0.10117 and 0.10118). The event of interest—the time at which the epidemic ends—lies beyond the *effective forecast horizon* for this system. This horizon is determined by how quickly neighboring trajectories diverge, i.e., how long it takes for small differences in initial conditions (differences lost in the rounding error in this example) to become amplified to the point where they drive the observed trajectory. (The rate of divergence concept can be precisely defined and quantified by the system's Lyapurnov exponent (*ibid*)).

Similar limits to predictability of future events hold in models of weather forecasting, as popularized in the notorious "butterfly effect". In general, sensitivity to exact initial conditions in deterministic chaotic models implies that there is a hard limit to how far ahead currently available information can be used to inform and improve predictions. As soon as differences in initial conditions that are smaller than the resolution of measurement grow to the point where they cause large differences in observed trajectories, more accurate predictions cannot be made by using the measured initial conditions than by ignoring them.

These examples illustrate limitations on ability to predict the timing and magnitude of future catastrophic events. Being unable to predict even the correct order of magnitude of the time and size of the next such event makes effective risk management preparation and planning more difficult than for more routine risks.

Decision Analysis Can Omit Crucial Details in Describing Catastrophes

In light of the limited predictability of catastrophic events in natural, man-made, and hybrid (e.g., social-ecological) systems, how much precaution is desirable? How much productive capacity should each generation devote to protecting against potential catastrophes? How does the answer vary across different locations, where the frequencies and predictability of different disasters differ widely? The usual tools of expected utility theory and decision analysis may not produce convincing answers when disasters threaten entire families, communities, societies, species, or ecosystems.

Imagine someone trying to assess his von Neumann-Morgenstern (NM) utilities for outcomes that include deaths of different family members (e.g., both parents but no children, one parent and all children, etc.) In such difficult settings, it may not even be clear that fewer deaths are always preferable to more, or whether different lives should be valued differently, depending on who else survives. Assessments of von Neumann-Morgenstern utilities for the possible dreadful outcomes of disasters may thus be highly uncertain and speculative. Such critiques of the practical applicability of expected utility theory in extreme, unusual, or unfamiliar settings are not new (see e.g., Chaps. 2 and 13 of Luce and Raiffa 1957), but are especially relevant for catastrophic risks.

Moreover, some traditional analytic formulations of trade-offs among potentially fatal outcomes do not capture everything that might matter to those involved. It is common to quantify probability distributions for number of fatalities, while omitting such details as whether victims die while trying to help others, or while trying to save themselves at the expense of others, and so forth. This does not deny the compelling logic of decision analysis, or that successful applications are still possible when outcomes include fatalities. Rather, it is a warning that standard presentations of risk information (especially about events that cause multiple fatalities) may seem to be sufficiently complete to support well-informed choices among alternatives, but not actually be so.

Example: Risk Curves for Frequency and Severity Do Not Show Risk Equity

A standard probabilistic risk assessment (PRA) tool for describing catastrophic risks is the "risk curve," or "F-N" curve, giving the average annual frequency of at least x fatalities from a catastrophic event, for all $x \geq 0$. (This is usually displayed as a complementary cumulative frequency distribution function on logarithmic axes.) These curves have been proposed as part of a quantitative definition of risk (Kaplan and Garrick 1981). However, they do not reveal information about distributions of individual risks (Thompson Jr. 1988; Gadjos et al. 2009) and have been criticized as inconsistent with expected utility theory (Evans and Verlander 1997).

To illustrate, suppose that a catastrophic event, if it occurs, will kill either 1 person (with probability 0.2), 2 people (with probability 0.3), or 3 people (with probability 0.5). One way for this risk curve to be generated would be for a "small" event to kill person A, a "medium" event to kill A and B, and a "large" event to kill A, B, and C. If these three sizes have respective conditional probabilities (given that the catastrophic event occurs) of 0.2, 0.3, and 0.5, then the corresponding individual risks for A, B, and C from the event are: (1, 0.8, 0.5). That is, A dies if the event occurs at all, C dies only if the event is large (probability 0.5), and B dies only if it is medium or large (probability 0.8). A different way to generate the same risk curve would be for the medium sized event to kill B and C, rather than A and B. In this case, the individual risks for A, B, and C would be (0.7, 0.8, 0.8), instead of (1, 0.8, 0.5). Individual A is no longer sure to die in case the catastrophic event occurs. Some decision-makers may prefer the more equitable distribution of individual risks, (0.7, 0.8, 0.8), to the less equitable distribution, (1, 0.8, 0.5). (The former may be viewed as being derived from the latter by transferring 0.3 fatality risk in the event of a catastrophe from the highest-risk individual, A, to the lowest-risk individual, C.) However, information about the distribution of individual risks is lost when only the risk curve, with respective probabilities (0.2, 0.3, 0.5) for 1, 2, or 3 deaths, is used to summarize risk. Decision-makers presented with risk curves might not know whether the omitted information would change the choices suggested by the risk curves (Thompson 1988; Gadjos et al. 2009).

Ordinary choices about how to respond to disasters may depend on details that are lost in succinct decision-analytic formulations. Moral intuitions about duty, loyalty, teamwork, and other moral and social norms (Elster 2007) shaping choices in a community cannot necessarily be easily summarized in expected utility calculations. A member of a community in the path of a hurricane might assess quite differently the NM utilities of "dying because I risked my life to stay and help" and "dying because I was among those too stubborn or foolish to leave when we had the chance." Which narrative description to use (if either) in deciding whether to flee might not be clear, but could profoundly affect the choice.

Examples such as these do not necessarily reveal flaws in the normative theory of decision analysis. One might, in principle, include additional attributes or constraints (e.g., to augment the traditional F-N curve with information about risk equity and more detailed narrative descriptions of fatalities), if one were aware that these things were missing and considered them important. But it is difficult to objectively map social and moral descriptions of responses to catastrophes (using evaluative terms and concepts such as teamwork, courage, responsibility, and heroism; or, from a different perspective, groupthink, foolhardiness, stubbornness, and irrationality) into more neutral conventional decision-analytic descriptions of acts, states, probabilities, consequences, and utilities. The meanings and values of acts and their consequences, to those directly involved in disaster preparation, response, and recovery, may depend on how they are construed in social and moral terms. It is not always clear how these rich descriptions can or should be translated into the language of expected utility maximization. These examples suggest that the expected utilities of disaster risk management actions or their consequences may depend greatly on how they are described (e.g., using admiring or pejorative terms), and no objectively "correct" description may be known.

Emergent Precautionary Choice Behaviors Can Be Incoherent and Unpredictable

When many individuals must decide for themselves how to prepare for, or respond to, a catastrophe, and their decisions influence each other, then new challenges to normative decision theory arise. There may be no mutually consistent set of individual choices. Patterns of individual choices may then evolve in ways that are difficult to predict, as each individual adapts his own choice to the choices of others, without necessarily approaching any stable configuration.

Example: Coherent Individual Preferences Can Create Incoherent Group Choices

Suppose that each of three people in a community must choose among the following three possible individual behaviors: (A) Invest heavily in defenses against a possible catastrophe (e.g., purchase insurance and take expensive precautions); (B) Invest lightly in defenses against a possible catastrophe (e.g., purchase insurance but do not take expensive precautions); (C) Do not invest in precautions. For purposes of illustrating how difficult prediction can be, suppose that each individual has the same preference ordering for his own investment decision, given what others are doing. Part of that preference ordering is as follows:

1. If the other two member of the community both invest heavily, then I prefer to invest lightly (e.g., because their good examples only partly offset my temptation to free-ride);
2. If one other member invests heavily and the other not at all, then I prefer to invest heavily (following the good example of the heavy investor, or because additional investment still has high marginal benefits);
3. If one other member invests heavily and the other lightly, then I prefer to invest not at all (perhaps feeling less peer pressure to resist the temptation of free-riding, and being conscious of diminishing marginal benefits).

(A possible interpretation is that each individual is torn between the temptation to become a free rider and the desire to contribute to defensive investments of the community; what he does depends in part on the examples set by others. Elster (2007) surveys why rational people have preference orderings that reflect social norms.) This preference ordering implies that the community of three individuals has *cyclic preferences*. There is a cycle of states (each state consisting of three components describing how many individuals invest heavily, lightly, or not at all, respectively) such that each state is reachable from its predecessor by one individual changing his behavior to one which he prefers, given the choices of others. Figure 5.2 shows this cycle (bold states), embedded in a larger preference pattern that leads to it from any starting state.

Starting from state (2, 1, 0), for example, the third preference rule above implies that an individual who has been investing heavily will prefer to invest not at all

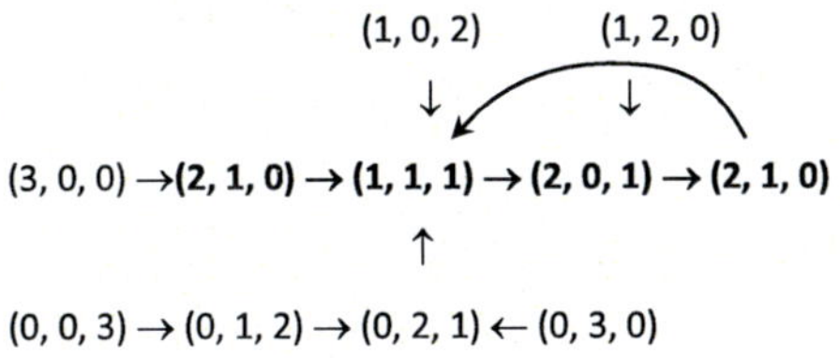

Fig. 5.2 Cyclic community preferences among defensive behaviors. (x, y, z) means that x people invest heavily, y people invest lightly, z people invest not at all. $x + y + z = 3$

instead. When he follows this preference, the state becomes (1, 1, 1). Now, rule 2 implies that the light investor in (1, 1, 1) will prefer to become a heavy investor, moving the community to state (2, 0, 1). Finally, rule 1 implies that the non-investor, confronted with heavy investment by the rest of the community, will increase his own investment to light, moving the state back to (2, 1, 0) and the beginning of a new cycle.

Perhaps the most fundamental prescription of normative decision analysis for individuals is that *one should choose an act to which no other available act is preferred.* Such an act always exists for an individual with a well-defined preference ordering for a finite set of alternative acts (Luce and Raiffa 1957, pp. 25–6). But the example shows that, for groups, there may be no act to which no other available act is preferred. Thus, concepts such as "the community's optimal choice," "collective preference," "general will," or "social utility" for the level of defensive or protective investment are not necessarily well defined. When each individual's own preferred choice depends on what others choose, there is not necessarily any mutually consistent set of choices. Psychologists have shown that individuals in small groups do in fact look to each other's behaviors to decide what is appropriate under conditions of risk and uncertainty (e.g., Cialdini 2001). The existence of cyclic preferences when preferences for acts are contingent on the behaviors of others makesthe outcome of such self-organizing behavior hard to predict. A decision theory that takes such social and moral norms into account (Elster 2007) is needed to understand not only how people respond to catastrophes, and to each others' behavior, but also to understand how they *should* respond, when these norms affecthow decisions are evaluated.

Cyclic preferences can exist for any number of players and choices greater than 2. [With only two choices, an equilibrium always exists (Schelling 1978).] However, even when each player has only 2 choices, allowing each player to watch his neighbors and adopt the locally most rewarding choice in each period can lead to *spatial chaos* (Nowak and May 1992). Even the smallest fluctuation, such as a change (or delay) in one player's choice on one play, is eventually amplified and propagated until it dramatically alters the evolution of the entire system On an infinite grid, the eventual fate of the system is *undecidable*: no algorithm exists that can predict whether one choice (such as a particular choice of investment level) will survive and spread at the expense of the other, from any given starting configuration (Grim 1997). Thus, allowing individual choices to respond to each other, as seems realistic, can induce these two very strong forms of unpredictability (spatial chaos and undecidability) for how the system of interacting individuals will self-organize to determine final levels of collective investment.

Example: Dynamic Inconsistency of Majority Preferences for Costly Precautions

A plausible normative rule for individual decision-making is: *Do not choose an act now that you are certain to regret choosing later.* If you know now that, in future, you will wish that you had chosen *A* instead of *B* now, then you should indeed choose *A* instead of *B* now. We illustrate next how this *dynamic consistency* principle (McClennen 1990) can fail for groups.

Consider a large number of communities trying to decide whether to continue to pay to maintain their expensive local disaster preparedness programs. It costs each community some amount per year to maintain its program, and the program generates benefits (in the form of reduced losses) whenever a certain type of disaster (e.g., hurricane, flood, etc.) strikes a prepared community. If the time T until the next disaster were known, then a community's optimal decision rule would have the form: Pay for the program if and only if $T < T^*$, where T^* years is the *break-even point* past which the costs (e.g., net present value) of continuing maintenance exceed the benefits. Now, suppose that T is a random variable, with a mean greater than T^* and a median less than T^*. (For example, if T is exponentially distributed, then its median is about 69% of its mean.) Members of the community might prefer to halt the disaster preparedness program, since its expected costs outweighs its expected benefits ($E(T) > T^*$), even though it is more likely than not that the actual benefits from the program will exceed its actual costs ($\Pr(T < T^*) > 0.5$). Suppose there are many communities, all in the same situation, but widely enough separated so that their random times until the next disaster are statistically independent. If the number of such communities is sufficiently large, then it becomes a statistical certainty that most of them (and hence the majority of individuals, if the communities are of similar sizes) will obtain a net benefit from maintaining their preparedness programs, even though none expects to so before the fact. In other words, probabilistic risk assessment undertaken before a catastrophe occurs would show that most people will eventually wish that they had invested in preparing for it, even if no one currently favors doing so. Politicians and institutions that consider themselves accountable to, and acting in the best interests of, "the people" must then decide whether it is their *ex ante* or *ex post* preferences that they should serve, if probabilistic risk analysis reveals that pre- and post-catastrophe majority preferences conflict.

That *ex ante* and *ex post* majority preferences may differ in ways that are predictable in advance raises ethical questions about the purpose of democratic government and the proper role of centralized decision-making in such cases. Should leaders choose the act that most people prefer now (i.e., let preparedness programs lapse), or the act that most people will prefer later in retrospect (i.e., maintain the presently unpopular preparedness programs, because fewer people will regret this choice later than would regret the decision to let them lapse)? This tension between present preferences and predictable future preferences pits different normative principles of collective decision-making for catastrophe risk management against each other. Majority rule based on present preferences, as well as social utility theories that aggregate individual utility functions (Hammond 1992), and risk-cost-benefit analyses that compare expected costs and benefits of continued maintenance, would all prescribe halting the programs. They would let communities bear the costs of disaster, rather than the greater average costs of maintaining preparedness. However, majority rule based on predictable future preferences would prescribe maintaining the expensive programs. Governments that fail to do so may become widely unpopular in the wake of disasters in locations where preparedness programs were allowed to lapse, even if letting them lapse was the majority (or unanimous) preference of individuals *ex ante*. This is especially so if hindsight reveals that maintaining the programs would have been cost-beneficial as of the date of a disaster (as will usually be the case, in the above example, since the median time until a disaster is less than the break-even time).

These examples have illustrated that defining coherent "social preferences" for risk management decisions can be difficult or impossible. Self-organizing behavior, in which each individual's preferred choice depends on the choices of others, as well as top-down centralized decision-making that attempts to implement majority preferences (while recognizing that these may be dynamically inconsistent), can both lead to inconsistencies in defining the "socially preferred" level of risk management investments.

Challenges to Normative Group Decision Theory for Risk Management

Results from experimental economics, neuroeconomics and behavioral economics suggest that individuals make poor decisions about catastrophe risk management, for a host of well-documented psychological, organizational, and economic reasons (Michel-Kerjan and Slovic 2010; Gul and Pesendorfer 2008; Thaler and Sunstein 2008). Catastrophes that occur seldom within an individual lifetime can prompt both under-preparation and over-reaction, as judged in retrospect. Economic analyses suggest that there is often too little individual concern and investment in risk management before the fact, and perhaps too much public concern and investment after the fact (Michel-Kerjan and Slovic 2010).

Following a catastrophe, emotionally charged, vivid outcomes and media images promote probability neglect and distort perceptions of the true risks of similar catastrophes. Before hand, however, predictable—and potentially correctable—incentives and psychological biases lead to misallocations of catastrophe risk management concern and resources, as judged in hindsight or by normative economic models (Michel-Kerjan and Slovic 2010). Among these are: (a) overconfidence in ability to control adverse outcomes when they occur; (b) indecision, procrastination, and excessive aversion to acting on uncertain probabilities ("ambiguity aversion"); (c) distorted incentives to take care (such as agency effects, moral hazard, free riding); (d) imperfect learning and social adaptation heuristics (e.g., herd-following, groupthink); (e) distributed responsibility and control in planning, coordinating, and implementing disaster preparedness measures and responses; and (f) difficulties in forecasting, pooling, diversifying, and insuring catastrophic risks.

Confronted with these challenges, it would be useful if present decision science provided a normative reference model for how communities or societies *should* invest in protection against rare and catastrophic events. But this contribution is beyond the current state-of-the-art of decision science for disaster risk management, modeled as a process of group deliberation and collective choice among alternative courses of action with uncertain consequences. Indeed, the main results from collective choice theory are impossibility theorems, showing that no group decision process satisfies several simultaneous desiderata (e.g., Mueller 2003). Separate formation of group beliefs and preferences based on the beliefs and preferences of group members leads to recommendations that violate basic normative criteria for group decision-making. Among these are that a group decision process should depend on the input from more than one member of the group and should not select one risk management intervention if everyone prefers a different one (Hylland and Zeckhauser 1979; Nehring 2007).

Example: Aggregating Individual Beliefs Can Lead to Group Risk Management Decisions that No One Likes

Suppose that members of a community must decide whether to pay for an expensive levee to protect against possible flooding in the event of a hurricane. Each individual believes that the benefits of the proposed levee will exceed its costs if and only if two conditions hold: (a) A hurricane powerful enough to cause flooding in the absence of the levee occurs within some time frame of interest; and (b) The levee does not fail during the hurricane. Everyone agrees that they should pay for the levee if and only if probabilistic risk assessment (PRA) shows that the joint probability of events (a) *and* (b) exceeds 20%. They agree to use the mean probabilities assessed by community members (or of experts who serve the community), for each event, (a) and (b), to aggregate different beliefs. Suppose that half of the probably judgments are relatively pessimistic: they assign probability 0.8 to event (a) (hurricane occurrence) and probability 0.2 to event (b) (levee performs). The other half are more optimistic: they assess a probability of only 0.2 for event (a) and a probability of 0.8 for event (b). (For example, the pessimistic group might consist of groups with more fear of natural disasters and less trust in engineering solutions than the optimistic group.) The average probability for event (a) is (0.8 + 0.2)/2 = 0.5, and the average probability for event (b) is (0.2 + 0.8)/2 = 0.5, so these group probability assessments imply that the joint probability of events (a) and (b) is 0.5*0.5 = 0.25. Since this is above the agreed-to decision threshold of 0.2, the levee would be built. On the other hand, every individual computes that the joint probability of events (a) and

(b) is only $0.8*0.2 = 0.16$. Since this is below the decision threshold of 0.2 required for projected benefits to exceed costs, no individual wants the levee to be built. Thus, aggregating individual beliefs about events and applying PRA to decide what to do leads to a decision that no one agrees with. Nehring (2007) generalizes such examples, showing that using a more sophisticated approach than averaging cannot overcome the problem: there is *no possible* way to aggregate individual beliefs and use them to make group decisions that guarantees avoiding making decisions that no one wants (other than such trivial ones as selecting a single individual as a "dictator" and ignoring everyone else's beliefs). For *any* aggregation and decision rule that treats individuals symmetrically (e.g., using geometric means, instead of arithmetic means, which would resolve the above example; or using rules that do not involve any form of averaging) there are choice problems for which the group's decision is not favored by any of its members.

Traditional normative decision science does not provide clear concepts for defining what the "best" risk management decision is in such settings. Principles such as unanimity, or Pareto-consistency of group preferences with individual preferences, may have to be replaced to develop a more useful approach to collective decision-making about how to defend against uncertain hazards.

This concludes our survey of challenges to normative decision theory as a practical guide for use in managing risks of catastrophic and extreme events. We turn next to some recent alternatives that forego (for now) the advantages of rigorous axiomatic foundations and mathematical clarity found in decision analysis. Instead, they seek principles for successful risk management in the experiences of communities that have fared more or less well in preparing for, and recovering from, disasters.

Toward a New Foundation for Disaster Risk Management: Building Disaster-Resilient Communities

Confronted with the practical need for a better basis for disaster risk management, many disaster aid researchers and practitioners have started to formulate ideas about what constitutes a *disaster-resilient community*, and how to encourage the deliberate creation and maintenance of such communities (Adger et al. 2005; Norris et al. 2008). In contrast to axiomatic models and theories of rational individual and collective choice, such resilience theories typically take communities, rather than individuals, as the actors of principle interest. Collective decision-making skills before, during, and after a disaster affect the probabilities of different outcomes. As stated in a National Research Council study on Facing Hazards and Disasters (NRC 2006), "Note that what is being discussed here are *group-level* deliberations and decisions, not individual ones. Actions under conditions of uncertainty and urgency such as those that accompany disaster warnings should not be conceptualized in individualistic terms." How communities respond to hazards, warnings, and imminent dangers depends on the psychology and sociology of mutual influence, formation of collective beliefs, and creation of new norms for novel or unfamiliar situations (Elster 2007). Expected utility calculations for individuals shed little light on these aspects of group dynamics and community-level decisions.

Community resilience theories and indices usually focus less on optimal planning of responses to possible future scenarios than on the capacities of communities to respond well to crises when they occur. Individuals are no longer viewed as autonomous bundles of preferences and beliefs, but as parts of an interrelated and interdependent community. Their values, skills, and capacities to respond in a crisis are developed largely through social interactions and networks. The healthy functioning of the community over a range of stresses, including responses to and recovery from occasional rare catastrophes, depends largely on how well its members can adapt together to changing circumstances.

Example: Resilient Response to the North Sea Flood of 1953

On the night of Saturday, January 31, 1953, an 18-foot rise in sea level, due to a combination of high tides and storm surge, overwhelmed dikes in the Netherlands, eventually causing 1836 deaths and the evacuation of about 72,000 people in the Netherlands. The disaster could have been—and almost was—thousands of times more deadly, had it not been for the resilient responses of those involved. Most of South Holland and all of North Holland, including about three million people sleeping in Rotterdam, were saved, even thoughthe sea broke through the final dike protecting them. A local mayor and a river ship captain agreed to steer the captain's ship into the gap, plugging the break in the dike. Within hours, a volunteer network of amateur radio operators organized themselves into a working emergency communications network, which they staffed for the next 10 days. Within a week, over 30,000 volunteers had mobilized to repair the dikes and distribute aid pouring in from other areas and countries, and the flooded region was well on its way to recovery (SEMP 2006).

This experience illustrates key features of a resilient response. They include prompt local improvisation (the improvised use of a river boat to dam the break in the ruptured dike probably saved millions of lives); quick and effective self-organization of experts (the local amateur radio operators) to help meet the needs of the crisis; massive but well-managed deployment of volunteers in a well-organized structure (working with dike engineers and personnel to quickly repair the many miles of damaged sea walls and dikes); and effective use of the abundant disaster relief aid provided by others from outside the disaster area.

Current research on "community-based disaster risk management" (Pandey and Okazaki 2005) seeks to help communities at risk from various types of natural disasters (including cyclones and earthquakes, as well as floods) to exhibit similar self-reliance, cooperation, and effective use of volunteers and aid, as well as community-based risk analysis and planning before disaster strikes, to help reduce risks and losses from natural disasters.

Several principles for successful effective community organization have been proposed and empirically tested. Principles with substantial empirical support include: clearly defining boundaries for communities and for the systems they manage; establishing local authority (rights) and collective choice procedures for participants to modify their own operating rules in light of information about local conditions; instituting low-cost and accessible conflict-resolution mechanisms; and monitoring performance of agreed-to duties, with accountability to other members of the community (Cox et al. 2010). In this view, being well prepared to cope with disasters has little to do with optimization and much to do with ability to work and adapt effectively together in the face of dangers and surprises. Community resilience is bolstered by economic development and infrastructure; effective coordination in cooperative problem solving; competent and effective leadership; and "social capital" and support networks that help to create, reinforce, and express values and norms that encourage individuals to trust and help each other (and to be trustworthy and helpful), even under pressure. Key resources promoting community resilience are sometimes described in terms of various types of "capitals," such as social, economic, ecosystem, education, and infrastructure that can help to prepare for, respond to, and recover from disasters.

No analytic framework comparable in elegance and precision to SEU theory yet exists for disaster resilience theory. Defining key terms and concepts such as "resilience," "vulnerability," "restoration," and "capacity" is an ongoing challenge for complex and interlinked systems (Haimes 2009). Academic theories of disaster-resilient communities, and techniques for fostering them, are in their infancy. Yet, the approach appears promising as a possible start for a new branch of decision science. This branch would overcome some of the challenges and limitations of traditional normative decision theories. It would also integrate and explain some empirical findings on how people prepare for (or fail to prepare for) and respond to disaster risks. For example,

1. Postulating that communities or groups, rather than individuals, are the basic units of risk management decision-making suggests some testable hypotheses. One is that individuals should have brain mechanisms that support effective community decision-making. For example, they may derive satisfaction from successful cooperation and adaptation even in the face of obvious incentives and uncertainties that would tempt merely individually rational individuals to cheat. This prediction is consistent with recent findings from brain science, neuroeconomics, and experimental psychology showing that "irrational trust" and "irrational altruism" indeed generate high levels of satisfaction in most people. They enable groups of real individuals to substantially out-perform purely rational players (e.g., those following the prescriptions of game theory and individual decision theory) in situations where incentives to cheat are strong (Elster 2007). Investing in precautionary measures and filling needed roles during a crisis are examples of such situations.

2. Considering relatively long-lived (compared to individuals) communities as the most relevant units of risk management decision-making for rare, catastrophic events highlights the importance of institutions, norms, and mores that help to bridge the interests and coordinate the decisions of multiple generations. It is commonly observed that, "Human beings are rule-following animals by nature; they are born to conform to the social norms that they see around them, and they entrench those rules with often transcendent meaning and value" (Fukuyama 2011). Norm-based community decision-making is facilitated and enforced by individual motivations and public institutions that reflect social concepts such as duty, rights, responsibility, law, honor, and stewardship. Conversely, shame for violating norms or earning widespread disapproval for behavior can help to harmonize individual decisions with community needs over time. This may reduce the exploitation of future generations and excessive discounting of far-future consequences that plague many models of individually rational decision-making (Elster 2007). Successive generations within a community or society may be viewed as investing in maintaining and improving the common resource of social capital and local decision-making skills and infrastructures needed to cope with occasional rare catastrophes. Most will never experience such catastrophes, but all have an in interest in being able to respond to and recover from them if necessary. On the negative side, our increasingly interconnected world may generate new potential catastrophes, interdependencies, and vulnerabilities to natural disasters (e.g., in supply chains) faster than entrenched mores and institutions adapt to cope with them. Then dysfunctional, outmoded, but stable institutions and habits can reduce ability to effectively manage disaster risks (Elster 2007; Fukuyama 2011).
3. Acknowledging that individuals are intrinsically social creatures, with preferences and values formed largely through interactions with others (Cialdini 2001), helps to resolve the apparent paradox that altruism and cooperation are much higher in many laboratory experiments—especially for participants from high-trust, high social capital cultures—than predicted by traditional models of rational choice (Axelrod 1990; Elster 2007).
4. Resilient community theory views risk management as *building community capacities* to adapt to, respond to, and recover from adverse events. By emphasizing building capacity to act effectively when needed, rather than focusing solely on choosing acts (as in the expected utility formalism), resilient community theory sidesteps some of the difficulties involved in trying to predict and prepare for specific disasters. It also avoids many of the paradoxes and pitfalls, discussed earlier, that arise in normative models of how groups should choose acts. Specifically,

   (a) Rather than focusing on predicting disasters, resilience theory seeks to build capacity to respond to them effectively, whenever they may occur. Building resilience in communities, and in other vulnerable systems (Haimes 2009), simultaneously reduces the potential for harm from many types of catastrophes, even without knowing which one is most likely to occur next.

Increasing the different capitals needed for resilience—for example, through investments in infrastructure, ecosystems, social capital, and economic development—may also improve the lives of community members in other ways, making it possible to combine risk management with other priorities.

(b) Ability to respond effectively is not primarily based on envisioning scenarios and identifying optimal actions for them before the fact, but on reacting well to events as they unfold. This reduces the need to envision and describe a full set of hypothetical consequences well enough to assess utilities.

(c) Group decision-making is no longer conceived of as either a mathematical procedure for aggregating individual preferences and beliefs, or as a decentralized process through which individuals adjust their own choices in response to the choices of other. Rather, participation in community decisions is guided by principles such as those already described (e.g., clear local authority, accountability to the community, and so forth) (Cox et al. 2010).

The resilience approach complements, but does not substitute for, traditional decision analysis and probabilistic risk assessment in aiding risk management. For example, admonishments to make investments that build disaster-resistant communities do little to answer specific questions, such as whether or how much a community should invest in building and maintaining a levee to protect against possible (but unlikely) catastrophic flooding. How best to use risk assessment and decision analysis to build resilience remains to be worked out. Instead of focusing on the usual constructs of individual decision theory—that is, states (and their probabilities), consequences (and their utilities), and acts (and their expected utilities)—community resilience approaches focus on the shared infrastructures and institutions and the moral, psychological and sociological variables that bind individuals into communities. These shape and coordinate their expectations and behaviors before, during, and after emergencies.

Emphasizing resilience also changes the kinds of questions that risk managers pay most attention to. For example, a common first step in analyzing a community's disaster preparedness is to identify different disaster scenarios (or states of nature) that might occur, along with alternative plans and courses of action to be followed if they occur. (This approach is well illustrated in the United Nations online educational game *Stop Disasters!*) A less analytic but more pragmatic first step might be to use realistic drills and exercises to find out how well the members of a community can generate options and make effective decisions, under time pressure and uncertainty, when disaster strikes. (After all, although it is unlikely that *ad hoc* innovations, such as plugging a broken dike with a river boat, would appear on any *a priori* listing of plausible scenarios and acts, such improvisation may be essential to prevent disasters from escalating or from costing more lives than necessary.)

Similarly, hazard forecasting and probabilistic risk assessment of state or scenario probabilities may have limited value for hard-to-predict events such as fire, flood, earthquake, landslide, storm, terrorist attack, infrastructure failure, or combinations of these. It may save more lives to focus instead on assessing and improving the ability of a community to marshal and deploy its resources (including local expertise, volunteers, and aid received following a disaster) effectively across a wide range of challenging conditions. A community that can sustain essential life-line services (e.g., providing clean water, food, shelter, power and fuel, communications, and emergency medical treatment) for most of its members under a wide variety of disruptive events is more likely to be able to manage and limit the harm from unforeseen events than a less capable community. Assessing and improving the robustness of these capacities (and of their supporting infrastructures, institutions, and mores) to a wide range of disruptions, may be more directly useful than better decision analysis and planning for improving risk management and reducing disaster-related fatalities and losses.

Decision analysis and community resilience offer complementary perspectives and recommendations on disaster risk management. *Ex ante* decision analysis and planning can help to identify and put in place the most crucial resources, emergency shelter and evacuation plans, first-responder training and response plans, and supporting infrastructure investments. Community-resilience and community-based disaster risk management approaches can help to assure that these resources and investments are used effectively when needed. Both approaches are useful in overcoming well-documented psychological biases, such as over-confidence, illusion of control, and optimism biases (Hilton et al. 2011), that might otherwise undermine needed preparation and resilience-building efforts. Neither is the only practical way forward, or guaranteed to work well in all cases. Rather, community-based approaches draw attention to a different set of variables from traditional analysis, emphasizing how community members influence each other and collaborate with each other (and with outsiders) in carrying out plans to deal with anticipated catastrophes; in creating and implementing new plans on the fly, if needed; and in coordinating and managing responses to, and recovery from, both foreseen and unforeseen contingencies, using the resources and opportunities at hand. This attention to community-level variables and performance provides a promising complementary conceptual framework to that of individual decision theory, as a basis for assessing and improving catastrophe risk management skills.

Bistability and the Evolution and Collapse of Social Cooperation

When Hurricane Katrina ravaged New Orleans, some early news footage and media stories presented sensational accounts of looting, violence, shooting at police, and the apparent collapse of civil order. Although the more extreme stories later proved false, the possibility of social collapse in the wake of catastrophes is a risk worth understanding. What factors explain the resilience of social order and the rule of law in communities hard hit by disasters? Conversely, when is the risk of disorder greatest?

Although full answers to these questions are hard to come by, game theory offers some simple models and insights into conditions for sustaining high trust and cooperation in populations of interacting individuals (e.g., Elster 2007). These include the following explanations for cooperative behavior, which progress from simple formal models of cooperation among purely rational players to the roles of mores and morals in sustaining cooperation—or in allowing it to collapse—in more realistic (evolved) populations.

- *Trustworthy behavior as a strategic equilibrium.* Suppose that each individual, in each of many pair-wise transactions with others, can either behave trustworthily or not. Suppose that incentives are such that each player does best (gains maximum payoff) by acting trustworthily when the other player does so too; does second best by acting untrustworthily when the other player acts trustworthily; does third best when neither acts trustworthily, and does worst by acting trustworthily when the other does not. An example of such an incentive pattern, in standard bi-matrix game notation (in which player 1 chooses a row, player 2 chooses a column, and the resulting cell of the table identified by this pair of choices shows the payoffs to players 1 and 2 in that order) is as follows:

Stag Hunt Game

	Player 2 cooperates	Player 2 defects
Player 1 cooperates	3, 3	0, 2
Player 1 defects	2, 0	1, 1

[It is conventional to refer to acting trustworthily in such games as "Cooperate" and to acting untrustworthily as "Defect," and we follow this usage. In evolutionary game theory, it is common to call the former "Dove" and the latter "Hawk.") The incentive pattern can then be summarized as: (*cooperate*, *cooperate*) > (*defect*, *cooperate*) > (*defect*, *defect*) > (*cooperate*, *defect*), where the inequalities reflect the relative sizes of the payoffs to the player who uses the first strategy in each pair, if the other player uses the second. Thispattern of incentives is known in game theory as the Stag Hunt Game (Helbing and Johansson 2010).] With these incentives, both (*cooperate*, *cooperate*) and (*defect*, *defect*) are pure-strategy Nash equilibria. The first is better for both players: each gains the maximum possible payoff by cooperating, e.g., 3 instead of 1, in the above example. Unfortunately, however, mutual defection is also a self-sustaining equilibrium: a player loses by being the only one to deviate from it (since (*defect*, *defect*) > (*cooperate*, *defect*), e.g., 0 is less than 1 in the above example). When players in a populations are repeatedly paired at random to play this game, evolutionary game theory shows that both of these two possible pure-strategy equilibria, (*cooperate*, *cooperate*) and (*defect*, *defect*), are also *evolutionarily stable*, as each a best response to itself. Moreover, there is a *cooperation threshold* such that, once the fraction of the population which acts trustworthily exceeds that threshold, all players maximize their own expected payoffs in each random encounter by acting trustworthily. Conversely, if the fraction of the population which acts trustworthily falls below the threshold, then all players maximize their own expected payoff by acting untrustworthily. Thus, the system is *bistable*: it can settle in either a high-trust, Pareto-efficient, Nash equilibrium (corresponding to (3, 3) in the abaove game), or in a lower-trust, Pareto-inefficient, Nash equilibrium (corresponding to (1, 1) in the above game). Interpretively, "social capital" (e.g., the fraction of the population that acts trustworthily in each random transaction) must reach a certain critical level in order to become self-sustaining. Otherwise, those who act trustworthily become "suckers," losing on average compared to those who do not. For such a bistable system, a shock that sends the level of social capital below the threshold can cause the high-trust equilibrium to become extinct as untrustworthy behavior becomes the new norm.

- *The Folk Theorem.* In repeated games (including repeated Prisoner's Dilemma), Pareto-efficient outcomes (and other individually rational outcomes) can be sustained by a mutual agreement to punish defection by withholding future cooperation. [This is the "Folk Theorem" for repeated games (Fudenberg and Maskin 1986).] However, this threat only works, thereby sustaining cooperation, if future repetitions of the game are sufficiently likely to occur, and if future payoffs are discounted (compared to immediate payoffs) at a sufficiently low rate. Empirical studies of poor communities and individuals show that their futures are typically both highly uncertain and heavily discounted (Banerjee and Duflo 2011). These conditions undermine the possibilities for creating self-reinforcing, credible (sub-game perfect equilibrium), mutually beneficial agreements to cooperate, using as incentives the threat of future exclusion from the benefits of cooperation if one defects. Under such conditions, the benefits of high-trust equilibria (or high "social capital") and sustained cooperation may be difficult or impossible to achieve. If a community or society that has been enjoying the fruits of sustained cooperation finds that continuity of future business relationships (or other repeated relationships) is no longer highly likely, ability to secure and maintain ownership of future gains from cooperation is no longer assured, or discount rates are higher than they used to be, then the foundations for rational sustained cooperation are weakened. Continued cooperation may no longer be rational in the face of temptations to achieve short-term gains by defecting, despite the possibility of future losses from cooperation foregone.
- *Mechanism design.* In order to induce players to reveal private information (e.g., truthful revelations of willingness to pay for a shared public good) for use in a public choice decision rule (or "mechanism"), it is in general necessary to make credible threats that put positive probability on outcomes that no one wants (e.g., refusing to let some players, who declared a low willingness-to-pay, use a public good if it is purchased anyway; or destroying a contested piece of property that can be allocated to at most one player). Carrying out these threats is clearly *ex post* Pareto-inefficient, yet failing to do so would undermine the credibility of the threats, and hence the capacity to use them to elicit Pareto-improving choices (e.g., about what to reveal) *ex ante*. Designing collective choice mechanisms to achieve multiple goals simultaneously, such as *ex post* Pareto efficient outcomes, balanced budget (for mechanisms that determine taxes and subsidies when a

public good is bought, or a public bad (negative externality) is endured), and providing positive individual incentives to participate, is possible only if individual preferences are sufficiently similar, or are otherwise restricted (unless one simply makes one of the individuals a dictator, and seeks to maximize his or her expected utility). Such trade-offs (known in the public choice literature as *impossibility theorems*, such as the Gibbard-Satterthwaite, Green-Laffont and Myerson-Satterthwaite impossibility theorems) set sharp limits to the rational design of public choice mechanisms for perfectly rational players (Othman and Sandholm 2009). They highlight that the collective choice mechanisms themselves may have to impose risks of violating some desirable goals or constraints if they are to achieve others. For example, they may risk *ex post* Pareto-inefficient outcomes (i.e., everyone would have preferred a different choice), loss of freedom (rational individuals must be forced to participate, or to abide by one person's preferences), and/or budget imbalance (more is purchased than can be collected from fees collected from voluntary participants).

- *Evolution and dissolution of cooperation.* Fortunately, real people are more altruistic and cooperative than game-theoretic models of purely rational players ("*homo economicus*") predict. We come equipped with a range of rule-following and cooperative tendencies and pro-social moral and social impulses that help societies and communities function better than might otherwise be expected. For example Gintis et al. (2003) have argued that *strong reciprocity*, meaning "a predisposition to cooperate with others and to punish those who violate the norms of cooperation, at personal cost, even when it is implausible to expect that these costs will be repaid" is an evolutionarily stable strategy (ESS) in simple models. In these models, individuals with a predisposition to cooperate, but also with a predisposition to punish (even at personal cost) those who do not cooperate, can gain more on average (and thus reproduce more successfully) than more selfish individuals. Institutions and social learning, as well as genetic predispositions, can increase gains from cooperation by punishing defectors (Henrich 2006), although costly punishment creates risks, especially when information about defection is imperfect. Although strong reciprocity makes possible more altruism and cooperation than could be justified by simpler mechanisms such as kin selection and reciprocal altruism, it also makes societies of strong reciprocators vulnerable to sudden drops in cooperation if perceptions that some members have defected (perhaps stoked by rumor, gossip, or, media reports, regardless of accuracy) lead to costly cycles of punishment and revenge (Guala 2012). Dynamic models of evolving social interactions within and between groups with different preferences or beliefs, with imitative learning within groups, indicate the possibility of *phase transitions*, in which social cooperation abruptly breaks down, societies become increasingly polarized, or revolutions and fragmentation occur as groups abandon cooperation with the rest of society to pursue their own interests (Helbing and Johansson 2010).

Summary and Conclusions

This chapter has illustrated challenges for the application of single-person decision theory to extreme and catastrophic events, and has suggested some alternatives. Events with exponential inter-occurrence times, heavy-tailed consequence distributions, or chaotic dynamics may impose an inherently low value of information on *any* forecast or risk assessment that attempts to predict future losses (or how alternative current control actions would change them) from currently available information. Decisions with hard-to-envision or hard-to-describe acts or consequences (such as losses of life) may omit details which could change decisions. Collective risk management decision processes for deciding whether and where to deploy expensive countermeasures are vulnerable to the possibility that *ex post* and *ex ante* majority preferences will be predictably different, as well as to the possibility that any procedure for aggregating individual beliefs and/or preferences can produce collective decisions that no one favors. Individual precautionary choices or behaviors that influence each other may lead to unpredictable outcomes, with no possible configuration of individual choices being stable.

Treating communities, rather than individuals, as the units of value- and preference-formation, risk management capacity development, and decision-making during crises, offers a promising complement to traditional decision analysis and risk management for rare and catastrophic events. Some of the same ideas may also improve management of more routine risks, including those prevalent in poor communities and countries, and risks associated with breakdowns in trustworthiness and cooperation. Helping communities to improve their catastrophe risk management skills may reduce losses due to a wide range of disasters, as well as to failures to routinely achieve the benefits of sustained cooperation. This remains true even if the nature, timing and magnitude of the next catastrophic event or disaster remain very uncertain, and even if community members cannot envision, evaluate, or agree on the probabilities of different adverse events and the utilities of different possible consequences. Identifying what needs to be done to enable a community to reduce average annual losses from a wide range of potential catastrophic risks may avoid many of the difficult technical challenges required for decision and risk analysis of any one of them in isolation.

Turning these insights and hopes into a practical discipline for demonstrably improving community resilience and reducing catastrophic losses is challenging. However, there is great practical need for such a discipline. Basing risk management decisions on improved understanding of how community members can and should work together to manage risks opens up risk management to a wealth of insights from social sciences and political economy. Making communities, rather than individuals, the protagonists in efforts to model and improve risk management decisions appears to be a fertile source for more realistic and useful principles of social risk management.

References

Abaimov SG, Turcotte DL, Shcherbakov R, Rundle JB (2007) Recurrence and interoccurrence behavior of self-organized complex phenomena. Nonlinear Process Geophys 14:455–464

Adger WN, Hughes TP, Folke C, Carpenter SR, Rockström J (2005) Social-ecological resilience to coastal disasters. Science 309(5737):1036–1039

Aguirre J, Valejo JC, Sanjuan MAF (2001) Wada basins and chaotic invariant sets in the Hénon-Heiles system. Phys Rev E Stat Nonlinear Soft Matter Phys 64(6):066208. ftp://ftp.ma.utexas.edu/pub/mp_arc/c/01/01-452.pdf

Axelrod R (1990) The evolution of co-operation. London; New York; Penguin; Basic Books. 1984

Bak P, Tang C, Wiesenfeld K (1988) Self-organized criticality. Phys Rev A 38:364–374

Banerjee AV, Duflo E (2011) Poor economics: a radical rethinking of the way to fight global poverty. PublicAffairs, New York, NY

Buchanan M (2001) Ubiquity: why catastrophes happen. Crown, Random House, Inc

Camargo S, Lopes SR, Viana RL (2010) Extreme fractal structures in chaotic mechanical systems: riddled basins of attraction. J Phys Conf Ser 246. http://iopscience.iop.org/1742-6596/246/1/012001/pdf/1742-6596_246_1_012001.pdf. last accessed 8/5/2011

Cialdini RB (2001) Influence: Science and Practice, 4th edn. Allyn and Bacon, Boston

Cox M, Arnold G, Villamayor Tomás S. 2010. A review of design principles for community-based natural resource management. Ecol Soc 15(4):38. http://www.ecologyandsociety.org/vol15/iss4/art38/ (Last accessed 8/4/2011)

Elster J (2007) Explaining social behavor: more nuts and bolts for the social sciences. Cambridge University Press, New York

Evans AW, Verlander NQ (1997) What is wrong with criterion FN-lines for judging the tolerability or risk? Risk Anal 17(2):157–167

Fudenberg D, Maskin E (1986) The Folk Theorem in repeated games with discounting or imperfect information. Econometrica 54(3):533–555

Fukuyama F (2011) The origins of political order: from prehuman times to the french revolution. Farrar, Straus and Giroux, New York

Gadjos T, Weymark JA, Zoli C 2009. Shared destinies and the measurement of risk equity. Ann Oper Res. http://www.arts.cornell.edu/poverty/kanbur/InequalityPapers/Weymark.pdf (Last accessed 8/5/2011)

Gintis H, Bowles S, Boyd R, Fehr E (2003) Explaining altruistic behavior in humans. Evol Hum Behav 24:153–172

Grim P (1997) The undecidability of the spatialized Prisoner's dilemma. *Theory and Decis* 42(1): 53–80

Guala F (2012) Reciprocity: weak or strong? What punishment experiments do (and do not) demonstrate. Behav Brain Sci 35(1):1–15

Guikema S (2020) Artificial intelligence for natural hazards risk analysis: potential, challenges, and research needs. Risk Anal 40(6):1117–1123. https://doi.org/10.1111/risa.13476

Gul F, Pesendorfer W (2008) The case for mindless economics. In: Caplin A, Shotter A (eds) The foundations of positive and normative economics. Oxford University Press

Haimes YY (2009) On the definition of resilience in systems. Risk Anal 29(4):498–501

Hammond PJ (1992) Harsanyi's utilitarian theorem: a simpler proof and some ethical connotations. In: Selten R (ed) Rational Interaction: Essays in Honor of John Harsanyi. Springer, Berlin

Helbing D, Johansson A (2010) Cooperation, norms, and revolutions: a unified game-theoretical approach. PLoS One 5(10):e12530. https://doi.org/10.1371/journal.pone.0012530

Henrich J (2006) Cooperation, punishment, and the evolution of human institutions. Science 312(5770):60–61

Hilton D, Régner I, Cabantous L, Charalambides L, Vautier S (2011) Do positive illusions predict overconfidence in judgment? A test using interval production and probability evaluation measures of miscalibration. J Behav Decis Mak 24(2):117–139

Howard RA (1966) Information value theory. IEEE Trans Syst Sci Cybern SSC-2:22–26
Hylland A, Zeckhauser RJ (1979) The impossibility of Bayesian group decision making with separate aggregation of beliefs and values. Econom Econ Soc 47(6):1321–1336
Kaplan S, Garrick BJ (1981) On the quantitative definition of risk. Risk Anal 1(1):11–27. http://josiah.berkeley.edu/2007Fall/NE275/CourseReader/3.pdf
Luce RD, Raiffa H (1957) Games and decisions. Wiley, New York
Mandelbrot B (1964) Random walks, fire damage amount and other Paretian risk phenomena. Oper Res 12(4):582–585
McClennen EF (1990) Rationality and dynamic choice: foundational explorations. Cambridge University Press, New York
Michel-Kerjan E, Slovic P (eds) (2010) The irrational economist: making decisions in a dangerous world. PublicAffairs, New York
Mueller DC (2003) Public Choice III. Cambridge University Press, New York
National Research Council (NRC) (2006) National Committee on Disaster Research in the Social Sciences: future challenges and opportunities, national research council. Facing Hazards and Disasters: understanding human dimensions. National Academies Press, Washington, DC
Nehring K (2007) The impossibility of a paretian rational: a Bayesian perspective. Econ Lett 96(1): 45–50
Norris FH, Stevens SP, Pfefferbaum B, Wyche KF, Pfefferbaum RL (2008) Community resilience. Am J Commun Psychol 41(1–2):127–150
Nowak MA, May RM (1992) Evolutionary games and spatial chaos. *Nature* 359:826–829
Othman A, Sandholm T (2009) How pervasive is the Myerson-Satterthwaite impossibility? Proceedings of the 21st international joint conference on artifical intelligence, IJCAI'09. Morgan Kaufmann Publishers Inc., San Francisco, CA, pp 233–238
Pandey P, Okazaki K (2005) Community based disaster management: empowering communities to cope with disaster risks. http://unpan1.un.org/intradoc/groups/public/documents/UN/UNPAN020698.pdf
Sae-jie W, Bunwong K, Moore EJ (2010) Qualitative behavior of SIS epidemic model on time scales. Latest trends on applied mathematics, simulation, modelling, pp 159–164. http://www.wseas.us/e-library/conferences/2010/Corfu/ASM/ASM-25.pdf. ISBN: 978-960-474-210-3
Schelling TC (1978) Micromotives and macrobehavior. W.W. Norton, New York, NY
Sheu S-H, Chiu C-H, Hsu T-S (2011) An age replacement policy via the Bayesian method. Int J Syst Sci 42(3):469–477
Solow AR (2005) Power laws without complexity. Ecol Lett 8(4):361–363
Suburban Emergency Management System (SEMP) (2006) The Catastrophic 1953 North Sea Flood of the Netherlands. www.semp.us/publications/biot_reader.php?BiotID=317 (Last accessed 08/07/2011)
Thaler RH, Sunstein C (2008) Nudge: improving decisions about health, wealth and happiness. Penguin Books, New York
Thompson WA Jr (1988) Point process models with applications to safety and reliability. Chapmand & Hall, London
Vandermeer J (2004) *Wada basins* and *qualitative unpredictability* in *ecological* models: a graphical interpretation. *Ecol Model* 176:65–74. http://sitemaker.umich.edu/jvander/files/wada_basins.pdf

Chapter 6
Learning Aversion in Benefit-Cost Analysis with Uncertainty

Introduction: Benefit-Cost Analysis (BCA) Fundamentals

For most of the past century, economists have sought to apply methods of benefit-cost analysis (BCA) (Portney 2008) to help policy makers identify which proposed regulations, public projects, and policy changes best serve the public interest. BCA provides methods to evaluate quantitatively, in dollar terms, the total economic costs and benefits of proposed changes. In versions commonly used by regulators and analysts, BCA prescribes that decisions should be made to *maximize the expected net present value* (NPV) of resulting time streams of net benefits (i.e., monetized benefits minus costs), with delayed and uncertain impacts being appropriately discounted to yield a net present value for each option being evaluated (e.g., Treasury Board of Canada Secretariat 1988). In the United States, OMB Circular A-4, which provides guidance to US Federal Agencies conducting regulatory impact analyses, explicitly requires that "the risk assessment methodology must allow for the determination of expected benefits in order to be comparable to expected costs." Similarly, in law-and-economics analyses of negligence torts, the Learned Hand Rule prescribes a duty to take care to prevent or reduce risk if the cost of doing so is less than the expected benefit (Grossman et al. 2006). Comparing expected costs to expected benefits is the mainspring of applied BCA. However, this chapter will argue that this principle, and hence traditional BCA calculations and recommendations based on it, are not well suited to guide public policy choices when costs or benefits are highly uncertain. Under those conditions, comparison of expected costs and benefits should be augmented with additional principles, such as reducing predictable rational regret, to improve BCA.

In the current practice of regulatory BCA, benefits are typically measured as the greatest amounts that people who want the changes would be willing to pay (WTP) to obtain them. In principle, costs are measured by the smallest amounts that people who oppose the changes would be willing to accept (WTA) as full compensation for

L. A. Cox Jr., *AI-ML for Decision and Risk Analysis*, International Series in Operations Research & Management Science 345,
https://doi.org/10.1007/978-3-031-32013-2_6

them (Portney 2008), although in practice it is more usual to focus on compliance costs. Recommending alternatives with the greatest expected NPV helps to adjudicate the competing interests of those who favor and those who oppose a proposed change.

Arguably, seeking to maximize net social benefit in this fashion promotes a society in which everyone expects to gain from public decisions on average and over time, even though not everyone will gain from every decision. Hence, BCA offers a possible approach to collective choice that appears to meet minimal standards for justice (it might be favored by everyone from an initial position behind Rawls's veil of ignorance) and economic efficiency (those who favor an adopted change gain more from it than those who oppose it lose). At first glance, BCA appears to have developed a decision-making recipe that circumvents the daunting impossibility theorems of collective choice theorists, flowing from the seminal work of Arrow (Arrow 1950; Mueller 2003; Man and Takayama 2013), but extended to collective choices under probabilistic uncertainty (e.g., Hylland and Zeckhauser 1979; Nehring 2007; Othman and Sandholm 2009), which imply that no satisfactory way exists in general to use available information about individual preferences to guide economically efficient social choices while protecting other desirable properties such as voluntary participation and budget balance. For that is precisely what BCA seeks to do.

However, this chapter argues that, whatever its conceptual strengths and limitations might be for *homo economicus*, or purely rational economic man, BCA for real-world regulations or projects with risky outcomes often leads to *predictably regrettable* collective choices in practice (and does not really succeed in bypassing impossibility results in principle). More useful recommendations can be developed by seeking to *minimize expected rational regret*, taking into account known biases in real human judgment, evaluation of alternatives, and decision-making, rather than by focusing on maximizing expected NPV, especially when probabilities for different costs and benefits are unknown or uncertain. This criterion is also better suited to the needs of real decision-makers with realistically imperfect information about the costs and benefits of proposed changes than is the principle of maximizing expected NPV.

The remainder of this chapter is structured as follows. The next section discusses aspirations and motivations for BCA and its promise and limitations for improving collective choices in societies of *homo economicus*. We then recall some impossibility results from collective choice theory for purely rational agents that limit what BCA or other choice procedures can accomplish in such societies. Key aspects of how real people make decisions, including many "predictably irrational" ones (Ariely 2009) are discussed, and we argue that well-documented decision heuristics and biases of the type discussed in Chap. 1 invalidate the usual normative prescriptive use of elicited or inferred WTP and WTA amounts in many practical applications. This is because both WTP and WTA amounts are sensitive to details of framing, context, perceptions of fairness and rights, feelings about social obligations and entitlements, and other factors discussed in Chaps. 1 and 5 that depart from the simplified economic models (e.g., quasi-linear preferences with additively separable costs and benefits) envisioned in the usual foundations of BCA. Psychological phenomena such as ambiguity aversion (reluctance to bet on unknown or highly uncertain subjective probabilities) imply several forms of what we will call *learning aversion*, i.e., refusal to use available information to improve decision-making. Simple examples illustrate mechanisms of learning-aversion for organizations as well as individuals. We argue that, in following the prescriptions of BCA, real people and organizations (whether individuals, companies, regulatory agencies, or legislators and policy-makers) typically spend too much to get too little, for a variety of reasons rooted in decision psychology and political theory. The chapter concludes by considering how to do better by using active information-seeking and learning to try to reduce predictable regret.

Aspirations and Benefits of BCA

A welcome element of common sense and benign rationality seem to infuse basic BCA prescriptions, such as *Don't take actions whose costs are expected to exceed their benefits;* or *Take actions to produce the greatest achievable expected net benefits*. People may argue about how best to quantify costs and benefits, includinghow to evaluate opportunity costs, delayed or uncertain rewards, real options, and existence values. They may disagree about how best to characterize uncertainties and information, models, and assumptions. But submitting proposed courses of action to the relatively objective-seeming tests of quantitative BCA comparisons has appealed powerfully to many scholars and some policy makers over the past half century. It is easy to understand why. Without such guidance, collective decisions—even those taken under a free, democratic rule of law—may harm all involved, as factional interests and narrow focusing on incremental changes take precedence over more dispassionate and comprehensive calculations for identifying which subsets of changes are most likely to truly serve the public interest.

Example: Majority Rule Without BCA Can Yield Predictably Regrettable Collective Choices

Table 6.1 shows five proposed changes that a small society, consisting of individuals 1–3 ("players," in game theory terminology) is considering adopting. The proposed changes, labeled A–E, are shown in the rows, of the table. These might represent proposed regulatory acts, investment projects, initiatives, mandates, etc. The table presents resulting changes in annual incomes for each player if each measure is adopted, measured in convenient units, such as thousands of dollars per year. For simplicity, the impacts of the different measures are assumed to be independent of each other. For example, project A, if implemented would cost player 1 three units of income, perhaps in the form of a tax on player 1's business or activities. It would

Table 6.1 A hypothetical example of changes in annual incomes (e.g., in thousands of dollars) for each of three people from each of five alternatives

Proposed change	Player 1's income change	Player 2's income change	Player 3's income change
A	−3	1	1
B	1	−3	1
C	1	1	-3
D	3	−1	−1
E	0	0	0

produce benefits valued at one unit of income for each of players 2 and 3. Thus, its costs are narrowly concentrated but its benefits are widely distributed. Conversely, project D would impose a tax, or other loss of income, of one unit of income on each of players 2 and 3, but would produce three units of income for player 1. E is the *status quo*.

If the collective choice process used in this small society is direct majority rule, with each participant voting for or against each proposed change, A–E, then which proposed changes will be approved? Assuming that each voter seeks to maximize his own income (or minimize his own loss), measures A–C will be adopted, since a majority (two out of three) of the players prefer each of these to the *status quo*. Summing the changes in incomes for all three of the adopted measures A–C shows that each player would receive a net loss of 1 unit of income from these three adopted collective decisions. Thus, applying simple majority rule to each proposed change A–E creates a *predictably regrettable outcome*: it is clear that changes A–C will be adopted (the outcome is *predictable*) and it is clear that this will make all voters worse off than they would have been had they instead rejected the changes and maintained the *status quo*. The adopted changes are, in this sense, jointly *regrettable*.

The problem illustrated here is familiar: each voter is willing to have "society" (as embodied in the collective choice process) spend other people's money to increase his own benefit. Yet, when each faction (a coalition, or subset of players, such as players 2 and 3, for change A) has the political power to adopt a measure that achieves gain for all its members at the expense of its non-members, the portfolio of alternatives that end up being adopted harms everyone, in the sense that everyone would have preferred the *status quo*. Political theorists have recognized this possibility for centuries; it loomed large in Federalist Paper Number 10, and in concerns about tyranny of the majority.

BCA seeks to remedy this ill by subjecting each alternative to a cost-benefit test. A familiar example is theKaldor-Hicks potential compensation test (Kaldor 1939; Hicks 1939): *Do the gainers gain more than the losers lose?* Would those who prefer adoption of a proposed alternative still prefer it if they had to fully compensate those who preferred the *status quo*? This question makes sense under the usual assumptions of quasi-linear preferences (utility can be expressed as benefits minus costs) and if utility is assumed to be transferable and proportional to money. Although these assumptions, in turn, may be difficult to defend, they suffice to illustrate some key points about strengths and limitations of BCA even under such idealized conditions. Alternatives A–C in Table 6.1 fail this test, but alternative D—which would not be selected by majority rule—passes. For example, if a tax of one income unit taken from each of individuals 2 and 3 allows individual 1 to gain a benefit (such as socially subsidized healthcare) evaluated as equivalent to three income units, it might be deemed an alternative worth considering further, since individual 1 could (at least in principle) pay one unit of income to each of individuals 2 and 3 and still be better off (by one income unit) than before the change. BCA practitioners often apply such tests for potential Pareto improvements to determine whether a proposed change is worth making (Feldman 2004).

Of course, taking from some to benefit others, especially if potential compensation remains only a theoretical possibility, raises questions about rights and justice (e.g., is enforced wealth transfer a form of theft? Would individuals voluntarily choose to adopt procedures that maximize estimated net social benefits, if they made the choice from behind the veil of ignorance in Rawls's initial position?) Moreover, it is well known that potential compensation criteria can lead to inconsistencies when a proposed alternative to the *status quo* increases one good, e.g., clean air, but reduces another, e.g., per-capita income. Those who prefer a change in the *status quo* might still do so if they had to fully compensate those who prefer it; and yet those who prefer the *status quo* might still do so if they had to fully compensate those who do not (Feldman 2004). Thus, potential compensation tests are not free of conceptual and practical difficulties. Nonetheless, the idea that a proposed change should not be adopted unless its benefit (defined as the sum of willingness-to-pay (WTP) amounts from those who want it) exceeds its cost (defined as the sum of willingness-to-accept (WTA) amounts needed to fully compensate those who don't) provides a plausible and much-cited screen for eliminating undesirable proposals (Portney 2008).

Limitations of BCA for Purely Rational People, *Homo Economicus*

BCA was developed by economists, and is most applicable to societies of purely rational individuals, sometimes called *homo economicus*, who conform to the usual (Savage-style) axioms for rational behavior via expected utility-maximizing choices (Gilboa and Schmeidler 1989; Smith and von Winterfeldt 2004). As discusssed in Chap. 1, such individuals differ markedly and in many respects from real people, as they are not prey to numerous well-documented heuristics and biases that shape decisions in practice (Kahneman 2011). However, even perfect individual rationality does not necessarily promote effective collective choice. Numerous impossibility results in game theory and the theory of collective choice reveal the difficulty of constructing collective choice procedures ("mechanisms") that will produce desirable (e.g., Pareto-efficient) results based on voluntary participation by rational people. Tradeoffs must be made among desirable characteristics such as budget balance (a mechanism should not run at a net loss), *ex post* Pareto-efficiency (a mechanism should not select an outcome that every participants likes worse than one that was rejected), voluntary participation, and nondictatorship (a mechanism should reflect the preferences of more than one of the participants) (e.g., Mueller 2003; Man and Takayama 2013; Othman and Sandholm 2009). Similar tradeoffs, although less well known, hold when collective decisions must be made by rational individuals with different beliefs about outcomes (Hylland and Zeckhauser 1979; Nehring 2007), as well as when they have different preferences for outcomes. The following example, which we also discussed in Chap. 5, illustrates this point.

Example: Pareto-Inefficiency of BCA with Disagreements About Probabilities

Suppose that members of a society (or an elected subset of members representing the rest) must collectively decide whether to implement a costly proposed regulation to further reduce fine particulate air pollution in order to promote human health and longevity. Each individual believes that the benefits of the proposed regulation will exceed its costs if and only if (a) Air pollution at current levels causes significantly increased mortality risks; and (b) The proposed regulation would reduce those (possibly unknown) components of air pollution that, at sufficiently high exposure concentrations and durations, harm health. Each individual favors the regulation if and only if the joint probability of events (a) *and* (b) exceeds 20%. That is, the product of the probabilities of (a) and (b) must exceed 0.2 for the estimated benefits of the proposed regulation to exceed its costs (as these two events are judged to be independent).

As a mechanism to aggregate their individual beliefs, the individuals participating in the collective choice have agreed to use the arithmetic averages of their individual probabilities for relevant events, here (a) and (b), They will then multiply the aggregate probability for (a) and the aggregate probability for (b) and pass the regulation if and only if the resulting product exceeds 0.2. (Of course, many other approaches to aggregating or reconciling expert probabilities can be considered, but the point illustrated here with simple arithmetic averaging holds generally.)

Individual beliefs can be described by two clusters with quite different world views and subjective probability assessments. Half of the community ("pessimists") fear both man-made pollution and our inability to control its consequences: they believe that air pollution probably does increase mortality risk, but that not enough is known for a regulation to reliably target and control the unknown components that harm human health. Specifically, they assign probability 0.8 to event (a) (exposure causes risk) and probability 0.2 to event (b) (regulation reduces relevant components of exposures). The other half of the community ("optimists") is skeptical that that exposure increases risk, but believe that, if it does, then it is probably the components targeted by the regulation that do so (i.e., fine particulate matter rather than sulfates or something else). They assess a probability of only 0.2 for event (a) and a probability of 0.8 for event (b). Note that both sets of beliefs are consistent with the postulates that all individuals are perfectly rational, since the axioms of rationality do not determine how prior probabilities should be set (in this case, reflecting two different world views about the likely hazards of man-made pollution and our ability to control them).

Using arithmetic averaging to combine the subjective probability estimates of participating individuals (assumed to be half optimists and half pessimists), the average probability for event (a) is (0.8 + 0.2)/2 = 0.5, and the average probability for event (b) is likewise (0.2 + 0.8)/2 = 0.5. These group probability assessments imply that the collective joint probability of events (a) and (b) is 0.5*0.5 = 0.25. Since this is above the agreed-to decision threshold of 0.2, the regulation would be passed. On the other hand, every individual computes that the joint probability of events (a) and (b) is only 0.8*0.2 = 0.16. Since this is below the decision threshold of 0.2 required for projected benefits to exceed costs, no individual wants the regulation passed. Thus, aggregating individual beliefs about events leads to a decision that no one agrees with—a regrettable outcome.

The point illustrated by this example is not that one should not average probabilities, or that other mechanisms might work better. To the contrary, an impossibility theorem due to Nehring (2007) demonstrates that *no* method of aggregating individual beliefs and using them to make group decisions can avoid selecting dominated decisions (other than such trivial procedures as selecting a single individual as a "dictator" and ignoring everyone else's beliefs). For *any* aggregation and decision rule that treats individuals symmetrically, one can construct examples in which the group's decision is not favored by any of its members. (For example, using a geometric mean instead of an arithmetic means would resolve the specific problem in this example, but such a procedure would also select dominated choices in slightly modified versions of the example.) Thus, the general lessons that when probabilities of events are not known and agreed to, and opinions about them are sufficiently diverse, then calculations (collective decision mechanisms) that combine the probability judgments of multiple experts or participants to determine what acts should be taken in the public interest risk producing regrettable collective choices with which no one agrees.

Example: Impossibility of Pareto-Efficient Choices with Sequential Selection

A possible remedy for the Pareto-inefficient outcomes in the preceding example would be to elicit from individuals their final, holistic evaluations of collective actions. For example, each individual might estimate his own net benefit from each alternative action (pass or reject the proposed regulation, with a proposed tax or other measure to pay for it if it is passed), and then society might take the action with the largest sum of estimated individual net benefits. This would work well in the preceding example, where everyone favors the same collective choice (albeit for different reasons, based on mutually inconsistent beliefs). But it leaves the resulting decision process squarely in the domain of other well-known impossibility theorems that apply when individuals directly express preferences for alternatives.

As an example, suppose a society of three people (or a Congress of three representatives of a larger society) makes collective choices by voting among various proposed regulatory alternatives as the relevant bills are brought forward for consideration. Suppose that the legislative history is such that, in the following list of possible alternatives, the choice between A and B comes to a vote first (e.g., because advocates for PM2.5 reduction organize themselves first or best), and that later the winner of that vote is run off against alternative C (perhaps because O3 opponents propose their bill later, and it is assumed that the current cost-constrained political environment will allow at most one such pollution reduction bill to be passed in the current session). Finally (maybe in the next session, with an expanded regulatory budget, or perhaps as a rider to an existing bill), alternative D is introduced, and run off against whichever of alternatives A–C has emerged as the collective choice so far. Here are the four alternatives considered:

A: Do not require further reductions in any pollutant
B: Require further reductions in fine particulate matter (PM2.5) emissions only
C: Require further reductions in ozone (O3) only
D: Require further reductions in both PM2.5 and O3.

Individual preferences are as follows (with "$>$" interpreted as "is preferred to"):

1. $A > D > C > B$
2. $B > A > D > C$
3. $C > B > A > D$

For example, individual 1 might believe that further reducing air pollution creates small (or no) health benefits compared to its costs, but believes that, if needless costs are to be imposed, they should be imposed on both PM2.5 and O3 producers (with a slight preference for penalizing the latter, if a choice must be made). Individual 2 believes that PM2.5 is the main problem, and that dragging in ozone is a waste of cost and effort; individual 3 believes that ozone is the main problem.

Applying these individual preferences to determine majority votes, it is clear that B will be selected over A (since B is preferred to A by both of individuals 2 and 3). Then, B will lose to C (since 1 and 3 prefer C to B). Finally, D will be selected over C (since 1 and 2 prefer D to C). So, the predictable outcome of this sequence of simple majority votes is that alternative D will be the society's final collective choice, i.e., require further reductions in both pollutants. But this choice is clearly Pareto-inefficient (and, in that sense, regrettable): everyone prefers option A (no further reduction in pollutants), which was eliminated in the first vote, to option D (further reductions in all pollutants), which ended up being adopted.

A central theme of collective choice theory for societies of rational individuals is that such perverse outcomes occur, in the presence of sufficiently diverse preferences, for all possible collective choice mechanisms (including those in which BCA comparisons are used to compare pairs of alternatives), provided that non-dictatorship or other desired properties hold (e.g., Mueller 2003; Man and Takayama 2013).

How Real People Evaluate and Choose Among Alternatives

As detailed in Chap. 1, psychologists, behavioral economists, marketing scientists, and neuroscientists have demonstrated convincingly that most people (including experts in statistics and decision science) depart systematically from purely rational, subjective expected utility (SEU)-maximizing decision-making (Kahneman 2011), perhaps especially when preferences and choices are elicited, rather than when they are observed in market transactions. To a very useful first approximation, most of us can be described as making rapid, intuitive, emotion-informed judgments and evaluations of courses of action ("System 1" judgments, in the current parlance of decision psychology), followed (time and attention permitting) by slower, more reasoned adjustments ("System 2" thinking) (*ibid*). Much of System 1 thinking, in turn, can be understood in terms of the *affect heuristic*, according to which gut reaction—a quick, automatically generated feeling about whether a situation, choice, or outcome is good or bad—drives decisions. For most decisions and moral judgments, including those involving how to respond in risky situations, the alternative choices, situations, or outcomes are quickly (perhaps instinctively) categorized as "bad" (to be avoided) or "good" (to be sought). Beliefs, perceptions, and System 2 rationalizations and deliberations then tend to align behind these prompt evaluations. This approximate account, while over-simplified, successfully explains many of the departures of real preferences and choice behaviors from those prescribed by expected utility theory, and is consistent with evidence from neuroeconomics studies of how the brain processes risks, rewards, delays, and uncertainties (including unknown or "ambiguous" ones) in arriving at decisions. Although there is a vast literature about relations among biases, the core relations on which we focus can be summarized succinctly as: *WTP* ← *Affect heuristic* → *Learning aversion* → *Overspending* → *Rational regret*, where each arrow indicates that the outcome pointed to is positively causally contributed to by the one that points into it. These components are explained next.

Decision biases make WTP amounts, both elicited and revealed, untrustworthy guides for quantifying the benefits of many risk-reducing measures, such as health, safety, and environmental regulations (Casey and Delquie 1995). Important, systematic departures of elicited WTP from normative principles include the following:

- *Affect heuristic.* Among many examples, people (and other primates) are willing to pay more for a small set of high-quality items than for a larger set that contains the same items, with some lower-quality one added as well (Kralik et al. 2012). More generally, in contrast to the prescriptions of SEU theory, expanding a choice set may change choices even if none of the added alternatives is selected, and may change satisfaction with what is chosen (Poundstone 2010).

- *Proportion dominance.* Willingness-to-pay is powerfully, and non-normatively, affected by use of proportions. For example, groups of subjects typically are willing to pay more for a safety measure described as saving "85% of 150 lives" in the event of an accident than for a measure described as saving "150 lives" (Slovic et al. 2002, 2004) (Similarly, one might expect that many people would express higher WTP for saving "80% of 100 lives" than for saving "10% of 1000 lives," even though all would agree that saving 100 lives is preferable to saving 80.) The high percentages act as cues triggering positive-affect evaluations, but the raw numbers, e.g., "150 lives," lack such contextual cues, and hence do not elicit the same positive response. This aspect of choice as driven by contextual cues is further developed in Ariely's theory of arbitrary coherence (Ariely 2009).
- *Sensitivity to wording and framing.* Describing the cost of an alternative as a "loss" rather than as a "cost" can significantly *increase* WTP (Casey and Delquie 1995). The opportunity to make a small, certain payment that leads to a large return value with small probability, and else to no return, is assessed as more valuable when it is called "insurance" than when it is called a "gamble" (Hershey et al. 1982). Describing the risks of medical procedures in terms of mortality probabilities instead of equivalent survival probabilities can change preferences among them (Armstrong et al. 2002), since the gain-frame and loss-frame trigger loss-averse preferences differently, in accord with Prospect Theory.
- *Sensitivity to irrelevant cues.* A wide variety of contextual cues that are logically irrelevant can nonetheless greatly affect WTP (Poundstone 2010). For example, being asked to write down the last two digits of one's Social Security Number significantly affects how much is willing to pay for consumer products (with higher SSNs leading to higher WTP amounts) (Ariely 2009). The "anchoring and adjustment" heuristic (Kahneman 2011) allows the mind to anchor on irrelevant cues (as well as relevant ones) that then shape real WTP amounts and purchasing behaviors (Poundstone 2010).
- *Insensitivity to probability.* If an elicitation method or presentation of alternatives gives different salience to attributes with different effects on affect (e.g., emphasizing amount vs. probability of a potential gain or loss), then choices among the alternatives may change (the phenomenon of *elicitation bias*, e.g., Champ and Bishop 2006). Similarly, although rational (System 2) risk assessments consider the probabilities of different consequences, System 1 evaluations may be quite insensitive to the magnitudes of probabilities (e.g., 1 in a million vs. 1 in 10,000), and, conversely, overly sensitive to the change from certainty to near-certainty: "When consequences carry sharp and strong affective meaning, as is the case with a lottery jackpot or a cancer ... variation in probability often carries too little weight. ... [R]esponses to uncertain situations appear to have an all or none characteristic that is sensitive to the possibility rather than the probability of

strong positive or negative consequences, causing very small probabilities to carry great weight." (Slovic et al. 2002)

- *Scope insensitivity.* Because the affect heuristic distinguishes fairly coarsely between positive and negative reactions to situations or choices, but lacks fine-grained discrimination of precise degrees of positive or negative response, WTP amounts that are largely driven by affect can be extraordinarily insensitive to the quantitative magnitudes of benefits involved. As noted by Kahneman and Frederick (2005), "In fact, several studies have documented nearly complete neglect of scope in CV [contingent valuation stated WTP] surveys. The best-known demonstration of scope neglect is an experiment by Desvouges et al. (1993), who used the scenario of migratory birds that drown in oil ponds. The number of birds said to die each year was varied across groups. The WTP responses were completely insensitive to this variable, as the mean WTP's for saving 2,000, 20,000, or 200,000 birds were $80, $78, and $88, respectively. ... [Similary], Kahneman and Knetsch (see Kahneman 1986) found that Toronto residents were willing to pay almost as much to clean up polluted lakes in a small region of Ontario as to clean up all the polluted lakes in Ontario, and McFadden and Leonard (1993) reported that residents in four western states were willing to pay only 28% more to protect 57 wilderness area than to protect a single area."
- *Perceived fairness, social norms, and moral intensity.* How much individuals are willing to pay for benefits typically depends on what they think is fair, on what they believe others are willing to pay, and on whether they perceive that the WTP amounts for others reflect moral convictions or mere personal tastes and consumption preferences (e.g., Bennett and Blaney 2002). The maximum amount that a person is willing to pay for a cold beer on a hot day may depend on whether the beer comes from a posh hotel or a run-down grocery store, even though the product is identical in either case (Thaler 1999).
- *Ambiguity aversion*, i.e., reluctance to take action based on beliefs about events with unknown objective probabilities (and willingness to pay to reduce uncertainty about probabilities before acting) implies each of the following limitations (Al-Najjar and Weinstein 2009,

  - *Decisions do not ignore sunk costs*, as normative theories of rational decision-making would prescribe;
  - *Dynamic inconsistency*, i.e., people will make plans based on assumptions about how they will behave if certain contingencies occur in the future, and then not actually behave as assumed.
  - *Learning aversion,* i.e., unwillingness to receive for free information that might help to make a better (SEU-increasing) decision.Many other anomalies

(e.g., preference reversal, endowment effect, *status quo* bias, etc.) drive further gaps between elicited WTP and WTA amounts, and between both and normatively coherent preferences. Taken together, they rule out any straight-forward use of WTP values (elicited or inferred from choices) for valuing uncertain benefits. Indeed, once social norms are allowed as important influencers of real-world WTP values (unlike the WTPs in textbook BCA models of quasi-linear individual preferences), coherent (mutually consistent) WTP values need not exist at all. For example, if Mr. Smith is willing to pay either at most x = $100 for a public good, if no one else is willing to pay anything, or else at most whatever Mr. Jones will pay (if Mr. Jones is willing to pay more than zero), and if Mr. Jones is willing to pay at most y = half of whatever Mr. Smith is willing to pay (perhaps feeling that this is his fair share to obtain the social benefit revealed by Mr. Smith's WTP), then there is no well-defined set of individual WTP amounts (since x = y, y = x/2 has no solution greater than zero).

In addition to the preceding biases, we propose that *learning aversion*, meaning reluctance to seek or use information that might change a decision for the better, plays a major role in decisions about what risks to take, consistent with the idea that System 2 is generally lazy, and insufficiently willing to exert more effort to produce better results (Kahneman 2011). Although the term "learning aversion" (Louis 2009) is not widely used in decision science, we believe it is central to understanding how to avoid premature action and to improve the practice and outcomes of BCA. For example, among ten well-documented "decision traps," or barriers to effective decision-making by individuals and organizations, discussed in a popular book (Russo and Schoemaker 1989), most involve failing to take sufficient care to collect, think about, appropriately use, and deliberately learn from relevant information that could improve decisions. These include not keeping track of decision results, failing to make good use of feedback from the real world, failing to collect relevant information because of overconfidence in one's own judgment, and trusting too much in the most readily available ideas and information. All exemplify failures to learn effectively from experience.

Possible contributors to learning-aversion include *over-confidence* (which, in turn, might reflect a predilection to consider only information and interpretations that support the views with which one is already endowed); *hyperbolic discounting* (which implies that the immediate costs of learning may overwhelm even much larger future benefits of being able to make better decisions); and ambiguity aversion. Ambiguity aversion, as axiomatized by Gilboa and Schmeidler (1989) and others (Maccheronia et al. 2006) implies that a decision-maker will sometimes refuse free information that could improve decisions (Al-Najjar and Weinstein 2009). For example, in principle, an ambiguity-averse decision-maker might refuse sufficiently informative, free genetic information that is highly relevant for decisions on lifestyle, healthcare planning, and insurance purchasing (Hoy et al. 2014). Empirically, fuller disclosure of scientific uncertainties to women facing cancer treatment choices does not necessarily improve the quality of their decisions by any measure evaluated, but does significantly reduce their subsequent (post-decision) satisfaction with the decisions that are eventually made (Politi et al. 2011).

The comparison of expected costs to expected benefits in BCA can facilitate learning-averse decision-making. Its golden rule is to choose the action (from among those being evaluated) that maximizes the *expected* discounted net benefit. But there is no requirement in BCA that expected values must be calculated from adequate information, or that a final BCA comparison among alternatives should be postponed until enough information has been collected so that the expected costs of collecting more (including the expected costs of delayed decisions) outweigh the expected benefits from making a better-informed comparison. In this respect, BCA differs from other normative frameworks, including stochastic dynamic programming (SDP), decision analysis with explicit value-of-information (VOI) calculations, and optimal statistical decision models (such as the Sequential Probability Ratio Test). These frameworks explicitly require continuing to collect more information, before making a final comparison and decision, until it is optimal to stop doing so. They provide methods to calculate (or approximate) optimal stopping rules and decision boundaries for determining when to stop collecting information and take action (DeGroot 1970). Within the framework of BCA, a similar effect could in principle be achieved by making "Collect more information before making a final decision" one of the options that should always be evaluated. Thus, improving the practice of BCA does not necessarily require rejecting the basic framework, but enriching it to include more options, such as learning more before acting, or experimenting some before adopting a widespread policy. Augmenting standard BCA calculations with VOI or SDP calculations can help to incorporate the value of resolving some uncertainties into optimal decision-making. Since learning-averse individuals (Hoy et al. 2014) and organizations (Russo and Schoemaker 1989) typically do not collect enough information (as judged in hindsight) before acting, prescriptive disciplines should explicitly encourage optimizing information collection and learning as a prelude to evaluating, comparing, and choosing among final decision alternatives (Russo and Schoemaker 1989). Helping users to overcome learning aversion is therefore a potentially valuable direction for improving the current practice of BCA.

Learning Aversion and Other Decision Biases Inflate WTP for Uncertain Benefits

Decision biases have a potentially surprising implication: Real people typically overestimate highly uncertain benefits and under-estimate highly uncertain costs, and hence are willing to pay too much for projects or other proposed changes with unknown or highly uncertain benefits and/or costs. Intuitively, one might expect exactly the reverse: that ambiguity aversion would reduce the perceived values or net benefits of such projects. But in fact, ambiguity aversion (and other drivers of learning aversion) mainly cut off information collection and analyses needed for careful evaluation, comparison, and selection of alternatives, leading to premature and needlessly risky decisions (Russo and Schoemaker 1989; Kahneman 2011). Then, overconfidence and optimism biases take over. From the perspective of obtaining desirable outcomes, members of most decision-making groups spend too much time and effort convincing each other that their decisions are sound, and increasing their own confidence that they have chosen well. They spend too little effort seeking and using potentially disconfirming information that could lead to a decision with more desirable outcomes (Russo and Schoemaker 1989). Moreover, in assessing the likely future outcomes of investments in risky projects, individuals and groups typically do *not* focus on the worst plausible scenario (e.g., the worst-case probability distribution for completion times of future activities), as theoretical models of ambiguity aversion suggest (Gilboa and Schmeidler 1989). To the contrary, they tend to assign low subjective probabilities to pessimistic scenarios, and to base plans and expectations on most-favorable, or nearly most-favorable, scenarios (e.g., Newby-Clark et al. 2000).

This tendency toward overly-optimistic assessment of both uncertain benefits (too high) and uncertain costs or delays (too low) has been well documented in discussions of optimism bias and corollaries such as the planning fallacy (Kahneman and Tversky 1979; Kahneman 2011). For example, it has repeatedly been found that investigators consistently over-estimate the benefits (treatment effects) to be expected from new drugs undergoing randomized clinical trials (e.g., Djulbegovic et al. 2011; Gan et al. 2012); conversely, most people consistently underestimate the time and effort needed to complete complex tasks or projects, such as new drug development (Newby-Clark et al. 2000). These psychological biases are abetted by statistical methods and practices that routinely produce an excess of false positives, incorrectly concluding that interventions have desired or expected effects that, in fact, they do not have, and that cannot later be reproduced (Nuzzo 2014; Sarewitz 2012; Lehrer 2012; Ioannidis 2005). Simple Bayesian calculations suggest that more than 30% of studies with reported P values of ≤ 0.05 may in fact be reporting false positives (Goodman 2001). Indeed, tolerance for, and even encouragement of, a high risk of false-positive findings (in order to reduce risk of false negatives and to continue to investigate initially interesting hypotheses) has long been part of the culture of much of epidemiology and public health investigations supposed to be in the public interest (e.g.. Rothman 1990).

Learning aversion and related decision biases such as overconfidence and optimism bias (Kahneman 2011) and sunk-cost effects (Navarro and Fantino 2005) contribute to a willingness to take costly actions with highly uncertain benefits and/or costs. These biases favor premature decisions to pay to achieve uncertain benefits, even in situations where free or inexpensive additional investigation would show that the benefits are in fact almost certainly much less than the costs.

Example: Overconfident Estimation of Health Benefits from Clean Air Regulations

Overconfidence and confirmation biases can be encoded in the modeling assumptions and analytic procedures used to develop estimates of cost and benefits for BCA comparisons. For example, the U.S. EPA (2011a, b) estimated that reducing fine particulate matter (PM2.5) air pollution in the United States has created close to

2 trillion dollars per year of annual health benefits, mainly from reduced elderly mortality rates. This is vastly greater than the approximately 65 billion dollars per year that EPA estimates for compliance costs, leading them to conclude that "***The extent to which estimated benefits exceed estimated costs*** and an in-depth analysis of uncertainties ***indicate that it is extremely unlikely the costs of 1990 Clean Air Act Amendment programs would exceed their benefits under any reasonable combination of alternativeassumptions or methods identified during this study***" (emphasis in original). However, the benefits calculation used a quantitative approach to uncertainty analysis based on a Weibull distribution (assessed using expert guesses) for the reduction in mortality rates per unit reduction in PM2.5. The Weibull distribution is a continuous probability distribution that is only defined over non-negative values. Thus, the quantitative uncertainty analysis implicitly assumes a 100% certainty that reducing PM2.5 does in fact cause reductions in mortality rates (the Weibull distribution puts 100% of the probability mass on positive values), in direct proportion to reductions in PM2.5 pollutant levels, even though EPA's qualitative uncertainty analysis states correctly that such acausal relation has not been established. An alternative uncertainty analysis that assigns a positive probability to each of several discrete uncertainties suggests that "EPA's evaluation of health benefits is unrealistically high, by a factor that could well exceed 1000, and that it is therefore very likely that the costs of the 1990 CAAA [Clean Air Act Amendment] exceed its benefits, plausibly by more than 50-fold. The reasoning involves re-examining specific uncertainties (including model uncertainty, toxicological uncertainty, confounder uncertainty, and uncertainty about what actually affects the timing of death in people) that were acknowledged qualitatively, but whose discrete contributions to uncertainty in health benefits were not quantified, in EPA's cost-benefit analysis" (Cox 2012). If this analysis is even approximately correct, then EPA's highly confident conclusion results from an uncertainty analysis that disregarded key sources of uncertainty. It implicitly encodes (via the choice of a Weibull uncertainty distribution) overconfidence and confirmation biases that may have substantially inflated estimated benefits from Clean Air Act regulations by assuming that reducing PM2.5 concentrations causes reductions in mortality rates, while downplaying (by setting its subjectively assessed probabilityto zero) the possibility that this fundamental assumption might be wrong.

In the political realm, the costs of regulations or other proposed expensive changes can also be made more palatable to decision-makers by a variety of devices, long known to marketers and politicians and increasingly familiar to behavioral economists, which exploit decision biases (Poundstone 2010). Among these arepostponing costs by even a little (to exploit hyperbolic discounting, since paying now provokes an adverse System 1 reaction that paying even slightly later does not); emphasizing annual costs instead of larger total costs; building in an annual rate increase so that increases become viewed as part of the *status quo*, and hence acceptable without further scrutiny; paying from unspecified, obscure, or general funds (e.g., general revenues) rather than from specific accounts owned by the decision-maker(s), so that trade-offs, opportunity costs and outgoing payments are less salient; adding comparisons to alternatives that no one would want, to make the recommended one seem more acceptable; creating a single decision point for committing to a stream of expenses, rather than instituting multiple review and decision points (e.g., a single yes/no decision, with a limited time window of opportunity, on whether to enact a costly regulation that will last for years, rather than a contingent decision for temporary funding with frequent reviews to ask whether it has served its purpose and should be discontinued); considering each funding decision in isolation, so that the proposal can be evaluated based on its affect when viewed outside the context of competing uses to which the funds could be put ("narrow framing," discussed by Kahneman 2011); framing the cost as protecting an endowment, entitlement, or option, i.e., as paying to avoid losing a benefit, rather than as paying to gain it; and comparing expenditures to those of others (e.g., to how much the European Union or Japan is spending on something said to be similar). These and related techniques are widely used in marketing and advertising, as well as by business leaders and politicians seeking to "sell" programs to the public (Gardner 2009). They are highly effective in persuading consumers to spend money that, in retrospect, they might feel would have been better spent on something else (Ariely 2009; Poundstone 2010).

Assuming No Risk Aversion Inflates the Estimated Value of Public Projects with Uncertain Benefits

In a collective choice context, learning aversion may be strategically rational if discovering more information about the probable consequences of alternative choices could disrupt a coalition's agreement on what to do next (Louis 2009). But collective learning aversion may also arise because of free-rider problems or other social dilemmas (Weber et al. 2004) that create gaps between private and public interests.

Example: Information Externalities and Learning Aversion in Clinical Trials

In clinical trials, a well known dilemma arises when each individual seeks his or her own self-interest, i.e., the treatment that is expected to be best for his or her own specific case, given presently available information. If everyone uses the same treatment, then the opportunity to learn about potentially better (but possibly worse) treatments may never be taken. Given a choice between a conventional treatment that gives a 51% survival probability with certainty and a new, experimental treatment that is equally likely to give an 80% survival probability or a 20% survival probability, and that will give the same survival probability (whichever it is) to all future patients, each patient might elect the conventional treatment (since

51% > 0.5*0.2 + 0.5*0.8 = 50%). But then it is never discovered whether the new treatment is in fact better. The patient population continues to endure an individual survival probability of 51% for every case, when an 80% survival probability might well be available (with probability 50%). The same remains true even if there are many possible treatment alternatives, so that the probability that at least some of them are better than the current incumbent approaches 100%. Ethical discussions of the principle of clinical equipoise (should a physician prescribe an experimental treatment when there is uncertainty about whether it performs better than a conventional alternative, especially when opinions are divided?) recognize that refusal to experiment with new treatments (possibly due to ambiguity-aversion) in each individual case imposes a costly burden from failure to learn on the patient population as a whole, and on each member of it when he or she must choose among options whose benefits have not yet been well studied (Gelfand 2013). The principle that maximizing expected benefit in each individual case can needlessly reduce the expected benefit for the entire population is of direct relevance to BCA, as discussed further in the next example.

Example: Desirable Interventions with Uncertain Benefits Become Undesirable When They Are Scaled Up

Many people who would be willing to pay $1 for a 50-50 chance to gain $3 or nothing (expected net value of $1.50 expected benefit - $1 cost = $0.50) might baulk at paying $100,000 for a 50–50 chance to gain $300,000 or nothing. Indeed, for risk-averse decision-makers, scaling up a favorable prospect with uncertain benefits by multiplying both costs and benefits by a large enough factor can make the prospect unacceptable. (As an example, for a decision-maker with exponential utility function evaluating a prospect with normally distributed benefits having mean M and variance V, the certainty equivalent of n copies of the prospect, where all of n of them share a common uncertainty and the same outcome, has the form $CE = nM - kn^2V$, where k reflects subjective relative risk aversion. Since the first term grows linearly and the second term grows quadratically with the scaling factor n, the certainty equivalent is negative for sufficiently large n.) Now consider a local ordinance, such as a ban on coal-burning, that has uncertain health benefits and known implementation costs, such that its certainty equivalent is assessed as positive for a single county. If the same ban is now scaled up to n counties, so that the same known costs and uncertain benefits are replicated n times, then the certainty equivalent will be negative for sufficiently large n. A bet worth taking on a small scale is not worth taking when the stakes are scaled up too many times. Yet, top-down regulations that apply the same action (with uncertain benefits) to dozens, hundreds, or thousands of counties or individuals simultaneously, based on assessment that $CE > 0$ for each one, implies that essentially the same bet is being made many times, so that the total social CE will be negative if the number of counties or individuals is sufficiently large. This effect of *correlated uncertainties* in reducing the net benefits of regulations with uncertain benefits that are widely applied is omitted from BCA calculations that only consider expected values.

BCA usually assumes risk neutrality for proposed public projects with uncertain costs and benefits, as implied by the rule of maximizing expected net present value. Yet, most members of the public are risk averse for gains in financial and other domains, implying that, for them, the value (meaning the certainty equivalent, or selling price value) of an uncertain gain is *less* than its expected value. The use of risk neutrality, or expected values, in public finance decisions for public projects in a society of risk-averse individuals is supported by the Arrow-Lind theorem (Arrow and Lind 1970), which justifies risk neutrality, under certain conditions, when individuals can diversify risk away. But in cases such as those just illustrated, where many individuals face the same (completely correlated) risks, the conditions of the theorem do not hold. As noted by Foldes and Rees (1977), "The assumption [of the Arrow-Lind theorem] that the share of the net benefits of an investment accruing to any person becomes negligible as population tends to infinity is unacceptable in at least three cases: for public goods, where the benefit is not 'shared' but increases with the population; for projects whose scale must be adjusted roughly in proportion to the size of population (such as the construction of a grid system of electricity distribution); and for projects whose benefits accrue wholly or in part to a section of the population which is 'small' in the sense of thetheorem." Our examples in which scaling up small uncertain net benefits (essentially by making the same bet, that net benefits of an action are positive, simultaneously for many individuals or at many locations) changes the certainty equivalent from positive to negative correspond to the first two of these exceptions: benefits (such as longer life or avoided deaths) that are not shared, and projects (such as restrictions on point source emissions) that are geographically distributed in the population. Under these conditions, the Arrow-Lind theorem no longer warrants use of risk neutrality in evaluating uncertain prospects, and the fact that risk aversion implies that the value of an uncertain benefit is less than its expected value again becomes relevant for public decision-making. By ignoring risk aversion, BCA risks accepting projects whose certainty equivalents are negative for those who pay for them. Of course, how aggregate certainty equivalents or project acceptability should be determined when individuals have different degrees of risk aversion, or different utility functions, invites the unwelcome complexities and impossibility results of collective choice theory. But simply assuming risk neutrality is not a satisfactory simplification if it encourages systematically approving large-scale projects with widely geographically distributed costs and benefits that would be rejected if uncertainty about net benefits were explicitly considered, in addition to expected values.

Doing Better: Using Predictable Rational Regret to Improve BCA

The preceding discussion suggests that decision biases can lead to both individual and collective decision processes that place too little value on collecting relevant information, rely too heavily on uninformed or under-informed judgments (which tend to be over-optimistic and over-confident), and hence systematically over-value prospects with uncertain costs and benefits, creating excessive willingness to gamble on them. One result is *predictable disappointment*: consistent over-investment in uncertain and costly prospects that, predictably, will be seen in retrospect to have (on average) cost more and delivered less than expected. A second possible adverse outcome is *predictable regret*: investing limited resources in prospects with uncertain net benefits when, predictably, it will be clear in hindsight that the resources could have been better spent on something else. Standard BCA facilitates these tendencies by encouraging use of current expected values to make choices among alternatives, instead of emphasizing more complex, but potentially less costly (on average), optimal sequential strategies that require waiting, monitoring, and inaction until conditions and information justify costly interventions (Stokey 2009). This section considers how to do better, and what "better" means.

A long-standing tradition in decision analysis and normative theories of rational decision-making complements the principle of maximizing expected utility with various versions of minimizing expected rational regret (e.g., Loomes and Sugden 1982; Bell 1985). Formulations of rational regret typically represent it as a measure of the difference between the reward (e.g., net benefit, in a BCA context) that one's decision *actually* achieved and the greatest reward that *could* have been achieved had one made a different (feasible) decision instead (Hart 2005; Hazan and Kale 2007). Adjusting decision rules to reduce rational regret plays a crucial role in current reinforcement learning algorithms, as well as in neurobiological studies of human and animal learning, adaptation, and decision-making, within the general framework of computational reinforcement learning (e.g., Li and Daw 2011; Schönberg et al. 2007). (By contrast, related concepts such as elation or disappointment (Delquié and Cillo 2006) reflect differences between expected or predicted rewards and those actually received. They do not necessarily attribute the difference to one's own decisions, or provide an opportunity to learn how to make more effective decisions.)

Intuitively, instead of prescribing that current decisions should attempt to maximize prospective expected reward (or expected discounted net benefits), rational regret-based theories prescribe that they should be made so that, even in hindsight, one has no reason to change the decision process to increase average rewards. In effect, instead of the advice "Choose the alternative with the greatest expected value or utility," normative theories of regret give the advice "Think about how, in retrospect, you would want to make decisions in these situations, so that no change in the decision procedure would improve the resulting distribution of outcomes. Then, make decisions that way." In this context, a no-regret rule (Chang 2007) is one that, even in retrospect, one would not wish to modify before using again, since no feasible modification would lead to a preferred distribution of future consequences. Equivalently, if various options are available for modifying decision rules to try to improve the frequency distribution of rewards that they generate, then a no-regret rule is one that cannot be improved upon: it is a fixed point of the decision rule-improvement process (Hazan and Kale 2007). These concepts apply to what we are calling rational regrets, i.e., to regrets about not making decisions that would have improved reward distributions.

Example: Rational vs. Irrational Regret

Suppose that a decision maker's reward (or "payoff," in the game-theoretic terminology often used) is determined by her choice of an act, A or B, together with a random state of nature (e.g., the outcome of one toss of a fair die, with faces 1-6 being equally likely, revealed only after the choice has been made. Possible payoffs range between 1 and 6.1 dollars, as described by the following table.

	State:					
Decision	1	2	3	4	5	6
Act A:	1	2	3	4	5	6.1
Act B:	2	3	4	5	6	1

Expected utility theory unequivocally prescribes choosing act A, since its probability distribution of rewards stochastically dominates that of act B, as $6.1 > 6$, even though act B yield a higher payoff than A 5/6 of the time. Minimizing rational regret also prescribes choosing act A, since any decision rule that prescribes choosing B in this situation (always or sometimes) will yield a payoff frequency distribution that is inferior to (stochastically dominated by) the payoff distribution from always choosing act A. In this simple case, choosing act A and then observing that choosing B would have yielded a higher reward provides no reason for a rational decision-maker to deviate from the optimal strategy of always choosing act A. Thus, minimizing rational regret recommends A, not B.

Other concepts of regret and regret-avoidance are linked to personality psychology. These include making decisions with low potential for regret to protect damaging already low self-esteem (Josephs et al. 1992), as well as preferring to avoid learning outcomes in order to avoid possible regrets. From a biological perspective, it has been proposed that the *emotion* of regret, when used as an error signal to adaptively modify decision rules in individual decision-making, is a "rational emotion" that helps us to learn and adapt decision-making effectively to uncertain and changing environments (e.g., Bourgeois-Gironde 2010). Although psychological and biological aspects of regret are important for some kinds of decision-making under risk, it is primarily proposed concepts of rational regret, as just discussed, that we believe can be most useful for improving the practice of BCA. The rest of this section explains how.

Does the shift in perspective from maximizing prospective expected net benefits to minimizing expected retrospective regret make any practical difference in what actions are recommended? Not for *homo economicus*. For an ideally rational SEU decision-maker, the principle of maximizing SEU, while optimally taking into account future plans (contingent on future events) and the value of information, is already a no-regret rule. But for real decision-makers (whether individuals or groups) who are not able to formulate trustworthy, crisp, agreed-to probabilities for the consequences of each choice, the shift to minimizing regret has several powerful practical advantages over trying to maximize expected net benefits. Among these are the following:

- *Encourage prospective hindsight analysis.* A very practical aid for reducing overconfidence and optimism bias is for decision-makers to imagine that a contemplated project or investment ends badly, and then to figure out what could have caused this and how it might have been prevented. Such "prospective hindsight" or "premortem" exercises have been used successfully in business to help curb under-estimation of costs and over-estimation of benefits when both are highly uncertain (Russo and Schoemaker 1989). In the domain of regulatory benefit-cost analysis, they prompt questions such as: Suppose that, 20 years from now, we rigorously assess the health benefits and economic costs actually achieved by extending Clean Air Act amendments, and find that the costs were on the order of a trillion dollars (EPA 2011a, b), but that the projected benefits of reduced mortality rates caused by cleaner air never materialized. How might this have happened? Could it have been discovered sooner? What might we do now or soon to prevent such an outcome? When such questions are asked on a small scale, such as for an individual study that mistakenly (but confidently) projected large, obvious, life-saving benefits from banning coal-burning in Dublin (Clancy et al. 2002; Harvard School of Public Health 2002), only to discover a decade later that the reduction reflected a historical trend that had no apparent relation to the ban (HEI 2013), they lead to simple answers, such as to use a control group (people outside the affected area) to determine whether the bans actually produce the effects projected for and attributed to them by those who believe they are beneficial (HEI 2013; Campbell and Stanley 1966). On a national level, a similar openness to the possibility of errors in projections, and vigilance in frequently testing uncertain assumptions against data as the effects of expensive regulations become manifest, might likewise be used to anticipate and prevent the BCA failure scenarios imagined in premortem exercises. In the U.S., for example, learning from the experiences of cities, counties, or states (such as California) who are early adopters of policies or initiatives that are later proposed for national implementation provides opportunities to check assumptions against data relatively early, and to modify or optimally slow-roll (Stokey 2009) the implementation of national-level policies as needed to reduce expected regret.

- *Increase feedback and learning.* Studies of individual and organizational decision-making document several types of failures to learn from real-world feedback based on the gaps between what was expected and what actually occurred, or between what was achieved and what could have been achieved by better decisions (if this is known) (Russo and Schoemaker 1989). Formal models of how to adaptively modify decision processes or decision rules to reduce regret—for example, by selecting actions next time a situation is encountered in a Markov decision process, or in a game against nature (with an unpredictable, possibly adversarial, environment) using probabilities that reflect cumulative regret for not having used each action in such situations in the past—require explicitly collecting and analyzing such data (Robards and Sunehag 2011; Hazan and Kale 2007). Less formally, continually assessing the outcomes of decisions and how one might have done better, as required by the regret-minimization framework, means that opportunities to learn from experience will more often be exploited instead of missed.
- *Increase experimentation and adaptation.* An obvious possible limitation of regret-minimization is that one may not know what would have happened if different decisions had been made, or what probabilities of different outcomes would have been induced by different choices (Jaksch et al. 2010). This is the case when relevant probabilities are unknown or ambiguous. It can arise in practice when no states or counties have been early (or late) adopters of a proposed national policy, and so there is no comparison group to reveal what would have happened had it not been adopted. In this case, formal models of regret reduction typically require exploring different decision rules to find out what works best. Such learning strategies (called "on-policy" learning algorithms, since they learn only from experience with the policy actually used, rather than from information about what would have happened if something different had been tried) have been extensively developed and applied successfully to regret reduction in machine learning and game theory (Chang 2007; Yu et al. 2009; Robards and Sunehag 2011). They adaptively weed out the policies that are followed by the least desirable consequences, and increase the selection probabilities for policies that are followed by preferred consequences. Many formal models of regret-minimization and no-regret learning strategies (e.g., Chang 2007; Jaksch et al. 2010 for Markov decision processes) have investigated how best to balance exploration of new decision rules and exploitation of the best ones discovered so far. Under a broad range of conditions, such adaptive selection (via increased probabilities of re-use) of the decision rules that work best empirically soon leads to discovery and adoption of optimal or near-optimal ('no-regret") decision rules (i.e., maximizing average rewards) (Chang 2007; Robards and Sunehag 2011; Hazan and Kale 2007). Of course, translating these mathematical insights from the simplified world of formal decision models (e.g., Markov decision processes with initially unknown transition and reward probabilities and costs of experimentation) to the real world requires caution. But the basic principle that the policies that will truly maximize average net benefits per period (or discounted net benefits, in other formulations) may initially be unknown, and

that they should then be discovered via well-designed and closely analyzed trials, has powerful implications for the practice of BCA and policy making. It emphasizes the desirability of conducting, and carefully learning from, pilot programs and trial evaluations (or natural experiments, where available) before rolling out large-scale implementations of regulations or other changes having highly uncertain costs or benefits. In effect, the risk of failure or substantially sub-optimal performance from programs whose assumptions and expectations about costs and benefits turn out to be incorrect can be reduced by small-scale trial-and-error learning, making it unnecessary to gamble that recommendations based on BCA using current information will turn out to coincide with those that will be preferred in hindsight, after key uncertainties are resolved.

- *Asymptotic optimization of decision rules with initially unknown probabilities for consequences.* In formal mathematical models of no-regret reinforcement learning with initially unknown environments and reward probabilities, swift convergence of the prescriptions from empirical regret-minimization algorithms to approximately optimal policies holds even if the underlying process tying decisions to outcome probabilities is unknown or slowly changing (Yu et al. 2009). This makes regret-minimization especially relevant and useful in real-world applications with unknown or uncertain probabilities for the consequences of alternative actions. It also provides a constructive approach for avoiding the fundamental limitations of collective choice mechanisms that require combining the subjective probabilities (or expected values) of different participants in order to make a collective choice (Hylland and Zeckhauser 1979; Nehring 2007). Instead of trying to reconcile or combine discrepant probability estimates, no-regret learning encourages collecting additional information that will clarify which among competing alternative policies work best. Again, the most important lesson from the formal models is that adaptively modifying policies (i.e., decision rules) to reduce empirical estimates of regret based on multiple small trials can dramatically improve the final choice of policies and the final results produced (e.g., average rewards per period, or discounted net benefits actually achieved). From this perspective, recommending *any* policy based on analysis and comparison of its expected costs and benefits to those of feasible alternatives will often be inferior to recommending a process of trials and learning to discover what works best. No-regret learning (Chang 2007) formalizes this intuition.

In summary, adjusting decision processes to reduce empirical estimates of regret, based on actual outcomes following alternative decisions, can lead to much better average rewards or discounted net benefits than other approaches. Real-world examples abound of small-scale trial and error leading to successful adaptation in highly uncertain business, military, and policy environments (e.g., Harford 2011).

Conclusions

This chapter has argued that a foundational principle of traditional BCA, choosing among proposed alternatives to maximize the expected net present value of net benefits, is not well suited to guide public policy choices when costs or benefits are highly uncertain. In principle, even if preferences are aligned (so that familiar collective choice paradoxes and impossibility results caused by very different individual preferences do not arise)—for example, even if all participants share a common goal of reducing mortality risks—there is no way (barring such extremes as dictatorship) to aggregate sufficiently diverse probabilistic beliefs to avoid selecting outcomes that no one favors (Hylland and Zeckhauser 1979; Nehring 2007). BCA does not overcome such fundamental limitations in any formulation that requires combining probability estimates from multiple participants to arrive at a collective choice among competing alternatives—including using such probabilities to estimate which alternative has the greatest expected net benefit.

Reviewers of an earlier version of this work raised the sensible question of whether the problems that we have noted for BCA are really fundamental, implied even in principle by the nature and logic of BCA, or whether they merely reflect imperfections of execution. Does not willingness to use expected costs and benefits with available information reflect the sorts of pragmatic compromises and imperfections that arise when methods that are sound and useful in principle are applied to messy real-world problems with realistic data and knowledge gaps? It is a fair question. The answer developed here is that BCA can be improved in many cases, in principle as well as in practice, by abandoning maximization of expected net benefits, or expected net present value, as a normative principle. Uncertainty often *should* be taken into account—deliberately exploited, even—in ways that require using more information than just a comparison of expected costs and benefits. Some of those ways include deliberate experimentation and learning, avoidance of large simultaneous bets on the same unresolved uncertainty (e.g., about whether changing exposure to a particular drug or chemical or food or educational practice will really create net benefits), and disciplining decision procedures to help reduce the effects of psychological biases on evaluations of expected costs and benefits.

In practice, as discussed in Chap. 1, a variety of well-known decision biases conspire to make subjectively assessed expected value calculations and WTP estimates untrustworthy, with highly uncertain benefits often tending to be over-estimated, and highly uncertain costs tending to be under-estimated. Biases that contribute to unreliable expected net benefit and WTP estimates range from the affect heuristic, which we view as fundamental, to optimism, over-confidence, and confirmation biases, ambiguity aversion, and finally to what we have called learning aversion. These biases make it predictable that projects and proposals with highly uncertain costs and benefits will tend to be over-valued, leading to potentially regrettable decisions, meaning decisions that, in retrospect, and upon rational review, one would want to have made differently. Similar results have been demonstrated for groups and for individuals (Russo and Schoemaker 1989). The net result is a proclivity to gamble on excessively risky proposals when the benefits and costs are highly uncertain.

To help overcome these difficulties, we have proposed shifting to a different foundation for BCA calculations and procedures: minimizing rational regret. Regret minimization principles been developed in both decision analysis (e.g., Loomes and Sugden 1982; Bell 1985) and extensively in more recent machine learning, game theory, and neurobiological models of reinforcement learning (Hart 2005; Chang 2007; Hazan and Kale 2007; Li and Daw 2011; Schönberg et al. 2007). We propose that seeking information and making decisions deliberately to reduce predictable rational regret can help to correct for many of the biases that predictably lead to exaggerated WTP values for proposed projects with good intentions (positive affect) but very uncertain consequences. Although the idealized mathematical models and analyses of these fields are not necessarily directly applicable to real-world BCA settings, they do suggest several practical principles that have proved valuable in improving real-world individual and collective decisions when potential costs and benefits are uncertain enough so that the best course of action (given clarity on goals) is not clear. In particular, we propose that BCA under such conditions of high uncertainty can be improved by greater use of prospective hindsight (or "premortem") analyses to reduce decision biases; explicit data collection and careful retrospective evaluation and comparison of what was actually achieved to what was expected, and to what could have been achieved by different choices (when this can be determined); and deliberate learning and adaptation of decision rules based on the results of multiple small-scale trials in settings for which this is practical.

Not all of these principles are applicable in all BCA situations, of course. Whether to build a bridge in a certain location cannot be decided by multiple small-scale trials, for example. But for many important health, safety, and environmental regulations with substantial costs and substantial uncertainty about benefits, learning from experiences on smaller scales (e.g., from the changes in mortality rates following different histories of pollution reductions in different counties) can powerfully inform and improve BCA analyses that are intended to guide larger-scale (e.g., national) policy-making. The main proposed shift in emphasis is from *guessing* what will work best (in the sense of maximizing the expected NPV of net benefits, as assessed by experts or other participants in the decision-making process), and then perhaps betting national policies on the answer, to *discovering* empirically what works best, when it is practical to do so and when the answer is initially highly uncertain.

References

Al-Najjar NI, Weinstein J (2009) The ambiguity aversion literature: a critical assessment. Econ Philos 25 (Special Issue 03): 249–284

Ariely D (2009) Predictably irrational: the hidden forces that shape our decisions. Revised and Expanded Edition. HarperCollins, New York, NY

Armstrong K, Schwartz JS, Fitzgerald G, Putt M, Ubel PA (2002) Effect of framing. Med Decis Making 22(1):76–83

Arrow KJ (1950) A difficulty in the concept of social welfare. J Polit Econ 58(4):328–346

Arrow KJ, Lind RC (1970) Uncertainty and the evaluation of public investment decisions. Am Econ Rev 60:364–378

Bell DE (1985) Putting a premium on regret. Manag Sci 31(1):117–120. https://doi.org/10.1287/mnsc.31.1.117

Bennett R, Blaney RJP (2002) Social consensus, moral intensity and willingness to pay to address a farm animal welfare issue. J Econ Psychol 23(4):501–520

Bourgeois-Gironde S (2010) Regret and the rationality of choices. Philos Trans R SocLond B Biol Sci 365(1538):249–257. https://doi.org/10.1098/rstb.2009.0163

Campbell DT, Stanley JC (1966) Experimental and quasi-experimental designs for research. Rand McNally, Chicago

Casey JT, Delquie P (1995) Stated vs. implicit willingness to pay under risk. Organ Behav Hum Decis Process 61(2):123–137

Champ PA, Bishop RC (2006) Is willingness to pay for a public good sensitive to the elicitation format? Land Econ 82(2):162–173

Chang YC (2007) No regrets about no-regret. Artif Intell 171:434–439

Clancy L, Goodman P, Sinclair H, Dockery DW (2002) Effect of air-pollution control on death rates in Dublin, Ireland: An intervention study. Lancet 360(9341):1210–1214

Cox LA Jr (2012) Reassessing the human health benefits from cleaner air. Risk Anal 32(5):816–829

DeGroot MH (2004) Optimal statistical decisions. Wiley Classics Library (Originally published (1970) by McGraw-Hill)

Delquié P, Cillo A (2006) Disappointment without prior expectation: a unifying perspective on decision under risk. J Risk Uncertainty 33(3):197–215. https://doi.org/10.1007/s11166-006-0499-4

Djulbegovic B, Kumar A, Magazin A, Schroen AT, Soares H, Hozo I, Clarke M, Sargent D, Schell MJ (2011) Optimism bias leads to inconclusive results-an empirical study. J ClinEpidemiol 64(6):583–593. https://doi.org/10.1016/j.jclinepi.2010.09.007

EPA (2011a) The benefits and costs of the clean air act from 1990 to 2020: Summary Report. U.-S. EPA, Office of Air and Radiation. Washington, DC. http://www.epa.gov/air/sect812/aug10/summaryreport.pdf

EPA (2011b) The benefits and costs of the clean air act from 1990 to 2020. Full report. U.S. EPA, Office of Air and Radiation, Washington, DC. http://www.epa.gov/oar/sect812/feb11/fullreport.pdf

Feldman AM (2004) Kaldor-Hicks compensation. In: Newman P (ed) The New Palgrave dictionary of economics and the law, vol 2, E-O, pp 417–412. http://www.econ.brown.edu/fac/allan_feldman/AMF%20Significant%20Published%20Papers/Kaldor-Hicks%20Compensation.pdf

Foldes LP, Rees R (1977) A note on the arrow-lind theorem. Am Econ Rev 67(2):188–193

Gan HK, You B, Pond GR, Chen EX (2012) Assumptions of expected benefits in randomized phase III trials evaluating systemic treatments for cancer. J Natl Cancer Inst 104(8):590–598. https://doi.org/10.1093/jnci/djs141

Gardner D (2009) The science of fear: how the culture of fear manipulates your brain. Penguin Group, New York, NY

Gelfand S (2013) Clinical equipoise: actual or hypothetical disagreement? J Med Philos 38(6): 590–604. https://doi.org/10.1093/jmp/jht023

Gilboa I, Schmeidler D (1989) Maxmin expected utility with a non-unique prior. J Math Econ 18: 141–153

Goodman SN (2001) Of P-values and Bayes: a modest proposal. Epidemiology 12(3):295–297. No abstract available

Grossman PZ, Cearley RW, Cole DH (2006) Uncertainty, insurance, and the Learned Hand formula. Law, Probab Risk 5(1):1–18

Harford T (2011) Adapt: why success always starts with failure. New York, NY, Farra, Straus and Giroux

Hart A (2005) Adaptive heuristics. Econometrica 73(5):1401–1430. http://www.math.huji.ac.il/~hart/papers/heurist.pdf

Harvard School of Public Health (2002) Press release: "Ban on coal burning in dublin cleans the air and reduces death rates". www.hsph.harvard.edu/news/press-releases/archives/2002-releases/press10172002.html

Hazan E, Kale S (2007) Computational equivalence of fixed points and no regret algorithms, and convergence to equilibria. Adv Neural Inf Process Syst 20:625–632

Health Effects Institute (HEI) (2013) Did the Irish coal bans improve air quality and health? HEI Update, Summer, 2013. http://pubs.healtheffects.org/getfile.php?u=929. Last Retrieved 1 February 2014

Hershey JC, Kunreuther HC, Schoemaker PJH (1982) Sources of bias in assessment procedures for utility functions. Manag Sci 28(8):936–954

Hicks J (1939) The foundations of welfare economics. Econ J 49(196):696–712

Hoy M, Peter R, Richter A (2014) Take-up for genetic tests and ambiguity. J Risk Uncertain 48:111–133. https://doi.org/10.1007/s11166-014-9186-z

Hylland A, Zeckhauser RJ (1979) The impossibility of bayesian group decision making with separate aggregation of beliefs and values. Economet Econom Soc 47(6):1321–1336

Ioannidis JPA (2005) Why most published research findings are false. PLoS Med 2(8):e124. https://doi.org/10.1371/journal.pmed.0020124

Jaksch T, Ortner R, Auer P (2010) Near-optimal regret bounds for reinforcement learning. J Mach Learn Res 11:1563–1600

Josephs RA, Larrick RP, Steele CM, Nisbett RE (1992) Protecting the self from the negative consequences of risky decisions. J PersSoc Psychol 62(1):26–37

Kahneman D (2011) Thinking, fast and slow. Farrar, Straus, and Giroux, New York, NY

Kahneman D, Frederick S (2005) A model of heuristic judgment. In: Holyoak KJ, Morrison RG (eds) The Cambridge handbook of thinking and reasoning. Cambridge University Press, New York, pp 267–293

Kahneman D, Tversky A (1979) Intuitive prediction: biases and corrective procedures. TIMS Stud Manag Sci 12:313–327

Kaldor N (1939) Welfare propositions in economics and interpersonal comparisons of utility. Econ J 49(195):549–552

Kralik JD, Xu ER, Knight EJ, Khan SA, Levine WJ (2012) When less is more: Evolutionary origins of the affect heuristic. PLoS One 7(10):e46240. https://doi.org/10.1371/journal.pone.0046240

Lehrer J (2012) Trials and errors: why science is failing us. Wired January 28, 2012. http://www.wired.co.uk/magazine/archive/2012/02/features/trials-and-errors?page=all

Li J, Daw ND (2011) Signals in human striatum are appropriate for policy. J Neurosci 31(14): 5504–5511. https://doi.org/10.1523/JNEUROSCI.6316-10.2011

Louis P (2009) Learning aversion and voting rules in collective decision making. mimeo, UniversitatAutonoma de Barcelona

Loomes G, Sugden R (1982) Regret theory: an alternative theory of rational choice under uncertainty. Econ J 92(368):805–824

Man PTY, Takayama S (2013) A unifying impossibility theorem. Economic Theory 54(2): 249–271. http://www.shino.info/papers/alternatives_6.pdf

Maccheronia F, Marinacci M, Rustichini A (2006) Ambiguity aversion, robustness, and the variational representation of preferences. Econometrica 74(6):1447–1498

Mueller DC (2003) Public Choice III. Cambridge University Press, New York

Navarro AD, Fantino E (2005 Jan) The sunk cost effect in pigeons and humans. J Exp Anal Behav 83(1):1–13

Nehring K (2007) The impossibility of a paretian rational: a Bayesian perspective. Econ Lett 96(1): 45–50

Newby-Clark IR, Ross M, Buehler R, Koehler DJ, Griffin D (2000) People focus on optimistic scenarios and disregard pessimistic scenarios while predicting task completion times. J Exp Psychol Appl 6(3):171–182

Nuzzo R (2014) Scientific method: Statistical errors. P values, the 'gold standard' of statistical validity, are not as reliable as many scientists assume. Nature 506:150–152. https://doi.org/10.1038/506150a

Othman A, Sandholm T (2009) How pervasive is the Myerson-Satterthwaite impossibility? Proceedings of the 21st international joint conference on artifical intelligence, IJCAI'09. Morgan Kaufmann Publishers, San Francisco, CA, pp 233–238

Politi MC, Clark MA, Ombao H, Dizon D, Elwyn G (2011) Communicating uncertainty can lead to less decision satisfaction: a necessary cost of involving patients in shared decision making? Health Expect 14(1):84–91. https://doi.org/10.1111/j.1369-7625.2010.00626.x

Portney PR (2008) Benefit-cost analysis. In Henderson DR (ed) The concise encyclopedia of economics. Library of Economics and Liberty. Last Retrieved 1 February 2014. http://www.econlib.org/library/Enc/BenefitCostAnalysis.html

Poundstone W (2010) Priceless: the myth of fair value (and how to take advantage of it). Scribe Publications

Robards M, Sunehag P (2011) Near-optimal on-policy control. https://www.academia.edu/86513073/Near_Optimal_On_Policy_Control

Rothman KJ (1990) No adjustments are needed for multiple comparisons. Epidemiology 1:43–46

Russo JE, Schoemaker PJH (1989) Decision traps: ten barriers to brilliant decision-making and how to overcome them. Doubleday, New York

Schönberg T, Daw ND, Joel D, O'Doherty JP (2007) Reinforcement learning signals in the human striatum distinguish learners from nonlearners during reward-based decision making. J Neurosci 27(47):12860–12867

Sarewitz D (2012) Beware the creeping cracks of bias. Nature 485:149

Slovic P, Finucane M, Peters E, MacGregor DG (2002) Rational actors or rational fools: implications of the affect heuristic for behavioral economics. J Socioecon 31(4):329–342

Slovic P, Finucane M, Peters E, MacGregor D (2004) Risk as analysis and risk as feelings: some thoughts about affect, reason, risk, and rationality. Risk Anal 24(2):311–322

Smith JE, von Winterfeldt D (2004) Decision analysis. Manag Sci 50(5):561–574

Stokey NL (2009) The economics of inaction: stochastic control models with fixed costs. Princeton University Press, Princeton, NJ

Thaler RH (1999) Mental accounting matters. J Behav Decis Mak 12:183–206

Treasury Board of Canada Secretariat (1988) Benefit-cost analysis guide (Draft). Last Retrieved 1 February 2014. http://classwebs.spea.indiana.edu/krutilla/v541/Benfit-Cost%20Guide.pdf

Weber M, Kopelman S, Messick DM (2004) A conceptual review of social dilemmas: applying a logic of appropriateness. Personal Soc Psychol Rev 8:281–307

Yu JY, Mannor S, Shimkin N (2009) Markov decision processes with arbitrary reward processes. Math Oper Res 34(3):737–757

Part III
Ways Forward

Chapter 7
Addressing Wicked Problems and Deep Uncertainties in Risk Analysis

Introduction: How to Make Good Decisions with Deep Uncertainties?

Some of the most troubling risk management challenges of our time are characterized by *deep uncertainties*. Well-validated, trustworthy, risk models giving the probabilities of future consequences for alternative present decisions are not available; the relevance of past data for predicting future outcomes is in doubt; experts disagree about the probable consequences of alternative policies—or, worse, reach an unwarranted consensus that replaces acknowledgment of uncertainties and information gaps with groupthink—and policy makers (and probably various political constituencies) are divided about what actions to take to reduce risks and increase benefits. For such risks, there is little or no agreement even about what decision models to use, and risk analysts may feel morally obliged not to oversimplify the analysis by imposing one (Churchman 1967; Rittel and Webber 1973). Passions may run high and convictions of being right run deep, in the absence of enough objective information to support rational decision analysis and conflict resolution (Burton 2008).

Examples of risk management with deep uncertainties include deciding where, when, and how to prepare for future effects of climate change (and, perhaps, of efforts to mitigate it); managing risks from epidemics and new or deliberately spread pathogens; protecting valuable but vulnerable species, habitats, and ecosystems from irreversible loss; testing and reducing new interdependencies in financial systems to reduce risks of catastrophic failure; designing and managing power grids and energy and traffic networks to increase their resilience and reduce their vulnerability to cascading failures; and trying to anticipate and defend against credible threats from terrorists, cybercriminals, bank fraud, and other adversarial risks. The final section will return to these motivating challenges, after we have reviewed technical concepts and methods that can help to meet them.

L. A. Cox Jr., *AI-ML for Decision and Risk Analysis*, International Series in Operations Research & Management Science 345,
https://doi.org/10.1007/978-3-031-32013-2_7

		Level 1	Level 2	Level 3	Level 4	
				Deep Uncertainty		
Determinism	**Context**	A clear enough future 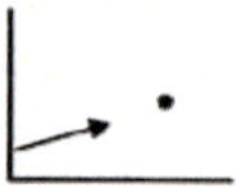	Alternate futures (with probabilities) 	A multiplicity of plausible futures 	Unknown future 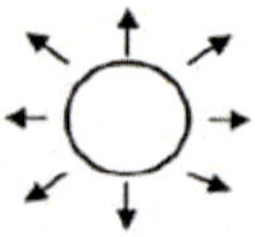	Total ignorance
	System model	A single system model	A single system model with a probabilistic parameterization	Several system models, with different structures	Unknown system model; know we don't know	
	System outcomes	A point estimate and confidence interval for each outcome	Several sets of point estimates and confidence intervals for the outcomes, with a probability attached to each set	A known range of outcomes	Unknown outcomes; know we don't know	
	Weights on outcomes	A single estimate of the weights	Several sets of weights, with a probability attached to each set	A known range of weights	Unknown weights; know we don't know	

Fig. 7.1 A suggested taxonomy of uncertainties (Walker et al. 2010)

Figure 7.1 (Walker et al. 2010) summarizes some uncertainties about matters of fact and value that separate deep uncertainties (right two columns, Levels 3 and 4) from the more tractable uncertainties encountered in statistics and scenario analysis with known probabilities (left two columns, Levels 1 and 2). (The "weights on outcomes" row at the bottom alludes to value weights, and allows for uncertain preferences or utilities.)

Although these challenges are formidable, the underlying risks are too important to ignore and too complex to dispose of easily. Policy makers will continue to turn to risk analysts for help. Risk analysts, in turn, need to be familiar with the best available methods for improving risk management decisions under such trying conditions. This chapter summarizes recent progress in ideas and methods that can help. There has been great progress in technical methods for assessing and managing risks with deep uncertainties in recent years, usually using multiple models and scenarios. These methods are not yet widely used in risk analysis, compared to older methods that select a single statistical or simulation model and then perform sensitivity analyses on its results. The following sections seek to create an expository bridge from statistical methods and concepts that many risk analysts might already be familiar with (such as resampling techniques for robust statistical inference) to newer ideas from machine learning, robust optimization, and adaptive control that may be less familiar, but that are promising for dealing with deep uncertainties in risk analysis.

Principles and Challenges for Coping with Deep Uncertainty

There is no shortage of advice for managing risks with deep uncertainties. We should design fault-tolerant, survivable, and resilient organizations, systems and infrastructure. We should experiment with possible improvements, learn quickly, effectively and humbly from our own and others' mistakes and experiences (including accident precursors and unexpected events), and actively seek feedback and local "on the ground" information so that we can adapt flexibly to unforeseen circumstances and performance glitches. We should loosen or decouple the tight couplings and dependencies in existing complex systems and infrastructure—from oil rigs to financial systems—that set the stage for swiftly cascading failures and "normal accidents" (Harford 2011). By adopting a vigilant, risk-aware mind set and culture, we can, perhaps, build highly reliable organizations (HROs) around the five principles of preoccupation with failure, reluctance to simplify interpretations of data and anomalies, sensitivity to operations, commitment to resilience, and deference to expertise rather than to authority (Weick and Sutcliffe 2007).

The practical problem is thus not finding logical principles for managing risks with deep uncertainties, but figuring out how best to implement them in detail. Risk analysts who, however rightly, respond to deep uncertainty by advocating greater learning and flexibility, or by promoting the virtues of adaptation and resilience to communities, institutions, and organizations, may be unsure how to bring them about, or how much good they will do if implemented. The following sections review methods that can help to improve risk management decisions when correct models are unknown and learning and adaptation to new data are essential.

Point of Departure: Subjective Expected Utility (SEU) Decision Theory

Traditional decision and risk analysis make extensive use of models to predict the probable consequences of alternative risk management decisions. The paradigmatic analysis of decisions using subjective expected utility (SEU) theory, the gold standard for normative models of rational decision-making with Level 1 uncertainties, proceeds as follows (Gilboa and Schmeidler 1989; Pinker 2021):

- Identify a *choice set A* of alternative risk management acts. The decision problem is posed as choosing among the acts in *A*. Acts may represent not only alternative actions, such as resource allocations, but also rules for making decisions over time, such as alternative regulatory standards, adaptive feedback control policies, decision rules, collective choice rules, liability allocation rules, investment strategies, intervention trigger rules, etc., depending on who is choosing what.

- Identify a set C of *possible consequences*. Choices of acts from A are to be made in an effort to make preferred consequences in C more likely and undesired ones less likely.
- *Quantify preferences*. This is typically done by assessing a von Neumann-Morgenstern utility $u(c)$, between 0 and 1, for each consequence c in C, such that the decision-maker is indifferent between receiving consequence c with certainty and receiving the most-preferred consequence in C with probability $u(c)$, and otherwise receiving the least-preferred consequence in C.
- *Optimize decisions*. Expected utility (EU) theory prescribes selecting an act in A that will maximize the expected value of $u(c)$, called *expected utility*. This prescription is typically justified by normative axioms for "rational" decision-making. It is implemented with the help of a probabilistic consequence (or risk) model, $Pr(c \mid a)$, giving the probability of each consequence c if each act a is selected. Specifically, the expected utility of act a is $EU(a) = \Sigma_c Pr(c \mid a)u(c)$, with the sum replaced by an integral if the consequence set is continuous.
- *Model and assess uncertainties*. If no well-validated empirical model, $Pr(c \mid a)$, is available, then use subjective probability judgments to complete a subjective expected utility (SEU) model. For example, suppose that the consequence of choosing act a depends on what else happens that is not directly controllable by the decision-maker (i.e., not in the choice set A). These other inputs—which, together with a, determine the consequence—lie in a set S of possible *scenarios*, or *states* of nature. If $c(a, s)$ denotes the consequence that occurs when act a is chosen and state s occurs, then the expected utility of act a can be expressed as $EU(a) = \Sigma_s u[c(a, s)]Pr(s)$. (More generally, if the consequence of a pair (a, s) is not deterministic, e.g., due to stochastic elements, then a conditional probability model for consequences, $Pr(c \mid a, s)$, can be used to compute expected utility via the formula $EU(a) = \Sigma_c u(c)Pr(c \mid a) = \Sigma_c u(c)[\Sigma_s Pr(c \mid a, s)Pr(s)]$.) If necessary, subjective probabilities $Pr(s)$ for the states can be developed or elicited, e.g., based on willingness to bet on each s compared to other events with known probabilities (perhaps after calibration training). Subjective expected utility (SEU) theory shows that a decision-maker with preferences satisfying certain axioms should behave as if she had coherent subjective probabilities $Pr(s)$ and should choose acts that maximize EU calculated from these probabilities

Thus, probabilistic consequence models, $Pr(c \mid a)$ (perhaps built up from components, such as $c(a, s)$ or $Pr(c \mid a, s)$, and $Pr(s)$) play a crucial role in enabling rational decision-making via the SEU paradigm. A completely known EU *decision model* for supporting a risk management decision can be summarized by a quadruple $M = \{A, C, u(c), Pr(c \mid a)\}$. When the correct decision model is known and agreed on, EU provides a compelling normative framework for deciding what to do.

Four Major Obstacles to Applying SEU to Risk Management with Model Uncertainty

Fruitful though the SEU framework is, it cannot easily be applied to some of the most important risk management decisions that trouble modern societies, due to deep uncertainties. One obstacle to its practical application is uncertainty about *what alternatives are available* in the choice set A. A common problem is premature focusing on only a few salient options, which are not necessarily the best that could be devised. A second obstacle is uncertainty about the full range of *possible consequences* in C. The challenges of "unknown unknowns" or failures of imagination for potential consequences—e.g., failure to correctly envision and account for all the important consequences that an act might make more probable, or lack of confidence that all such important consequences have been identified—raise the concern that surprising "black swan" outcomes may occur which were not considered when the decision was being made, but which would have changed the decision if they had been considered. The oft-bemoaned law of unintended consequences expresses this concern.

A third obstacle is that, even if A and C are both known, the *correct risk model* $Pr(c \mid a)$ for consequence probabilities for different acts may not be known (perhaps because the underlying state or scenario probabilities, $Pr(s)$, are not well known). Different stakeholders may have conflicting beliefs about $Pr(c \mid a)$, and hence conflicting beliefs about which act will maximize expected utility.

Finally, uncertainties or conflicts about *values and preferences* to be encoded in the utility function $u(c)$ used to evaluate different consequences—for example, arising from differences in willingness to take risks to achieve potential rewards, or because the preferences of future generations for consequences of current decisions are not well known—can make the expected utilities of different acts uncertain. Any of these obstacles can inhibit uncontroversial application of SEU theory to risk management problems, pushing a risk management problem to the right in Fig. 7.1. If a completely known SEU model for supporting a risk management decision is denoted by $M = \{A, C, u(c), Pr(c \mid a)\}$, then the preceding difficulties can be viewed as instances of decision-making when the correct model, M, is unknown or disputed.

Decision-making without knowledge of, or agreement about, the basic assumptions needed to structure a decision problem by specifying a unique decision model, M, has been studied under headings such as deep uncertainty (Lempert and Collins 2007), severe uncertainty (Ben-Haim 2001), model uncertainty, and wicked decision problems (Rittel and Webber 1973). Constructive proposals to help guide risk management decision-making when relevant data are available, but a unique correct decision model is not known, are described next. Then we address the challenges of deeper uncertainty that arise when neither trustworthy predictive models nor relevant data are available at first, and it is necessary to learn and adapt as one goes. Finally, we will consider practical applications of these techniques.

Ten Tools of Robust Risk Analysis for Coping with Deep Uncertainty

Table 7.1 summarizes ten tools that can help us to better understand deep uncertainty and make decisions even when correct models are unknown. They implement two main strategies: finding *robust decisions* that work acceptably well for many models (those in the uncertainty set); and *adaptive risk management*, or learning what to do

Table 7.1 Methods for decision-making with unknown models

Method	Model generation	Optimization/Adaptation	Combination
Expected utility/ SEU theory	One model specified	Maximize expected utility (over all acts in the choice set, *A*)	None
Multiple priors, models, or scenarios; robust control, robust decisions	Identify multiple priors (or models or scenarios, etc.) e.g., all models close to a reference model (based on relative entropy)	Maximize the return from the worst-case model in the uncertainty set	Penalize alternative models based on their dissimilarity to a reference model.
Robust optimization	Use decision maker's risk attitude, represented by a coherent risk measure, to define the uncertainty set	Optimize objective function while satisfying constraints, for all members of uncertainty set	None
Average models	Use multiple predictive (e.g., forecasting) models	None	Simple average or weighted majority
Resampling	Create many random subsets of original data and fit a model to each	Fit models using standard (e.g., least squares, maximum likelihood) statistical criteria	Create empirical distribution of estimates
Adaptive boosting (AdaBoost)	Iteratively update training data set and fit new model	Re-weight past models based on predictive accuracy	Use weights to combine models
Bayesian Model Averaging (BMA)	Include all models that are consistent with data based on likelihood	Condition model probabilities on data	Weight models by their estimated probabilities
Low-regret online decisions	Set of experts, models, scenarios, etc. is given, $\{M_1, M_2, \ldots, M_n\}$	Reduce weights of models that make mistakes	Weighted majority or selection probability.
Reinforcement Learning (RL) for MDPs: UCRL2	Uncertainty set consists of confidence region around empirical values	Approximately solve Bellman equations for most optimistic model in uncertainty set to determine next policy	Update from episode to episode based on new data
Model-free reinforcement learning (RL) for MDPs: SARSA	No model used (model-free learning)	Approximately solve Bellman equations for unknown model	Update value estimates and policies based on new data

by well-designed and analyzed trial-and-error. Each is discussed in the following paragraphs, which also explain the different columns for generating, optimizing/ adapting, and combining multiple model results.

Use Multiple Models and Relevant Data to Improve Decisions

When the correct model linking acts to their probable consequences is unknown, but relevant data are available, good risk management decisions can often be made by combining predictions from multiple models that are consistent with available knowledge and data (e.g., as judged by statistical criteria discussed later). We will call the set of alternative models considered the *uncertainty set.*

A "good" decision, given the information available when it is made, can be defined as one to which no other choice is clearly preferable, e.g., by stochastic dominance (Buckley 1986), giving clearly higher probabilities of preferred outcomes and lower probabilities of undesired outcomes, as assessed by all models in the uncertainty set. Alternatively, a "good" decision procedure might be defined as one that, despite all uncertainties, performs almost as well as some ideal procedure (e.g., optimal decision-making with perfect information, or the best-performing of all the models in the uncertainty set), as assessed in hindsight by the difference in rewards that they generate (often referred to as the *regret* for using the inferior model). Both approaches have led to strikingly successful procedures for using *multiple models*, or *model ensembles*, to let data inform decisions, when the correct model is unknown. We will refer to both as methods for *robust risk analysis*, i.e., risk analysis that delivers recommendations that are robust to deep (and other) uncertainties, especially about the correct probabilistic relation between acts and their consequences.

Several practical options are available for generating plausible models or scenarios (using various definitions of "plausible" or "consistent with data," as discussed below); optimizing decisions within and across these multiple possibilities; and combining the different decision recommendations into a final decision recommendation in a way that allows some performance guarantees for the quality of the result. The essence of robust risk analysis, for a large class of decision procedures, can be summarized as follows:

1. *Generate:* Generate or select multiple plausible models or scenarios, given available data and knowledge.
2. *Optimize/improve:* Find the best decision for each considered model or scenario. This may be interpreted as the decision "recommended" or "voted for" by that model or scenario. Alternatively, if optimization of decisions is not clearly defined or is not practical, but criteria and methods for improving models and decisions are available, then improve upon the ones considered so far until no further clear improvements can be made.

3. *Combine:* Use the multiple decision recommendations to recommend a final risk management decision, by using some *combination rule* (such as majority voting) to combine the individual decision recommendations from step 2.

The robustness of the final decision recommendation can be defined and characterized in various ways: not only by the fraction of models that support it (or by upper bounds for the probability of models that do not), but also by upper bounds for the *difference in average reward* (e.g., expected utility or disutility) from following it *vs.* from making the best decisions possible if the correct model were known. The latter criterion leads to *low-regret* and *reinforcement learning* decision strategies for managing uncertain risks. The following paragraphs review methods for model generation, improvement, and combination to support robust risk analysis.

Robust Decisions with Model Ensembles

A crucial contribution to decision-making with deep uncertainty (Lempert and Collins 2007; Bryant and Lempert 2010) is the generation and analysis of many (e.g., thousands of) scenarios or models. Uncertainty about the correct decision model can be treated as just one more source of uncertainty, with each scenario in the uncertainty set now specifying a decision model to be used, as well as the values of other quantities that lie outside the choice set but that, together with the choice of act, affect consequences. If scenario probabilities are known, then expected utility can be maximized with respect to these probabilities. Even if the probabilities of different scenarios are not known, a decision that performs well by some criterion (e.g., that is undominated or that yields close to some provable upper bound on expected utility, given the information available when it is made) for most scenarios is likely to also do so in reality, if reality is well described by at least some scenarios in the uncertainty set, and if this set is much more likely than the set of scenarios not considered—something that might be easier to assess than the individual scenario probabilities.

If one or a few decisions are "best" (e.g., maximizing scenario-specific expected utilities) or "good" for all or most of the considered scenarios, then these decisions are, in this sense, *robust to uncertainty* about which scenario in the uncertainty set is correct (if any). By contrast, if such ensemble analysis reveals that different choices are best for substantial fractions of the plausible scenarios, then it will be clear that no robust decision exists that makes the choice of decision immune to uncertainty about the correct scenario, and that more information is therefore needed before a decision recommendation can be made that is robust, in this sense, to remaining uncertainties.

Example: Robust Decisions with Model Uncertainty

Tables 7.2 and 7.3 present two very different views of a risk management decision problem. In this example, a *perceived threat* of concern to some stakeholders (e.g., crop blights from climate change, genetically modified organisms in food, nanoparticles in air, electromagnetic radiation from cell phones, etc.) is assumed to have been identified, but it is not yet known whether complete scientific knowledge would reveal that the exposures or activities of concern actually cause the harms that people worry about (abbreviated as "Perceived threat is real") or not ("Perceived threat is not real.") (More generally, one might be uncertain about the size of the threat, but these two states suffice to illustrate the basic challenge.) The alternative risk management acts being considered are to intervene now, perhaps by limiting exposures as a precautionary measure; or to wait for more information before deciding whether to intervene.

The tables show the expected disutility (scaled from 0 to 100) for each act-state pair. For simplicity, we assume that everyone agrees that the best choice of act is the one that minimizes expected disutility (equivalent to maximizing expected utility). However, perhaps due to the affect heuristic, optimistic stakeholders who think that the threat is probably not real ($p \leq 0.1$) also tend to think that its disutility, should it occur after all, will be modest (even though there is no logical reason that probability and severity must be positively correlated). Conversely, those who perceive the probability of a threat as being relatively high ($p = 0.4$) also tend to perceive the severity of the threat (its disutility if it occurs) as being relatively great. Tables 7.2 and 7.3 are intended to capture these perceptions. Each constitutes one scenario.

The pessimists in Table 7.3 are shown as having crisp probabilities for the states (probability that the threat is real = 0.4), but the optimists in Table 7.2 have only imprecisely specified probabilities ($0 \leq p \leq 0.1$). Simple expected utility calculations show that acting now is less desirable than waiting, for the scenario in Table 7.2, if the threat probability is $p < 1/3$ (since then $20p + 10(1—p) < 40p$); hence, they prefer to wait. Similarly, the pessimists described by Table 7.3 prefer to wait if

Table 7.2 Decision problem for an optimistic scenario

	Perceived threat is real, $p \leq 0.1$	Perceived threat is not real, $1—p \geq 0.9$	Expected disutility
Act now	Disutility = 20	Disutility = 10	≥10
Wait for more information	Disutility = 40	Disutility = 0	≤4

Table 7.3 Decision problem for a pessimistic scenario

	Perceived threat is real, $p = 0.4$	Perceived threat is not real, $1—p = 0.6$	Expected disutility
Act now	Disutility = 90	Disutility = 10	42 = 36 + 6
Wait for more information	Disutility = 100	Disutility = 0	40

$p < ½$, which it is ($p = 0.4$ in this scenario). Hence, both scenarios prescribe waiting. Even if many other scenarios lie between these two extremes (i.e., with scenario-specific probabilities and disutilities lying between those in Tables 7.2 and 7.3), and even if we are ignorant of the respective probabilities of these scenarios, or even of what all the scenarios are, waiting for more information is a robust optimal decision with respect to this uncertainty set. (However, if the pessimists who see the world as in Table 7.3 become slightly less pessimistic, by changing their assessment of the disutility of acting now if the perceived threat is real from 90 to 80, then neither decision would be robustly optimal.)

Example: Robustness, Multiple Models, Ambiguous Probabilities, and Multiple Priors

Expected utility theory has been extended to allow for uncertain or "ambiguous" probabilities and models, and to consider ambiguity aversion as well as risk aversion. Instead of evaluating expected utility with respect to a unique "best guess" prior probability distribution (or measure), an uncertainty set of multiple priors, all of which are considered plausible, can be used to represent ignorance of the true probability distribution. Then, axioms for decision-making with uncertain probabilities imply that a decision-maker should choose the act that *maximizes the minimum expected utility* obtained by using any of these plausible probability distributions (or measures) (Gilboa and Schmeidler 1989). More generally, Maccheroni et al. (2006) presented conditions under which a decision-maker should choose the act in A that *maximizes the minimum penalized expected utility*, where different probability distributions or measures in the uncertainty set carry different penalties based on their plausibility. Symbolically, such "variational preferences" prescribe choosing an act from choice set A to maximize the minimized value (over all members p of the uncertainty set) of the weighted sum $E_p[u(c \mid a)] + \alpha(p)$, where $E_p[u(c \mid a)]$ is the usual expected utility of act a if probability measure p is used to compute expected values, and $\alpha(p)$ is the penalty for using p. ($\alpha(p) = 0$ if p is known to be correct, and is larger for less plausible probability distributions or measures.) Robust decision-making in this sense—maximizing the minimum expected reward (or credibility-penalized expected utility) over an uncertainty set of alternative probabilities—connects to a tradition of *robust control* in control engineering (Hansen and Sargent 2001, 2008), in which controls are sought that perform well for all models not too dissimilar to a known reference model that is considered plausible but not necessarily correct. The measure of dissimilarity is typically based on information-theoretic metrics [such as relative entropy or Kullback-Leibler divergence between the reference model and the model being weighted (Laeven and Stadje 2011). Robust control of stochastic systems with somewhat misspecified models (not too dissimilar from the reference model) is mathematically equivalent to a special case of decision-making with multiple priors (Hansen and Sargent 2008).

Example: Robust Optimization and Uncertainty Sets Using Coherent Risk Measures

One of the most useful paradigms for decision-making is *constrained optimization*, in which the choice set A consists of all values of one or more decision variables satisfying a set of constraints, and the decision-maker seeks a set of values for the decision variables to maximize or minimize some objective function (e.g., average production of net benefits, or average cost of losses per unit time, respectively). For example, the decision variables might be the amounts invested in risky stocks or opportunities, the constraint might be that the amount invested must not exceed a total budget available to invest, and the objective function might be the expected value of the resulting portfolio. More generally, a *robust linear optimization* problem (Bertsimas and Brown 2009) seeks to maximize a weighted sum of decision variables (the linear objective function, e.g., the value of a risky portfolio), while keeping other weighted sums of the decision variables (e.g., the costs or resources required to implement the decision) acceptably small (the constraints), when it is only known that the values of the weights and constraints belong to some uncertainty set of alternative possibilities, but the probabilities of different sets of weights and constraints are not known. Standard methods for solving deterministic constrained optimization problems, such as linear programming, which are suitable when the optimization problem is known with certainty, can give highly infeasible solutions when the problem data are uncertain; therefore, *robust optimization* methods must be used instead to address these model uncertainties (Ben-Tal et al. 2009).

Any coherent risk measure representing the decision-maker's aversion to risk of violating a budget (or other linear) constraint can be expressed as an equivalent robust linear optimization problem with a convex uncertainty set that is derived directly from the coherent risk measure (Bertsimas and Brown 2009). For example, if the conditional value at risk (CVaR) risk measure is used to specify that the expected value of cost in the worst (most costly) $x\%$ of cases must be no greater than some level b, then the corresponding uncertainty set can be generated by finding a set of probability measures that *represent* the CVaR measure of risk as minimizing expected values over that set. (Any coherent risk measure has such a minimum-expected-value-over-a-set-of-probabilities representation.) The uncertainty set for the corresponding robust optimization problem is then just a convex set (a polytope) of weighted averages of the probability measures that represent the coherent risk measure. The set of decisions that create "acceptable" risks of violating the linear constraint compared to the *status quo* according to a coherent risk measure is identical to the set of decisions that satisfy the constraint for all sets of weights in the uncertainty set.

Robust linear optimization problems can be solved via linear programming (due to the polytope shape of the uncertainty set). Both linear and nonlinear robust optimization problems can be computationally advantageous compared to non-robust formulations, and the gap between the maximized expected utility or return from the correct model (if it were known) and the robust model is often surprisingly small (Ben-Tal et al. 2010; Bertsimas et al. 2011).

Averaging Forecasts

During the 1980s and 1990s, forecasting experts in time series econometrics and management science participated in several competitions (the "M-competitions") to discover empirically which forecasting models and methods worked best (e.g., minimizing mean squared error between forecast and subsequently revealed true values) in over 1000 different economic and business time series. One finding was that a simple arithmetic average of forecasts made by different methods usually out-performed any of the individual forecasts being averaged (Makridakis and Hibon 2000). Averaging tends to reduce the error from relying on any single model (even the single best one), when even the best-fitting model is unlikely to be perfectly correct, and even relatively poorly-fitting models are likely to contribute some information useful for prediction. This is similar to Condorcet's centuries-old observation on majority voting with probabilistic knowledge: when each voter independently has a greater than 50% probability of correctly identifying which of two competing answers to a question is correct (assuming that one of them is), then majority rule in a large population of such voters has a probability close to 100% of selecting the correct answer—possibly very much greater than the probability for any of the individuals (de Condorcet 1785). Even if the voter opinions are not completely statistically independent, a similar conclusion often holds, as discussed later (e.g., for resampling, boosting, Bayesian model averaging, and online decisions). Note that this argument does not require knowing the probabilities that the different voters will be correct. Replacing voters with models and votes with model-based forecasts or probabilistic predictions provides heuristic motivation for the benefits of averaging predictions across multiple models.

Since these early experiments, a variety of *model ensemble methods* have been developed that seek to make predictions and decisions that are robust to some model uncertainties, in the sense that they work well for a large set of alternative plausible models, and do not depend on assuming that any specific model (e.g., the best-fitting one) correctly describes or predicts the real situation.

Resampling Data Allows Robust Statistical Inferences in Spite of Model Uncertainty

One way to generate multiple models to contribute to an ensemble prediction is to identify the "best" models (e.g., by traditional statistical criteria such as maximum likelihood or least squares or maximum a posteriori probability, or minimum expected loss) for each of many randomly sampled subsets of the data. It is common in applied risk assessment that the correct statistical model for fitting a curve (e.g., a dose-response function) or estimating a quantity of interest (e.g., an odds ratio) from data, is unknown. Then, modern computational statistical *resampling methods*—such as the bootstrap, jackknife, model cross-validation, and bagging—can create many random sub-samples of the original data; fit a (possibly nonparametric) model or estimate to each sub-sample; and average these sample-specific estimates to obtain a final estimate (e.g., Molinaro et al. 2005). The empirical distribution of the sample-specific estimates around the final estimate indicates how far from the final estimate the unknown true model might fall. Resampling can reduce bias from over-fitting, leading to wider confidence intervals for model-based estimates (because model uncertainty is considered), and correspondingly fewer false-positives for significant effects, than selecting a single "best" model. It allows robust statistical inferences and model-based predictions, within limits (set in part by the model-fitting strategies used for the random samples, as well as by how the multiple samples are generated) even when the correct model is uncertain.

Adaptive Sampling and Modeling: Boosting

Instead of resampling data purely randomly, it turns out to be profoundly useful, for statistical classification problems, to construct deliberately biased samples that overweight data points that cannot yet be predicted well, and then to iteratively improve models by fitting them to these deliberately biased training sets. On each iteration, a new statistical model is developed by fitting it to a new training set. Predictions from successive models are combined via a weighted-majority decision rule in which each model's "vote" (predicted class) is weighted based on its relative performance in correctly classifying data points in the training set. If the data points are then weighted based on how well they are predicted by the current best model, and these weights are used to determine the inclusion probability for each data point in the next training sample (with the least-well-predicted points receiving higher ("boosted") probabilities of being included), then a few hundred or thousand iterations can often generate an excellent statistical classifier, starting from even a weak initial predictive model that classifies data points with only slightly greater than random accuracy. Such *adaptive boosting* (AdaBoost) algorithms have proved highly successful in applications that require classifying cases into two or more classes. Examples include classification of credit applicants as "good" or "bad" credit risks (or into more than two credit risk categories) (Zhou and Lai 2009); diagnosis of patients based on symptoms and markers (Tan et al. 2009); prediction of which companies are most likely to go bankrupt over a stated time interval (Cortés et al. 2007); predicting toxicities of organic compounds (Su et al. 2011) and detection of intrusion in computer networks (Hu et al. 2008).

Bayesian Model Averaging (BMA) for Statistical Estimation with Relevant Data But Model Uncertainty

One of the best-developed model ensemble methods is Bayesian Model Averaging (BMA) for statistical inference when the correct statistical model is uncertain. BMA seeks to weight model outputs (e.g., inferences, predictions, or decision recommendations) according to their probabilities of being correct, based on consistency with data. Like resampling methods, BMA creates many models (e.g., by considering all 2^n subsets of n candidate predictors in different regression models), but it weights each model based on its likelihood in light of the data, rather than fitting different models to different subsets of the data. (If there are too many plausible models to make it practical to generate and fit all of them, then sampling only those that are most consistent with the data, according to some statistical criterion, in the ensemble of considered models may yield a computationally tractable compromise.) BMA typically assesses consistency with the data by statistical criteria such as likelihood (i.e., model-predicted probability of the observed data), or likelihood penalized by model complexity, as reflected in degrees of freedom or number of constraints on data—the Bayesian Information Criterion (BIC). For example, a hypothesized causal model for multi-factorial disease causation might be considered "consistent with data" if it implies a likelihood or BIC value for the observed data that is not much less than (e.g., is within an order of magnitude of) the maximum value for any model.

Given a model ensemble, BMA estimates the probability that a statistical property of interest holds (e.g., that a particular exposure variable is a significant predictor of a particular adverse health effect) or that a model-based prediction or conclusion is true (e.g., that the risk created by a given exposure exceeds a specified level), by considering the weighted fraction of the models in the ensemble that have that property or make that prediction, with each model weighted to reflect its conditional probability given the data (via a "Bayes factor" that reflects the likelihood of the data, given the model, in accord with Bayes' rule). An intuitive motivation is that the conditional probability that any conclusion, *X*, is true, given some set of observations that we will call *Data*, can be written (tautologically, via the law of total probability) as:

$$Pr(X \mid Data) = Pr(X \mid M_1)\text{Pr}(M_1 \mid Data) + \ldots + Pr(X \mid M_n)\text{Pr}(M_n \mid Data)$$

where $M_1, M_2, \ldots, M_n$ are any set of mutually exclusive and collectively exhaustive hypothesized models; *Data* represents any available observations; and $Pr(M_j \mid Data)$ is proportional to the likelihood of the data if model M_j is correct, $Pr(Data \mid M_j)$. Various approximations made for computational tractability and convenience, such as only sampling from a large set of possible models, and only considering models with tractable priors (glossed over in this brief overview) and with likelihood function values within an order of magnitude or so of the maximum-likelihood one, lead to different detailed BMA algorithms, appropriate for different types of statistical models ranging from regression models to Bayesian networks and causal graphs (Hoeting et al. 1999).

A substantial literature documents cases for which BMA-based statistical predictions or conclusions are less biased and more realistic than corresponding predictions or conclusions based on any single (e.g., best-fitting or maximum-likelihood) model. A typical result, as with resampling methods, is that confidence intervals for parameters estimated by BMA are wider, and type-1 errors (false positives) for falsely discovering what seem to be statistically "significant" results correspondingly less common, that when inferences are obtained from any single model, including the "best" model according to some model-selection criterion (Hoeting et al. 1999). This can have important implications for risk assessment results when model uncertainty is important. For example, when BMA is used to assess the statistical association between fine particulate matter (PM2.5) and mortality rates in some time series data sets, effects previously reported to be significant based on model selection (with model uncertainty ignored) no longer appear to be significant (Koop and Tole 2004).

Learning How to Make Low-Regret Decisions

Resampling, boosting, and BMA methods are useful when they can fit multiple models to data that are known to be relevant for predicting future consequences of present decisions. If relevant data are initially unavailable, however, or if the relevance of past data to future situations is uncertain, then a different strategy is needed. This section considers what to do when data will be collected only as decisions are made, and various different models (or experts or hypotheses or causal theories, etc.), with unknown probabilities of being correct, are available to inform decisions. This deeper uncertainty forces adaptive decision-making as relevant data become available, rather that pre-determining the best course of action from available relevant data.

For systems with quick feedback, where the loss (or reward) for each act is learned soon after it is taken, some powerful approaches are now available for using multiple models to improve decisions. These situations can be modeled as *online decision* problems, in which what to do in each of many sequentially presented cases must be decided without necessarily knowing the statistical characteristics of the cases—which may be changing over time, or selected by one or more intelligent adversaries, or influenced by continually adapting agents or speculators in a market.

Suppose that $\{M_1, M_2, \ldots, M_n\}$ are the different models (or theories, experts, scenarios, prediction algorithms, etc.) being considered, but their prior probabilities of being correct are unknown. Decision opportunities and feedback on resulting consequences arrive sequentially. For example, the risk manager may be confronted with a series of cases that require prompt decisions (such as stock market investment opportunities, patients to be treated, chemicals to be tested and classified, new drug applications or loan applications to be approved or rejected, etc.) If the correct model were known, then it could be used to make decisions that would maximize the total reward earned from each decision, assuming that each choice of act for a case results in a consequence that can be evaluated by the decision-maker as having some value, which we call the "reward" for that decision in that case. In practice, the correct model is usually not known, but, in online decision problems, the risk manager learns the actual consequence and reward soon after each decision; if the different models are specific enough, then the consequences and rewards that would have been received if each model had been used to make the decision may also be known.

The *cumulative regret* for using one model rather than another can be defined and quantified as the difference between the cumulative reward that would have been earned by following the decision recommendations from the second model instead of the first, if this difference is positive; equivalently, it is the cumulative loss from using the first model instead of the second. A good (or, more formally, *low-regret*) sequence of decisions, with respect to the ensemble $\{M_1, M_2, \ldots, M_n\}$ has an average regret per decision that approaches zero, compared to the best decisions that, in retrospect, could have been made using any of the models in the ensemble. In other words, a low-regret decision sequence does almost as well, on average, as if the decision-maker had always use the best model, as judged with the advantage of hindsight. Practical low-regret decision algorithms focus on homing in quickly on correct (or low-regret) decision rules, while keeping regret small during the learning period.

Somewhat remarkably, low-regret decision strategies are often easy to construct, even if the probabilities of the different models in the ensemble are unknown (Cesa-Bianchi and Lugosi 2006). The basic idea is to weight each model based on how often it has yielded the correct decision in the past, and to make decisions at any moment recommended by a *weighted majority* of the models. After each decision is made and its outcome is learned, models that made mistaken recommendations are penalized (their weights are reduced). Thus, the model ensemble produces recommendations that adapt to the observed performances of the individual models, as revealed in hindsight. (An alternative is to use the model weights to create *probabilities* of selecting each model as the one whose recommendation will be followed for the next case; such probabilistic selection (with the wonderful name of a "follow the perturbed leader" (FPL) strategy) also produces low-regret decision sequences (Hutter and Poland 2005). A further variation (Blum and Mansour 2007) is to adjust the weight on each model only when it is actually used to make a decision; this is important if the consequences that would have occurred had a different model been used instead are not known.) In each of these cases, weighted majority or FPL algorithms produce low-regret decision sequences; moreover, performance guarantees can be quantified, in the form of upper bounds for the average regret using the model ensemble algorithm compared to always using the best model (if it were known in advance).

If the environment is stationary (offering fixed but unknown probabilities of consequences for different decisions), then the low-regret strategies effectively learn, and then exploit, its statistical properties. If the environment changes over time, then low-regret strategies can be transformed to yield adaptive low *adaptive regret* strategies. These replace cumulative regret measures with measures of performance on successive intervals, to make the decision sensitive to changes in the underlying process (Hazen and Seshadhri 2007). Risk analysts and policy analysts often recommend using *efficient adaptation* in light of future information to cope with deep uncertainty. Model ensemble decision algorithms provide one constructive framework to implement such recommendations.

Example: Learning Low-Regret Decision Rules with Unknown Model Probabilities

To understand intuitively how low-regret online decisions are possible, consider the extremely simple special case in which one must decide which of two possible decisions to make for each of a sequence of cases (e.g., invest or decline to invest in a new business opportunity; approve or deny a chemical product for consumer use; sell or hold a stock, administer or withhold an antibiotic in the treatment of a sick patient who might have a viral infection, etc.) After each decision is made, one of two possible outcomes is observed (e.g., business succeeds or fails, chemical product proves safe or hazardous, stock price moves up or down, patient would or would not have benefitted from the antibiotic, respectively). The decision-maker evaluates the results, assigning a "reward" (or loss) value to each outcome. The correct model for deciding what to do (or for predicting the outcome of each decision in each case) is uncertain. It belongs to some finite uncertainty set of alternative competing models $\{M_1, M_2, \ldots, M_n\}$ (perhaps developed by different experts or research groups or constituencies), but initially the risk manager knows nothing more about which model is correct (e.g., there is no experience or available knowledge to assign meaningful probabilities to the individual models, or even to assign Dempster-Shafer beliefs to subsets of models, within the uncertainty set).

Despite this ignorance of the correct model, a low-regret sequence of decision can still be constructed, as follows (Cesa-Bianchi and Lugosi 2006).

1. Assign all the models in the uncertainty set the same initial weight, $1/n$.
2. As each case arrives, make the decision recommended by the weighted majority of models (i.e., sum the weights of all models in the ensemble that recommend each decision, and choose the decision with the maximum weight. In this simple example, with equal initial weights, this is the same as choosing the simple majority decision.) Resolve ties arbitrarily.
3. As long as the ensemble-based recommendation is correct (reward-maximizing) for each case in hindsight, make no changes; but when the ensemble recommendation is mistaken, reduce the weights of all of the models that made the mistaken recommendation to zero.

Since majority rule is used, each new mistake eliminates at least half of the surviving models; thus successive eliminations will lead to all decisions being made by the correct model (or to a subset of models that agree with the correct model), after a number of mistakes that is at most logarithmic in the number of models in the uncertainty set. After that, regret will be zero, and hence average regret will approach zero as the correct model continues to be used.

For more realistic and complex cases, this simple procedure must be modified to achieve low-regret decisions. When there is no guarantee that the correct model is in the uncertainty set, and if only the consequences of the selected decisions are revealed (but not the consequences that other decisions would have produced), then the weights of models that contribute to incorrect (positive-regret) decisions are reduced only partially at each mistake, rather than jumping all the way to zero. Moreover, rather than making deterministic recommendations, the weights of models in the ensemble are used to set probabilities of selecting each possible act. Nonetheless, for a variety of sequential decision problems (including ones with more than two possible outcomes and more than two possible acts to choose among for each case), such refinements allow efficient adaptive learning of decision rules that perform almost as well on average as if the best model in the ensemble (as evaluated with 20–20 hindsight) were always used.

Reinforcement Learning of Low-Regret Risk Management Policies for Uncertain Dynamic Systems

The online risk management decision problems considered so far, such as deciding whether to approve loans, administer antibiotics, sell stocks, etc., are perhaps less exciting than the grand challenges of risk management under deep uncertainty mentioned in the introduction. However, key ideas of low-regret decision making can be generalized to a broad class of *reinforcement learning* (RL) decision problems and algorithms that encompass many more complex risk management decision problems of practical interest. State-of-the-art RL algorithms also show how to generate continuous uncertainty sets based on observations, and how to apply mathematical optimization to the resulting infinite ensembles of models to make low-regret decision in both stationary and changing environments.

Many risk management decision problems with deep uncertainties involve trading off relatively predictable immediate gains against uncertain future rewards or losses. Examples include extracting valuable non-renewable resources with uncertain remaining reservoirs (such as oil or minerals); managing forests, vulnerable habitats, fisheries, or other renewable resources having uncertain population dynamics and extinction thresholds; attempted control of climate change with uncertain damage thresholds and points of no return; and medical use of antibiotics whose use increases, to an unknown extent, the risk of future antibiotic-resistant infections. In each case, a decision about how much benefit to extract now, given the present (perhaps uncertain) state of the word, yields an immediate reward, but it may also cause a transition to a new, possibility inferior state offering different (perhaps lower or even zero) rewards for future actions.

For purposes of quantitative analysis, the usual formulation of such a problem is a *Markov Decision Process* (MDP). In an MDP, choosing act a when the state of the system is s yields an immediate reward, $r(a, s)$ and also affects probabilities of transitions to each possible next state, $Pr(s' \mid a, s)$, where s' = a possible next state and s = present state when act a is taken. (For stochastic rewards, the immediate reward may be the mean of a random variable with a distribution that depends on a and s.) A *decision rule*, or *policy* for an MDP specifies the probability of taking each act when in each state. (The set of such policies constitutes the choice set, A, in the standard expected utility formulation of decision theory discussed earlier, and the decision-maker seeks to identify the best policy.) An *optimal policy* maximizes the *value* of the stream of rewards starting from each state; this value is usually denoted by $Q(s)$, and is defined as the expected sum of the immediate reward and the discounted value of future rewards, assuming that decisions now and in the future are consistently optimized. If β is the one-period discount factor, then optimal values, denoted by $Q^*(s)$, satisfy the following equation (the Bellman equation):

$$Q^*(s) = \max_{a \text{ in } A} \{r(a, s) + \beta\, \Sigma_{s'} Q^*(s') Pr(s' \mid a, s)\}$$

In words, the optimized reward starting from state s is the maximized (over all possible current acts) sum of the immediate reward plus the maximized expected discounted future reward starting from the next state. This system of equations (one for each s) can be solved for the optimal policy, by standard algorithms from operations research (such as linear programming, value iteration, policy iteration, and stochastic dynamic programming) or by reinforcement learning (RL) algorithms (such as Q-learning or temporal difference learning) that use successive empirical estimates of the optimal value function, based on the observed history of states, acts, and rewards so far, to gradually learn an optimal, or nearly optimal policy (Sutton and Barto 2005).

Robust low-regret risk management policies for MDPs (Regan and Boutilier 2008) generate low regrets even when the reward distributions and state transition probabilities are initially not known, but must be estimated from observations; and even when they may change over time, rendering what has been learned so far no longer useful. These complexities move the decision toward the right in Fig. 7.1—the domain of deeper uncertainties. Practical applications of RL algorithms to date have ranged from controlling hazardous chemical production processes to maximize average yield under randomly changing conditions, while keeping the risk of entering dangerous process states within specified bounds (Geibel and Wysotzk 2005), to devising stop light control policies to reduce jams and delays in urban traffic (Gregoire et al. 2007).

As discussed in Chap. 3, experiments and brain-imaging (functional MRI) studies of human subjects suggest that RL also has neural correlates, with the human brain processing differences between anticipated and obtained rewards for different policies under risk, and subsequently adapting perceptions and behaviors, in ways that can be interpreted in terms of RL algorithms (e.g., Kahnt et al. 2009). For example, whether subjects successfully learn which of four risky reward processes generates the highest average, based on repeated trial and error learning, appears to be predicted by the strength of physiologically measurable signals involved in reinforcement learning (Schönberg et al. 2007), although other experiments show that learning is also affected by mental models (possibly incorrect) of processes generating data (Green et al. 2010).

Example: Reinforcement Learning of Robust Low-Regret Decision Rules

If a decision-maker must make choices in an unknown MDP model, with only the sets of possible states and acts (S and A) known, but rewards and state transition probabilities resulting from taking act a in state s having to be estimated from experience, then a low-regret strategy can be constructed using the following principle of *optimism in the face of uncertainty* (Jaksch et al. 2010):

1. Divide the history of model use into consecutive *episodes*. In each episode, a single policy is followed. The episode lasts until a state is visited for which the act prescribed by the current policy has been chosen as often within the current episode as in all previous episodes. (The new episode thus at most doubles the cumulative number of occurrences of any state-act pair.) When an episode ends, the data collected is used to update the uncertainty set of considered models, as well as the policy to be followed next, as described next.
2. At the start of each episode, create a new uncertainty set of plausible MDP models from confidence intervals around the empirically observed mean rewards and transition probabilities.
3. Choose an optimistic MDP model (one yielding a high average reward) from the uncertainty set. Solve it via operations research optimization techniques to find a near-optimal policy.
4. Apply this policy until the episode ends (see Step 1). Then, return to Step 2.

Analysis of a detailed algorithm (UCRL2, for upper confidence reinforcement learning) implementing these steps shows a high probability (depending on the confidence levels used in step 2 to generate uncertainty sets) of low regret, compared to the rewards that would have been achieved if optimal policies for each of the true MDPs had been used (Jaksch et al. 2010). This result holds when any state can be reached from any other in finite time by appropriate choice of policies, and even when the true but unknown underlying MDP (i.e., reward distributions and transition probabilities) can change at random times (or in any other way that is oblivious to the decision-maker's actions), provided that the number of changes allowed in an interval is finite. Intuitively, the UCRL2 algorithm seeks the best return by exploring different plausible models, starting with those that would yield the best returns if correct. As data accumulates, confidence intervals around estimated model parameters shorten. When the current model no longer appears best, exploration switches to a different model. The UCRL2 algorithm learns efficiently and can adapt to changes in the underlying unknown MDP quickly enough so that the policies it recommends are unlikely to spend long yielding returns much lower than those from the best policies given perfect information.

Example: Model-Free Learning of Optimal Stimulus-Response Decision Rules

Rather than solving the Bellman equations directly, RL algorithms use data to approximate their solution increasingly well. For example, the *SARSA* (state-act-reward-state-act) RL algorithm updates the estimated value (the sum of immediate and delayed rewards) from taking act a in state s, denoted by $Q(s, a)$, via the equation:

$$\begin{aligned}\text{new } Q(s,a) \text{ value} &= \text{previous } Q(s,a) \text{ value} \\ &+ \alpha[\text{change in estimated value of } Q(s,a)]\end{aligned}$$

where α is a *learning rate* parameter and the change in the estimated value of $Q(s, a)$ is the difference between its new value (estimated as the sum of the most recently observed immediate reward and the previously estimated discounted value starting from the observed new state) and its previously estimated value:[change in estimated value of $Q(s, a)$] = $[r(s, a) + \beta Q(s', a')]$—$Q(s, a)$.

(Here, a' is the act taken in the observed next state, s', according to the previously estimated value function $Q(s, a)$; and $r(s, a) + \beta Q(s', a')$ is the estimated value just received when act a was taken in state s.) The difference between this estimate of value just received and the previous estimated value $Q(s, a)$ expected from taking act a in state s provides the feedback needed to iteratively improve value estimates and resulting policies. The change in the estimated value of $Q(s, a)$ is zero only when its previously estimated value agrees with its updated value based on the sum of observed immediate reward and estimated delayed reward starting from the observed next state, i.e., only when $Q(s, a) = r(s, a) + \beta Q(s', a')$. When this condition holds for all states, the Bellman equation is satisfied, and the observed sequence of state-act-reward-state-act (SARSA) data $(s, a, r(s, a), s', a')$ have been used to learn the optimal policy. Detailed implementations of this idea (e.g., incorporating randomization to assure that all act-state pairs will eventually be tried with non-zero probability; and specifying the act to be selected in each state, typically as the "epsilon-greedy" one that chooses an act at random with small probability, and otherwise chooses the one that maximizes the current estimated expected value of

$r(s, a) + \beta Q(s', a')$, perhaps with statistical regression or nonparametric smoothing models and Monte-Carlo simulation of a random sample of future trajectories used to approximate $Q(s', a')$ for large state spaces) yield practical RL algorithms for a variety of sequential decision problems with random transitions and immediate and delayed losses or rewards (Szepesvari 2010).

Many RL algorithms learn by comparing the estimated rewards received using the current policy to the best estimated rewards that could have been received (as predicted by a model) had a different policy been used instead, and revising the current policy based on this difference (which can be interpreted as a measure of regret). By contrast, the SARSA algorithm uses only the *observed data* on what was done and what reward was experienced (the SARSA data) to update the value estimates for state-act pairs and to gradually learn an optimal policy. No model of the underlying MDP (or other process) is required. In effect, the learner maintains estimated values for an ensemble of different stimulus-response (i.e., state-act) pairs; updates these value estimates based on the experienced differences between obtained and expected rewards; and uses them to decide what to do as each new state occurs. Such adaptive learning is suitable even when no model is available, and will converge to the optimal policy for the underlying MDP, if one exists, under quite general conditions, even if the unknown MDP itself occasionally changes (Yu et al. 2009).

Recent work has started to extend RL algorithms to *partially observable* MDPs (POMDPs) in which the state at each moment (e.g., the size of a fishery stock) is not known with certainty, but must be inferred from statistical information (e.g., sampling). State-of-the-art RL algorithms for POMDPs balance *exploration* of new or under-investigated decision rules (each of which maps histories of observed information, acts, and rewards to decisions about what act to take next) and *exploitation* of known high-performing decision rules. Similar to SARSA, this approach can learn optimal or nearly-optimal polices for the underlying POMDP, if one exists, even without a model of the process (Cai et al. 2009; Ross et al. 2011).

Ongoing extensions and refinements of these ideas—especially, *multi-agent (social) learning* and *evolutionary optimization* algorithms, in which the (perhaps fatal) experiences of some agents help to inform the subsequent choices of others (Waltman and van Eck 2009)—will bring further improvements in ability to solve practical problems. However, the techniques summarized in Table 7.1 already suffice to support many valuable applications.

Applying the Tools: Accomplishments and Ongoing Challenges for Managing Risks with Deep Uncertainty

Conceptual frameworks and technical tools such as those in Table 7.1 have practical value insofar as they help to improve risk management decisions with deep uncertainties. This section sketches applications of robustness and adaptive risk management methods to practical risk management problems with deep uncertainties, and highlights some key challenges.

Before seeking sophisticated solutions to difficult problems, of course, it is well to cover the basics: Pay attention to what doesn't work, and stop doing it; if possible, encourage many independent experiments on a small scale to find out what works better; identify, reward, and spread successes; don't bet too heavily on unvalidated models or assumptions (Harford 2011). The increasing capabilities of technical methods should not lead to neglect of such useful commonsense advice.

Planning for Climate Change and Reducing Energy Waste

In *robust decision-making* (RDM), participants develop multiple scenarios—perhaps with the help of computer-aided scenario generation and an experienced facilitator (Bryant and Lempert 2010)—to identify potential vulnerabilities of proposed decisions, such as where to build a road to connecting villages. These scenarios help participants to identify cost-effective ways to change the proposed decision to decrease vulnerabilities (e.g., potential loss of the road due to flooding or mud slides), and to develop increasingly robust decision options.

RDM has been advocated as a practical way to help multiple stakeholders in communities and developing countries engage in planning for climate change and infrastructure development (Lempert and Kalra 2008). Some limitations are that a robust decision may not exist, and the most relevant and likely scenarios, as viewed in hindsight, may not be identified during planning. [For example, empirical surprises, such as larger-than-predicted effects of "global dimming," might not be considered among the scenarios, leading to an ensemble of predictions with uncertain or debated credibility (Srinivasan and Gadgil 2002).] However, practical experience suggests that RDM can be helpful in envisioning and planning for possible futures (Bryant and Lempert 2010).

While scenario-based planning methods such as RDM, can help plan large-scale adaptation to envisioned potential changes, adaptive risk management methods can also guide smaller, immediate changes that significantly reduce energy waste and pollution by increasing the efficiency energy-consumption in uncertain environments. For example, RL algorithms have been used to design more efficient building energy conservation programs (subject to comfort constraints) (Dalamagkidis et al. 2007); devise more efficient use and coordination of stop lights to greatly reduce time spent by vehicles in urban traffic (Balaji et al. 2010); and optimize dynamic power use by devices (Wang et al. 2011). These applications reduce energy consumption without decreasing quality of life, by adaptively reducing wastes of energy.

Sustainably Managing Renewable Resources and Protecting Ecosystems

Sustainable management and harvesting of renewable resources can be formulated in terms of Markov Decision Processes (MDPs) (or generalizations, such as semi-Markov decision processes, in which the times between state transitions may have arbitrary distributions; or POMDPs). When the resources extend over large areas, with sub-areas developing differently over time, then the spatially distributed control problem of managing them can be factored into many local MDPs, represented as the nodes of a network, with local dependencies between the MDPs indicated by edges between nodes. Such graph-based MDPs (GMDPs) represent a variety of spatially distributed control problems in forestry and agriculture (Forsell and Sabbadin 2009).

As an example, in a large commercial forest consisting of many stands of trees, a decision must be made about when to harvest each stand, taking into account that random severe wind storms (perhaps every few decades) pose a risk of wiping out most of the commercial value of a stand that is blown down before it is harvested, but that neighboring stands can provide some shelter to each other, and hence reduce risk of wind damage (Forsell and Sabbadin 2009). If the probability distributions for rewards (e.g., based on market values of the crop over time) and state transition probabilities (e.g., based on statistics for wind storm arrival times and severities) were known in advance (Level 1 uncertainty), then a state-of-the-art way to devise a value-maximizing harvesting policy would be to use *simulation-optimization*. Simulation-optimization tries one or more initial policies (perhaps a mix of randomly generated and historical ones), simulates the consequences of each policy many times via Monte Carlo simulation using the known probability distributions, and iteratively improves policies until no further increases in the reward (e.g., average simulated net present value) can be found. Coupled with design-of-experiment principles for adaptively exploring the set of policies, together with sophisticated optimization steps (e.g., evolutionary optimization routines), current simulation-optimization algorithms can solve a wide range of forestry management problems under Level 1 uncertainty. These include multicriteria decisions in which the utility derived from biodiversity, carbon sequestration, and standing forests as well as the market value of timber, are taken into account (Yousefpour and Hanewinkel 2009).

Simulation-optimization is impossible under deep uncertainty, however, because the probability distributions of consequences for different policies are unknown. Instead, current algorithms for risk management of GMDPs with unknown probabilities use collaborative multiagent reinforcement learning (RL) algorithms. Each "agent" (typically identified with one node of the GMDP) makes decisions about one part of the problem (e.g., when to harvest one specific stand in a commercial forest). Each agent must coordinate with its neighbors to achieve optimal results. This is well within the capabilities of current multiagent RL algorithms for spatially distributed management of agricultural and forest resources (Forsell and Sabbadin 2009).

Similar RL algorithms have been developed to adaptively manage risks of forest fires, which again pose locally linked risks that increase with time since last harvest

(Chades and Bouteiller 2005); and to protect and conserve biodiversity in Costa-Rican forests over time, by adaptively coordinating and optimizing the reservation of sub-areas that will not be commercially exploited, in order to preserve habitats and species (Sabbadin et al. 2007). Partially observable MDPs (POMDPs) are now starting to be used to optimize allocation of scarce conservation resources to multiple conservation areas, when the presence and persistence of threatened species in each area is uncertain (McDonald-Madden et al. 2011). Thus, current applications of RL can help to protect forests and other ecosystems, as well as to manage commercial forests and other resources over long periods in the presence of uncertain, and possibly changing, risks.

Managing Disease Risks

Like the spatial spread of wind damage, forest fires, and habitat loss or gain, many contagious diseases also have strong spatial, as well as temporal, dependencies. Stopping the spread of an epidemic requires deciding not only how to act (e.g., vaccine vs. quarantine), but also where and when and with what intensity. The stakes are high: failing to quickly contain a potential epidemic or pandemic can impose enormous economic and health costs. For example, one estimate of the economic consequences of delaying detection of a foot-and-mouth disease (FMD) outbreak in a California cattle herd from 7 days to 22 days is about $66 billion (with over half a billion of additional loss, and 2000 additional cattle slaughtered, for each extra *hour* of delay after 21 days) (Carpenter et al. 2011). Managing such risks in real time, with constantly changing spatiotemporal disease data and uncertainties about where and when new cases may be discovered, requires a new generation of risk management tools to inform intervention decisions far more quickly than traditional methods. RL algorithms are being developed to meet this need.

For several decades, simulation-optimization has been applied to design epidemic risk management plans for both animal and human contagious diseases, when infectious disease control models (e.g., for mass dispensing of stockpiled medical countermeasures) involve only Level 1 or Level 2 uncertainties (Lee et al. 2010). For epidemic models with deeper uncertainties, RL optimization of policies is now starting to be used. For example, RL algorithms applied to a stochastic simulation model of the spread of an H1N1 influenza pandemic and its consequences—from illnesses and deaths, to healthcare expenses and lost wages, to shortages of vaccines, antiviral drugs, and hospital capacity—have recently been proposed to coordinate and optimize risk mitigation measures (early response, vaccination, prophylaxis, hospitalization, and quarantine applied at different times and locations) to create a cost-effective overall risk management strategy (Das et al. 2007).

In livestock, the spread of highly contagious foot-and-mouth disease (FMD) can be controlled by a combination of vaccination and culling. Both over-reaction and under-reaction cost animal lives and economic losses; therefore, adroit and flexible risk management that exploits information as it becomes available is very valuable.

Recent research suggests that adaptive risk management of FMD epidemics substantially outperforms traditional pre-specified control strategies (in which observed cases trigger automatic culling and/or vaccination within a set area around affected farms), saving unnecessary loss of animal life and more quickly suppressing FMD (Ge et al. 2010).

Robust, ensemble, and adaptive risk management techniques are also starting to be used to improve medical screening, diagnosis, prediction, and treatment of a variety of diseases. Examples include the following:

- *Earlier detection of Alzheimer's.* Ensemble prediction methods can dramatically improve ability to detect and predict some medical conditions from data. The challenging task of using brain imaging data to automatically identify women with mild Alzheimer's disease is one where AdaBoost appears to substantially improve accuracy (Savio et al. 2009), and detection of Alzheimer's in brain MRIs by model ensemble methods that incorporate AdaBoost compare favorably even to manually created "gold standard" classifications (Morra et al. 2010).
- *Improving HIV treatment using reinforcement learning.* A model-free RL algorithm has been proposed for using clinical data to decide adaptively when to cycle HIV patients off of harsh drug therapies, as part of a structured treatment interruption program designed to reduce risk of acquisition of drug resistance, as well as alleviating side effects (Ernst et al. 2006). The RL algorithm works directly with clinical data (e.g., observed levels of CD4+ T cell counts), with no need for an accurate model of HIV infection dynamics.
- *Treating depression.* RL algorithms that estimate value functions (the Q functions in the Bellman equation) despite missing data (e.g., caused by incomplete compliance and non-response bias in the patient population) have been used to adaptively refine treatments of depressed patients by adjusting the combination of antidepressants administered over time, based on patient responses, to achieve quicker and more prevalent relief of symptoms (Lizotte et al. 2008).
- *Managing ischemic heart disease (IHD) and other dynamic diseases.* The problems of managing various dynamic diseases over time based on inconclusive observations, have been formulated as MDPs and POMPDs (e.g., Schaefer et al. 2004; Alagoz et al. 2010). For example, for IHD, the physician and patient must decide when to administer or change medication, schedule stress tests or coronary angiograms, perform angioplasty or coronary artery bypass graft surgery, etc., based on time-varying information of uncertain relevance that may range from reports of chest pain to EKG readings. This disease management process has been formulated as a POMD (Hauskrecht and Fraser 2000), and uncertainty sets and practical solution algorithms for imprecisely known POMDs have been developed (Itoh and Nakamura 2007; Ni and Liu 2008).
- *Optimizing treatment of lung cancer patients in clinical trials.* Treatment of patients with advanced lung cancer typically requires switching among different lines of chemotherapy. RL algorithms are now being developed to approximately optimize the treatment of individual patients even when not enough is known to model the progression of cancers in detail (Zhao et al. 2009). The authors note

that, "reinforcement learning has tremendous potential in clinical research because it can select actions that improve outcomes by taking into account delayed effects even when the relationship between actions and outcomes is not fully known."

- *Predicting toxicity of chemicals.* Ensemble learning and prediction methods, including AdaBoost and its generalizations, have recently been shown to improve prediction of mechanisms of toxicity for organic compounds (e.g., phenols) based on molecular descriptors (Niua et al. 2009) and to out-perform other QSAR methods (Svetnik et al. 2005).
- *Better targeting of radiation therapy under uncertainty.* Robust optimization of intensity-modulated proton beam therapy spares more healthy tissues and organs than conventional optimization methods (e.g., based on probabilistic margins of error), while providing excellent coverage of the target tissue despite range and setup uncertainties (Fredriksson et al. 2011; Inaniwa et al. 2011). Multiobjective evolutionary optimization algorithms have also been developed to automatically identify undominated choices for beam angles and intensities in radiation therapy treatment planning (Fiege et al. 2011).
- *Reducing schizophrenia hospitalization episodes.* Model ensemble predictors incorporating AdaBoost have been used recently to improve prediction of schizophrenia relapses in patients participating in a weekly remote patient monitoring and disease management program (via a PC-to-phone platform), increasing specificity of predictions from 0.73 to 0.84, while keeping sensitivity at 0.65 (Hrdlicka and Klema 2011).

These examples suggest the potential for robust and adaptive methods to improve health risk management under uncertainty. This potential is only starting to be realized, since the methods are still relatively new, but it seems certain that many more practical applications in medical decision and risk analysis will be seen over the next few years.

Maintaining Reliable Network Infrastructure Service Despite Disruptions

Quickly containing and recovering from cascading failures in a power grid is somewhat analogous to quickly suppressing a spreading epidemic. In both, observations and control opportunities are spatially distributed; costly preemptive measures can be taken at different places (e.g., vaccinating as-yet uninfected flocks, or shedding power loads before generators are knocked off-grid); and a quick, effective response can potentially avert orders-of-magnitude larger losses. It is therefore perhaps unsurprising that multiagent reinforcement learning (MARL) algorithms (especially hierarchies and teams of RL controllers, each using an RL algorithm) are now being studied as effective risk management tools for increasing network resilience and responding to catastrophic failure events. For example, a two-level hierarchical control framework has recently been proposed to manage power generation and distribution in interconnected power grids under changing load and hydrothermal energy supply conditions (Zhou et al. 2011). Model-free RL (via Q-learning) is used both to figure out how best to implement high-level commands at generation units, and what high-level commands to give them to meet changing demands reliably and cheaply across the interconnected areas under normal conditions.

In the event of a catastrophic failure event that disables one or more generators (e.g., a storm, accident, or attack), decentralized (multiagent) Q-learning can again be used to quickly detect and prevent cascading failures and rapidly restore power grid systems (Ye et al. 2011). Under such a contingency, adaptive load-shedding, i.e., selective deliberate dropping of electric power, keeps the network stable, preventing the spread of blackouts, and minimizing power losses to customers as failures are isolated, power is rerouted, and service is automatically restored (Jung et al. 2002).

Similarly, multiagent distributed RL algorithms facilitate quick automated rerouting of data packet traffic in telecommunications networks following loss of fibers or switching centers, helping to make these networks highly resilient to equipment and link failures. Although vehicles cannot be rerouted as easily as data packets or electric power, control of urban traffic flow by applying similar distributed RL algorithms to traffic lights can reduce average delays and expedite passage of emergency equipment, when traffic networks and communications networks are interlinked (Kuyer et al. 2008).

Adversarial Risks and Risks from Intelligent Agents

Methods of ensemble, robust, and adaptive risk analysis do more than provide useful concepts and detailed algorithms for coping with model uncertainty (including ambiguous beliefs and preferences) in a variety of practical applications. They also shed light on some key theoretical questions in risk analysis, for example, by providing performance guarantees for how quickly adaptive low-regret risk management policies learned from data converge to approximately the best possible policy, or by giving upper bounds on the size of the cumulative difference in rewards obtained from the policy used vs. those that would have been obtained from the perfect-information optimal policy, or some other reference policy. Mathematical analysis shows that risks from intelligent adversaries cannot necessarily be managed effectively by using the same concepts and methods as for risks from non-intelligent sources: the same performance guarantees do not hold for systems that respond intelligently to a decision maker's choices as for systems that do not (Yu et al. 2009).

This does not mean that the methods are not useful for detecting and mitigating vulnerabilities to deliberate attacks. Indeed, RL algorithms for POMDPs have been shown to improve the performance of early detection systems for anthrax outbreaks, and proposed for use in reducing the consequences of possible bioterrorist attacks

(Izadi and Buckeridge 2007). RL algorithms are also used successfully to detect fraud in health insurance and auto insurance data (Lu et al. 2006; see background in Bolton and Hand 2002), and cost-sensitive modifications of AdaBoost (AdaCost and asymmetric boosting) are effective in detecting credit card fraud (Fan et al. 1999; Masnadi-Shirazi and Vasconcelos 2007). AdaBoost and RL algorithms are also used to detect intrusions into computer systems and networks (Chen and Chen 2009; Hu et al. 2008). Thus, methods of robust risk analysis, including ensemble and adaptive learning techniques, are becoming well established as tools for managing risks from intelligent adversaries.

However, the behaviors of systems of interacting intelligent agents (including software agents running their own RL algorithms, as well as humans) can be unpredictable, and low-regret policies (compared to the best that could be done with perfect information and coordination among agents on the same team) cannot necessarily be learned from data in the presence of intelligent adversaries (Yu et al. 2009). Moreover, while single-agent RL methods can be constrained to operate safely (avoiding acts that might cause harm) while still learning optimal control laws for engineering systems with nonlinear responses and random disturbances (e.g., in robotics or industrial process control) (Perkins and Barto 2002), interacting adaptive controllers in multi-agent systems can settle into behavioral patterns that do not converge at all, or that lead to a clearly dominated equilibrium (Busoniu et al. 2008).

Multi-agent reinforcement learning (MARL) algorithms are a hot research area (Dickens et al. 2010), with promising applications both for broad classes of decision problems, such as POMDs (Osada and Fujita 2005), and also for practical problems such as automated trading in finance (Busoniu et al. 2008) or detection and response to cyberterrorist distributed denial of service attacks in data networks (Xu et al. 2007). However, much remains to be understood about how intelligent agents should and do coordinate, cooperate, compete, and conflict in networks and other environments before effective risk management tools can be created for the deep uncertainties created by the interaction of multiple agents.

Conclusions

For decades, the field of health, safety, and environmental risk analysis has defined itself largely in terms of providing useful answers to a few fundamental questions, such as: What can go wrong? How likely is it to happen? If it does happen, what are the consequences likely to be? What should we do about it? What should we say about it, how, to whom? (The first three of these questions are from Kaplan and Garrick 1981; the remaining two incorporate elements of risk management decision-making and risk communication that have been emphasized more recently.) Tools for robust risk analysis, including model ensemble, robust optimization, and adaptive learning and decision-making methods, now make it practical to refine some of these questions and to pose new ones, as follows.

- Instead of (or in addition to) asking "*What can go wrong?*" one might ask "*Is there a clearly better risk management policy than the one I am now using?*" The latter question implicitly acknowledges that not everything that might plausibly go wrong can necessarily be anticipated. What can be addressed, even with very imperfect information (e.g., in a POMDP with imprecise or unknown parameters), is whether some other policy mapping observed conditions to acts, or to probabilities of acts, would be clearly better than the current one, by any of various criteria for comparing policies in the presence of deep uncertainty (e.g., stochastic dominance, expected utility with imprecise probabilities, minimum expected utility with ambiguous probabilities, robust optimization, or measures of regret.)
- Instead of asking "*How likely is to happen?*" one can ask "*How probable should I make each of my next possible actions?*" The probabilities of different scenarios or states or events are often unknown when decisions must be made, and they depend in part on what acts we take now and later. For example, the probability of an accident at a nuclear power plant over some time horizon depends largely on the acts and policies chosen by its operators. The probability of survival over time for a patient depends on what the physician (and, perhaps, the patient) do, now and later. In general, asking how likely something is to happen requires specifying what we will do, now and later. What can be answered, therefore, is not necessarily how likely different future events are, but what one will do now and what policy, mapping observations to probabilities of acts, one will use to determine what to do later. Adaptive learning policies such as SARSA and UCRL2 typically prescribe probabilities for acts, to balance the two goals of maximizing rewards based on current estimates ("exploiting" what is known now) and searching for possibly better polices ("exploring" what is still uncertain).
- Instead of asking "*If it does happen, what are the consequences likely to be?*" one can ask "*Would a different choice of policy give me lower regret (or higher expected utility of consequences), given my uncertainties?*" Even though the probabilities of consequences of events, given a choice of acts, (and hence the immediate and delayed rewards from different act-state pairs) may be unknown, or estimated only within some ranges, low-regret policies can still be developed using adaptive learning algorithms. Robust optimization can sometimes identify recommended acts even if the consequences are highly uncertain. It is therefore not necessary (and may not be possible) to predict consequences of possible future events in order to recommend low-regret or robust risk management policies. As a practical matter, decision-makers can choose policies, not events or consequences. Robust risk analysis therefore focuses on improving these choices, recognizing that event and consequence probabilities may be too uncertain to specify.

Robust risk analysis methods, including model ensemble, robust optimization, and adaptive learning and decision algorithms, shift the emphasis of the questions that define risk analysis from passive (*What might happen, and how likely is it?*) to more active (*How should I act, now and in the future?*) Risk managers are viewed not only as helping to create the future through their current decisions, but also as being able to act intelligently on the basis of future information to mitigate and control risks in ways that perhaps cannot be anticipated with the more limited information available today.

Many of the future challenges for robust risk analysis will focus on changing from a single decision-maker perspective (*What should I do?*) to a multi-agent perspective (*What should we do, how might they respond, and how should we respond to their responses?*) Understanding how multiple adaptive agents collectively affect and respond to a variety of risks, from economic and financial, to sociopolitical, to war and terrorism, remains an outstanding challenge for the next wave of advances in robust risk analysis concepts and methods.

References

Alagoz O, Hsu H, Schaefer AJ, Roberts MS (2010) Markov decision processes. Med Decis Making 30(4):474–483

Balaji PG, German X, Srinivasan D (2010) Urban traffic signal control using reinforcement learning agents. Intell Transp Syst IET 4(3):177–188

Ben-Haim Y (2001) Information-gap decision theory. Academic, San Diego, CA

Ben-Tal A, Bertsimas D, Brown DB (2010) A soft robust model for optimization under ambiguity. Oper Res 58(4, Part 2 of 2):1220–1234

Ben-Tal A, El Ghaoui L, Nemirovski A (2009) Robust optimization. Princeton University Press

Bertsimas D, Brown DB, Caramanis C (2011) Theory and applications of robust optimization. SIAM Rev 53(3):464–501

Bertsimas D, Brown DB (2009) Constructing uncertainty sets for robust linear optimization. Oper Res 57(6):1483–1495

Blum A, Mansour Y (2007) From external to internal regret. J Mach Learn Res 8:1307–1324

Bolton RJ, Hand DJ (2002) Statistical fraud detection: a review. Stat Sci 17(3):235–255. https://projecteuclid.org/journals/statistical-science/volume-17/issue-3/Statistical-Fraud-Detection-AReview/10.1214/ss/1042727940.full

Bryant B, Lempert RJ (2010) Thinking inside the box: A participatory, computer assisted approach to scenario discovery. Technol Forecast Soc Change 77(1):34–49

Buckley JJ (1986) Stochastic dominance: an approach to decision making under risk. Risk Analysis 6(1):35–41

Burton R (2008) On being certain: believing you are right even when you're not. St. Martin's Press, New York, NY

Busoniu L, Babuska R, Schutter BD (2008) A comprehensive survey of multiagent reinforcement learning. IEEE Trans Syst Man Cybern-Part C Appl Rev 38(2):156–172. www.sciweavers.org/publications/comprehensive-survey-multiagent-reinforcement-learning

Cai C, Liao X, Cari L (2009) Learning to explore and exploit in POMDPs. Adv Neural Inf Process Syst 22:198–206. http://people.ee.duke.edu/~lcarin/LearnE2_NIPS09_22_FINAL.pdf

Carpenter TE, O'Brien JM, Hagerman AD, McCarl BA (2011) Epidemic and economic impacts of delayed detection of foot-and-mouth disease: a case study of a simulated outbreak in California. J Vet Diagn Investig 23(1):26–33

Cesa-Bianchi N, Lugosi G (2006) Prediction, learning, and games. Cambridge University Press

Chades I, Bouteiller B (2005) Solving multiagent Markov decision processes: a forest management example. In MODSIM 2005 international congress on modelling and simulation

Chen Y, Chen Y (2009) Combining incremental Hidden Markov Model and Adaboost algorithm for anomaly intrusion detection. In: Chen H, Dacier M, Moens M, Paass G, Yang CC (eds) Proceedings of the ACM SIGKDD workshop on cybersecurity and intelligence informatics (Paris, France, June 28–28, 2009), CSI-KDD '09. ACM, New York, NY, pp 3–9. https://doi.org/10.1145/1599272.1599276

Churchman CW (1967) Wicked problems. Manag Sci 14(4):B141–B142

de Condorcet NC (1785) Essai sur l'Application de l'Analyse a la Probabilite des Decisions Rendues a la Pluralite des voix. Paris

Cortés EA, Gámez M, Rubio NG (2007) Multiclass corporate failure prediction by Adaboost. Int Adv Econ Res 13(3):301–312

Dalamagkidis D, Kolokotsa D, Kalaitzakis K, Stavrakakis GS (2007) Reinforcement learning for energy conservation and comfort in buildings. Build Environ 42:2686–2698. http://www.tuc.gr/fileadmin/users_data/elci/Kalaitzakis/J.38.pdf

Das TK, Savachkin AA, Zhu Y (2007) A large scale simulation model of pandemic influenza outbreaks for development of dynamic mitigation strategies. IIE Trans 40(9):893–905. http://wwweng.usf.edu/~das/papers/das_r1.pdf

Dickens L, Broda K, Russo A (2010) The dynamics of multi-agent reinforcement learning. In In Coelho H, Studer R, Wooldridge M (eds) Frontiers in artificial intelligence and applications, vol 215. Proceedings of the 2010 conference on ECAI 2010: 19th European conference on artificial intelligence. http://www.doc.ic.ac.uk/~lwd03/ecai2010.pdf

Ernst D, Stan G-B, Gongalves J, Wehenkel L (2006) Clinical data based optimal STI strategies for HIV: a reinforcement learning approach. 45th IEEE conference on decision and control, 13–15 Dec, San Diego, CA, pp 667–672. http://www.montefiore.ulg.ac.be/~stan/CDC_2006.pdf

Fan W, Stolfo S, Zhang J, Chan P (1999) Adacost: misclassification cost-sensitive boosting. In: Proceedings of the 16th international conference on machine learning, pp 97–105

Fiege J, McCurdy B, Potrebko P, Champion H, Cull A (2011) PARETO: a novel evolutionary optimization. Med Phys 38(9):5217–5229

Forsell GF, Sabbadin R (2009) Reinforcement learning for spatial processes. World IMACS/MODSIM congress, Cairns, 13–17 July 2009. http://www.mssanz.org.au/modsim09/C1/forsell.pdf

Fredriksson A, Forsgren A, Hårdemark B (2011) Minimax optimization for handling range and setup uncertainties in proton therapy. Med Phys 38(3):1672–1684

Ge L, Mourits MC, Kristensen AR, Huirne RB (2010) A modelling approach to support dynamic decision-making in the control of FMD epidemics. Prev Vet Med 95(3–4):167–174

Geibel P, Wysotzk F (2005) Risk-sensitive reinforcement learning applied to control under constraint. J Artif Intell Res 24:81–108

Gilboa I, Schmeidler D (1989) Maxmin expected utility with a non-unique prior. J Math Econ 18:141–153

Green CS, Benson C, Kersten D, Schrater P (2010) Alterations in choice behavior by manipulations of world model. Proc Natl Acad Sci U S A 107(37):16401–16406

Gregoire PL, Desjardins C, Laumonier J, Chaib-draa B (2007) Urban traffic control based on learning agents. In Intelligent transportation systems conference. ITSC 2007 IEEE: 916–921. Seattle, WA. Print ISBN: 978-1-4244-1396-6. Digital Object Identifier: 10.1109/ITSC.2007.4357719

Hauskrecht M, Fraser H (2000) Planning treatment of ischemic heart disease with partially observable Markov decision processes. Artif Intell Med 18(3):221–244. http://veryoldwww.cs.pitt.edu/~milos/research/AIMJ-2000.pdf

Hansen LP, Sargent TJ (2001) Robust control and model uncertainty. Am Econ Rev 91:60–66

Hansen LP, Sargent TJ (2008) Robustness. Princeton University Press, Princeton, NJ

Harford T (2011) Adapt: why success always starts with failure. New York, NY, Farra, Straus and Giroux

Hazen E, Seshadhri C (2007) Efficient learning algorithms for changing environments. ICML '09 Proceedings of the 26th annual international conference on machine learning. http://ie.technion.ac.il/~ehazan/papers/adap-icml2009.pdf

Hoeting JA, Madigan D, Raftery AE, Volinsky CT (1999) Bayesian model averaging: a tutorial. Stat Sci 14(4):382–401. http://mpdc.mae.cornell.edu/Courses/UQ/2676803.pdf

Hrdlicka J, Klema J (2011) Schizophrenia prediction with the adaboost algorithm. Stud Health Technol Inform 169:574–578

Hu W, Hu W, Maybank S (2008) AdaBoost. IEEE Trans Syst Man Cybern B Cybern 38(2): 577–583

Hutter M, Poland J (2005) Adaptive online prediction by following the perturbed leader. J Mach Learn Res 6:639–660. http://jmlr.csail.mit.edu/papers/volume6/hutter05a/hutter05a.pdf

Inaniwa T, Kanematsu N, Furukawa T, Hasegawa A (2011) A robust algorithm of intensity modulated proton therapy for critical tissue sparing and target coverage. Phys Med Biol 56(15):4749–4770

Itoh H, Nakamura K (2007) Partially observable Markov decision processes with imprecise parameters. Artif Intell 171(8–9):453–490

Izadi MT, Buckeridge DL (2007) Optimizing anthrax outbreak detection using reinforcement learning. IAAI'07 Proceedings of the 19th national conference on Innovative applications of artificial intelligence, vol 2. AAAI Press http://www.aaai.org/Papers/AAAI/2007/AAAI07-286.pdf

Jaksch T, Ortner R, Auer P (2010) Near-optimal regret bounds for reinforcement learning. J Mach Learn Res 11:1563–1600

Jung J, Liu CC, Tanimoto S, Vittal V (2002) Adaptation in load shedding under vulnerable operating conditions. IEEE Trans Power Syst 17:1199–1205

Kahnt T, Park SQ, Cohen MX, Beck A, Heinz A, Wrase J (2009) Dorsal striatal-midbrain connectivity in humans predicts how reinforcements are used to guide decisions. J Cogn Neurosci 21(7):1332–1345

Kaplan S, Garrick BJ (1981) On the quantitative definition of risk. Risk Anal 1(1):11–27. http://josiah.berkeley.edu/2007Fall/NE275/CourseReader/3.pdf

Koop G, Tole L (2004) Measuring the health effects of air pollution: to what extent can we really say that people are dying from bad air? J Environ Econ Manag 47:30–54. See also: http://citeseerx.ist.psu.edu/viewdoc/summary?doi=10.1.1.164.6048

Kuyer L, Whiteson S, Bakker B, Vlassis N (2008) Multiagent reinforcement learning for urban traffic control using coordination graphs. In ECML 2008: Proceedings of the nineteenth European conference on machine learning, pp 656–671

Laeven R, Stadje MA. Entropy coherent and entropy convex measures of risk. Tilburg University CentER Discussion Paper 2011-031. http://arno.uvt.nl/show.cgi?fid=114115

Lee EK, Chen CH, Pietz F, Benecke B (2010) Disease propagation analysis and mitigation strategies for effective mass dispensing. AMIA Annu Symp Proc 13(2010):427–431. http://www.ncbi.nlm.nih.gov/pubmed/21347014

Lempert RJ, Collins MT (2007) Managing the risk of uncertain threshold response: comparison of robust, optimum, and precautionary approaches. Risk Anal 27(4):1009–1026

Lempert R, Kalra N (2008) Managing climate risks in developing countries with robust decision making. World Resources Report, Washington, DC. Available online at http://www.worldresourcesreport.org/files/wrr/papers/wrr_lempert_and_kalra_uncertainty.pdf

Lizotte DJ, Gunter L, Laber E, Murphy SA (2008) Missing data and uncertainty in batch reinforcement learning, NIPS-08 Workshop on Model Uncertainty and Risk in RL. http://www.cs.uwaterloo.ca/~ppoupart/nips08-workshop/nips08-workshop-schedule.html

Lu F, Boritz JE, Covvey HD (2006) Adaptive fraud detection using Benford's law. Advances in artificial intelligence: 19th conference of the Canadian society for computational studies of intelligence http://bit.csc.lsu.edu/~jianhua/petrov.pdf

Maccheroni F, Marinacci M, Rustichini A (2006) Ambiguity aversion, robustness, and the variational representation of preferences. Econometrica 74:1447–1498

Makridakis S, Hibon M (2000) The M3-Competition: results, conclusions and implications. Int J Forecast 16:451–476. http://www.forecastingprinciples.com/files/pdf/Makridakia-The%20M3%20Competition.pdf

Masnadi-Shirazi H, Vasconcelos N (2007) Asymmetric boosting. In: Proceedings of the 24th international conference on machine learning, p 609{619}

McDonald-Madden E, Chadès I, McCarthy MA, Linkie M, Possingham HP (2011) Allocating conservation resources between areas where persistence of a species is uncertain. Ecol Appl 21(3):844–858

Molinaro AM, Simon R, Pfeiffer RM (2005) Prediction error estimation: a comparison of resampling. Bioinformatics 21(15):3301–3307

Morra JH, Tu Z, Apostolova LG, Green AE, Toga AW, Thompson PM (2010) Comparison of AdaBoost and support vector machines for detecting Alzheimer's disease through automated hippocampal segmentation. IEEE Trans Med Imaging 29(1):30–43

Ni Y, Liu Z-Q (2008) Bounded-parameter partially observable Markov decision Processes. In Proceedings of the eighteenth international conference on automated planning and scheduling

Niua B, Jinb Y, Lua WC, Li GZ (2009) Predicting toxic action mechanisms of phenols using AdaBoost Learner. Chemometr Intell Lab Syst 96(1):43–48

Osada H, Fujita S (2005) CHQ: a multi-agent reinforcement learning scheme for partially observable Markov decision processes. IEICE Trans Inf Syst E88-D(5)

Perkins TJ, Barto AG (2002) Lyapunov design for safe reinforcement learning. J Mach Learn Res 3: 803–883. http://jmlr.csail.mit.edu/papers/volume3/perkins02a/perkins02a.pdf

Pinker S (2021) Rationality: what it is, why it seems scarce, why it matters. Viking, an imprint of Penguin Random House LLC. New York, NY

Regan K, Boutilier C (2008) Regret-based reward elicitation for Markov decision processes. NIPS-08 workshop on model uncertainty and risk in RL. http://www.cs.uwaterloo.ca/~ppoupart/nips08-workshop/nips08-workshop-schedule.html

Rittel H, Webber M (1973) Dilemmas in a general theory of planning. Policy Sci 4:155–169. [Reprinted in Cross N (ed) Developments in design methodology. Wiley, Chichester, 1984, pp 135–144.] http://www.uctc.net/mwebber/Rittel+Webber+Dilemmas+General_Theory_of_Planning.pdf

Ross S, Pineau J, Chaib-draa B, Kreitmann P (2011) POMDPs: a new perspective on the explore-exploit tradeoff in partially observable domains. J Mach Learn Res 12:1729–1770

Sabbadin R, Spring D, Bergonnier E (2007) A reinforcement-learning application to biodiversity conservation in costa-rican forest. In: Oxley, L. and Kulasiri, D. (eds) MODSIM 2007 international congress on modelling and simulation. Modelling and simulation society of Australia and New Zealand, December 2007, pp. 2189–2195. https://www.mssanz.org.au/MODSIM07/papers/41_s34/AReinforcement_s34_Sabbadin_.pdf

Savio A, García-Sebastián M, Graña M, Villanúa J (2009) Results of an Adaboost approach on Alzheimer's disease detection on MRI. Bioinspired Applications in Artificial And Natural Computation Lecture Notes in Computer Science, vol 5602/2009:114–123. www.ehu.es/ccwintco/uploads/1/11/GarciaSebastianSavio-VBM_SPM_SVM-IWINAC2009_v2.pdf

Schaefer AJ, Bailey MD, Shechter SM, Roberts MS (2004) Handbook of operations research/management science applications in health care. Kluwer Academic, Boston, MA. Modeling medical treatment using Markov decision processes, pp 593–612. http://www.ie.pitt.edu/~schaefer/Papers/MDPMedTreatment.pdf

Schönberg T, Daw ND, Joel D, O'Doherty JP (2007) Reinforcement learning signals in the human striatum distinguish learners from nonlearners during reward-based decision making. J Neurosci 27(47):12860–12867

Srinivasan J, Gadgil S (2002) Asian brown cloud - fact and fantasy. Curr Sci 83:586–592

Su Q, Lu W, Niu B, Liu X (2011) Classification of the toxicity of some organic compounds to tadpoles (Rana Temporaria) through integrating multiple classifiers. Mol Inform 30(8):672–675

Sutton RS, Barto AG (2005) Reinforcement learning: an introduction. MIT Press. http://rlai.cs.ualberta.ca/~sutton/book/ebook/the-book.html

Svetnik V, Wang T, Tong C, Liaw A, Sheridan RP, Song Q (2005) Boosting: an ensemble learning tool for compound classification and QSAR modeling. J Chem Inf Model 45(3):786–799

Szepesvari C (2010) Reinforcement learning algorithms. Morgan & Claypool Publishers

Tan C, Chen H, Xia C (2009) Early prediction of lung cancer based on the combination of trace element analysis in urine and an Adaboost algorithm. J Pharm Biomed Anal 49(3):746–752

Walker WE, Marchau VAWJ, Swanson D (2010) Addressing deep uncertainty using adaptive policies introduction to section 2. Technol Forecast Soc Chang 77(6):917–923

Waltman L, van Eck NJ (2009) Robust evolutionary algorithm design for socio-economic simulation: Some comments. Comput Econ 2009(33):103–105. http://repub.eur.nl/res/pub/18660/RobustEvolutionary_2008.pdf

Wang Y, Xie Q, Ammari A (2011) Deriving a near-optimal power management policy using model-free reinforcement learning and Bayesian classification DAC '11 Proceedings of the 48th design automation conference. ACM New York, NY

Weick KE, Sutcliffe KM (2007) Managing the unexpected: resilient performance in an age of uncertainty, 2nd edn. Wiley

Xu X, Sun Y, Huang Z (2007) Defending DDoS attacks using hidden Markov models and cooperative reinforcement learning. Proceedings, PAISI'07 Proceedings of the 2007 Pacific Asia conference on Intelligence and security informatics. Springer Berlin, Heidelberg

Yousefpour R, Hanewinkel M (2009) Modelling of forest conversion planning with an adaptive simulation-optimization approach and simultaneous consideration of the values of timber, carbon and biodiversity. Ecol Econ 68(6):1711–1722

Ye D, Zhang M, Sutato D (2011) A hybrid multiagent framework with Q-learning for power grid systems restoration. IEEE Trans Power Syst 26(4):2434–2441

Yu JY, Mannor S, Shimkin N (2009) Markov decision processes with arbitrary reward processes. Math Oper Res 34(3):737–757

Zhao Y, Kosorok MR, Zeng D (2009) Reinforcement learning design for cancer clinical trials. Stat Med 28(26):3294–3315

Zhou B, Chan KW, Yu T (2011) Q-Learning approach for hierarchical AGC scheme of interconnected power grids. The Proceedings of international conference on smart grid and clean energy technologies energy procedia, vol 12, pp 43–52

Zhou L, Lai KK (2009) Adaboosting neural networks for credit scoring. Adv Intell Soft Comput 56(2009):875–884. https://doi.org/10.1007/978-3-642-01216-7_93

Chapter 8
Muddling-Through and Deep Learning for Bureaucratic Decision-Making

Introduction: Traditional Benefit-Cost Analysis (BCA) and Decision Analysis

As discussed in Chaps. 1, 5, 6, and 7, traditional benefit-cost analysis and decision analysis typically involves multiple steps such as the following (Raiffa 1968; Clemen and Reilly 2014; Howard and Abbas 2015; Pinker 2021):

1. *Identify alternative feasible choices*, decision rules, or courses of actions. This "choice set," or set of decision alternatives, may be specified explicitly as a discrete set of alternatives, such as whether or not to fund a public project, or implicitly via constraints on the allowed values of decision variables, such as quantities of limited resources available to be allocated.
2. *Identify preferences and value trade-offs* for possible outcomes. These may be formally represented via a net benefit function or via a (possibly multi-attribute) von Neumann-Morgenstern utility function or social utility function to be maximized (Keeney and Raiffa 1976).
3. *If the outcomes for each choice are uncertain, estimate the probabilities of different outcomes for each choice* (e.g., its risk profile); and
4. *Optimize choices* subject to feasibility constraints (e.g., on available time, budget, or limited resources) to identify and recommend a feasible choice that maximizes expected net benefit, expected utility, or expected social utility of outcomes

These steps are all well-established parts of prescriptive decision analysis for a single decision-maker and benefit-cost analysis for a social decision-maker (Howard and Abbas 2015; Raiffa 1968).

In 1957, political economist Charles Lindblom of Yale University pointed out that almost none of these steps can be applied in practice to the decisions and uncertainties faced by real government decision-makers, or by decision-makers in other bureaucracies. Preferences and value trade-offs may be unknown and difficult

L. A. Cox Jr., *AI-ML for Decision and Risk Analysis*, International Series in Operations Research & Management Science 345,
https://doi.org/10.1007/978-3-031-32013-2_8

or impossible to articulate, quantify, and justify. Lindblom wrote, "Typically the administrator chooses—and must choose—directly among policies in which [different] values are combined in different ways. He cannot first clarify his values and then choose among policies," as multiattribute utility theory prescribes. Even identifying possible outcomes for each feasible choice may be impracticable if the number of possible choices is immense or possible outcomes are unknown. In addition, real-world bureaucratic and organizational decisions are almost never made by a single decision-maker. Rather than seeking to extend or refine normative decision analysis to overcome what he perceived as its fatal practical limitations for large-scale, multi-person organizational decision-making over time, Lindblom instead described a method of *successive limited comparisons* that he contrasts with the "rational-comprehensive" normative approach favored in benefit-cost analysis, decision analysis, operations research, and optimal control engineering. The rational-comprehensive approach seeks to solve decision optimization problems such as.

$$max_{a \in A} R(a) \tag{8.1}$$

where

- a is a decision variable or policy (e.g., a vector or a time series of decision variables, or a feedback control decision rule mapping observations to actions)
- A is the set of feasible alternative decisions (the "choice set")
- $R(a)$ is the reward (expected utility or net benefit) from choosing a. In many traditional economic, policy, and operations research analyses, the reward function be maximized is assumed to be known. In statistical design of experiments and machine learning, it may have to be discovered. If the reward received depends both on the decision-maker's choice a and also on other variables not controlled by the decision-maker, collectively referred to as the *state* and modeled as a random variable s, then $R(a)$ is the expected reward from choosing a given the probability distribution of s. When there are many players, R is often taken to be a weighted sum of individual utility functions (Gilboa et al. 2004)
- $max_{a \in A}$ indicates that an act a in A is to be selected to maximize $R(a)$.

Lindblom wrote that "the attention given to, and successes enjoyed by operations research, statistical decision theory, and systems analysis" have strengthened a "tendency to describe policy formulation even for complex problems as though it followed [this] approach," emphasizing "clarity of objective, explicitness of evaluation, a high degree of comprehensiveness of overview, and, wherever possible, quantification of values for mathematical analysis. But these advanced procedures remain largely the appropriate techniques of relatively small-scale problem-solving where the total number of variables to be considered is small and value problems restricted."

In contrast, for large-scale real-world decision problems faced by most bureaucracies, Lindblom considers the rational-comprehensive approach in Eq. (8.1) to be impracticable because the net benefit or reward function R is not known or agreed to; choice set A may be too large to enumerate or search effectively, or unknown and costly to develop; and often no single centralized authority is capable of, authorized to, or accountable for identifying and implementing the best choice in A. Instead of clarifying values and objectives in advance, goals and actions to achieve them are selected together as opportunities arise. The test of a "good" policy is not that it is the best means to desired ends, or that it maximizes some measure of expected net benefit, utility, or collective welfare, but that people will agree to it (possibly for different, and perhaps conflicting, private reasons). Important possible outcomes, feasible alternative policies, and affected values and trade-offs are neglected in favor of relatively simple comparisons between the current policy and a proposed incremental modification of it. A succession of such modifications may, if all goes well, produce gradually improving policies; this is the process that Lindblom refers to as successive limited comparisons, or, more colloquially, as *muddling through.* He states that "Making policy is at best a very rough process. Neither social scientists, nor politicians, nor public administrators yet know enough about the social world to avoid repeated error in predicting the consequences of policy moves. A wise policy maker consequently expects that his policies will achieve only part of what he hopes and at the same time will produce unanticipated consequences that he would have preferred to avoid. If he proceeds through a succession of incremental changes, he avoids serious lasting mistakes in several ways" including learning from experience and being able to correct missteps fairly quickly. Of course, this view is optimistic if a single misstep could lead to disaster, ruin, or the destruction of the decision-making organizations, but Lindblom does not dwell on these grim possibilities. To model and evaluate muddling through approach more formally, however, we will have to consider possibilities for *safe learning*, i.e., surviving and avoiding disastrous decisions during learning (Garcia and Fernandez 2015). Lindblom proposes muddling through not only as a descriptive theory of bureaucratic decision-making, but also as a normative one: "Why then bother to describe the method in all of the above detail? Because it is in fact a common method of policy formulation and is, for complex problems, the principal reliance of administrators as well as of other policy analysts. And because it will be superior to any other decision-making method available for complex problems in many circumstances, certainly superior to a futile attempt at superhuman comprehensiveness." In short, muddling through by successive incremental adjustments of policy is proposed as both more desirable and more widely practiced than the rational-comprehensive approach.

Since Lindblom's essay, revolutions have occurred in computer science, game theory, collective choice theory, automated and adaptive control, artificial intelligence, robust optimization and risk analysis, machine learning, computational statistics and data science, and the intersection of these fields with political economy, law-and-economics, and management science. It is timely to reexamine the extent to which Lindblom's critique of rational-comprehensive techniques for risk management decision support still applies; the extent to which the ferment of ideas and technical developments in artificial intelligence and other fields dealing with multi-agent control has overcome his objections; how both the strengths and the limitations of muddling through can be understood better, and the technique applied more successfully, in light of progress since 1957; and whether there are circumstances in which muddling through provides a viable alternative or complement to decision analysis. The following sections undertake such a reexamination.

Developments in Rational-Comprehensive Models of Decision-Making

An individual, team, organization, or artificial intelligence that repeatedly makes decisions to achieve some overall purposes or goals must repeatedly decide *what* to do next—e.g., what subgoals or tasks to undertake next—and *how* to do it, e.g., which agents should do what, and how much planning should be local and autonomous instead of centralized or hierarchical. In teams with no central coordinator, such as robot soccer teams of cooperating autonomous agents, cooperating swarms of drones, or search-and-rescue teams with autonomous agents and limited communication, the agents may have to infer and adapt to each other's plans on the fly as they observe each other's behaviors and messages (Hunt et al. 2014; Zhao et al. 2016). In bureaucracies or other organizations where policies are formulated and adapted via muddling through, success or failure in achieving stated goals may depend on who may propose what when, how decisions are made about which proposals to adopt, and how these changes and their consequences are linked to incentives and rewards for those participating in policy making and administration.

In the face of such complexities, the simple prescriptive model of optimization-based rational-comprehensive decision-making in (8.1) has been generalized and extended in the following ways.

- *Non-cooperative game theory* (Luce and Raiffa 1957) replaces the reward function $R(a)$ in (8.1) with a set of reward functions (also called "payoff functions"), one for each participant (called a "player" or "agent"). Each player has its own choice set of feasible alternatives to choose among, often called *strategies* in game theory, or *policies* in decision analysis, machine learning, and artificial intelligence. Player i now seeks to choose a_i from A_i to maximize $R_i(a_i, a_{i'})$, where a_i denotes the strategy selected from A_i by player i; $a_{i'}$ denotes all the strategies selected by the other players; and $R_i(a_i, a_{i'})$ is the reward to player i from choosing strategy a_i when the other players choose $a_{i'}$. There is no single net benefit, social welfare or public interest to be maximized. Rather, each player seeks to act to maximize its own reward, given the actions of the rest. A *Nash equilibrium* is a set of choices such that no player can improve its own reward by unilaterally modifying its own choice, given the choices of the other players. Each player's choice is a best response to the choices of the rest. A set of choices by the players is *Pareto-efficient* if no other set of choices would give all players equal or greater rewards, and at least some of them greater rewards. In practical applications such as deciding how to manage air pollution, antibiotic resistance, or climate change, a common challenge is that each player benefits if everyone else exercises restraint to avoid making the current problem worse, but each player also maximizes its own benefits by being unrestrained itself, whatever the other players are doing. In such cases, the unique Nash equilibrium is that no one exercises self restraint, even though all would gain if all would do so; hence, it is not Pareto efficient. A variety of "folk theorems" of game theory prove that both Pareto efficiency and multi-period versions of Nash equilibrium can be achieved if players are sufficiently patient (i.e., they do not discount delayed rewards too steeply) in repeated games with discounted rewards and uncertain time horizons, where the players have a chance to observe each other's behaviors and make choices repeatedly over time. The trick is to have players make choices that punish those who do not cooperate in sustaining a Pareto-efficient outcome (Fudenberg and Maskin 1986; Fudenberg et al. 1994; Hörner and Olszewski 2006).

- *Cooperative game theory* further generalizes the multi-player choice problem by allowing players to form coalitions and to bargain or negotiate with each other. For example, in the *treaty participation* game model of international cooperation (or lack of it) to limit emissions in hopes of limiting undesired climate change, a coalition of signatories might choose emissions levels to maximize their collective benefits, while non-signatories choose emissions levels to maximize their individual benefits (Barrett 2013). The final levels of cooperation and emissions achieved in multistage games of coalition formation and decision-making about emissions depend on factors such as whether coalitions, once formed, are exclusive; whether players (e.g., countries) can make and enforce conditional agreements such as that some will reduce their emissions more if and only if others do; whether binding commitments can be made and enforced; how steeply participants discount future rewards and penalties compared to current ones; and whether the timing of catastrophic consequences from failure to muster sufficient cooperation is known or uncertain (Heitzig et al. 2011; Wood 2011; Barrett 2013).
- *Team theory* (Marschak and Radner 1972) focuses on design of costly communication and agent decision rules (and, in some versions, on allocation of limited resources among the agents) for the special case of cooperating agents in an organization where all of the agents have identical preferences and goals. That is, they all seek to maximize the same reward function of their joint choices, but local observations, actions, and communications are costly. Team theory has been applied to distributed control of systems by agents with sensors and actuators at different locations, as well as to organizational design, design of compensation systems, and dynamic allocation of tasks, roles, and responsibilities within teams of cooperating agents.
- *Mechanism design:* Institutions, social and moral norms, legal constraints and liabilities, regulations and their enforcement, wages and contractual incentives, outcome-sharing rules in principal-agent relationships and investment syndicates, and reputations in repeated transactions and long-term relationships all help to shape the rewards (positive or negative) and feedback that players receive for their choices and behaviors. Game theory studies how agents make choices in response to incentives. *Mechanism design theory* (Nisan 2007) studies the inverse

problem of how to design incentives, or the rules determining rewards in the games in which agents participate, to elicit choices that satisfy desired properties. These may include Pareto efficiency, self-enforcing stability (e.g., Nash equilibrium and its multi-period extensions), implementability using information that can actually be obtained and incentives (e.g., payments) that can actually be provided, and voluntary participation. Although important impossibility theorems show that successful mechanism design satisfying most or all of these properties is impossible if preferences are arbitrary, many positive results are available when preferences satisfy restrictions (e.g., risk neutrality and "quasilinear preferences" with utility linear in money) commonly assumed in traditional benefit-cost analyses.

- *Organizational design and law-and-economics:* Within bureaucracies and other hierarchical organizations (e.g., principal-agent relationships), as well as in the more specialized contexts of designing contracts and auctions, mechanism design can be applied to design incentive systems to promote revelation of local information, elicit desired behaviors despite private information, and optimize delegation and tradeoffs between centralization and decentralization, taking into account costs of communication, monitoring, and control and inefficiencies due to remaining private information (Mookherjee 2006). As a prominent application of the mechanism design perspective, the modern theory of law and economics (Miceli 2017) explains how systems of laws establishing tort liability rules for hazardous activities, remedies for breach of contracts, property rights to internalize externalities, product liability and implicit warranty principles, and so forth can be designed to maximize the expected net economic benefit from voluntary transactions, usually assuming risk-neutral participants with quasilinear preferences. Practical designs that explain many aspects of observed legal practice account for market imperfections such as private and asymmetric information (e.g., a consumer may not know how much care a manufacturer has taken to keep a product safe, or the manufacturer may not know how much care the consumer will exercise in using the product safely), costs of litigation, misperceptions of risk by buyers, and incentives for socially valuable research and disclosure of information by sellers.

Modern Algorithms for Single- and Multi-Agent Decision-Making

The intersection of computer science with decision models and algorithms has tremendously advanced the design and practical application of algorithms for solving large-scale single-person and team decision optimization problems, as well as games and collective choice problems, in recent decades. Current state-of-the-art algorithms are briefly described next.

- *Monte Carlo Tree Search (MCTS).* Decision trees and game trees showing possible sequences of actions (choice nodes) and uncertainty resolutions (chance nodes, with probabilities for each branch) leading to rewards (utilities) at the ends (leaf nodes) of the tree are perhaps the best known rational-comprehensive models of normative decision analysis for small problems (Raiffa 1968; Luce and Raiffa 1957). For large problems, recent *Monte Carlo Tree Search* (MCTS) algorithms (Munos 2014; Silver et al. 2016, 2018) sample possible future paths and rewards to avoid enumerating all possibilities. This decouples "rational" decision-making, based on optimizing current decisions based on predicted future reward probabilities, from "comprehensive" modeling of the causal relationship between choices and reward probabilities by selecting only the most promising choice nodes in a tree for further simulation and evaluation. MCTS can be combined with reinforcement learning (RL) techniques discussed next (Vodopivec et al. 2017) and applied to more general settings, such as those in which it is costly to observe the reward (Schulze and Evans 2018), as is the case for many social policy interventions.
- *Reinforcement learning (RL) of high-reward policies through trial and error learning* (Sutton and Barto 1998, 2018). Decision-makers (agents) often initially do not know how their choices affect reward probabilities, or expected benefits, but must discover the immediate and longer-term costs and benefits of alternative policies or choices from experience. Denote true expected value starting in state s and acting optimally thereafter by an (initially unknown) *value function*, $V(s)$ and let $Q(a, s)$ denote an estimate of the value from taking each feasible action a when in each state s and then acting optimally (e.g., to maximize the discounted sum of future rewards) ever after. The initial estimates of these values may be random guesses, but they are updated in light of experience by adjusting current estimates by an amount proportional to the difference between expected and experienced rewards. The constant of proportionality is interpreted as the *learning rate*. For example, *Q-learning* uses the current estimate $Q(a, s)$ to select which action to take next in the current state s. Then the resulting reward is used to update the estimate of $Q(a, s)$ based on the difference between estimated and observed rewards. In many settings, estimated $Q(a, s)$ values converge and the policy of selecting a to maximize $Q(a, s)$ is then the optimal policy, while the estimated value of $Q(a, s)$ when that policy is used is the true value function, $V(s)$. This procedure is similar to value iteration in classical stochastic dynamic programming, but without the requirement that the reward function and state transition probabilities be initially known. It converges to yield optimal policies under certain conditions for Markov decision processes (MDPs), in which the actions taken affect next-state probabilities as well as probability distributions of current rewards) (Krishnamurthy 2015). The main conditions are that learning rates be kept small enough and that the MDPs are ergodic, involving no irreversible choices or fatal outcomes that would limit or prevent future exploration and adaptation (Bloembergen et al. 2015; Krishnamurthy 2015; Xu et al. 2017).
- *RL using policy gradient algorithms.* RL can also be based on algorithms that emphasize adjusting polices directly rather than estimating values for different

actions as in benefit-cost analysis. As usual, a *policy* in RL is a decision rule mapping observations (e.g., the current state) to actions. In most RL algorithms, however, this mapping is randomized: thus, a policy RL specifies the *probability* of taking each feasible action when in each state (or, more generally, given current information, which may include imperfect observations of the current state). Polices are updated to favor selecting actions with higher expected values. The tension between exploring further in hopes of finding a more valuable policy and exploiting what has been learned so far by selecting the actions with the highest expected values is managed carefully by choosing action-selection probabilities to avoid premature convergence to sub-optimal policies. For example, a simple and effective policy in many settings is to select each action with a probability equal to the currently estimated probability that it is the best (value-maximizing) action; this is called Thompson sampling (Schulze and Evans 2018). Such randomized sampling schemes prevent jumping to possibly erroneous conclusions about what works best in clinical trials and similar sequential decision optimization settings (Villar et al. 2015). Adjustments of policies continue until expected and experienced average rewards no longer differ. For large classes of adaptive decision problems under uncertainty, the policies arrived at by such successive incremental adjustments are the optimal policies that would be obtained by classical operations research methods (Bloembergen et al. 2015; Krishnamurthy 2015; Xu et al. 2017). Table 8.1 lists important refinements and enhancements used in practice to make RL quicker and more robust to data limitations. Table 8.2 summarizes methods for safe learning that have proved effective in applications ranging from learning to control helicopters and quadcopters (e.g., allowing them to hover or navigate safely in cluttered environments) to learning to manage power grids and other networked infrastructures, without risking costly accidents and failures during learning. Table 8.3 summarizes variations and extensions of *multi-agent reinforcement learning (MARL)* in which multiple agents act, learning, and perhaps communicate about how to control a system or accomplish a task. MARL can greatly increase speed of learning and average rewards generated per unit time, under certain conditions (Omidshafiei et al. 2017; Gupta et al. 2017).

MARL algorithms and architectures that incorporate MCTS and enhancements to speed convergence, safe learning, and communication and control hierarchies represent the current state-of-the-art in machine learning models and methods for solving large-scale and distributed decision and control problems under uncertainty, including problems with sparse and delayed feedback. Although most MARL algorithms are designed for cooperating agents, Bowling and Voloso (2001) showed that convergence to Nash equilibria can also be achieved in a variety of noncooperative Markov games (generalizations of MDPs to multiple agents) if each agent uses RL but manages its learning rate to take large steps when the agent's experienced rewards are less than expected ("learn fast when losing") and small steps otherwise (when it is "winning" by receiving higher than expected rewards). The resulting WoLF ("win or learn fast") principle has been incorporated into many

Table 8.1 Some enhancements to Reinforcement Learning (RL) algorithms

Enhancement	Main ideas
Policy gradient RL algorithms	Directly modify policies, without first estimating a value function for the states, by estimating the gradient (slope) of the reward as a function of policy parameters and adjusting those parameters incrementally to ascend the estimated slope (Arulkumaran et al. 2017)
Actor-critic architectures	Interpret the policy at any time as an "actor" and the value function as a "critic" that evaluates how well the current policy is working. Separating these two roles helps to speed convergence (Grondman et al. 2012).
Model-based RL	Fit statistical models of reward probabilities and state transition probabilities to observed state-act-reward-next-state data. Use the models to speed learning of high-reward policies (if the models are usefully accurate) (Clavira et al. 2018).
Model-free RL	Use empirically observed rewards to estimate state or action value functions (via iteratively updated Q values). Powerful statistical and machine learning techniques for approximating unknown functions from data, such as deep neural networks, can obtain most of the advantages of model-based RL while avoiding the potential pitfalls from using incorrect models (Mnih et al. 2015; Andrychowicz et al. 2018).
Reward shaping	Modify the original reward function received from the environment to encourage quicker learning and discovery of better policies (Mannion et al. 2017)
Experience replay	Use Monte Carlo simulation from frequency distributions of past experiences (e.g., state-action-reward-next state sequences) to reduce computational burden and augment sparse training data (Andrychowicz et al. 2018).
Deep learning control of the learning rate	Use deep learning neural networks to automatically adjust the learning rate parameter using an actor-critic architecture in which one neural network adjusts the parameter and another provides feedback on how well the adjustments appear to be working (Xu et al. 2017).
Meta-learning	Estimate crude high-level models of rewards and value functions relatively rapidly. Refine and improve them and use them to guide actions via RL as new observations are made. Such a hierarchy of modeling allows relatively rapid and effective adaptation to new conditions in non-stationary, including graceful compensation for and recovery from partial system failures (Lemke et al. 2015; Clavira et al. 2018).
Inverse RL and imitation learning.	Use observed data on state and action sequences leading to success or failure in a task to infer successful policies for choosing actions to take in each state to accomplish it successfully. This makes it possible for agents to learn quickly from humans or other more experienced and higher-performing agents how to do complex tasks (Shiarlis et al. 2016).
Hybrids of above techniques	Example: Interleaving updates of the estimated value function with sampling from the experience replay buffer and adjustment of policies to increase expected reward ("policy gradient ascent" for rewards or "policy gradient descent" for losses, using a step size determined by the current learning rate parameter).

Table 8.2 Some principles for safe learning, i.e., learning without risking catastrophic failures

Safe learning principle	Main ideas
Risk-sensitive learning and control	Modify the reward function to consider variance in return; probabilities of ruin or large loss, such as crash of an autonomous vehicle; and risk-sensitive control policies (Garcia and Fernandez 2015)
Imitation learning with safe instruction	Use imitation learning from demonstrations supplied by instructors to assure that only safe examples are imitated (Garcia and Fernandez 2015)
Knowledge-based constraints on exploration	Use knowledge-based constraints supplied by instructors to assure that that only safe changes in policies are explored during learning(Garcia and Fernandez 2015)
Maintain system stability while learning and exploring modified policies	Apply feedback control theory for dynamic systems to maintain stability of the system while collecting data. Use the collected data to learn to improve control performance and to expand the safe region of the state space, i.e., the set of states for which safe control policies are available (Bernkamp et al. 2017). Keeping changes in control policies small enough to avoid destabilizing the system while learning is effective for systems that are known to have well-behaved dynamics, without large (e.g., discontinuous jump) responses to small changes in controls.
Use model uncertainty to constrain exploration	Create uncertainty zones around regions of potentially high loss (e.g., around pedestrians with unpredictable behaviors) based on model uncertainty estimates, and avoid them during learning (Lütjens et al. 2018).
Safe policy improvement using a known safe policy as default when model uncertainty is high	Engage in *safe policy improvement* by using known safe (i.e., catastrophe-avoiding) default policies when model uncertainty about effects of changing the policy is high. Explore for possible improvements in policies when model uncertainty is low (Petrik et al. 2016)
Safe policy improvement using statistical confidence bounds to limit the risk from policy modifications	Use statistical confidence bounds (e.g., derived from importance sampling and probability inequalities) for performance of modified policies to avoid those that pose unacceptable risks (Thomas et al. 2015).

subsequent MARL algorithms for cooperative learning. It gives agents who are lagging in learning to contribute to the team's success time to catch up, while agents who are ahead of the rest continue to explore relatively cautiously (via relatively small incremental adjustment steps) for even better policies. In practice, MARL algorithms have been applied successfully to obtain high-reward policies for difficult

Table 8.3 Some MARL variations and extensions

Setting	Main ideas, results, and applications
MARL for non-cooperative stochastic games	Convergence to Nash equilibria occurs under certain conditions if each agent uses RL and manages its learning rate appropriately (Hu and Wellman 1998, 2003). (However, Nash equilibria may be Pareto-efficient.)
Collective choice MARL	Agents initially know only their own preferences. They negotiate by proposing joint actions to each other to improve their own payoffs. Accepted proposals are binding and generate mutual gains. This cooperative negotiation leads to Pareto-superior outcomes than non-cooperative MARL in many games (Hu et al. 2015).
MARL for teams without communication among agents	Teams of cooperating agents with the same goal (i.e., cooperating to maximize the same reward function) can learn to behave effectively in many settings even without explicit communication, by observing, modeling, and adjusting to each other's behaviors (Gupta et al. 2017).
Decentralized MARL for distributed control of a system by a team of cooperating and communicating agents	Decentralized cooperative learning by a team of agents based on explicit communication (e.g., over an unreliable communication network), with agents sharing experiences (data, estimated value functions, or policies), improves learning of distributed control policies to maximize average reward. Applications include control of power grids, mobile sensor networks, and autonomous vehicles (Zhang et al. 2018)
Hierarchical MARL (HMARL)	MARL systems with hierarchical organizations of agents, as well as other techniques such as reward shaping, speed convergence to high-reward policies in many settings (Mannion et al. 2017).
Decentralized multi-level HMARL	In a multi-level hierarchy of agents, supervisory agents abstract and aggregate information from their subordinates, share it with their peers, pass summaries upward to their own supervisors, and pass supervisory suggestions and constraints on next actions down to their subordinates. This approach has been found to improve convergence of MARL learning in tasks requiring distributed control, such as network routing (Zhang et al. 2008).
Two-level HMARL	A central controller coordinates learning among the agents. Local agents manage different parts of a system, such as a supply chain network. They send to the central controller

(continued)

Table 8.3 (continued)

Setting	Main ideas, results, and applications
	information about their current policies (e.g., represented as deep neural networks for mapping observations to actions) and observations on local costs (e.g., arising from inventory ordering, holding, and stockout costs). The central controller sends feedback to the agents (e.g., weights for the best policies learned so far by each agent) to coordinate their learning. In experimental supply chains, such two-level hierarchical MARL systems discovered policies that substantially reduce costs (e.g., by 80%) compared to the performance of human managers (Fuji et al. 2018).
Hierarchy of tasks assigned to a hierarchy of agents	Hierarchical deep MARL can be used to decompose a learning task into a hierarchy with high-level learning of policies over multi-step goals and low-level controllers learning policies for taking the actions or steps needed to complete those goals. This task decomposition architecture combined with experience replay proved effective for learning high-reward policies in complex and rapidly changing test environments, such as managing a team of cooperating agents in a simulated basketball attack/defense game, even in the presence of sparse and delayed rewards. (Tang et al. 2018)

distributed decision and control problems such as job shop scheduling among multiple agents (Gabel and Riedmiller 2007); coordination of military force attacks in increasingly large-scale and realistic war game simulations (e.g., StarCraft battles) (Usunier et al. 2016); and self-organizing control of swarms of drones to perform missions or to cooperate in choosing locations to obtain full visual coverage of a complex and initially unknown environment (Pham et al. 2018). Safe MARL (SMARL) and Hierarchical MARL (HMARL) algorithms have demonstrated promising performance in controlling autonomous vehicles (Shalev-Shwartz et al. 2016) and teams of robots performing challenging tasks such as urban search and rescue in complex and uncertain environments (Cai et al. 2013), respectively. Such results suggest the potential for MARL principles and their extensions to contribute to improved control of complex distributed systems in important practical business, military, and industrial engineering applications.

Discussion: Implications of Advances in Rational-Comprehensive Decision Theory for Muddling Through

A key insight from machine learning is that policy gradient algorithms and other RL and MARL techniques that take successive incremental steps guided by experience—and in this sense muddle through—end up solving dynamic optimization problems. This finding addresses the "rational" component of Lindblom's critique by showing that *muddling through and optimization are not opposed*: muddling through provides one way to solve optimization problems. Likewise, RL's ability to solve adaptive dynamic optimization problems without requiring initial knowledge of the optimization problems being solved—specifically of how different choices affect reward probabilities and next-state transition probabilities in dynamic systems or environments—renders the "comprehensive" knowledge requirement no longer necessary. Sampling-based approximate optimization algorithms such as MCTS further reduce the need for a comprehensive examination and evaluation of decision options. In short, rather than being thesis and antithesis, as Lindblom framed them, optimization and muddling through have undergone a useful synthesis in modern machine learning via RL and MARL.

However, fully automated RL and MARL techniques for quickly discovering optimal or near-optimal policies remain elusive. Computational complexity results for decentralized control of Markov decision processes (MDPs) and their generalizations suggest that some of these limitations are intrinsic for MARL (although not for single-agent RL with MDPs) (Papadimitiou and Tsitsiklis 1985), and hence that discovery of high-reward policies will always be time-consuming unless there is some measure of centralized control (Bernstein et al. 2000). Of course, real organizations do not simply implement computer science algorithms, and it would be simplistic to read into the complexities of human organizational design and behavior all the limitations (or only the limitations) of RL and MARL algorithms. Nonetheless, understanding how and why these algorithms fail in some settings suggests important pitfalls to avoid in organizations that rely on muddling through, insofar as they follow the same basic principles. Conversely, success factors that turn out to be necessary for effective RL or MARL machine learning of high-reward policies in relatively simple environments may help to suggest necessary (although not sufficient) conditions for effective organizational learning within and among human organizations. The following paragraphs summarize key lesson and some comparisons with observed real-world decision processes for human organizations.

1. *Collect accurate, relevant feedback data and use it to improve policies.*

After each new action is taken, RL evaluates the reward received and compares it to the reward that was expected so that the difference can be used to correct erroneous expectations and update the current policy. This requires that the effects of actions be evaluated and compared to prior expectations or predictions, and also that policies then be adjusted in light of the data. In the real world, policy-making and policy-administering bureaucracies frequently violate each of these requirements. For example, finding that investments in a costly course of action have yielded lower-than-expected returns may provoke those who originally chose it to escalate their commitment to it (Molden and Hui 2011; Schultze et al. 2012). Possible psychological and political explanations for escalating commitment range from loss aversion to seeking to manage the impressions of others, but clearly such resistance to modifying or abandoning previous choices in light of experience inhibits effective learning (Cox 2015; Tetlock and Gardner 2015). In business as well as government, data needed to evaluate and compare actual to predicted performance of a policy are often not even collected, or are ignored or misinterpreted if they are collected (Russo and Schoemaker 1989). In social policy application areas as diverse as education, criminal justice, and healthcare, changes in policy are often implemented without any clear predictions about expected changes in rewards or careful evaluations of actual changes in rewards (Tetlock and Gardner 2015). These failures of design and analysis prevent the crucial learning from experience that is essential to effective muddling through. The remedy is to collect, retain, candidly communicate, and use accurate data on predicted and observed outcomes from implemented policies to improve them over time.

2. *Explore via experiments to discover How to cause desired changes in outcome probabilities.*

It is tempting for a policy analyst or policy maker steeped in the rational-comprehensive tradition criticized by Lindblom to create the best possible model of how one believes the world works and then to choose the action or policy that maximized expected utility according to this model, as in Eq. (8.1). But in reality, the causal relationship between choices of policies and resulting conditional probabilities of different consequences and rewards is often initially highly uncertain. Prudent and effective policy-making require acknowledging and coping with this *model uncertainty*, rather than selecting and using a single model. RL and MARL algorithms do this via randomized selection of actions (e.g., using Thompson sampling or other randomized sampling schemes) (Schulze and Evans 2018) to discover which policies work best and to avoid becoming stuck in local optima, but it is counter-cultural among people who believe that one should know and not guess about the best course of action before taking it (Tetlock and Gardner 2015), and among decision analysts who believe that one should solve an expected utility optimization problem and then make deterministic decisions based on the results. Neither belief fully acknowledges or responds constructively to the reality emphasized by Lindblom, that current knowledge is often simply insufficient to permit confident identification of the best policy, and that experimentation is the only practical way to discover how to do better. Fortunately, use of randomized controlled trials (RCTs) in social policy experimentation and evaluation of interventions has become increasingly accepted and practiced recently, in areas ranging from disrupting poverty (Tollefson 2015) to preventing delinquency (de Vries et al. 2018) to improving oral health of fifth grade students (Qadri et al. 2018) to reducing child abuse by intervening with substance-abusing parents (Barlow et al. 2018). For collective learning and decision problems, such as controlling air pollution health effects, RCTs may not be practicable or ethical, but natural experiments and quasi-experiments provide valuable opportunities to learn from observed responses to unplanned or non-random interventions (Boogaard et al. 2017; Henneman et al. 2017).

3. *During collective learning, agents should advance slowly when doing better than expected, but retreat quickly when doing worse.*

The "win or lose fast" (WoLF) principle from MARL provides a useful heuristic for coordinating the rates at which agents on a team adjust their individual policies to prevent collective instability, so that they can eventually find and exploit a coordinated set of individual policies for maximizing team reward. In practice, destabilized policy-making processes in human organizations can manifest as "policy churn," in which new policies are proposed before old ones are well implemented and evaluated by the teams of agents implementing them (Monios 2016). Teachers implementing education reform programs; bankers implementing new risk management regulations and requirements; medical staff implementing new infection control protocols in hospital wards; and workers in bureaucracies implementing policy changes have all been frustrated by policy churn that encourages costly activity and change without providing the opportunities for careful and thorough evaluation and improvement needed to improve outcomes. Perhaps fear of constant deflections and resulting lack of progress explains some of the previously discussed reluctance to systematically collect and use feedback data to evaluate and improve policies. Conversely, desire to show action and strong leadership, or to obscure the results of previous ineffective choices, might provide incentives for policy churn. In any case, the study of RL and MARL performance suggests that deliberately controlling step sizes and adjustment rates for policy updates might facilitate productive incorporation of feedback data into policy updates for a group of cooperating agents without destabilizing their learning and improvement process.

4. *Separate actors and critics*

The RL idealization of frequent small adjustments made without significant costs, delays, or uncertainties in implementation is too simple to describe most real-world decision processes. Nonetheless, some RL and MARL principles may still be useful for human organizations. One of the most useful may be that decision and evaluation of decision performance should be kept distinct processes. Reasons abound in individual and group psychology for keeping those who make decisions about policy adjustments (analogous to "actors" in actor-critic RL algorithms) separate from those who evaluate the performance of the policies and provide feedback and suggestions for improving them (the "critics"). Among these reasons are confirmation bias, motivated reasoning, groupthink, and other heuristics and biases (Cox 2015). RL suggests an additional reason, rooted in statistics: in deep learning RL algorithms, training one network to decide what to do next and a separate one to evaluate how well it is working has been found to prevent overly optimistic assessments of policy performance due to overfitting, i.e., using the same data to both select estimated value-maximizing actions and estimate the values from taking those actions (vanHesselt et al. 2015). The principle of separating the processes for choosing which changes to make and evaluating how well they perform can also be applied usefully to choice of learning rates (i.e., choosing how much to modify current policies in light of feedback) as well as to choice of policies (Xu et al. 2017). Possible future advances include deliberately diversifying the learning rates of different agents on the same team to obtain the advantages of both rapid exploration of new policies and thorough exploitation and refinement of old ones. This is an old concept in organizational science (e.g. March 1991), but is still being developed in MARL research (Potter et al. 2001).

As a practical matter, separation of actors and critics can be applied fruitfully to major social learning and improvement initiatives, such as air pollution regulation, through accountability studies that revisit previous regulatory actions or other decisions to assess their results (Boogaard et al. 2017; Henneman et al. 2017). Use of such evaluation studies to evaluate and update previous policy decisions—ideally, in time to be useful in guiding policy decisions elsewhere—is clearly consistent with the principle of collecting and using relevant feedback data. Separation of actors and critics provides an additional principle for using feedback data to maximum advantage to improve polices and their results.

5. *Shape rewards to promote learning and improvement.*

Recently, it has been found that using causal (counterfactual) models to shape each agent's reward to reflect the estimated difference it has made—the difference between what was actually achieved and what would have been expected without each agent's contribution, or its marginal value, in microeconomic terms—can speed collective learning and optimization when each agent seeks to maximize its own reward (Devlin et al. 2014). This research uses mathematical rewards that are costless to implement, so that budget constraints such as that the sum of agent rewards must not exceed the collective reward of the team, do not apply. However, it seems plausible that, even in the presence of budget constraints, rewarding each agent according to its estimated marginal contribution (or its expected marginal contributions, or Shapley values in non-cooperative game theory) might promote joint learning about how to contribute more effectively, as well as having other properties of efficiency and fairness familiar from microeconomics and game theory. Of course, the asymmetric information about relative roles of chance and effort typical in principal-agent problems can inhibit accurate reward-shaping in practice, and causal modeling of individual marginal contributions to team performance is challenging. Nonetheless, research on how best to use reward shaping to provide feedback and encourage effective learning, as well as to create incentives, may be useful for human organizations as well as for MARL algorithms.

6. *Learn from the experiences and expertise of others*

Learning from each other by sharing valuable memories, experiences, and expertise (typically encoded as causal models or trained neural nets) helps teams of MARL agents discover high-reward joint policies for controlling large-scale systems and accomplishing tasks in complex, changing, uncertain environments. In applying such ideas to human organizations, it is valuable to recognize that the "agents" may themselves be organizations, such as different schools, hospitals, or companies; or similar government bureaucracies in different states or countries. States and counties implementing pollution-reducing regulations might learn from each others' experiences about which combinations of interventions and conditions (possibly involving copollutants, weather variables, and sociodemographic characteristics of the exposed population) generate the greatest public health benefits from pollution reduction measures. As usual, effective learning in human organizations must overcome challenges from various types of learning aversion that have no clear counterparts in machine learning (Cox 2015). For example, human bureaucracies may reorganize to visibly mimic organizational structures in more successful organizations whose reputations they covet, but without corresponding learning of the more effective policies that drive improved performance (Monios 2016). Players preoccupied with managing the perceptions and impressions of others to shape allocations of collective efforts and rewards to their own individual advantages may be unable to achieve Pareto efficiency or to maximize any measure of collective success or reward. These threats do not arise for teams of agents trying to cooperate in maximizing the same reward function. Our recommendation that agents should learn from each other in order to speed mastery of joint policies for obtaining high rewards from the environment is primarily applicable to such teams of cooperating agents.

Conclusions

In 1973, two professors of design and city planning offered the following sober assessment of the prospects for scientifically based social policy:

> The search for scientific bases for confronting problems of social policy is bound to fail, because of the nature of these problems. They are 'wicked' problems, whereas science has developed to deal with 'tame' problems. Policy problems cannot be definitively described. Moreover, in a pluralistic society there is nothing like the undisputable public good; there is no objective definition of equity; policies that respond to social problems cannot be meaningfully correct or false; and it makes no sense to talk about 'optimal solutions' to social problems unless severe qualifications are imposed first. Even worse, there are no 'solutions' in the sense of definitive and objective answers. (Rittel and Webber 1973).

We believe that subsequent developments warrant greater optimism. While it is true that sufficiently heterogeneous preferences may make it impracticable or impossible to define and measure a single indisputable public good to be optimized (see Chap. 5), it is also true that agents with at least some shared goals have already achieved impressive feats of cooperation and control using MARL principles, in applications as diverse as autonomous vehicle fleet and drone swarm control, search-and-rescue via teams of cooperating autonomous robots, distributed management of supply chains, and military gaming. Such applications are admittedly far less difficult than the wicked problems referred to by Rittel and Webber, but many of the differences are of scale rather than of kind: robot teams are already using RL, MARL, and HMARL to confront, with increasing competence, the difficulties of distributed decision-making with initially unclear roles and priorities, uncertain and changing environments, opportunistic revision of goals and plans, and local information that may be time consuming and expensive to share. Multiple practical applications have demonstrated the advantages of improving via small steps rather than trying to optimize in one big decision, and this insight from Lindblom's 1959 paper remains true for machine learning as well human organizations. It has been augmented by the discovery that successive incremental improvement based on feedback at each step and careful selection of step sizes is often an effective way to solve dynamic optimization problems when they can be clearly formulated, as well as an effective way to learn how to act when not enough is initially known to formulate a clear decision optimization problem.

As artificial intelligence and machine learning algorithms are tested and improved on increasingly challenging tasks, principles for learning how to manage risks and act effectively in a variety of centralized, decentralized, and hierarchical organizational structures have begun to emerge. We have discussed several based on recent work that uses deep neural networks to approximate value functions in RL, MARL, and HMARL algorithms. These principles are only the beginning of what may soon become a substantial flow from multi agent machine learning to human management science of useful principles for improving organizational design and performance in coping with realistically complex and uncertain collective decision and policy improvement challenges. These principles will doubtless require modifications and extensions for the human world, since human psychology for both individuals and groups differs greatly from RL and MARL agent programming. But the pace of discovery and progress in using machine learning to solve increasingly large, difficult, and important real-world problems of decision-making under uncertainty is now extremely rapid. Discovering how groups and teams of agents can organize, learn, decide, and adapt more effectively is becoming an experimental and applied science, as well as a theoretical one, in current artificial intelligence and machine learning. It seems likely that this research will produce insights and principles to help tame currently wicked problems and develop increasingly effective and beneficial polices in collective choice applications with high stakes for humans.

References

Andrychowicz M, Wolski F, Ray A, Schneider J, Fong R, Welinder P, McGrew B, Tobin J, Abbeel P, Wojcieh Z (2018) Hindsight experience replay. https://arxiv.org/pdf/1707.01495.pdf

Arulkumaran K, Deisenroth MP, Brundage M, Bharath AA (2017) A brief survey of deep reinforcement learning. IEEE Signal Process Magazine. Special Issue on Deep Learning for Image Understanding (Arxiv Extended Version). https://arxiv.org/pdf/1708.05866.pdf

Barlow J, Sembi S, Parsons H, Kim S, Petrou S, Harnett P, Dawe S (2018 Nov 3) A randomized controlled trial and economic evaluation of the parents under pressure program for parents in substance abuse treatment. Drug Alcohol Depend 194:184–194. https://doi.org/10.1016/j.drugalcdep.2018.08.044

Barrett S (2013) Climate treaties and approaching catastrophes. J Environ Econ Manag 66:235–250. https://doi.org/10.1016/j.jeem.2012.12.004i

Bernkamp F, Turchetta M, Schoellig AP, Krause A (2017) Safe model-based reinforcement learning with stability guarantees. In: 31st Conference on Neural Information Processing Systems (NIPS 2017), Long Beach, CA, USA. https://papers.nips.cc/paper/6692-safe-model-based-reinforcement-learning-with-stability-guarantees.pdf

Bernstein DS, Zinberstein S, Immerman N. (2000) The complexity of decentralized control of Markov decision processes. In: Uncertainty in Artificial Intelligence Proceedings 2000. https://arxiv.org/ftp/arxiv/papers/1301/1301.3836.pdf

Bloembergen D, Tuyls K, Hennes D, Kaisers M (2015) Evolutionary dynamics of multi-agent learning: a survey article. J Artif Intell 53:659–697. https://doi.org/10.1613/jair.4818

Boogaard H, van Erp AM, Walker KD, Shaikh R (2017) Accountability studies on air pollution and health: the HEI experience. Curr Environ Health Rep 4(4):514–522. https://doi.org/10.1007/s40572-017-0161-0

Bowling M, Veloso M (2001) Convergence of gradient dynamics with a variable learning rate. In Proceedings of the eighteenth international conference on machine learning, 2001. Morgan Kaufmann Publishers Inc., San Francisco, CA, pp 27–34. https://webdocs.cs.ualberta.ca/~bowling/papers/01icml-wolfiga.pdf (Last accessed 6/1/2023)

Cai Y, Yang SX, Xu X (2013, April) A combined hierarchical reinforcement learning based approach for multi-robot cooperative target searching in complex unknown environments. In: 2013 IEEE Symposium on Adaptive Dynamic Programming and Reinforcement Learning (ADPRL). IEEE, New York, pp 52–59

Clavira I et al (2018) Learning to adapt: meta-learning for model-based control. https://arxiv.org/abs/1803.11347

Clemen RT, Reilly T (2014) Making hard decisions, with the decision tools suite, 3rd edn. Duxbury Press, North Scituate, MA

Cox LA Jr (2015 Oct) Overcoming learning aversion in evaluating and managing uncertain risks. Risk Anal 35(10):1892–1910. https://doi.org/10.1111/risa.12511

Devlin S, Yliniemi L, Kudenko K, Tumer K (2014) Potential-based difference rewards for multiagent reinforcement learning. In: Lomuscio S, Bazzan H (eds) Proceedings of the 13th International Conference on Autonomous Agents and Multiagent Systems (AAMAS 2014), May 5–9, 2014, Paris, France. http://web.engr.oregonstate.edu/~ktumer/publications/files/tumer-devlin_aamas14.pdf

de Vries SLA, Hoeve M, Asscher JJ, Stams GJJM (2018 Sep) The long-term effects of the youth crime prevention program "new perspectives" on delinquency and recidivism. Int J Offender Ther Comp Criminol 62(12):3639–3661. https://doi.org/10.1177/0306624X17751161

Fudenberg D, Maskin E (1986) The Folk Theorem in repeated games with discounting or with incomplete information. Econometrica 54:533–554

Fudenberg D, Levine D, Maskin E (1994) The Folk Theorem with imperfect public information. Econometrica 62(5):997–1040

Fuji T, Ito K, Matsumoto K, Yano K (2018) Deep multi-agent reinforcement learning using DNN-weight evolution to optimize supply chain performance. In: Proceedings of the 51st Hawaii International Conference on System Sciences | 2018

Gabel T, Riedmiller M (2007) On a successful application of multi-agent reinforcement learning to operations research benchmarks. In: Proceedings of the 2007 IEEE Symposium on Approximate Dynamic Programming and Reinforcement Learning (ADPRL 2007). http://ml.informatik.uni-freiburg.de/former/_media/publications/gabelriedmiller07a.pdf

Garcia J, Fernandez F (2015) A comprehensive survey on safe reinforcement learning. J Mach Learn Res 16:1437–1480. http://www.jmlr.org/papers/volume16/garcia15a/garcia15a.pdf

Gilboa I, Samet D, Schmeidler D (2004) Utilitarian aggregation of beliefs and tastes. J Polit Econ 112(4):932–938. https://doi.org/10.1086/421173. https://www.jstor.org/stable/10.1086/421173

Grondman I, Busoniu L, Lopes GAD, Babuska R (2012) A survey of actor-critic reinforcement learning: standard and natural policy gradients. IEEE Trans Syst Man And Cybern C Appl Rev 42(6):1291–1307. http://busoniu.net/files/papers/ivo_smcc12_survey.pdf

Gupta JK, Egorov M, Kochenderfer M (2017) Cooperative multi-agent control using deep reinforcement learning. In: International Conference on Autonomous Agents and Multi-agent Systems. http://ala2017.it.nuigalway.ie/papers/ALA2017_Gupta.pdf

Heitzig J, Lessmann K, Zou Y (2011 Sep 20) Self-enforcing strategies to deter free-riding in the climate change mitigation game and other repeated public good games. Proc Natl Acad Sci U S A 108(38):15739–15744. https://doi.org/10.1073/pnas.1106265108

Henneman LR, Liu C, Mulholland JA, Russell AG (2017) Evaluating the effectiveness of air quality regulations: a review of accountability studies and frameworks. J Air Waste Manag Assoc 67(2): 144–172. https://doi.org/10.1080/10962247.2016.1242518

Hörner J, Olszewski W (2006) The Folk Theorem for games with private almost-perfect monitoring. Econometrica 74:1499–1544. https://doi.org/10.1111/j.1468-0262.2006.00717.x

Howard R, Abbas A (2015) Foundations of decision analysis. Pearson

Hu J, Wellman MP (1998) Multiagent reinforcement learning: theoretical framework and an algorithm. In: Proceedings of International Conference on Machine Learning (ICML'98), pp 242–250

Hu J, Wellman MP (2003) Nash Q-learning for general-sum stochastic games. J Mach Learn Res 4: 1039–1069

Hu Y, Gao Y, An B (2015) Multiagent reinforcement learning with unshared value functions. IEEE Trans Cybern 45(4):647–662

Hunt S, Meng Q, Hinde C, Huang T (2014) A consensus-based grouping algorithm for multi-agent cooperative task allocation with complex requirements. Cognit Comput 6(3):338–350

Keeney R, Raiffa H (1976) Decisions with multiple objectives: preferences and value tradeoffs. Wiley

Krishnamurthy V (2015) Reinforcement learning: stochastic approximation algorithms for Markov decision processes arXiv:1512.07669v1 [math.OC]. https://doi.org/10.48550/arXiv.1512.07669

Lemke C, Budka M, Gabrys B (2015) Metalearning: a survey of trends and technologies. Artif Intell Rev 44(1):117–130. https://doi.org/10.1007/s10462-013-9406-y. PMID: 26069389; PMCID: PMC4459543

Lindblom CE (1959) The science of "muddling through". Public Adm Rev 9(2):79–88. https://www.jstor.org/stable/973677?origin=JSTOR-pdf

Luce DR, Raiffa H (1957) Games and decisions. Wiley, New York

Lütjens B, Everett M, How JP (2018) Safe reinforcement learning with model uncertainty estimates. https://arxiv.org/abs/1810.087001

Mannion P, Duggan J, Howley E (2017) Analysing the effects of reward shaping in multi-objective stochastic games. In: May 2017 Conference: Adaptive and Learning Agents workshop (at AAMAS 2017). http://ala2017.it.nuigalway.ie/papers/ALA2017_Mannion_Analysing.pdf

March JG (1991) Exploration and exploitation in organizational learning. Organ Sci 2(1):71–87. Special Issue: Organizational Learning: Papers in Honor of (and by) James G. March. http://www.jstor.org/stable/2634940

Marschak J, Radner R (1972) Economic theory of teams. New Haven Yale University Press

Miceli TJ (2017) The economic approach to law, 3rd edn. Stanford University Press

Mnih V, Kavukcuoglu K, Silver D, Rusu AA, Veness J, Bellemare MG, Graves A, Riedmiller M, Fidjeland AK, Ostrovski G et al (2015) Human-level control through deep reinforcement learning. Nature 518(7540):529–533

Molden DC, Hui CM (2011 Jan) Promoting de-escalation of commitment: a regulatory-focus perspective on sunk costs. Psychol Sci 22(1):8–12. https://doi.org/10.1177/0956797610390386

Monios J (2016, October 13) Policy transfer or policy churn? Institutional isomorphism and neoliberal convergence in the transport sector. Environ Plan A Econ Space. https://doi.org/10.1177/0308518X16673367

Mookherjee D (2006) Decentralization, hierarchies, and incentives: a mechanism design perspective. J Econ Lit 44(2):367–390

Munos R (2014) From bandits to Monte-Carlo Tree search: the optimistic principle applied to optimization and planning. Found Trends Mach Learn 7(1):1–129. https://doi.org/10.1561/2200000038. https://hal.archives-ouvertes.fr/hal-00747575v5/document

Nisan N (2007) Introduction to mechanism design (for computer scientists). In: Nisan N, Roughgarden T, Tardos E, Vazirani V (eds) Algorithmic game theory. Cambridge University Press, Cambridge, MA

Omidshafiei S, Pazis J, Amato C, How JP, Vian J (2017) Deep decentralized multi-task multi-agent reinforcement learning under partial observability. https://arxiv.org/abs/1703.06182

Papadimitiou C, Tsitsiklis JN (1985) The complexity of Markov decision processes. https://dspace.mit.edu/bitstream/handle/1721.1/2893/P-1479-13685602.pdf?sequence=1

Petrik M, Chow Y, Ghavamzadeh M (2016) Safe policy improvement by minimizing robust baseline regret. https://arxiv.org/abs/1607.03842

Pham HX, La HM, Feil-Seifer D, Nefian A (2018) Cooperative and distributed reinforcement learning of drones for field coverage. https://arxiv.org/pdf/1803.07250.pdf

Pinker S (2021) Rationality: what it is, why it seems scarce, why it matters. Viking, an imprint of Penguin Random House LLC, New York

Potter M, Meeden L, Schultz A (2001) Heterogeneity in the coevolved behaviors of mobile robots: the emergence of specialists. In: Proceedings of the Seventeenth International Conference on Artificial Intelligence (IJCAI-2001)

Qadri G, Alkilzy M, Franze M, Hoffmann W, Splieth C (2018) School-based oral health education increases caries inequalities. Community Dent Health 35(3):153–159. https://doi.org/10.1922/CDH_4145Qadri07. https://www.ncbi.nlm.nih.gov/pubmed/30106523

Raiffa H (1968) Decision analysis: introductory lectures on choices under uncertainty. Addison Wesley

Rittel HWJ, Webber MW (1973) Dilemmas in a general theory of planning. Policy Sci 4(2): 155–169. http://urbanpolicy.net/wp-content/uploads/2012/11/Rittel+Webber_1973_PolicySciences4-2.pdf

Russo JE, Schoemaker PJH (1989) Decision traps: ten barriers to brilliant decision-making and how to overcome them. Doubleday, New York

Schultze T, Pfeiffer F, Schulz-Hardt S (2012 Jan) Biased information processing in the escalation paradigm: information search and information evaluation as potential mediators of escalating commitment. J Appl Psychol 97(1):16–32. https://doi.org/10.1037/a0024739

Schulze S, Evans O (2018) Active reinforcement learning with Monte-Carlo Tree search. https://arxiv.org/abs/1803.04926

Shalev-Shwartz S, Shammah S, Shashua A (2016) Safe, multi-agent, reinforcement learning for autonomous driving. https://arxiv.org/pdf/1610.03295.pdf

Shiarlis K, Messias J, Whiteson S (2016) Inverse reinforcement learning from failure. In: Proceedings of the 2016 International Conference on Autonomous Agents & Multiagent Systems (AAMAS'16). International Foundation for Autonomous Agents and Multiagent Systems, Richland, SC, pp 1060–1068. http://www.cs.ox.ac.uk/people/shimon.whiteson/pubs/shiarlisaamas16.pdf

Silver D, Huang A, Maddison CJ, Guez A, Sifre L, van den Driessche G, Schrittwieser J, Antonoglou I, Panneershelvam V, Lanctot M, Dieleman S, Grewe D, Nham J, Kalchbrenner N, Sutskever I, Lillicrap T, Leach M, Kavukcuoglu K, Graepel T, Hassabis D (2016) Mastering the game of Go with deep neural networks. Nature 529(7587):484–489. https://doi.org/10.1038/nature16961

Silver D, Hubert T, Schrittwieser J, Antonoglou I, Lai M, Guez A, Lanctot M, Sifre L, Kumaran D, Graepel T, Lillicrap T, Simonyan K, Hassabis D (2018) A general reinforcement learning. Science 362(6419):1140–1144. https://doi.org/10.1126/science.aar6404

Sutton RS, Barto AG (1998) Reinforcement learning: an introduction. MIT Press, Cambridge, MA

Sutton RS, Barto AG (2018) Reinforcement learning: an introduction, 2nd edn. MIT Press, Cambridge, MA

Tang H, Hao J, Lv T, Chen Y, Zhang Z, Jia H, Ren C, Zheng Y, Fan C, Wang L (2018) Hierarchical deep multiagent reinforcement learning. https://arxiv.org/pdf/1809.09332.pdf

Tetlock PE, Gardner D (2015) Superforecasting: the art and science of prediction. House LLC, New York

Thomas PS, Theocharous G, Ghavamzadeh M (2015) High confidence policy improvement. In: Proceedings of the 32nd International Conference on Machine Learning, Lille, France, 2015. JMLR: W&CP 37. https://people.cs.umass.edu/~pthomas/papers/Thomas2015b.pdf

Tollefson J (2015) Can randomized trials eliminate global poverty? Nature 524(7564):150–153. https://doi.org/10.1038/524150a. https://www.nature.com/news/can-randomized-trials-eliminate-global-poverty-1.18176

Usunier N, Synnaeve G, Lin Z, Chintala S (2016) Episodic exploration for deep deterministic policies: an application to StarCraft micromanagement tasks. https://arxiv.org/pdf/1609.02993.pdf

vanHesselt H, Guez A, Silver D (2015) Deep reinforcement learning with double Q-learning. https://arxiv.org/pdf/1509.06461.pdf

Villar S, Bowden J, Wason J (2015) Multi-armed bandit models for the optimal design of clinical trials: benefits and challenges. Stat Sci 30(2):199–215. https://doi.org/10.1214/14-STS504. https://arxiv.org/pdf/1507.08025.pdf

Vodopivec T, Samothrakis S, Ster B (2017) On Monte Carlo tree search and reinforcement learning. J Artif Intell Res 60:881–936. https://doi.org/10.1613/jair.5507

Wood PJ (2011 Feb) Climate change and game theory. Ann N Y Acad Sci 1219:153–170. https://doi.org/10.1111/j.1749-6632.2010.05891.x

Xu C, Qin T, Wang GLTY (2017) Machine learning reinforcement learning for learning rate control. https://arxiv.org/abs/1705.11159

Zhang C, Abdallah S, Lesser V (2008) MASPA: multi-agent automated supervisory policy adaptation. In: Computer Science Department University of Massachusetts Amherst UMass Computer Science Technical Report #08-03. https://pdfs.semanticscholar.org/418f/2ddfea52da09f21fea633e128ffccd00c8f6.pdf

Zhang K, Yang Z, Liu H, Zhang T, Basar T (2018) Fully decentralized multi-agent reinforcement learning with networked agents. In: Proceedings of the 35 th international conference on machine learning, Stockholm, Sweden, PMLR 80. http://proceedings.mlr.press/v80/zhang18n/zhang18n.pdf

Zhao W, Meng Q, Chung PW (2016 Apr) A heuristic distributed task allocation method for multivehicle multitask problems and its application to search and rescue scenario. IEEE Trans Cybern 46(4):902–915. https://doi.org/10.1109/TCYB.2015.2418052

Chapter 9
Causally Explainable Decision Recommendations Using Causal Artificial Intelligence

Introduction: Creating More Trustworthy AI/ML for Acting Under Risk and Uncertainty

How can the predictions, decisions and recommendations made by artificial intelligence and machine learning (AI/ML) systems be made more trustworthy, transparent, and intelligible? Enabling AI/ML systems to explain the reasons for their recommendations in terms that make sense to humans would surely help. This chapter reviews a set of concepts, principles and methods for creating AI/ML systems that recommend (or execute, for autonomous systems) appropriate actions—those we would want them to take, or at least understand and approve of their rationales for taking, if they were acting on our behalf. The key ideas are based on *causal models* of the relationship between actions and outcome probabilities. To a limited but useful extent, current causal models enable appropriate decisions even under unforeseen conditions and in response to new and unanticipated events. Such "causal artificial intelligence" (CAI) principles might also be useful in improving and explaining decision and policy recommendations in human organizations when risk, uncertainty, and novelty make the consequences of different courses of action hard to predict and thus make collecting information for predicting it valuable (Wu et al. 2017).

CAI builds on the intuition that systems that can explain the rationales for their inferences, predictions, recommendations, and behaviors in clear cause-and-effect terms are likely to be more trusted (and, perhaps, more trustworthy) than those that can't. It applies principles of information theory and closely related probabilistic and statistical concepts of conditional independence, probabilistic dependence (e.g., conditional probabilities), causality, uncertainty reduction, and value of information (VoI) to model probabilistic dependencies among variables and to infer probable consequences caused by alternative courses of action. To recommend best decisions despite incomplete causal knowledge and information, CAI seeks not only to

L. A. Cox Jr., *AI-ML for Decision and Risk Analysis*, International Series in Operations Research & Management Science 345,
https://doi.org/10.1007/978-3-031-32013-2_9

identify facts that make current observations less surprising, and in this sense to explain them; but also to identify actions, policies, and plans that make preferred future outcomes more probable, and in this sense explain how to achieve them. CAI applies statistical tests for conditional independence (or, conversely, mutual information) and other relationships (e.g., directed information flows and invariant causal prediction properties, reviewed later) among random variables to identify causal regularities that are consistent with multiple data sets, thereby enabling generalization from experience and prediction of probable consequences of courses of action in new settings (Heinze-Deml et al. 2018). Such causal generalization may be essential for acting effectively, as well as for explaining the basis for actions or recommendations, when confronting novel situations and unanticipated risks.

Current AI systems are typically most fragile and least trustworthy in novel situations, because they lack common-sense knowledge and ability to reason effectively (and causally) about likely consequences of actions when relevant prior data with stable predictive patterns are not available. CAI seeks to help bridge this crucial gap between experiential learning and the need to act effectively under novel conditions by applying causal generalizations from past observations. It recognizes the tight link between *causally effective plans*, meaning plans (i.e., sequences of actions contingent on events) that make preferred outcomes more probable; and *causal explanations* for preferred courses of action. Both exploit the fact that causes provide unique information about their direct effects (or joint distributions of effects): conditioning on levels of other variables does not remove the statistical dependency of effects on their direct causes. Causes and their direct effects have positive mutual information, and this information can be used to identify courses of action that make preferred outcomes more likely, and hence less surprising. Intuitively, causally effective decision-making can be thought of as mapping *observations*—signals received by an agent from its environment—to *decisions* and resulting behaviors that are calculated to change outcome probabilities to make preferred outcomes more likely. Decisions are implemented by control signals—typically transduced by effectors which may be unreliable or slow—sent into the agent's environment to change outcome probabilities. A completely specified probabilistic causal model predicts conditional probabilities of outcomes, given observations and actions. This provides the information needed to optimize actions. Even partial information about causal dependencies among variables can help to decide what additional information to seek next to formulate more causally effective policies. To act effectively, an agent must receive and process observations and transmit control signals quickly enough to keep up with changes in its environment.

Figure 9.1 summarizes key concepts and methods reviewed in subsequent sections and shows how they fit together. *Observations* (upper left) provide information about the underlying state (lower left) of the system or environment via an *information channel*, i.e., a probabilistic mapping from the state to observations. *Actions* (lower right) cause state transitions and associated costs or benefits, generically referred to as rewards (bottom) via a *causal model*, i.e., a probabilistic mapping from current state-action pairs to conditional probabilities of next-state and reward pairs. Table 9.1 list several specific classes of causal models discussed later, in rough

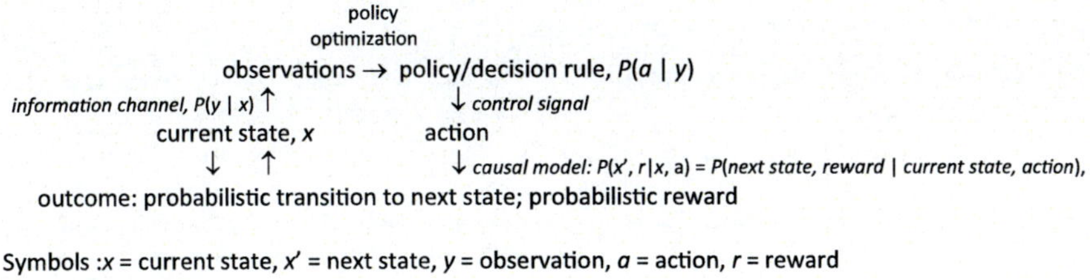

Fig. 9.1 Summary of key ideas and methods used in causal AI (CAI) to explain decisions. Symbols: x = current state, x' = next state, y = observation, a = action, r = reward

order of increasing generality and flexibility in representing uncertainty. Actions are selected by *policies*, also called decision rules, strategies, or control laws (upper right), which, at their most general, are probabilistic mappings from observations to control signals sent to actuators; these control signals are the decision-maker's or controller's choices. The mapping from choices to actions may be probabilistic if actuators are not entirely reliable; in this case, the capacity of the control channel and actuators to transfer information from control signals to future states of the system limits the possibilities for control. In many settings, actions are implemented via hierarchies of learned skills (i.e., abilities to complete tasks and subtasks); what can be done in a situation depends in part on the repertoire of skills that have been acquired (Shu et al. 2017). In Fig. 9.2, *observations* are explained by underlying states (and the information channels via which they are observed). By contrast, rational *decisions* are explained via optimization of decision-rules (i.e., policies). If knowledge of the causal model and the other components in Fig. 9.1 is inadequate to support full optimization, then reinforcement learning or other adaptive control methods, together with optimization heuristics, are used to improve policies over time. The vast majority of "explainable AI" (XAI) research to date has focused on explaining observations, as in, diagnostic systems; predictions; and prediction-driven recommendations (Mittelstadt et al. 2019). Such explanations emphasize the left side of Fig. 9.1, where observations are used to draw inferences about states, which can then be used to predict further observations. A principal goal of this chapter is to help extend XAI to more fully explain the rationales for recommended decisions and policies. Such explanations draw also on the right side of Fig. 9.1, using preferences for outcomes (i.e., rewards), choice sets (e.g., possible control signals), causal models, and optimization of policies as key explanatory constructs, in addition to observations and inferences.

The following sections seek to show how these intuitions can be sharpened and formalized and how they can be implemented in computationally effective CAI algorithms to support and explain decision recommendations for several widely applied classes of probabilistic causal models. One goal is to review how current causal AI/ML methods support and explain causally effective decisions—decisions that make preferred outcomes more probable—in practical applications, such as allocating a budget across advertising channels that interact in affecting consumer preferences and sales. We seek to present an accessible exposition, review and

Table 9.1 Some important probabilistic causal models

Probabilistic causal models, in order of increasing generality	Knowledge representation, inference, and decision optimization assumptions
Decision trees (Raiffa 1968), event trees (decision trees without decisions); fault trees, event trees, bow tie diagrams (Cox et al. 2018)	Event trees show possible sequences of events (realizations of random variables). Decision trees are event trees augmented with choice nodes and utilities at the leaves of the tree. Fault tree are trees of binary logical events and deterministic logic gates, supporting bottom-up inference from low-level events to top-level event (e.g., systems failure). Bow-tie diagrams integrate fault trees leading up to an event and event trees following it.
Bayesian networks (BNs), dynamic BNs, causal BNs Influence diagrams (IDs) are BNs with decision nodes and utility nodes (Howard and Matheson 1981)	Random variables (nodes) are linked by probabilistic dependencies (described by conditional probability tables, CPTs). In a DBN, variables can change over time. In a causal BN, changing a variable changes the probability distributions of its children in a directed acyclic graph (DAG). Bayesian inference of unobserved quantities from observed (or assumed) ones can proceed in any direction.
Markov decision process (MDP) optimization models, can be risk-sensitive	Markov transition assumptions, observed states, actions completed without delays
Partially observable MDPS (POMDPs)	States are not directly observed, but must be inferred from observations (signals, symptoms, data) via information channels, $P(observation = y \mid state = x)$
PO semi-MDPs (POSMDPs); behavior trees	Actions take random amounts of time to complete, and may fail
Discrete-event simulation models	Realistic lags and dependencies among events are modeled by state-dependent conditional intensities for individual-level transitions
Causal simulation-optimization models	Known models. Inference and optimization can be NP-hard and may require heuristics such as Monte Carlo Tree Search (MCTS).
Model ensemble optimization; reinforcement learning (RL) with initially unknown or uncertain causal models	Unknown/uncertain models. Probabilistic causal relationships between actions and consequences (e.g., rewards and state transitions) are learned via (heuristic-guided) trial and error.

synthesis of CAI ideas and advances from several decades of AI/ML progress, for a broad audience that might include decision analysts, risk analysts, decision science researchers, psychologists, and policy analysts, as well as AI/ML researchers. A second goal is to propose a framework clarifying the types of information and argument structure needed to provide convincing and responsible causal explanations for CAI-based decision and policy recommendations.

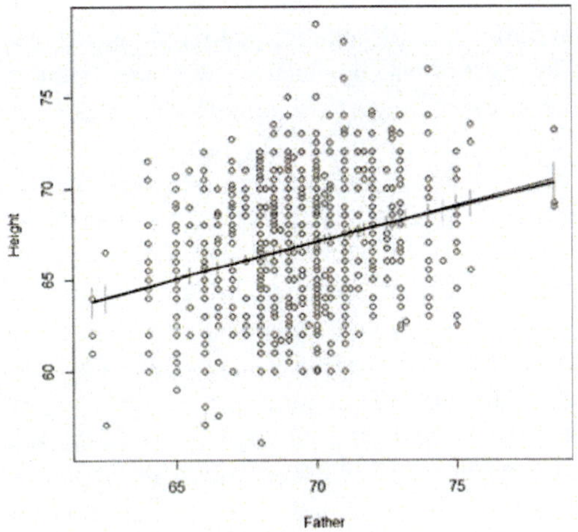


```
# Data Source and R code for Figure 2
#  Get dataset from web site, rename it as dataframe "df",attach it to current session
df <- read.csv(url("https://raw.githubusercontent.com/dspiegel29/ArtofStatistics/master/05-1-
sons-fathers-heights/05-1-galton-x.csv")); attach(df);
# Fit and plot simple linear regression model
plot(Father, Height); lines(Father, predict(lm(Height~Father)));
# Fit and plot nonparametric regression model
library(npreg); smod<- sm(Height~Father); plotci(Father, smod$fitted.values, smod$se.fit, bars =
TRUE, xlab = "Height", ylab = "Father", add=TRUE)
```

Fig. 9.2 Linear (black line) and Nonparametric Smoothing (grey line) regression models for adult offspring's Height vs. Father's Height in inches

To accomplish this exposition of CAI for explaining decisions, we first consider the traditional concept of explanation in statistics as the proportion of variance in a dependent variable in a regression model that is "explained" by differences in the independent variables on which it depends. A different concept is needed to explain decisions, since regression coefficients only address how to predict dependent variables from observed independent variables, but not the causal question of how intervening to exogenously change independent variables would change probability distributions of dependent variables (Pearl 2010). We therefore examine the structure of causal explanations for decisions in the causal models in Table 9.1. These models have been widely used in causal analytics, risk analysis, and applied AI/ML to represent probabilistic dependencies of outcomes (e.g., rewards and next states, in a Markov Decision Process) on actions (Cox et al. 2018). For each model, we discuss how the concepts in Fig. 9.1 can be used to recommend decisions and explain their rationales. The discussion includes relatively recent developments (e.g., integration of Thompson sampling into Monte Carlo Tree Search) while noting foundational works in prescriptive decision analysis and decision optimization (e.g., Bellman 1957; Howard 1960; Howard and Matheson 1981).

Our focus is on CAI concepts, principles and methods for causal explanation of decisions, but a variety of practical applications have been described in industrial engineering and industrial control, managerial economics (e.g., for forestry or fishery management), personalized medicine, supply chain management, logistics optimization, urban traffic control, robotics, autonomous vehicle and drone control, pest management in ecosystems, financial investments, game-playing, and other areas of applied risk analysis (e.g., Cox et al. 2018). Many of these applications have focused solely on decision optimization, rather than also on decision explanation, under risk and uncertainty. This motivates our focus on the structure of causal explanations under risk and uncertainty. However, to understand how CAI methods and explanations can improve the trustworthiness of causal inferences and intervention decision recommendations in practice, we recommend recent analyses and applications of computational causal methods in Systems Biology for cancer research. Although current CAI methods do not yet fully automate valid causal discovery with high reliability (Triantafillou et al. 2017), causal discovery and understanding of low-level (molecular-biological) pathways are increasingly able to inform, and build confidence in, high-level public health policies by helping target the right causal factors at the macro-level (e.g., diet, exposures) to be causally effective in reducing risks (Vineis et al. 2017). Explaining how interventions cause desired changes helps select effective interventions.

The Structure of Traditional Statistical Explanations

What should explainable AI (XAI) explain? Observations (why did an observed event happen?), predictions (what might happen next, and how probable is it?), and recommendations (what should we do now to make preferred outcomes more probable?) are natural candidates for diagnostic, predictive, and decision support (prescriptive) systems, respectively. Explaining actions, choices, and behaviors by AIs becomes more crucial as increasingly autonomous AI-directed systems interact more with people. Most of this chapter focuses on explaining *decision recommendations*: if an AI system recommends an action or policy, how can and should it explain and justify its recommendation? We first contrast this with explaining *observations*, the more usual focus of descriptive, diagnostic, predictive, and prescriptive systems that rely on stable patterns in data to explain some proportion of the variation in observed outcomes across cases.

It is natural in many cases to regard a system under study as having inputs, some of which may be controllable; and outputs, some of which are observed. Learning to explain observed output values in terms of combinations of input values that make them expected, or at least less surprising, is a fundamental task for statistics and machine learning. Statistical models posit a simple, powerful explanatory structure of the following form (Zhao and Hastie 2021):

$$output = f(inputs, noise) \tag{9.1}$$

Statistical regression models simplify this further, often to the form

$$\textit{expected value of output} = f(\textit{inputs}) + \textit{noise} \tag{9.2}$$

The expected value of a dependent variable (the "output" of the regression model—a quantity, response, or outcome whose value is to be predicted or explained) is expressed as a deterministic function of variables on which it depends (the "inputs" or independent variables). The actual value of the dependent variable for each case is modeled as a random variable, typically (but not necessarily) the sum of the predicted value and a random noise or error term having zero mean. The regression approach thus postulates that variability in observed output values is "explained" as the sum of predicted values and unpredictable error or noise. More generally, the mean of the conditional probability distribution of a dependent variable, given the values of the variables on which it directly depends (its parents in a directed acyclic graph (DAG) or probabilistic graphical model of dependencies among variables) constitutes a *response surface model* (RSM) for the dependent variable (output) as a function of the independent variables (inputs). (See Fig. 9.8 for an example.)

As an explanation for output values, a statistical model does not run deep. It does not reveal *why* the expected value of the output depends on the inputs, or *how* (or whether) changes in the inputs propagate along causal pathways of successive mechanisms to change the output so that the regression model is again satisfied. Such deeper explanations are the province of causal modeling. Statistical regression models merely describe observed patterns and quantify the proportion of variance in output that is "explained" by differences in inputs, given the model assumptions. Figure 9.2 shows an example. In a dataset recording heights of parents and their adult offspring, we might wonder how much of the variance in offspring heights (the *Height* variable on the vertical axis in Fig. 9.2) is explained by differences in father's height (*Father*, on the horizontal axis).

```
# Data Source and R code for Figure 2
# Get dataset from web site, rename it as dataframe "df",attach it to
current session
df <- read.csv(url("https://raw.githubusercontent.com/dspiegel29/
ArtofStatistics/master/05-1-sons-fathers-heights/05-1-galton-x.
csv")); attach(df);
# Fit and plot simple linear regression model
plot(Father, Height); lines(Father, predict(lm(Height~Father)));
# Fit and plot nonparametric regression model
library(npreg); smod<- sm(Height~Father); plotci(Father, smod$fitted.
values, smod$se.fit, bars = TRUE, xlab = "Height", ylab = "Father",
add=TRUE)
```

A simple linear regression model, represented by the black line in Fig. 9.2, shows that the proportion of variance in *Height* explained by *Father* is about 7.5% (the adjusted R-squared value, produced in R via summary (lm(Height ~ Father, data = df))), corresponding to the fit of the black line to the scatterplot in Fig. 9.2). Nonparametric (smoothing) regression produces an almost identical line and R^2 value (it is shown by the smooth grey curve in Fig. 9.2, which falls almost on top of the black line), but now the straight-line relation between *Father* and *Height* is a discovery rather than an assumption, as smoothing regression allows many other shapes. Various nonparametric smoothing (e.g., loess, kernel density, spline) and parametric (e.g., polynomial) regression models give closely similar fits; we relegate details to the underlying software packages, such as package *npreg* in R, which was used to produce the nonparametric regression curve in Fig. 9.2. The key point is just that differences in father's height "explain" (in the statistical sense) a small but significant (i.e., non-zero with high statistical confidence) proportion of observed variance in *Height* across offspring. Even such a partial explanation has practical value for prediction: it shows that father's height helps to predict offspring height. However, it does not address causal questions, such as how or whether the distribution of *Height* in a population would change if the distribution of *Father* heights were to change. For example, if it turned out that tall men tend to marry tall women, and mother's height alone affects offspring height, then the observed statistical association between *Father* and *Height* might be entirely explained by the mediating variable of mother's height. In this sense, statistical "explanation" of variance in *Height* by differences in *Father* has nothing necessarily to do with causal mechanisms or explanations.

```
#R code for Figure 3: library(car); scatter3d(Height ~ Father + Mother |
as.factor(Gender), fit = c("linear" )); scatter3d(Height ~ Father +
Mother | as.factor(Gender), fit = c("smooth"))
```

Including more independent variables, also called "explanatory variables," on the right side of a regression model can increase the proportion of explained variance in the dependent variable. For example, including mother's height increases the proportion of explained variance in offspring heights from 7.5% to 10.7% (adjusted $R^2 = 0.1069$, as shown by the R commands `attach(df); model <- lm (Height ~ Father + Mother); summary(model)`); and including offspring's sex increases the R^2 to 0.6385 (using `summary(lm(Height ~ Father + Mother + Gender))`). Figure 9.3 shows response surface models (RSMs) for this dataset, illustrating that offspring *Height* increases more steeply with *Father* height than with *Mother* height, and that *Gender* plays a larger role than either and does not interact with parent heights, insofar as the two sheets for men and women are approximately parallel.

The Structure of Explanations in Causal Bayesian Networks (BNs)

Bayesian network (BN) models and more general probabilistic graphical models (e.g., with some undirected edges) generalize RSMs by quantifying probabilistic dependencies among all of the variables in a model, with no need for just one dependent variable. In a BN, nodes represent variables and arrows between nodes

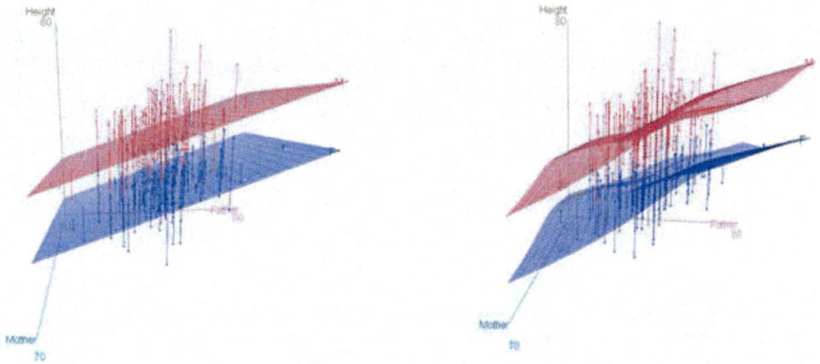

```
#R code for Figure 3: library(car); scatter3d(Height ~ Father + Mother | as.factor(Gender), fit =
c("linear" ));  scatter3d(Height ~ Father + Mother | as.factor(Gender), fit = c("smooth"))
```

Fig. 9.3 Linear (left) and smooth nonparametric (right) response surface models (RSMs) and data points for offspring heights

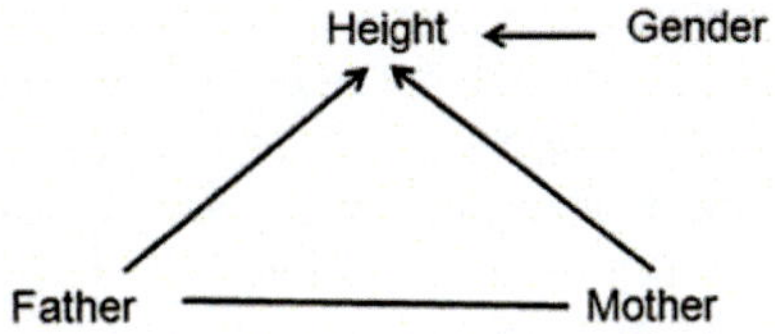

Fig. 9.4 A probabilistic graphical model for the *Height* dataset

represent statistical dependence relationships between variables. The response surface concept is generalized to that of a *conditional probability table* (CPT); we use this as a generic term to denote any probability model (not necessarily a table) specifying the *conditional probability distribution* of a random variable, given the values of its parents. (Nodes with no parents, e.g., input nodes with only outward-directed arrows, have marginal (unconditional) probability distributions, but we subsume these as special cases of CPTs.) A fully specified BN is a DAG with a CPT at each node. As in regression models and RSMs, part of the variation in a dependent variable is "explained" by differences in the values of other variables on which it depends, i.e., variables in its CPT.

Explaining Direct, Indirect (Mediated), and Total Effects

Figure 9.4 shows a probabilistic graphical model, with its arrows indicating that the value of *Height* depends on *Father*, *Mother*, and *Gender*. More specifically, the arrows signify that *Height* is not conditionally independent of any of the variables to which it is connected, even after conditioning on the other two; thus, for example, the hypothesis that the statistical effect of *Father* on *Height* is fully explained (or explained away) by *Mother* as a mediating variable is rejected in this model (since otherwise *Height* would be conditionally independent of *Father* given

Mother, and there would be no direct link between *Father* and *Height*). In general, two nodes (variables) are connected in such a probabilistic graphical model if and only if they are found to be informative about each other, i.e., the null hypothesis of conditional independence between them is rejected, implying that they have positive mutual information even after conditioning on other variables. Such graphs enable a slightly deeper level of causal explanation for observations than do regression models by distinguishing between, and quantifying, direct and indirect (mediated) causation. If a person is observed to be surprisingly tall or short, a partial explanation may be given in terms of the parent's heights: the observed height may be less surprising or more probable after conditioning on relevant information about (a) The parent's heights (explanatory inputs) and (b) The CPT for *Height* (i.e., knowledge of how inputs affect the conditional distribution of the *Height* output). This is similar to regression model-based explanations, with the CPT playing the role of the regression model and the parents of the outcome being the explanatory variables. However, the probabilistic graphical model goes beyond regression modeling by distinguishing between direct and indirect causes: a person's unusual height might be partially explained by the direct effect of his or her father's height, but also by the direct effect of his or her mother's height, which in turn may be correlated with, and in that sense partly explained by, father's height. Thus, the *total effect* of father's height on the expected value (or conditional probability distribution) of offspring's height consists of both a *direct effect* and an *indirect effect* mediated by mother's height (VanderWeele 2011).

In more detail, to be well-defined, counterfactual causal questions about how offspring height probabilities would differ if father's height were different must specify whether mother's height should be held fixed—and, if so, at what level. Holding mother's height fixed in considering how offspring height distribution varies with father's height isolates the *controlled direct effect* (or "partial" effect) of father's height on the conditional probability distribution of offspring height (VanderWeele 2011). Alternatively, if mother's height is allowed to vary realistically based on father's height, then we quantify the *total* (direct plus indirect, mediated by mother's height) effect of father's height. Distinctions among direct (or partial), indirect, and total effects of one variable on another, while holding some other variables fixed at specified levels, are causal concepts with no counterparts in statistical regression, as the concept of "*holding fixed*" is not a statistical (observational) construct, but a causal one (Pearl 2001, 2010). It is of fundamental importance for definitions of causality that consider *X* to be a direct cause of *Y* if and only if the value (or, more generally, the conditional probability distribution) of *Y* differs for different values of *X*, holding all other potential causes of *Y* fixed (Mill 1843).

Conditional Independence and Potential Causality in BNs

Arrows in a BN are *causally oriented* if direct causes point into their direct effects, but this begs the question of what is meant by a direct cause (or its direct effect) and how they are to be identified from data. Algorithms for causal discovery propose constructive answers to these questions and clarify what causal questions can and cannot be answered from observational data (Pearl 2010; Heinze-Deml and Meinshausen 2020). The undirected arc between *Father* and *Mother* in Fig. 9.4 signifies that they are informative about each other (e.g., because men and women of similar heights are more likely to marry), although perhaps neither can easily be viewed as a cause of the other. More generally, arrow directions in non-causal probabilistic graphical models are often best regarded as arbitrary, since mutual information between random variables is symmetric: if $P(x \mid y)$ differs from $P(x)$ (conditional probability does not equal marginal probability), then $P(y \mid x) = P(x \mid y) P(y)/P(x) = [P(y \mid x)/ P(x)]*P(y)$ must also differ from $P(y)$. Just as Bayes' Rule allows the joint probability $P(x, y)$ to be decomposed equally well as either $P(x \mid y) P(y)$, corresponding to the DAG model $Y \rightarrow X$; or as $P(y \mid x)P(x)$, corresponding to the DAG model $X \rightarrow Y$, so arrow directions in a fully specified Bayesian network (BN) only show one way to decompose a joint probability distribution into marginal distributions and CPTs. Other equally valid decompositions are usually possible. Thus, if suggested causal interpretations for arrows are provided, they must rest on principles other than conditional independence or mutual information, as these are symmetric.

Yet, there is a close relationship between conditional independence and most concepts of causality. Intuitively, a direct cause of a variable provides information about it that cannot be obtained from other direct causes. Thus, a variable is not conditionally independent of its direct causes, even though (by definition) it is conditionally independent of more remote ancestors, given the values of its parents. These considerations usually stop short of uniquely identifying the direct causes of each variable, because conditional independence tests alone may not be able to distinguish between the parents and children of a node. For example, the three DAGs $X \rightarrow Y \rightarrow Z$, $X \leftarrow Y \rightarrow Z$, and $Z \rightarrow Y \rightarrow X$ have the same conditional independence relationships among variables, with X and Z being conditionally independent, given Y. They are indistinguishable by conditional independence tests (i.e., they belong to the same "Markov equivalence class"). But the constraints imposed by conditional independence do provide a useful filter: if no variable is conditionally independent of its direct causes, then statistical tests for conditional independence can help identify *potential* direct causes, even if they cannot always distinguish direct causes from direct effects. Conditional independence tests are built into current causal discovery and BN structure-learning software packages, such as *CompareCausalNetworks* (Heinze-Deml and Meinshausen 2020) and *bnlearn* (Nagarajan et al. 2013). Although conceptual counterexamples can be contrived, e.g., by making X identical to Y in $X \leftarrow Y \rightarrow Z$, so that Z is conditionally independent of Y given X, even though Y rather than X is a direct cause of Z, practical workers might be willing to disregard such special cases, perhaps regarding them as having probability measure 0 or close to it in real datasets, in order to get the benefits of automated screening for potential direct causes based on an (almost) necessary, although not sufficient, condition. For example, in the above Markov equivalence class, conditional independence implies that X is not a direct cause of Z (since Y separates them, i.e., conditioning on Y renders them conditionally independent) and that X and Z are not both causes of Y (since conditioning on Y renders them independent rather than dependent), although it leaves ambiguous whether Y is a cause or an effect of Z.

Causal Discovery for Predictive, Interventional, and Mechanistic Causation

Causal discovery algorithms apply additional principles, beyond conditional independence, to disambiguate possible directions of causality (if any) between variables. Useful principles for orienting mutual-information links between variables incorporated in modern causal discovery packages include the following (Heinze-Deml and Meinshausen 2020):

- **V-structures (explaining away):** If X and Y are unconditionally independent, but become dependent after conditioning on a third variable Z, then Z is a common child (or possibly a common descendant, if there are other variables) of X and Y. This is called a "collider" or "v structure" DAG: $X \rightarrow Z \leftarrow Y$. For example, if *Mother* and *Father* heights are independent variables, but adult offspring's *Height* depends on both of them, then observing a tall person with a short father might make it more likely that the mother is tall, thus inducing a negative correlation between parent heights.
- **Directed information flow and predictive causation (Granger) vs. manipulative causation:** Information flows from causes to their effects over time. Parents' heights are potential causes of offspring heights, but not *vice versa.* This intuitively appealing idea leads to tests for *predictive causality*, such as classical Granger causality testing in time series analysis and its nonparametric generalization to transfer entropy (Wieczorek and Roth 2019). In these tests, X is defined as a predictive cause of Y if future values of Y are not conditionally independent of past and present values of X, given past and present values of Y itself. An intuitive motivation is that changes in causes help to predict changes in their effects (Wiener 1956). However, predictive causation is a relatively weak causal concept: in a dataset that does not contain smoking as a variable, nicotine-stained fingers might be a predictive cause of lung cancer without being a *manipulative (or "interventional") cause*, i.e., X being a predictive cause of Y does not necessarily imply that intervening to change the value (or distribution) of X would change the probability distribution of Y. Interventional causation, in turn, is weaker than *mechanistic causation*: in a world where fingers can only be kept clean by not smoking, nicotine-stained fingers could be a manipulative cause and a predictive cause of lung cancer without being a mechanistic cause. For causally effective decision-making, manipulative causation is necessary and sufficient. But it cannot be inferred from conditional independence and directed information flows alone.
- **Dynamic Bayesian networks (DBNs)** make the flow of information between variables over time explicit by replacing individual variables with time-stamped

values, creating a time series of values for each variable in successive periods. The conditional probability distribution for each variable in each period is allowed to depend on past values of other variables, possibly including its own lagged values. Causal discovery algorithms for learning DBNs from both experimental (Spencer et al. 2015) and observational data (Lo et al. 2015) have been developed (e.g., CaMML package, https://bayesian-intelligence.com/software/BI-CaMML-Quickstart-Guide-1.4/).

- **Effects are often simple (e.g., approximately linear) functions** of their direct causes and of random noise, as in the regression model in Eq. (9.1). This provides a basis for statistical tests for the direction of causality between variables if noise is not Gaussian (Shimizu and Bollen 2014; Yamayoshi et al. 2020).
- **Causal laws (CPTs) are invariant across applications contexts.** If a CPT $P(Y \mid Pa(Y))$ represents a causal law, then it should be the same across contexts, meaning that the same inputs (parent values) produce the same conditional probability distribution for the output, Y. The property of *invariant causal prediction* (ICP) stipulates that the CPT is the same across studies (causal generalization) and across interventions (transportability), where an intervention sets values of some of the parents (Peters et al. 2016; Heinze-Deml et al. 2017). This property can sometimes be tested if effects of interventions are available from multiple studies with different values of $Pa(Y)$. For example, adding a study identifier code to other potential predictors of Y lets a CART tree or other conditional independence tests check whether the distribution of Y is conditionally independent of the study ID (and hence the specific interventions, if any, in different studies) once values of $Pa(Y)$ are known (Heinze-Deml et al. 2017).

These principles and algorithms have helped to identify plausible structures for causal BNs in many domains (Lo et al. 2015). But they often give somewhat different answers. Some require strong assumptions, such as that there are no unmeasured common causes of the observed variables (e.g., hidden confounders) that, if known, would explain away the observed dependencies between measured variables. Even whether one variable is a direct or an indirect cause of another in a dataset may depend on what other variables are measured. For example father's height might be a direct cause of offspring's height in a data set consisting of *Father*, *Mother*, *Gender*, and *Height*, and yet the two might be conditionally independent in a dataset that also contained more mechanistically relevant information, such as offspring levels of human growth hormone during development until adulthood.

Knowledge-Based Constraints Help to Orient Arrows to Reflect Causal Interpretations

Current causal discovery algorithms are often unable to establish unique directions for arrows based on passive observations alone. For example, Fig. 9.5 shows the results of applying four current structure-learning algorithms to the heights dataset (hill-climbing (hc), tabu search, grow-shrink (gs), and incremental association

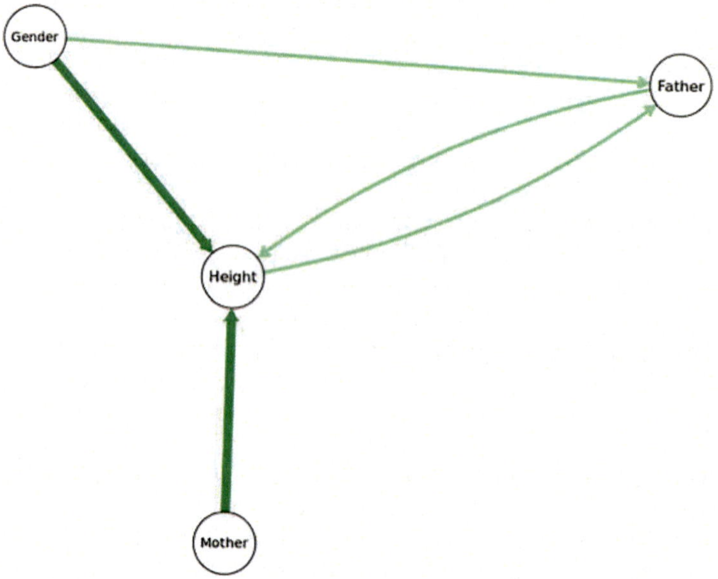

Fig. 9.5 Bayesian network (BN) structures learned by four different algorithms in the *bnlearn* package only partly agree. All 4 algorithms agree on the heavy arrows (darker color), but only 2 of 4 agree on each thin arrow (lighter color). (Continuous variables were discretized using the Hartemink discretization pre-processing option, and hc, tabu, gs, and iamb algorithms were applied)

Markov blanket (iamb); see Nagarajan et al. (2013) for details of these algorithms). Only two of them (hc and tabu) oriented an arrow from *Father* to *Height*; the other two algorithms reversed this orientation. Knowledge-based constraints can help orient arrows to reflect causal interpretations. For example, if it is known that a variable is a sink in a causal BN, meaning that it can have parents (representing direct causes) but not children (e.g., *Height* might be an effect of *Father*, *Mother*, or *Gender*, but not a cause of any of them), then classification and regression tree (CART) algorithms (Friedman et al. 2009) with the sink as a dependent variables provide useful heuristics for identifying its direct causes and estimating its CPT from data. Figure 9.6 shows an example for *Height*. The CART algorithm (implemented in the *rpart* package in R) identifies *Height* as depending directly on (i.e., not being conditionally independent of) *Father*, *Mother*, and *Gender*. (Their relative "importance" scores are estimated as 15, 6, and 80, respectively, based on their contributions to predicting *Height*; see the *rpart* package documentation and Friedman et al. 2009 for details of CART importance scores.) Likewise, imposing the knowledge-based constraint that *Height* is a sink, all four of the BN-learning algorithms used in Fig. 9.5 correctly identify *Father*, *Mother*, and *Gender* as parents of *Height*.

Figure 9.6 displays CPT information for *Height*, summarized as box plots for the empirical conditional distributions of *Height* in the leaf nodes at the bottom of the tree. Each leaf node yields a different conditional probability distribution for *Height*,

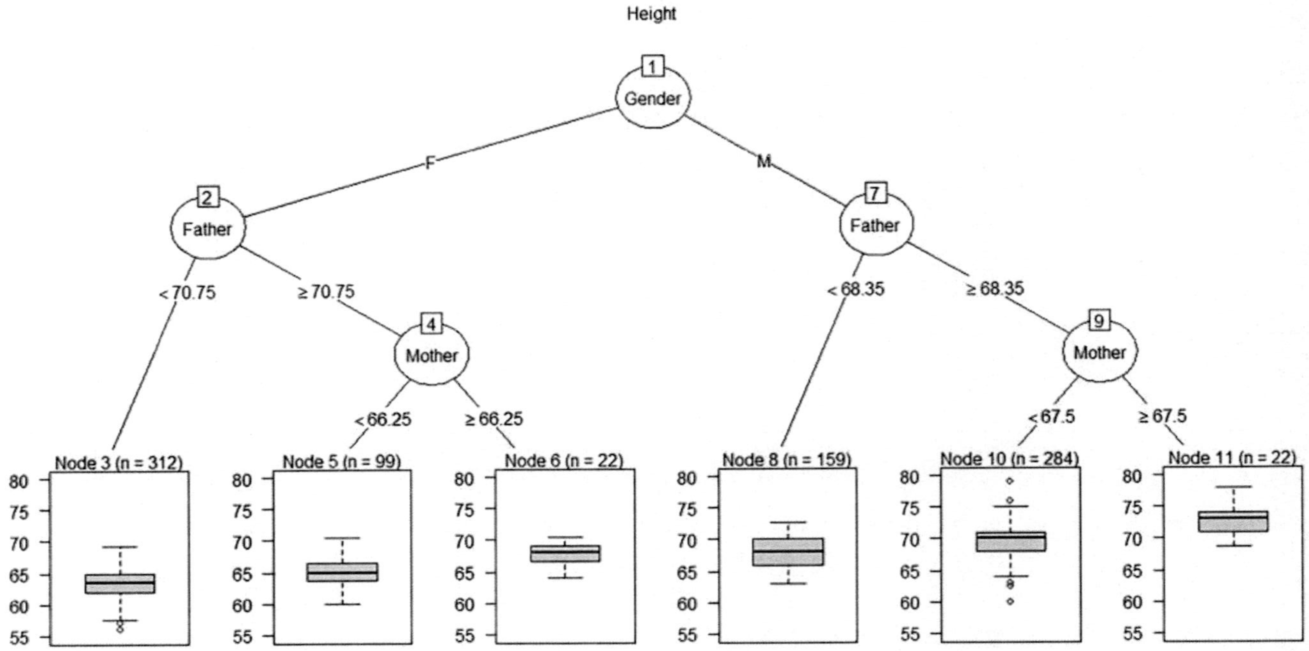


```
#R code for Figure 6
library(rpart); library(rpart.plot); library(partykit); tree <- rpart(Height ~ Father + Mother +
Gender);
plot(as.party(tree), type = "extended", main = "Height")
```

Fig. 9.6 A Classification and Regression Tree (CART) for *Height* showing box plots for the conditional distributions of *Height* given *Gender*, *Father*, and *Mother*. Dots in the boxes represent outliers

given the values or ranges of values shown on the branches for the "splits" (answers to binary questions) leading from the root node at the top of the tree to the leaf node. These splits define each leaf node and the information on which its probability distribution for *Height* is conditioned. For example, a male (M) with Father ≥68.35 inches and Mother ≥67.5 inches has a conditional distribution for *Height* summarized by the box plot in the right-most leaf node, with a median value of over 72 inches. There are 22 such cases in the dataset (displayed as n = 22 for node 11). BN-learning algorithms such as those in packages *bnlearn* and *CompareCausalNetworks* provide multiple methods for performing conditional independence tests and identifying the parents and estimating the CPTs of sink nodes, but the CART tree heuristic remains one of the simplest and most interpretable. If variables are not conditionally independent of their direct causes, then CART trees for sink nodes show their direct causes, provided that the differences they make in the conditional distribution of the sink variable are large enough to be detected by the tree-growing algorithm.

```
#R code for Figure 6
library(rpart); library(rpart.plot); library(partykit); tree <- rpart
(Height ~ Father + Mother + Gender);
plot(as.party(tree), type = "extended", main = "Height")
```

Structure of Most Probable Explanations (MPEs) in Bayesian Networks

In a probabilistic DAG model with causally oriented arrows, a *direct explanation* for an observed value states that *Y* has the value it does because the parents of *Y*, denoted by *Pa*(*Y*), have the values they do. These values entered the CPT for *Y*, which can be denoted as $P(Y \mid Pa(Y))$ (or more explicitly as $P(Y = y \mid Pa(Y) = x)$, where *y* denotes a specific value of *Y* and *x* a specific vector of values for *Pa*(*Y*)) to generate the observed outcome for *Y*. Alternatively, *indirect explanations* for the value of *Y* state that the parents of *Y* have the values they do because of their own parents and CPTs. One can therefore at least partially explain an *a priori* improbable value of *Y* by finding values (or combinations of values) of its parents that, when conditioned on, make the observed value of *Y* more probable. Such explanations in causal DAG models can be deepened recursively: values of variables are explained by explaining the values of their parents. This extends the concept of explanation beyond the statistical regression concept of proportion of variance of a dependent variable explained by differences in its explanatory variables. A causal DAG model makes probabilistic dependencies among the explanatory variables explicit, leading to a hierarchy of explanations rooted in the values of input variables (nodes with only outward-pointing arrows) and in chance at the CPTs. Explanations in such models are not limited to explaining factual observations. A standard application of Bayesian inference algorithms in BNs is to find values for unobserved variables to maximize functions of the observed data, such as the (log) likelihood function, the posterior probability, or a generalized Bayes factor of the observed values, thereby generating *most probable explanations* (MPEs), *maximum a posteriori* (MAP) explanations, and *most relevant explanations* (MREs) for the observed data, respectively (Yuan et al. 2011). These algorithms are also frequently applied to hypothetical, counterfactual, and potential future events, rather than to observed ones. For example, if it is posited that a complex system eventually fails, then these BN "explanations" (i.e., settings of the values of unobserved variables to maximize an objective function), given the hypothesized failure information, can be used to help envision how the failure might occur, which may help to prevent it. A combinatorially complex problem is to find the simplest such explanations (by various criteria, e.g., with irrelevant conditions pruned away). Exact and approximate methods have been developed for this purpose (Zhu and Yuan 2017). However all these methods are restricted to passive explanations: they explain some (real or imagined) observations in terms of other observations, rather than in terms of design decisions or other active decisions and interventions.

Explaining Predictions for Effects of Interventions: Adjustment Sets, Causal Partial Dependence Plots (PDPs) and Accumulated Local Effects (ALE) Plots for Known Causal BNs

In a causal BN, exogenously changing the value of one variable changes the conditional probability distributions of variables into which it points (its direct effects), via their CPTs. Changing their distributions, in turn, changes the distributions of their children. The effects of the initial change propagate along causally oriented arrows, leading to updated aleatory distributions for its descendants. The process is analogous to the propagation of evidence in a non-causal BN, but describes propagation of *changes* in aleatory probabilities in response to actions, rather than (or in addition to) propagation of *inferences* about epistemic probabilities in response to observations (Pearl 2010). In this setting, the statistical challenge of estimating how changing one variable would change the distributions of others can be met using a combination of graph-theoretic and statistical estimation methods. Specifically, to predict how changing X (via an exogenous intervention or manipulation) would change the distribution of one of its descendants, Y, it is necessary to identify an *adjustment set* of other variables to condition on (Textor et al. 2016). Adjustment sets generalize the principle that one must condition on any common parents of X and Y, to eliminate confounding biases; but must not condition on any common children, to avoid introducing selection biases (Wieczorek and Roth 2019). Appropriate adjustment sets can be computed from the DAG structure of a causal BN for both direct causal effects and total causal effects (Textor et al. 2016). Then, given a dataset of observed values for all variables for each of many cases, the quantitative effect of changes in X on the conditional mean (or, more informatively, the conditional distribution) of Y can be displayed via a *partial dependence plot* (PDP) (Zhao and Hastie 2021). This shows how the conditional distribution for Y (or its mean and uncertainty bands), as predicted by a machine-learning technique (e.g., random forest, support vector machines, gradient boosted machines, deep learning networks, etc.), changes for different values of X, holding the values of variables in the adjustment set fixed at the values they have in the cases in the dataset. We call the result a *causal PDP*. It shows how Y changes with X when sources of bias (e.g., possible confounders) are controlled for by holding their values fixed, but other variables are allowed to adjust realistically as X varies. For example, in a causal BN model $X \to Z \to Y$, the probability distribution of Z would change as described by its CPT in response to changes in the value of X. The distribution of Y would then change in response to the change in the distribution of its input, Z. If all relevant confounders are measured and controlled for this way, then the changes in Y's distribution are presumably caused by the differences in X values (Mill 1843; Wieczorek and Roth 2019).

In this causal BN framework, the predicted change in Y in response to a change in X is explained in terms of the following components:

(a) **Probabilistic causal laws**, represented by CPTs. These causal CPTs remain fixed ("invariant"), independent of the situation or interventions to which they are applied (Peters et al. 2016) and of the particular studies from which they were derived (Heinze-Deml et al. 2018). In this sense, a CPT in a causal BN encodes a *causal generalization* from specific studies and experiments. It describes how aleatory probabilities for a variable's values depend on the values of its parents. The invariance of this description across settings and interventions, termed *invariant causal prediction* (ICP), reflects the universality of causal laws (*ibid*). (In a non-causal BN, by contrast, CPTs are statistical description of conditional probabilities that may change across applications, perhaps reflecting different mixtures of aleatory CPTs based on different values of causally relevant latent variables.)

(b) **Initial conditions** to which these laws are applied. These conditions may differ across applications. For example, an exogenous intervention that sets X to a new value changes the conditions entering the CPTs of its children, initiating a change in their conditional distributions that then propagates along causally directed paths to change downstream distributions. The same intervention can have different effects in different populations if other conditions are also different between them. Thus, an infection control intervention that greatly reduces infection-related mortality rates in one hospital might not do so in another with a different mix of patient conditions, not because the same causal laws (aleatory CPTs) don't hold, but because the distributions of other variables that affect success (i.e., other causal parents of success) differ. In such cases, *transportability formulas* show how to adjust the findings from one set of conditions to apply to a new set of conditions, if there is enough common information between them so that this can be done (Bareinboim and Pearl 2016). For example, if the success of the infection control program varies by the age and sex of patients (and perhaps other covariates, which would be parents of the success indicator variable in a BN model), then the conditional success probabilities for different combinations of age and sex and any other causally relevant covariates can be applied to the empirical joint distributions of these variables at other hospitals to predict how well the program will succeed elsewhere, and to explain why.

(c) **Causal network structure**, i.e., the topology of the causal BN, showing the direct causes that each variable depends on, i.e., its parents in the causal BN, if any.

(d) **An adjustment set** specifying which variables are held fixed in calculating the causal effect of a change in *X* on changing the distribution of *Y*. There are often multiple valid adjustment sets, and comparing the PDPs for the estimated causal effects of changes in *X* on *Y* across alternative adjustment sets provides an internal validity consistency check for whether the assumed BN model provides a consistent description of causal effects (Textor et al. 2016).

(e) **Assumed values** for the variables that are held fixed. In PDPs, the assumed values are those that the variables have in the dataset used to estimate the PDP.

A causal PDP curve, plotting the conditional expected value of *Y* for each value of *X*, with uncertainty bands included if desired, provides a concise summary of how *Y* is predicted to change if *X* is changed, holding variables in the adjustment set fixed at specified levels. Rather than answering "*If we observe that* $Y = y$, *what is the most probable explanation?*" as for MPEs, a causal PDP answers "*If we set* $X = x$, *how will the expected value (or, more generally, the conditional distribution) of Y change?*" given stated levels of other variables. Fully explaining the answer requires the preceding five components, (a)-(e). The resulting ability to make (and to explain) predictions about effects of interventions provides essential information for optimizing decisions.

The graph-theoretic part of the causal BN explanatory framework provides rigorous conditions and algorithms for identifying adjustment sets and transportability conditions and formulas, when they exist (Textor et al. 2016; Bareinboim and Pearl 2016). The quantitative estimation and display of average causal effects by PDPs is less satisfactory, because averaging effects over different individual cases and hypothetical conditions obscures exactly what is (and should be) assumed about each case (Goldstein et al. 2015; Apley and Zhu 2020). The concept of "holding fixed" some variables to isolate the causal effects of others becomes problematic when the variables are tightly coupled, so that changing one entails changing others, rather than holding them fixed. For example, suppose that heart attack risk depends on both height and weight, and that weight also depends on height. To isolate the effect of weight on heart attack risk, a causal PDP would first include height in an adjustment set, and then modify the value of weight in each record of a dataset (i.e., each case), setting it equal to different values while holding heights fixed at the values they have in the records. The PDP would record the average predicted heart attack risk for each value of weight. The needed predictions are made by applying a predictive ML method such as random forest to each record, for each value of weight, holding other variables fixed. But in reality, weight and height are not independent. Averaging predicted risks over hypothetical cases with modified values of weight that are unlikely or impossible in light of the corresponding heights undermines the practical relevance of the resulting PDP by diluting it with predictions for unrealistic hypothetical cases that fall outside the observed range of observed data values. To avoid this limitation, *accumulated local effects* (ALE) plots have been developed that do not involve counterfactual conditions outside the range of the data, but instead quantify how small ("local") changes in one variable would affect the predicted value for another for each individual (again using a predictive model such as random forest) (Apley and Zhu 2020). The effect of a small change in X on the expected value of Y, evaluated at $X = x$, is averaged over cases with X approximately equal to x, rather than being averaged over all cases (by setting $X = x$ for all cases, while holding other variables fixed, as in PDPs). Thus, instead of asking how heart attack risk would differ for a 7 foot man if his weight were 100 pounds instead of 300 pounds, the ALE asks how heart attack risks would change for each individual if his or her weight were reduced by 1 pound. It displays the answer for individuals with different weights. Similarly, *individual conditional expectation* (ICE) plots (Goldstein et al. 2015) can be used to show how $E(Y \mid x)$ varies with x for each individual case, thus avoiding the ambiguities of aggregation. These ICE plots, in turn, can be clustered to reduce visual clutter.

Figure 9.7 illustrates these ideas, showing a PDP (upper left), ALE (upper right, y axis showing deviations from average), and ICE cluster plot (lower, y axis showing deviations from average) for the Heights dataset, using random forest as the prediction engine. (The plots were generated by the Causal Analytics Toolkit at http://cloudcat.cox-associates.com:8899/, which automatically optimizes the number of clusters in the ICE cluster plot.) We refer the interested reader to Goldstein et al. (2015) for ICE plots and to Apley and Zhu (2020) for ALE plots. Both also discuss PDPs. Apley and Zhu note that both PDPs and ALEs correctly recover the additive components in generalized additive models. To our knowledge, no software package currently disaggregates ALE curves to the individual level, analogous to ICE plots for PDPs, but this seems a natural next step.

These various approaches to defining, quantifying, and visualizing effects of one variable on another in a multivariate context are complicated mainly by conceptual ambiguities about what "effect" should mean when there are interactions among variables, and heterogeneity in their values across cases. For direct causal effects in individual cases, there is no such ambiguity in the causal BN framework: the causal CPT fully describes how each variable's aleatory probability distribution depends on, and varies with, the values of its parents. The CPT *describes* probabilistic dependency, rather than offering a deeper (e.g., mechanistic) explanation of why it exists and how it functions, but this description suffices to predict effects of interventions. For this purpose, a simple descriptive causal BN model $X \rightarrow Y$ for predicting or controlling Y via observations and interventions on X is as useful as a more refined causal BN model $X \rightarrow Z \rightarrow Y$, describing how Z mediates the causal relationship between X and Y (via $P(y \mid x) = \Sigma_z P(z \mid x)P(y \mid z)$). The model *Father* → *Height* is just as useful and valid as the more refined model *Father* → *HGH*→

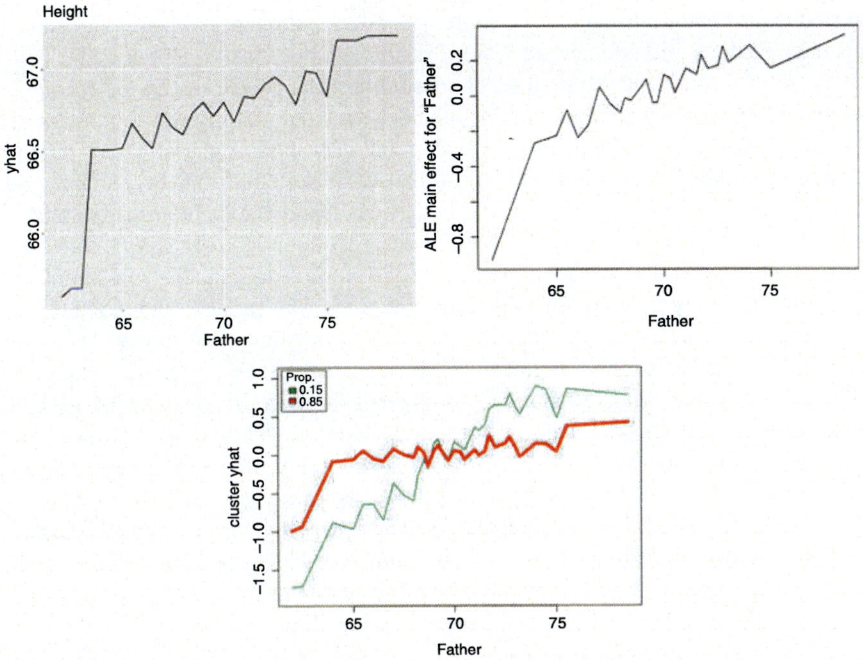

Fig. 9.7 PDP (upper left), ALE (upper right), and ICE cluster plots (lower left) for effect of Father height (x axes) on adult offspring heights (y axes). Plots are produced using the *pdp*, *ALEPlot*, and *ICEbox* packages in R, respectively, by the Causal Analytics Toolkit at http://cloudcat.cox-associates.com:8899/

Height, indicating that human growth hormone (*HGH*) mediates the probabilistic dependence of *Height* on *Father*, for purposes of predicting effects of *Father* on *Height*. More generally, the most useful level of detail to include in a causal explanation for supporting decisions describes how feasible changes in controllable decision variables change probabilities of outcomes. Causal BNs and their extensions to include decision variables and evaluation nodes (influence diagrams, discussed next) support this level of detail, while leaving open scientific questions of how underlying causal mechanisms mediate the dependencies among variables that are described by causal CPTs.

Structure of Explanations for Decision and Policy Recommendations in Influence Diagrams (IDs): Maximizing Expected Utility with a Known Causal Model

Causal BNs can be extended to produce AI decision-support advisory systems that recommend *decisions* (also called choices, actions, interventions, or manipulations) and *policies* (also called *decision rules* or *strategies*, defined as mappings from available information to actions, or, more generally, to probability distributions over actions). Actions and policies are viewed as *controllable causes*, i.e., decision variables with values selected from a set of possibilities (the choice set) by the decision-maker. The CPT for a random variable includes the values of any decision nodes on which it depends, i.e., with arrows that point into it in a causal DAG model. Decisions do not fit smoothly into the formalism of probability theory, as they are not events (subsets of a sample space), but they can readily be incorporated into BNs by introducing *decision nodes* and a sink node called the *value* or *utility node* that evaluates the outcomes at other nodes. The *utility node* represents the von-Neumann Morgenstern utility function for evaluating outcomes. The variables that point into it are those that matter to the decision-maker in evaluating the consequences of decisions. The resulting augmented BN is called an *influence diagram* (ID) (Howard and Matheson 1981; Shachter 1988). In an ID, any of several alternative actions can be taken at a *decision node* (or "*choice node*"), based on information about the values of nodes that point into it. It is often convenient to allow deterministic functions for some variables in IDs as well, e.g., body mass index is a deterministic function of height and weight. These may be regarded as special cases of CPTs giving probability 1 to a specific output value for each combination of values of its parents. Policies or decision rules, mapping information to (possibly randomized) choices of actions, are implemented at choice nodes.

In the ID framework, a decision rule sets the value of a choice node. The distributions of variables that depend on the choice node are then updated based on their CPTs, and these updates are propagated through the ID, ultimately leading to a conditional distribution and an expected utility (EU) value at the utility node. Decisions are optimized by algorithms that select decision rules to maximize expected utility. For tabular IDs, with a few possibilities at each node and explicit tables showing how conditional probability distributions of random variables depend on their parents, the same Bayesian inference algorithms used to calculate posterior distributions of chance nodes given any set of observations can be repurposed to infer decisions that maximize expected utility (Crowley 2004; see also Mauá 2016 for related work). More computationally efficient algorithms have also been developed for solving for optimal decisions in IDs (Shachter 1988; Crowley 2004). In an ID, the explanation for a recommended (optimized) decision rule is that it *maximizes expected utility*, given the information available when decisions are made. If the ID is known to be a trustworthy model of reality, then its decision recommendations are justified by the normative axioms of expected utility theory. If the ID is estimated from data, and has uncertain validity, then maximizing estimated EU may still be a valuable heuristic, but other considerations involving model uncertainty may also become important.

Figure 9.8 presents an example based on a dataset showing levels of sales for customers receiving different levels of Facebook, YouTube, and newspaper advertising. The left side of Fig. 9.8 shows the results of four BN structure-learning algorithms; all agree that *sales* depends on *YouTube* and *Facebook*, but not on *newspaper* advertising levels. The right side shows a nonparametric response surface for *sales* as a function of *YouTube* and *Facebook*. Its shape reveals that ads in each

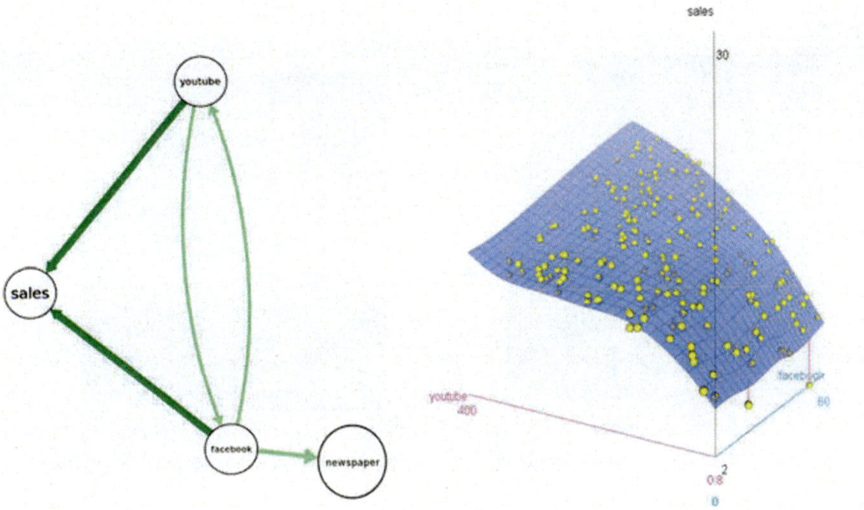


```
# R code for response surface
library(car); data("marketing", package = "datarium"); scatter3d(sales ~ youtube + facebook, fit
= c("smooth"), data = marketing)
```

Fig. 9.8 Left: Bayesian network (BN) structures learned by 4 algorithms in the *bnlearn* R package (hc, gs, tabu, iamb). Arrow thicknesses show how many algorithms (4 for darker-color arrows, 2 for lighter-color arrows) produce each arrow. Right: A response surface model for *sales* given *YouTube* and *Facebook*

channel increase expected sales most when the other channel also has relatively many ads. That is, there is cross-channel positive interaction (synergy). Given these data and information on a marketing budget and advertising prices for each channel, what decision recommendation should be made for how to allocate the budget to maximize resulting sales, and how should this recommendation be explained and justified?

```
# R code for response surface
library(car); data("marketing", package = "datarium"); scatter3d
(sales ~ youtube + facebook, fit = c("smooth"), data = marketing)
```

Figure 9.9 shows a standard managerial economics solution to this decision optimization problem. The solution is not related to standard ID solution algorithms for discrete decision variables, but instead applies nonlinear programming to choose the point satisfying the budget constraint that achieves the highest possible sales (the point, indicated by a star, where the budget line and an iso-sales contour are tangent). For simplicity, we assume that the goal is to maximize sales, or that the budget constraint is binding. The budget line is the set of feasible choices (the choice set for the decision problem) of all affordable combinations of *Facebook* and *YouTube* advertising that spend the budget, given the prices of advertising in each channel (which are divided into the budget to determine the intercepts of the line). Contours

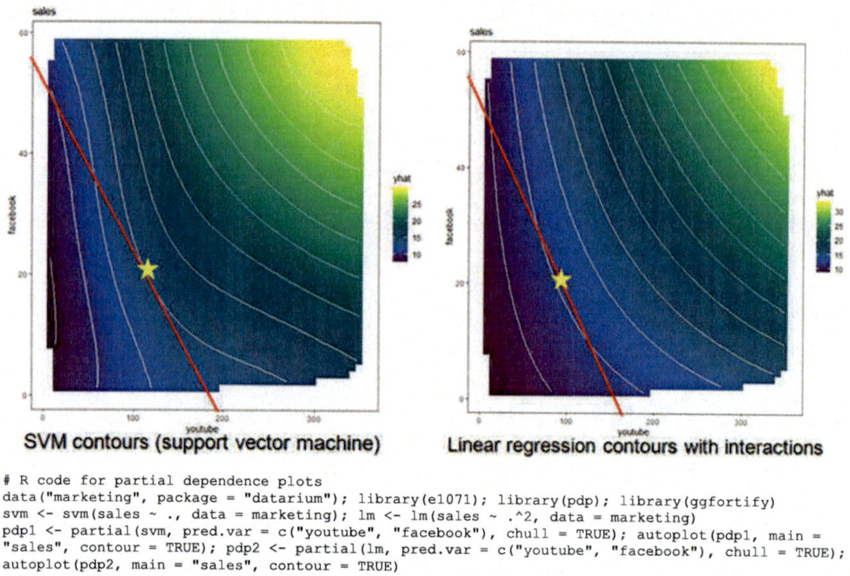

Fig. 9.9 Estimated causal partial dependence plot (PDP) contours for sales based on support vector machine (left) and quadratic regression (right), with superimposed budget lines and estimated optimal budget allocations (stars) for each estimated causal PDP

for equal predicted sales (the "iso-sales" contours, in microeconomics parlance), are estimated from the data by applying PDP estimation algorithms to the direct causes of sales, averaging over the fixed levels of other variables.

```
# R code for partial dependence plots
data("marketing", package = "datarium"); library(e1071); library
(pdp); library(ggfortify)
svm <- svm(sales ~ ., data = marketing); lm <- lm(sales ~ .^2, data =
marketing)
pdp1 <- partial(svm, pred.var = c("youtube", "facebook"), chull =
TRUE); autoplot(pdp1, main = "sales", contour = TRUE); pdp2 <- partial
(lm, pred.var = c("youtube", "facebook"), chull = TRUE); autoplot
(pdp2, main = "sales", contour = TRUE)
```

The left panel of Fig. 9.9 shows the contours estimated by a support vector machine (SVM) algorithm, and the right panel shows the corresponding contours estimated by multiple linear regression with second-order interactions. (In the right panel, the location of the sales-maximizing allocation is interpolated from the contours above and below the budget line, since none of the displayed contours is tangent to it.) Numerous other machine learning techniques could be used equally well to estimate these contours, e.g., random forest, multiple adaptive regression splines (MARS), gradient boosted machines (GBM), and so forth. They give somewhat different estimates of the contours, and hence somewhat different recommendations for the optimal budget allocation. These differences reflect *model uncertainty:* the true contours are unknown, and the estimated contours depend somewhat on the specific estimation technique used. However, the causal structure for decision recommendations is clear: select the feasible decision (or combination of decisions, if we view expenditures in each channel as distinct decision variables) to maximize the expected utility of consequences (here, the expected resulting sales). Moreover, to the extent that the relevant causal model can be estimated from data, it is also clear how the optimal decision recommendation should change as conditions change. If budget or advertising prices change (altering the slope and intercepts of the budget line), or if the contours of the PDP change (e.g., because of events that affect the advertising effectiveness of different channels, perhaps by changing the reputations, credibility, or market penetrations of the parent companies), then the new point of tangency between the budget constraint line and the iso-sales contours identifies the new recommended decision for altered circumstances.

In this way, the optimization-based framework can cope with limited novelty, offering defensible decision recommendations for what to do under new pricing or budget or effectiveness conditions, even if no historical data are available for the new conditions. The underlying causal model (the PDP contours, corresponding to a causal CPT for *sales* given values of its parent decision variables, *YouTube* and *Facebook,* and averaging over the values of all other variables that might affect individual purchase decisions) provides the knowledge needed to justify decision recommendations even under new conditions. In this way, causal AI avoids the need for new training data to make and explain decision recommendations, if causal CPTs satisfying ICP are available for predicting and optimizing effects of interventions under current, possibly novel, conditions. However, *model uncertainty* about the relevant causal CPTs, illustrated by the different estimates of contours in Fig. 9.9, leads to uncertainty about the best decision recommendation. It creates a need to learn more about the shape of the response surface (or causal CPT) over time to enable higher-value decisions. Reinforcement learning (RL) algorithms, discussed below, address this need.

Structure of Explanations for Decision Recommendations Based on Monte Carlo Tree Search (MCTS) and Causal Simulation

Making and explaining current decisions in terms of probabilities and utilities for their potential consequences epitomizes rational, deliberative, planning and decision-making for reason-driven "System 2" decisions, as contrasted with quick, pattern-driven "System 1" decisions (Kahneman 2011; Gershman 2017; see Chap. 1 for further discussion). When a known probabilistic causal model capable of predicting outcome probabilities for different actions is available, along with a utility function for evaluating different outcomes, then solving for expected utility-maximizing policies even in multi-period settings is conceptually similar to solving the simple nonlinear programming optimization problem illustrated in Fig. 9.9. The relevant optimization techniques for stochastic dynamic systems are more complicated, e.g., with stochastic dynamic programming instead of simple nonlinear programming, but the principle is similar in that current decisions are optimized to maximize expected utility, taking into account future decisions and outcome probabilities (Blondel and Tsitsiklis 2000). For example, a Markov Decision Process (MDP) is a multistate transition reward process, for which a policy specifies which feasible action to take in each state. Actions can affect rewards from transitions as well as next-state transition probabilities. MDPs with finite numbers of states and actions can be solved for optimal policies that maximize expected discounted rewards or average rewards per unit time by using either linear programming, dynamic programming, or well-known special-purpose iterative algorithms (Bellman 1957; Howard 1960). As in Fig. 9.9, re-optimizing when conditions change allows new optimal policies and resulting decision recommendations to be created as needed for new situations; they are justified by optimization.

The optimization-based paradigm for recommending and justifying decisions must be extended if the objective function to be optimized is unknown, either because it is difficult to compute (perhaps requiring lengthy simulation) or because of lack of causal information about how actions affect outcome probabilities. If uncertainty about an MDP is described by convex uncertainty sets for transition matrices—e.g., if the one-step transition probabilities are known to lie in certain intervals; or if the rows of the transition matrix are known only to lie within some relative entropy (Kullback-Leibler) "distance" from known reference distributions—then MDP solution algorithms can be extended to find the best policy for the worst-case model in the uncertainty set, providing a form of robust control for uncertain MDPs (Bagnell et al. 2001). (Extensions of expected utility theory that imply such maximin decision-making over convex uncertainty sets in decision analysis are discussed by Schervish et al. 2003.) However, the optimization paradigm breaks down if the optimization problem is unsolvable, e.g., because optimal policies are uncomputable. This happens in some partially observable MDPs (POMDPs), i.e., if the states of MDPs cannot be directly observed (so that policies mapping states to actions cannot be implemented), but must be inferred from observations that depend probabilistically on the underlying state—e.g., from signals observed via noisy information channels in engineering systems; from observed symptoms in patients with underlying unmeasured health conditions; from observed behaviors of customer with underlying unmeasured beliefs, preferences, and intents; and so forth. Existence of optimal policies—or even of policies meeting desired constraints, such as having at least a stated probability of achieving a goal state in finite time without first encountering any catastrophic state—is undecidable for some POMDPs with infinite planning horizons, and with either discounted or undiscounted total reward models (Madani et al. 2003). Although important special cases are decidable (Chatterjee et al. 2016), e.g., by converting the POMDPs to equivalent dynamic Bayesian networks (DBNs) and then optimizing policies by maximizing the likelihood of suitably normalized rewards (Toussaint et al. 2012), finding optimal or near-optimal policies in many POMDPs, even when it is possible in principle, is sufficiently computationally complex so that heuristics must be used (Chatterjee et al. 2016).

The rationale for heuristic-guided decision or policy recommendations in such cases is that *no better choice has been discovered by the time decisions must be made*. This "any time planning" principle applies more generally, well beyond the class of POMDP decision problems. For example, several recent state-of-the-art probabilistic planning heuristics apply sampling-based methods such as Monte Carlo Tree Search (MCTS), with deep learning or other function approximators of value functions, to search for and incrementally improve the best (highest expected value) plan that can be discovered with a given computational budget. When an action must be taken, the best plan discovered so far is used to recommend what to do next (e.g., Shen et al. 2019; Xiao et al. 2019; Parascandolo et al. 2020). Other causal simulation and simulation-optimization algorithms (Eskandari et al. 2011; Juan et al. 2015), e.g., particle filtering for efficient sampling of decision and outcome trajectories in POMDPs with hidden (latent) variables, differ in detail from MCTS, but they use similar principles of selective sampling and expansion (via simulation) of future trajectories for decisions and outcomes to estimate value functions and heuristically optimize immediate decisions (Du et al. 2020; Piché et al. 2019). We will refer to such methods generically as *causal simulation heuristics*, as they use causal models to envision a sample of future scenarios that are then used to evaluate and choose among currently feasible courses of action. Causal models describing probabilistic relationships between actions and consequences (next states and rewards) are crucial inputs to these methods. They are used under the guidance of MCTS, particle filtering, or other (e.g., hybrid MCTS-particle filtering) causal simulation heuristics to efficiently sample, simulate ("roll out"), and evaluate possible futures. The resulting information is used to improve plans specifying what to do now and what to do next if various contingencies occur (Piché et al. 2019). Re-planning and search for better plans occur as new information and additional time for planning are gained, with decisions at any time being guided by the current best plan. Causal simulation guided by entropy-based upper bounds on the value of the best policy (maximum entropy MCTS) converges quickly (exponentially fast) to the optimal next decision for some classes of sequential decision problems for MCTS (Xiao et al. 2019).

Structure of Explanations for Decision Recommendations Based on Reinforcement Learning (RL) with Initially Unknown or Uncertain Causal Models

If a causal model specifying how actions affect outcome probabilities (e.g., reward and state transition probabilities in an MDP) is not available, a key new principle is needed to guide decision-making: When causal models and optimal policies are initially uncertain, *actions are valued not only for the rewards and state transitions that they cause, but also for the value of information they reveal* about how to improve policies (Sledge and Príncipe 2018). Managing the famous exploration-exploitation tradeoff between applying the most promising policy discovered so far (exploitation), and deviating from it to discover whether a different policy might perform better (exploration) requires taking into account the value of information (VoI) produced by actions (*ibid*). For example, in a multiarm bandit (MAB) problem such as a clinical trial of several alternative treatments, where the efficacy of each treatment is initially unknown, a simple Bayesian heuristic starts with a uniform prior probability that each treatment is best. It updates these probabilities by conditioning on observed results (Russo et al. 2018). The convenient fact that the family of beta distributions, which includes the uniform, is a self-conjugate family for binomial sampling yields simple update formulas (posterior beta distributions) for MABs with fixed but initially unknown probabilities of success for each treatment. More complex update formulas and approximations have been developed for MABs with more complex reward distributions (Kauffmann et al. 2012). A simple but very effective technique for balancing exploration against exploitation in this setting is to probabilistically select, on each trial, each treatment with a probability equal to its current probability of being best. Thus, a treatment known to be best would always be selected; one known not to be best is never be selected; and treatments deemed more likely to be best are selected more frequently ("reinforced") as experience accumulates. This simple *reinforcement learning* (RL) heuristic (Jin et al. 2021), known as Thompson sampling (TS), has been extended to MDPs, with the number of periods (the "decision epoch" length) for which each trial policy is evaluated being tuned to be long enough to evaluate the policy without needlessly inefficient search (Russo et al. 2018). Variations of Thompson sampling are asymptotically optimal (converging in mean to optimal-value policies) even for many non-Markov and partially observable environments (Leike et al. 2016).

Performance guarantees for TS and other RL algorithms are usually given in terms of asymptotic optimality and sublinear growth of regret. The *regret* from using a policy *P* over any time interval is defined as the expected difference between the sum of rewards that would be obtained by acting optimally throughout the interval, as assessed by someone with perfect information about how actions affect outcome probabilities, and the sum of rewards actually obtained by following policy *P*. If regret grows sublinearly with the length of the interval, then the policy being used must be optimal (zero regret) or getting closer to it as time passes, since otherwise regret would increase at least in proportion to the length of the interval over which it is accumulated. Sublinear bounds and asymptotic zero-regret guarantees for the performance of RL algorithms, including TS, have been established for MDPs, MABs, and other classes of sequential and adaptive decision problems (Russo et al. 2018; Chowdhury and Oliveira 2021). Recent refinements include sampling each action to minimize the ratio of square of expected error (i.e., of squared one-period regret) per unit of information gained about how to optimize choices (i.e., per bit of mutual information between the optimal action and the next observation). This heuristic outperforms TS and other heuristics across a wide class of models, with a bound on regret that scales with the uncertainty (entropy) of the optimal action distribution (Russo and Van Roy 2018). If the pace of change is slow enough, then RL algorithms yield low regret policies even for MDPs with drifting (i.e., slowly changing, in a sense that can be made precise) parameter values (Cheung et al. 2019). If risk is important, imposing safety constraints on learning to avoid sampling actions with potentially catastrophic consequences still allows zero-regret policies to be learned in some MAB problems (Kim et al. 2017; Moradipari et al. 2020). Risk-averse dynamic programming methods that protect against large negative deviations from expected values have been developed for MDPs using average value-at-risk (VaR) (Carpin et al. 2016; Bauerle and Ott 2011). Other coherent risk measures, including conditional-value-at-risk (CVaR) and entropic-value-at-risk (EVaR), have recently been used in risk-averse planning and decision optimization algorithms for both MDPs and POMDPs (Ahmadi et al. 2020, 2021), as well as in model-free RL algorithms (Urpi et al. 2021).

Other refinements, extensions, and improvements in RL algorithms for various classes of decision problems continue to be made, as this is a very active area of ongoing research (Chowdhury and Oliveira 2021; Jin et al. 2021). For example, in nonlinear optimization problems such as optimizing cross-channel advertising expenditures in Fig. 9.9, the *smoothness* of the estimated response surface, as exhibited in its contours, allows valuable information about the best policy (feasible allocation of budget) to be gleaned from observations at nearby points. Collecting information in the vicinity of the current estimated optimum to better ascertain the shape of the response surface thereby enabling incremental adjustments of the decision variables to increase mean system performance, was one of the earliest approaches to decision-making with uncertain response surface models (Box 1957). MABs have been extended to allow for such spatial correlations among rewards at different locations in a decision space, and RL algorithms have been developed for safe exploration and optimization of policies with correlated rewards when some combinations of decision variables have low rewards (or high penalties) and must be avoided (Wu et al. 2017; Sui et al. 2015). RL methods have also been integrated into MCTS heuristics for sequential planning and decision-making in MDPs, POMDPs, and more general probabilistic environments, to guide efficient search for high-performance plans satisfying safety constraints (Bai et al. 2014; Kim et al. 2017). They have been applied to POMDPs and to semi-Markov decision processes (SMDPs) and partially observable SMDPs, in which executing an action (or the subtasks of which it is composed) takes an uncertain amount of time (Li 2012; Le et al. 2018). For this often realistic setting of tasks with uncertain durations and success probabilities, RL has also been integrated with behavior trees (de Pontes and Engel 2015) and with hierarchical POMDPs (Toussaint et al. 2012; Le et al. 2018) to enable AIs to adaptively select behaviors (sequences of tasks and subtasks) over time. Importantly, RL can be used to acquire and improve *skills*, i.e., learned sequences of actions for accomplishing tasks and subtasks (Moradi et al. 2010; Shu et al. 2017). Multiple skills and behaviors can be executed simultaneously, as when an autonomous robot executes in parallel routines for moving toward a target while maintaining balance and avoiding collisions. This concept of choosing behaviors from a set that can be expanded by acquiring and perfecting new skills extends the usual decision-analytic model of decision-making, in which a decision-maker optimizes over a fixed choice set of possible alternative choices, by allowing investment in skills that expand the repertoire of behaviors that can be successfully executed in response to a situation. What it knows how to do (i.e., the current choice set) then forms part of the explanation for what an agent (whether an AI, a person, or a team or organization) chooses to do at any time (Shu et al. 2017).

Throughout these variously extended and generalized classes of decision problems, however, the basic RL principle *adaptively select actions and policies to reduce expected regret, considering value of information as well as value of rewards* has proved extremely valuable, not only for MABs and for MDPs and POMDPs with initially unknown parameter values, but also for an increasingly wide variety of other dynamic stochastic systems with initial uncertainty about how actions affect transitions and rewards (Leike et al. 2016). Simply adding an "entropy regularization" term, reflecting current uncertainty (entropy) about the best policy for an MDP, to the standard objective function (e.g., a weighted sum of rewards) in RL to encourage more random exploration when the best policy is more uncertain, has been found to improve the robustness and convergence of RL for MDPs with continuous state and action spaces (Haarnoja et al. 2018). In effect, successful RL heuristics use adaptive exploration, guided by current uncertainty about the best (reward-maximizing) policy, to learn enough about the causal relationships between actions and consequences to enable low-regret policies to be discovered (Ortega and Braun 2014). RL heuristics thereby provide a basis for decision recommendations along the way that can be explained and justified as contributing to a balanced mix of exploration and exploitation, while meeting safety and risk constraints if necessary (Kim et al. 2017). Although they are guaranteed to eventually learn optimal policies with probability approaching 1 in certain stationary environments such as ergodic MDPs or MABs, RL algorithms may perform poorly in non-stationary environments, such as for "restless bandits" with reward distribution that change quickly compared to the convergence rate for learning low-regret policies (Raj and Kalyani 2017). More generally, in dynamic decision environments, the performance of an AI decision and control system is limited by the rates at which it can interact with and learn from its environments, as discussed next.

Limitations and Failures of Causally Explainable Decisions for Dynamic Systems

An AI system can fail to provide causally effective decision recommendations to achieve desired states and safety goals if it is overwhelmed by its environment—for example, if conditional state transition or reward probability distributions in response to its actions change too quickly for the AI learning and decision algorithms to keep up, or if the AI's repertoire of available actions is too limited to greatly change the course of events. Suppose that the state of a controlled system changes probabilistically not only in response to choices made by an AI controller, but also in response to stochastic inputs from the environment. The state of the system at any time is only partially observable by the AI controller via a noisy (probabilistic) information channel representing its limited sensing capabilities. In contrast to mathematical optimization formulations that treat choosing an action as a primitive that can be executed instantaneously, correctly, and with certainty, suppose that the AI controller can only select control signals to send to imperfect actuators (e.g., people or agents or mechanisms acting on its behalf). Implementation of its control signals as actions may involve errors and uncertainties. An intended action may be interrupted or fail before it is completed. Thus, the control channel is also probabilistic: a control signal sent by the AI creates a probability distribution over implemented actions (controllable inputs) entering the system and possibly its future state. In this context, the AI's ability to steer the system toward desired (highly valued) goal states, while avoiding undesired ones, is limited by the capacities of its sensing and control channels, i.e., the rate at which information about the state of the system can be received (Nair and Evans 2004 for feedback control stabilization of stochastic linear systems) and the rate at which information can be transferred from control signals to future states of the system, called the *empowerment* or *control capacity* of the controller (Achlioptas et al. 2017; Tiomkin et al. 2017; Ranade and Sahai 2019). Thus, a scalar stochastic linear system with actuator uncertainty can be stabilized around a desired state if and only if its control capacity—which is used to dissipate uncertainty about the state of the system—is greater than the rate at which uncertainty about the state accumulates in the absence of control (Ranade and Sahai 2019). If control capacity is below this limit, then even an arbitrarily "intelligent" AI, meaning one capable of instantly computing fully optimal policies with respect to the information it has and the control signals it can send, will lack power to implement causally effective policies to bring the system to (or close to) a desired state and maintain it there. Conversely, even if control and sensor capacities are high enough not to limit the performance of a controller, the difficulty or impossibility (undecidability) of computing optimal or approximately optimal polices for some problems, including some POMDPs with infinite horizons, limits the quality of decision recommendations that can be given for these problems (Madani et al. 2003; Leike and Hutter 2015). In short, the value of decision recommendations that any system can give depends greatly on the decision problem and on the available causal model (if any) linking actions to outcome probabilities. Such models range from relatively simple tabular influence diagrams, response surface, and optimization models (Figs. 9.7 and 9.8), for which optimal recommendations are readily computed and defended; to undecidable infinite-horizon POMDPs or computationally hard (NP-complete) stochastic optimal control problems (Blondel and Tsitsiklis 2000); to MDPs with initially unknown parameters, for which RL heuristics may provide practical approximate solutions (Table 9.1). This heterogeneity in decision optimization problem types and achievable results suggests that no single type of causal explanation for decision recommendations is best for all situations. Rather, rationales for decision recommendations should be matched to problem type.

Discussion: Explaining CAI Decision Recommendations

The different causal models we have reviewed lead to different detailed algorithms for decision optimization, but all draw on a small set of causal artificial intelligence (CAI) principles for making and defending (by explaining their rationales in causal terms) decision and policy recommendations when actions have uncertain consequences. This section attempts to distil those principles. In this context, a *decision* is a choice from a choice set of feasible alternatives. A *policy* is a decision rule specifying how to make a decision whenever one must be made, based on the information available then. A fully specified probabilistic causal model gives the probabilities of outcomes of actions. In influence diagrams (IDs), these are probabilities for the utility node, mediated by changes in the distributions of other variables caused by decisions. In Markov decision processes (MDPs), semi-Markov decision processes (SMDPs), and other stochastic optimal control models, actions can affect probabilistic state transition rates and reward distributions for occupying and/or transitioning between states. In partially observable MDPs (POMDPs) or partially observable SMDPs (POSMDPs), actions may also affect the information channels that map unobserved underlying states to conditional probabilities of observed signals. In all of these classes of causal models, *causally effective decision-making* recommends decisions and policies that make preferred outcomes more probable, where preferences are represented by utility functions, reward functions (including discount rates for future rewards), or goal states (perhaps with forbidden states and safety constraints), depending on the model. A *causal explanation* for a recommended decision or policy explains *why* it is recommended, i.e., why it is believed that it will make preferred outcomes more probable.

CAI principles and computational methods that have proved useful for supporting causally effective decision-making and control in a variety of applications include the following.

- **Expected utility (EU) optimization with known causal models:** *Choose a feasible action or policy to maximize expected utility.* This is practical when probabilistic causal models are known and optimization is tractable. Decision problems with known causal models represented by small (tabular) Bayesian networks or influence diagrams (IDs), decision trees, or response surfaces (Figs. 9.7 and 9.8) are amenable to EU optimization-based decision recommendations and causal explanations, with a fundamental rationale for recommended decisions being that they are implied by the normative axioms of EU theory.
- **Dynamic decision optimization and risk management with known dynamic causal models:** *Make current decisions to maximize a long-run objective function, taking into account potential future decisions and chance events.* Common objective functions include long-run average, total, or discounted rewards; and long-run expected utility. Dynamic optimization is practical for known causal models represented by small decision trees, influence diagrams based on dynamic Bayesian networks (DBNs), or MDPs. It is practical for some POMDPs, but computationally difficult for others, and impossible (undecidable) for some

infinite-horizon POMDPs. MDP and POMDP solution algorithms can be modified to incorporate risk aversion using a variety of coherent risk measures (Bauerle and Ott 2011; Ahmadi et al. 2021). When dynamic optimization is impracticable or impossible, heuristics must be used to guide choice.

- **Causal simulation and Monte Carlo Tree Search (MCTS) heuristics**: *Make current decisions based on the best plan available when decisions must be made.* MCTS and other causal simulation methods use causal models to simulate possible futures and incrementally improve plans, i.e., sequences of decisions contingent on future events, with increasing CPU time. MCTS does so by selecting (via adaptive sampling), simulating, and evaluating possible futures and keeping the best (most highly valued) plans discovered. Although full optimization of an objective function may be impossible with the time and resources provided, causal simulation uses the best plan discovered so far to recommend what to do when decisions must be made. It is practical if a probabilistic causal model, e.g., a POMDP or a discrete-event simulation model, is available to support its simulations ("roll outs") of possible futures.
- **Reinforcement Learning (RL) with initially unknown causal models:** *Make current decisions to reduce expected regret, taking into account the estimated value of information from actions*, as well as the estimated immediate state transitions and rewards that the actions cause. RL with entropy regularization is useful for discovering optimal or near-optimal (low-regret) policies in MDPs, even when relevant causal models linking actions to probabilities of consequences (immediate rewards and state transitions) are initially unknown (Haarnoja et al. 2018).
- **Information and control capacities:** *Make current decisions based on realistic assessments of what is possible to accomplish in future.* If an AI makes or recommends decisions in order to control a system that evolves dynamically in response to a mix of controlled and uncontrolled (stochastic) inputs, then the set of probability distribution over outcomes that it can reach and maintain in a given time interval via its control signals is limited by its control capacity—the rate at which it can transfer information to future states of the system—as well as by its capacity to sense and process relevant information needed for causally effective control. To the extent that these constraints are understood, or can be learned from experience, they can inform planning by showing what goals (construed as desired probability distributions over states reached within specified times) are realistically achievable. While this principle has been most thoroughly developed for discrete-time linear dynamical systems, the link between incoming and outgoing information rates and causally effective controllability of systems is applicable more generally (Touchette and Lloyd 2000) and is being clarified via recent work (e.g., Sinha and Vaidya 2017).

These principles are complementary. Current state-of-the-art AI/ML systems for discovering how to act effectively under uncertainty typically integrate several of them. For example, RL techniques such as Thompson sampling are now commonly used in MCTS to prioritize scenarios to simulate and evaluate. RL, MCTS, and numerical methods for approximate stochastic dynamic programming and policy optimization, have all benefitted from use of flexible nonparametric function approximators, such as deep neural networks, to estimate value functions for evaluating policies from limited samples of decision and outcome trajectories. AI/ML systems that integrate dynamic decision optimization, RL, and MCTS methods have proved useful in a wide range of applications, both when complex causal models (e.g., POMDPs) are available, and also in the absence of an *a priori* known causal model (model-free RL).

Applying CAI Principles to Explain Decision Recommendations

A powerful motivation for CAI is its potential to identify decisions and policies that improve average rewards (or improve other, risk-adjusted, objective function values) for decision and control under uncertainty, compared to methods that do not use causal models to help optimize decisions. We have emphasized an additional motivation: the potential for CAI to explain the causal rationales for its decision recommendations, thereby increasing their transparency and trustworthiness. However, the CAI methods we have discussed—EU maximization, dynamic optimization with or without risk aversion or safety constraints, and MCTS or optimal control, if causal models are known; RL if causal models are initially uncertain or unknown; and hybrids of these methods, assisted by function approximators, such as deep learning—are best understood by specialists. To what extent, then, can CAI facilitate humanly satisfying explanations that succeed in making decision recommendations more transparent, trusted, and acceptable for non-specialists? Although this question can only be fully answered empirically via social science research informed by the psychology of explanations (Miller 2019; Mittelstadt et al. 2019), the distinctions we have discussed among model-based optimization, causal simulation heuristics, experiential learning (RL), and capacity to effect change may be useful in setting expectations for the different kinds of information that can, should, and must be included to create convincing explanations for decision and policy recommendations.

Current thinking on "explainable AI" (XAI) focuses largely on explanations for AI/ML black-box *predictions*, with palatable explanations generated by simplifying complex black-box predictive models to yield simpler (e.g., linear or threshold), approximate predictive rules (Mittelstadt et al. 2019). By contrast, any rational explanation for a *decision* recommendation must include information about each of the following components:

- *A causal model describing how outcome probabilities depend on choices*. Examples of causal models discussed in previous sections include conditional probability tables (CPTs) in influence diagrams and in dynamic Bayesian networks (DBNs); response surface models; and MDPs, POMDPs, stochastic control models, and causal simulation models for dynamic systems. Examples of outcomes over which preferences are defined in these various causal models include trajectories (time sequences) of system states, outputs, or rewards; terminal states to which utilities are attached; or realizations of the utility random variable in an influence diagram model. If a unique causal model is not known, then model ensemble methods (e.g., different estimated causal models produced by different ML techniques, as in the slightly different PDP contour maps in Fig. 9.9) or reinforcement learning methods that maintain posterior distributions over possible models (e.g., beta distributions for the rewards from each arm of a MAB, or from each act in each state of a MDP, in Thompson sampling) may be used to reflect model uncertainty (Ouyang et al. 2017).

- *Preferences for outcomes.* Preferences are represented in several ways in the different causal models we have discussed, e.g., by a utility function; by an objective function for trajectories of states and outputs, in POPMDPs and stochastic control models, perhaps modified or constrained to reflect desired risk aversion; by estimated one-step transition rewards for state transitions and/or occupancy, together with a discount rate, in MDPs; or by goal states to be reached and taboo states to be avoided over some time horizon in stochastic optimal control models.
- *A choice set of feasible alternatives that the decision-maker can currently choose among.* While POMPDs, behavior trees, and control capacity models all acknowledge that a decision-maker may not be able to simply choose an action that is then implemented promptly and without error, all of them also recognize that decisions involve selecting from a feasible set of choices—even if what is selected is only which control signal to send or what task(s) to attempt next, rather than successful completion of an action.

Given these three main components, a CAI decision support algorithm applies optimization methods or heuristics to recommend decisions that are predicted by the causal model to make preferred outcomes more probable. If a causal model is known for which optimization is tractable, then CAI recommends decisions to maximize an objective function. For example, Fig. 9.9 illustrates how a choice set (the budget line), possible outcomes (sales levels), a causal model learned from data (the iso-sales contours expressed as functions of the decision variables, namely, advertising in each channel), and assumed preferences (higher contours are assumed to be preferred) come together in a simple traditional nonlinear programming optimization model. If optimization is intractable, then causal simulation heuristics such as MCTS are used. If no causal model is available, then reinforcement learning is used.

In each case, the optimization algorithms or heuristics for selecting a recommended decision are likely to be inscrutable to many users. Their results are not likely to be readily explicable except in simple cases such as Fig. 9.9. However, the three required inputs (causal model, preference model, choice set) can often be explained and visualized with the help of Bayesian networks and influence diagrams (or decision trees and tables, for very small problems); state transition diagrams and example trajectories for MDPs and POMDPs; and response surface models, partial dependence plots (PDPs), and accumulated local effects (ALE) plots, if there are only a few continuous decision variables. Moreover, if a trusted causal simulation model is available, then the recommendations from a CAI-based decision recommendation engine can be made credible, even if the optimization algorithms that lead to them are obscure, by simulating deviations from them and showing that non-recommended policies reduce the value of the objective function being maximized (or stochastically reduce it, i.e., shift its cumulative distribution leftward over repeated simulation runs). If a trusted causal model is not available, then sensitivity analyses and reinforcement learning methods that maintain and update posterior distributions for the best policies can be used instead to show how deviating from recommended decisions reduces expected utility, as assessed with currently available information.

From this perspective, explaining CAI-based decision recommendations to build trust and acceptance among users need not confront the challenge of explaining how the CAI algorithms work. Rather, they can focus on assuring that the choice sets, preference models, and causal models that the optimization algorithms or heuristics use to generate recommendations are well understood and agreed to, and that the resulting decision recommendations are credible, as shown by the demonstrable difficulty (or, for exact optimization methods, impossibility) of improving upon them. For decades, human decision analysts have emphasized precisely these features in communicating with and earning the trust of decision-makers: it is essential to be clear about the inputs and outputs and the structure of the analysis (choice sets, possible outcomes, preferences and value tradeoffs among outcomes, and conditional probabilities for outcomes given choices), but technical details for solution algorithms can be safely encapsulated in influence diagram software or other solvers, with the validity and robustness of results (i.e., decision recommendations) being clarified more through simulations and sensitivity analyses than by explaining how they were calculated (Howard 1988; Keeney and Raiffa 1976). To this traditional perspective from decision analysis on how to build trust in analyses and recommendations, we add the following points:

- In recommending a decision or policy, a CAI advisory system should disclose whether its recommendation is based on full optimization, heuristic search, or RL exploration.
- It should also disclose whether its recommendation is based on a well-validated causal model (e.g., a causal model with CPTs for which properties of transportability and invariant causal prediction (ICP) have been tested and verified); on a causal model that has been assumed but not yet validated (as in many traditional decision analyses that use subjectively elicited decision trees or IDs); on an uncertain model or ensemble of plausible causal models learned from data (e.g., with contours in Fig. 9.9 estimated using causal PDP or ALE plots based on ensemble ML techniques); or on reinforcement learning (RL) trajectories seeking to estimate from experience how decisions affect rewards or risk-adjusted objective functions.

- If the relevant causal model is uncertain (e.g., if it is still being learned by an RL heuristic that has not yet converged to a zero-regret answer), then the CAI systems should disclose current uncertainty about the best recommendation. It should note how, if at all, risk aversion and safe exploration constraints have been incorporated into the search for the best recommendation.
- If heuristics have been used to make a decision recommendation, then bounds on regret, or other performance guarantees, should be provided if possible. For example, such information is produced by several current state-of-the-art POMDP heuristics (et al. 2017).

While explaining details of CAI algorithms is probably unnecessary for justifying decision recommendations and explaining to what extent they are trustworthy, explaining the technical basis for the analyses that supports the recommendations—especially, the above aspects of causal model uncertainty and optimization uncertainty—can help to set realistic expectations for the reliability of the recommendations.

Conclusions: Explaining Recommended Decisions in Causal AI

Needing to show that following a CAI advisory system's decision recommendations is expected to cause probability distributions for possible outcomes that are preferred to the distributions caused by feasible alternatives creates a distinctive set of information requirements and a distinctive structure for causal explanations of decision recommendations that differ markedly from the information requirements and structure needed to explain observations or predictions. Explaining decision rationales requires not simply conditional probability distributions, but causal models of effects of possible (counterfactual) actions on probability distributions. It requires being able to use knowledge of choice sets, preferences (e.g., utility functions or goal states) for possible future outcomes, and estimated causal links (e.g., CPTs satisfying ICP) between current actions and probabilities of future outcomes to explain why the recommended course of action is best; as well as procedural knowledge of appropriate exact or heuristic optimization methods for finding the recommended course of action. In relatively simple situations, decision recommendations are then obtained by optimizing a known (or approximately known) objective function that serves as a causal model for predicting the value or expected utility of outcomes caused by different feasible settings of decision variables. Figure 9.9 illustrates this concept, with feasible allocations of available budget between marketing channels as the choice set; estimated iso-sales contours as the causal model; and sales as the outcome variable to be maximized. The recommended decision (the starred point of tangency between the budget constraint line and the estimated contours of expected sales caused by different actions, i.e., allocations) has a clear explanation in this example: deviating from the recommendation in either direction causes lower performance contours.

Such a causal model also readily explains how and why recommended decisions change as conditions change, even if the new conditions have never been encountered before. For example, a change in the budget or in advertising prices (changing the budget line) or in advertising effectiveness (changing the contours) that implies a new point of tangency in the example of Fig. 9.9 thereby immediately implies a new optimal budget allocation decision, even if no data have been collected for the new situation. In this way, causal models can provide defensible recommendations for new situations without requiring further training data, as long as the relevant causal laws are known, and are known to apply for the new conditions (transportability and ICP). This enables a CAI advisory system that monitors changes in causally relevant conditions (e.g., parents or ancestors of the utility node in an ID model) within the scope of known causal laws (causal CPTs), to exercise *automated vigilance* on behalf of those it advises, proactively recommending next optimal actions in response to observed changes in conditions, without having to wait for new training data or to learn new predictive patterns. The recommendations can be explained by showing how estimated optimal decisions change as conditions change, similar to comparative statics in economics. For more elaborate dynamic causal simulation models, explanations cannot necessarily be visualized as easily in Fig. 9.9, but simulating the consequences of deviating from recommended policies can provide convincing evidence that the recommendations should be followed, provided that the causal simulation model is trusted. To validate a causal model, in turn, an array of tools from causal analysis is available, as discussed in the section on the structure of explanations in causal Bayesian networks. These include structure-learning algorithms, CPT estimation algorithms, and conditional independence tests for implications of hypothesized causal structures; statistical tests for internal validity based on comparison of causal effect estimates across multiple distinct adjustment sets; and statistical tests for external validity based on transportability and invariant causal prediction (ICP) (Textor et al. 2016; Heinze-Deml et al. 2018). As with all statistical tests, passing these various validation tests does not assure that a causal model is valid, but provides evidence against the null hypothesis that its testable predictions are false. These methods have been most extensively developed in the context of causal Bayesian networks, but they apply also to other causal models, including MDPs and POMDPs, that can be represented by equivalent dynamic Bayesian networks (Toussaint et al. 2012).

If relevant causal laws (CPTs satisfying ICP) are known for only a limited range of conditions, however—perhaps only those in the vicinity of the currently estimated optimal policy, if it has been learned via RL—then changes that push observed conditions outside the scope of the currently known CPTs may require new rounds of RL before confident decision recommendations can be made. In Fig. 9.9, for example, if the response surface had been fit to a narrow range of data near the starred optimum, instead of to a relatively wide data cloud, the resulting causal PDP model might have produced a relatively narrow swath of well-estimated contours surrounded by unknown contours outside the range of observed data. To characterize the reliability of the basis for its recommended decisions, CAI advisory systems should therefore not only distinguish among exact optimization, heuristic (sub)-optimization, and exploratory RL heuristics, but they should also characterized remaining uncertainty about their recommendations whenever possible, e.g., by providing regret bounds or performance guarantees. Such uncertainty information differs from the statistical confidence intervals and Bayesian posterior distributions typically used in predictive analytics, insofar it requires counterfactual comparisons—for example, for the expected difference between achieved rewards and the maximum rewards that would have been achieved had the truly optimal policy been known and followed. Plausible explanations for why one action should be taken instead of another make such counterfactual comparisons necessary. Causal models of the effects actions on of outcome probabilities make them possible.

Although CAI is recent enough to make confident prediction of trends difficult, it has recently been identified as an important next step for the evolution of applied AI/ML by AI/ML developers, users and commentators (e.g., Dhar 2020; Heinze-Deml and Meinshausen 2020; Sgaier et al. 2020). The main reason is that it allows AI to go beyond pattern-driven inferences and predictions to calculate the probable effects of different interventions, or courses of action, and to optimize policies and behaviors accordingly—in effect, leading to smarter AI/ML. As explained by Sgaier et al. (2020), "Ultimately, knowing the 'why' behind complex problems helps us to understand how the world really operates and, in turn, to identify the right actions to achieve desired outcomes. We may yet find that an ounce of causal AI is worth a pound of prediction." For practitioners, the greatest practical value of CAI is probably that it helps to identify actions and policies that are causally effective in increasing the probabilities of achieving desired outcomes (Vineis et al. 2017). For developers, a strong contribution of CAI is that it improves the speed and accuracy of machine learning, especially in novel situations, by enabling efficient generalization and transfer of knowledge learned under previously encountered conditions to apply to new situations, as in the example of Fig. 9.9. CPTs satisfying the ICP property need not be re-learned every time an environment changes, or as new conditions or interventions are encountered, but instead can be used to "transport" previously acquired causal knowledge to the new situations (via transportability calculations) to predict effects of interventions without waiting to collect new data (Heinze-Deml et al. 2017).

In some applications areas, such as Systems Biology, users and developers have collaborated closely in developing and applying CAI (Triantafillou et al. 2017). CAI implementation and public-domain software have advanced enormously in the past decade, and uniform interfaces to multiple causal discovery packages are now readily available, such as the *CompareCausalNetworks* package in R (Heinze-Deml and Meinshausen 2020). Despite these advances, the situation illustrated in Figs. 9.4 and 9.7 (left panel), in which the structure of a causal model is not uniquely determined automatically from data alone, is still common in practice. As a result, current CAI technology is often used as a tool by knowledgeable experts to generate and test causal hypotheses and quantify causal relationships. Human experts apply common-sense causal knowledge (such as that age and sex can be causes, but not effects, of diseases; or that death can be an effect, but not a cause, of diseases and exposures) to constrain the search for causal models that are consistent with data (Nagarajan et al. 2013; Triantafillou et al. 2017). Whether used as a tool by human experts, or used by autonomous or human-guided AI to speed learning and help optimize decisions, CAI will be most likely to be trusted in future if it can give clear explanations for its conclusions. We have proposed that for decisions—as opposed to predictions, diagnoses, or causal discovery—the structure of convincing explanations involves the components in Fig. 9.2 for a wide range of causal models (Table 9.1); and making this structure clear can provide a useful and credible explanation for CAI-based decision recommendations, even if details of policy optimization heuristics are not discussed. Future systems that incorporate causal explanations together with CAI methods may help to increase the trustworthiness and acceptability of CAI in practical applications to proposing improved decisions and policies under uncertain and changing conditions.

We have discussed the structures and types of information and argument needed to give convincing causal explanations for recommended decisions or policies. The essence of such an explanation is a counterfactual comparison between the changes in outcome probabilities that the recommended decisions or policies are expected to cause, on the one hand, and those that alternative decisions would cause, on the other. This is precisely the type of comparison that optimization and RL heuristics use to make their recommendations. Hence, the same CAI methods can be used both to identify *what* decisions to recommend, as well as to explain *why* they are recommended. Emphasizing causally explainable decision recommendations improves their quality as well as their credibility. This is because, in contrast to simplification-based XAI for predictive analytics, CAI for prescriptive analytics emphasizes production of evidence that following the recommendations makes desired outcomes more probable. We propose that demonstrable effectiveness in trusted causal simulation models, rather than simplicity, is the essence of convincing explanations for decision recommendations from CAI-assisted prescriptive analytics.

References

Achlioptas D, Iliopoulos F, Vlassis N (2017). Stochastic control via entropy compression. ICALP 83:1–83:13. https://drops.dagstuhl.de/opus/volltexte/2017/7427/pdf/LIPIcs-ICALP-2017-83.pdf

Ahmadi M, Ono M, Ingham MD, Murray RM, Ames AD (2020). Risk-averse planning under uncertainty. In: 2020 American Control Conference (ACC), Denver, CO, USA, pp 3305–3312. https://doi.org/10.23919/ACC45564.2020.9147792

Ahmadi M, Rosolia U, Ingham MD, Murray RM, Ames AD (2021) Constrained risk-averse Markov decision processes. https://www.aaai.org/AAAI21Papers/AAAI-6211.AhmadiM.pdf; https://arxiv.org/abs/2012.02423

Apley DW, Zhu J (2020) Visualizing the effects of predictor variables in black box supervised learning models. J R Stat Soc Ser B 82(4):869–1164. https://arxiv.org/pdf/1612.08468.pdf

Bagnell JA, Ng AY, Schneider JG (2001) Solving uncertain Markov decision processes. Technical report. Carnegie Mellon University

Bai A, Wu F, Zhang Z, Chen X (2014) Thompson sampling based Monte-Carlo planning in POMDPs. In: Proceedings of the Twenty-Fourth International Conference on International Conference on Automated Planning and Scheduling (ICAPS'14). AAAI Press, pp 29–37

Bareinboim E, Pearl J (2016 Jul 5) Causal inference and the data-fusion problem. Proc Natl Acad Sci U S A 113(27):7345–7352. https://doi.org/10.1073/pnas.1510507113

Bauerle N, Ott J (2011) Markov decision processes with average-value-at-risk criteria. Math Methods Oper Res 74(3):361–379

Bellman R (1957) A Markovian decision process. J Math Mech 6(5):679–684

Blondel VD, Tsitsiklis JN (2000) A survey of computational complexity results in systems and control. Automatica 36(9):1249–1274

Box GEP (1957) Evolutionary operation: a method for increasing industrial productivity. J R Stat Soc Ser C (Appl Stat) 6(2):81–101. https://doi.org/10.2307/2985505

Carpin S, Chow Y, Pavone M (2016) Risk aversion in finite Markov decision processes using total cost criteria and average value at risk. In: 2016 IEEE International Conference on Robotics and Automation (ICRA), Stockholm, Sweden, pp 335–342. https://doi.org/10.1109/ICRA.2016.7487152

Chatterjee K, Martin C, Tracol M (2016) What is decidable about partially observable Markov decision processes with ω-regular objectives. J Comput Syst Sci 82(5):878–911. https://doi.org/10.1016/j.jcss.2016.02.009

Cheung WC, Simchi-Levi D, Zhu R (2019, May 23) Non-stationary reinforcement learning: the blessing of (more) optimism. Available at SSRN https://ssrn.com/abstract=3397818 or https://doi.org/10.2139/ssrn.3397818

Chowdhury SR, Oliveira R (2021) No-regret reinforcement learning with value function approximation: a kernel embedding approach. https://arxiv.org/pdf/2011.07881.pdf

Cox LA Jr, Popken DA, Sun X (2018) Causal analytics for applied risk analysis. Springer, New York

Crowley M (2004). Evaluating influence diagrams. www.cs.ubc.ca/~crowley/papers/aiproj.pdf

de Pontes Pereira R, Engel PM (2015) A framework for constrained and adaptive behavior-based agents. CoRR, abs/1506.02312. https://dblp.uni-trier.de/db/journals/corr/corr1506.html#PereiraE15

Dhar P (2020, October 29) Understanding causality is the next challenge for machine learning. IEEE Spectrum. https://spectrum.ieee.org/tech-talk/artificial-intelligence/machine-learning/understanding-causality-is-the-next-challenge-for-machine-learning

Du S, Hu W, Li Z, Shen R, Song Z, Wu J (2020) When is particle filtering efficient for POMDP sequential planning? https://arxiv.org/abs/2006.05975

Eskandari H, Mahmoodi E, Fallah H, Geiger CD (2011) Performance analysis of commercial simulation-based optimization packages: OptQuest and Witness optimizer. In: Jain S, Creasey RR, Himmelspach J, White KP, Fu M (eds) Proceedings of the 2011 Winter Simulation Conference. https://www.informs-sim.org/wsc11papers/212.pdf

Friedman J, Hastie T, Tibshirani R (2009) The elements of statistical learning. www.web.stanford.edu/~hastie/ElemStatLearn/

Gershman SJ (2017) Reinforcement learning and causal models. In: Waldmann MR (ed) The Oxford handbook of causal reasoning. Oxford University Press, Oxford. https://doi.org/10.1093/oxfordhb/9780199399550.013.20

Goldstein A, Kapelner A, Bleich J, Pitkin E (2015) Peeking inside the black box: visualizing statistical learning with plots of individual conditional expectation. J Comput Graph Stat 24(1):44–65. https://doi.org/10.1080/10618600.2014.907095

Haarnoja T, Zhou A, Abbeel P, Levine S (2018) Soft actor-critic: off-policy maximum entropy deep reinforcement learning with a stochastic actor. In: Proceedings of the 35th International Conference on Machine Learning, in Proceedings of Machine Learning Research 80:1861–1870. Available from http://proceedings.mlr.press/v80/haarnoja18b.html

Heinze-Deml C, Meinshausen N (2020) Package CompareCausalNetworks. https://cran.r-project.org/web/packages/CompareCausalNetworks/CompareCausalNetworks.pdf
Heinze-Deml C, Peters J, Meinshausen N (2017) Invariant causal prediction for nonlinear models. https://arxiv.org/pdf/1706.08576.pdf
Heinze-Deml C, Maathuis MH, Meinshausen N (2018) Causal structure learning. Annu Rev Stat Appl 5:371–391. https://www.annualreviews.org/doi/10.1146/annurev-statistics-031017-100630. (Full text is at https://arxiv.org/pdf/1706.09141.pdf)
Howard RA (1960) Dynamic programming and Markov processes (PDF). The MIT Press
Howard RA (1988). Decision analysis: practice and promise. Manag Sci 34(6):679–695. www.jstor.org/stable/2632123
Howard RA, Matheson JE (1981) Influence diagrams. In: Howard RA, Matheson JE (eds) Readings on the principles and applications of decision analysis, vol II. Strategic Decisions Group, Menlo Park, CA
Jin C, Liu Q, Miryoosefi S (2021) Bellman eluder dimension: new rich classes of RL problems, and sample-efficient algorithms. https://arxiv.org/pdf/2102.00815.pdf
Juan AA, Faulin J, Grasman SE, Rabe M, Figueira G. (2015) A review of simheuristics: extending metaheuristics to deal with stochastic combinatorial optimization problems, operations research perspectives, vol 2, pp 62–72, ISSN 2214-7160. https://doi.org/10.1016/j.orp.2015.03.001. https://www.sciencedirect.com/science/article/pii/S221471601500007X
Kahneman D (2011) Thinking, fast and slow. Farrar, Straus, and Giroux, New York
Kauffmann E, Korda N, Munos R (2012) Thompson sampling: an asymptotically optimal finite time analysis. In: Proceedings of the 24th International Conference on Algorithmic Learning Theory, pp 199–213
Keeney RL, Raiffa H. (1976) Decisions with multiple objectives: preferences and value tradeoffs. Wiley, republished by Cambridge University Press (1993), New York
Kim Y, Gur Y, Kochenderfer MJ (2017) Heuristics for planning with rare catastrophic events. In: Proceedings of the 2017 Winter Simulation Conference (WSC'17). IEEE Press, 247, pp 1–12
Le TP, Vien NA, Chung T (2018) A deep hierarchical reinforcement learning algorithm in partially observable Markov decision processes. IEEE Access 6:49089–49102. https://doi.org/10.1109/ACCESS.2018.2854283. https://arxiv.org/pdf/1805.04419.pdf
Leike J, Hutter M (2015) On the computability of AIXI. In: Proceedings of the Thirty-First Conference on Uncertainty in Artificial Intelligence (UAI'15). AUAI Press, Arlington, Virginia, USA, pp 464–473
Leike, LT, Orseau L, Hutter M (2016) Thompson sampling is asymptotically optimal in general environments. arxiv, abs/1602.07905
Li Y (2012) Reinforcement learning algorithms for semi-Markov decision processes with average reward. In: Proceedings of 2012 9th IEEE International Conference on Networking, Sensing and Control, Beijing, China, 2012, pp 157–162. https://doi.org/10.1109/ICNSC.2012.6204909
Lo LY, Wong ML, Lee KH, Leung KS (2015 Nov 25) High-order dynamic Bayesian network learning with hidden common causes for causal gene regulatory network. BMC Bioinformatics 16:395. https://doi.org/10.1186/s12859-015-0823-6
Madani O, Hanks S, Condon A (2003) On the undecidability of probabilistic planning and related stochastic optimization problems. Artif Intell 147(1–2):5–34
Mauá DD (2016) Equivalences between maximum a posteriori inference in Bayesian networks and maximum expected utility computation in influence diagrams. Int J Approx Reason 68 (C):211–229
Mill JS (1843) A system of logic, vol 1. John W. Parker, London, p 455
Miller T (2019) Explanation in artificial intelligence: insights from the social sciences. Artif Intell 267:1–38, ISSN 0004-3702. https://doi.org/10.1016/j.artint.2018.07.007
Mittelstadt B, Russell C, Wachter S (2019) Explaining explanations in AI. In: Proceedings of the Conference on Fairness, Accountability, and Transparency (FAT*'19). Association for Computing Machinery, New York, pp 279–288. https://doi.org/10.1145/3287560.3287574

Moradi P, Shiri ME, Entezari N (2010) Automatic skill acquisition in reinforcement learning agents using connection bridge centrality. In: Kim T, Vasilakos T, Sakurai K, Xiao Y, Zhao G, Ślęzak D (eds) Communication and networking. FGCN 2010. Communications in Computer and Information Science, vol 120. Springer, Berlin. https://doi.org/10.1007/978-3-642-17604-3_6

Moradipari A, Alizadeh M, Thrampoulidis C (2020) Linear Thompson sampling under unknown linear constraints. In: ICASSP 2020 - 2020 IEEE International Conference on Acoustics, Speech and Signal Processing (ICASSP), Barcelona, Spain, pp 3392–3396. https://doi.org/10.1109/ICASSP40776.2020.9053865

Nagarajan R, Scutari M, Lèbre S (2013) Bayesian networks in R: with applications in systems biology. Springer, New York. https://www.bnlearn.com/book-useR/

Nair GN, Evans RJ (2004) Stabilizability of stochastic linear systems with finite feedback data rates. SIAM J Control Optim 43(2):413–436. https://doi.org/10.1137/S0363012902402116

Ortega PA, Braun DA (2014) Generalized Thompson sampling for sequential decision-making and causal inference. Complex Adapt Syst Model 2:2. https://doi.org/10.1186/2194-3206-2-2

Ouyang Y, Gagrani M, Nayyar A, Jain R (2017). Learning unknown Markov decision processes: a Thompson sampling approach. In: Proceedings of the 31st International Conference on Neural Information Processing Systems (NIPS'17). Curran Associates Inc., Red Hook, NY, USA, pp 1333–1342

Parascandolo G, Buesing L, Merel J, Hasenclever L, Aslanides J, Hamrick JB, Hees N, Neitz A, Weber T (2020) Divide-and-conquer Monte Carlo Tree Search for goal-directed planning. https://arxiv.org/pdf/2004.11410.pdf

Pearl J (2001) Direct and indirect effects. In: Proceedings of the Seventeenth Conference on Uncertainty in Artificial Intelligence. Morgan Kaufmann, San Francisco, CA, pp 411–420

Pearl J (2010) An introduction to causal inference. Int J Biostat. 6(2):7. https://doi.org/10.2202/1557-4679.1203. https://www.ncbi.nlm.nih.gov/pmc/articles/PMC2836213/

Peters J, Bühlmann P, Meinshausen N (2016) Causal inference by using invariant prediction: identification and confidence intervals. J R Stat Soc Ser B 78(5):947–1012

Piché A, Thomas V, Ibrahim C, Bengio Y, Pal C (2019) Probabilistic planning with sequential Monte Carlo methods. ICLR

Raiffa H (1968) Decision analysis: introductory lectures on choices under uncertainty. Addison-Wesley, Reading, MA

Raj V, Kalyani S (2017) Taming non-stationary bandits: a Bayesian approach

Ranade G, Sahai A (2019) Control capacity. IEEE Trans Inf Theory 65(1):235–254. https://doi.org/10.1109/TIT.2018.286

Russo DJ, Van Roy B (2018) Learning to optimize via information-directed sampling. Oper Res 66(1):230–252. https://pubsonline.informs.org/doi/pdf/10.1287/opre.2017.1663

Russo DJ, Van Roy B, Kazerouni A, Osband I, Wen Z (2018) A tutorial on Thompson sampling. Found Trends Mach Learn 11(1):1–96. https://doi.org/10.1561/2200000070

Schervish M, Seidenfeld T, Kadane J, Levi I (2003) Extensions of expected utility theory and some limitations of pairwise comparisons. ISIPTA. https://www.cmu.edu/dietrich/philosophy/docs/seidenfeld/Extensions%20of%20Expected%20Utility%20and%20Pairwise%20Comparisons.pdf

Sgaier SK, Huang V, Summer GC (2020) The case for causal AI. Stanford Social Innovation Review. Summer. https://ssir.org/pdf/Summer2020-Feature-Sgaier-Causal-AI-Updated.pdf

Shachter RD (1988) Probabilistic inference and influence diagrams. Oper Res 36(4):589–604. https://doi.org/10.1287/opre.36.4.589

Shen W, Trevizan F, Toyer S, Thiebaux S, Xie L (2019) Guiding search with generalized policies for probabilistic planning. In: Proceedings of the Twelfth International Symposium on Combinatorial Search (SoCS 2019)

Shimizu S, Bollen K (2014) Bayesian estimation of causal direction in acyclic structural equation models with individual-specific confounder variables and non-gaussian distributions. J Mach Learn Res 15(1):2629–2652

Shu T, Xiong C, Socher R (2017) Hierarchical and interpretable skill acquisition in multi-task reinforcement learning. https://arxiv.org/abs/1712.07294

Sinha S, Vaidya U (2017) On information transfer in discrete dynamical systems. In: 2017 Indian Control Conference (ICC), Guwahati, India, 2017, pp 303–308. https://doi.org/10.1109/INDIANCC.2017.7846492

Sledge IJ, Príncipe JC (2018) An analysis of the value of information when exploring stochastic, discrete multi-armed bandits. Entropy 20(3):155. https://doi.org/10.3390/e20030155. https://www.mdpi.com/1099-4300/20/3/155#cite

Spencer SEF, Hill SM, Mukherjee S (2015) Inferring network structure from interventional time-course experiments. Ann Appl Stat 9(1):507–524. https://doi.org/10.1214/15-AOAS806

Sui Y, Gotovos A, Burdick JW, Krause A (2015) Safe exploration for optimization with Gaussian processes. In: International conference on machine learning (ICML). Lille, France

Textor J, van der Zander B, Gilthorpe MS, Liskiewicz M, Ellison GT (2016) Robust causal inference using directed acyclic graphs: the R package 'dagitty'. Int J Epidemiol 45(6): 1887–1894

Tiomkin S, Polani D, Tishby N (2017) Control capacity of partially observable dynamic systems in continuous time. arXiv preprint arXiv:1701.04984

Touchette H, Lloyd S (2000) Information-theoretic limits of control. Phys Rev Lett 84(6): 1156–1159. https://link.aps.org/doi/10.1103/PhysRevLett.84.1156

Toussaint M, Charlin L, Poupart P (2012) Hierarchical POMDP controller optimization by likelihood maximization. https://arxiv.org/ftp/arxiv/papers/1206/1206.3291.pdf

Triantafillou S, Lagani V, Heinze-Deml C et al (2017) Predicting causal relationships from biological data: applying automated causal discovery on mass cytometry data of human immune cells. Sci Rep 7:12724. https://doi.org/10.1038/s41598-017-08582-x

Urpi NA, Curi S, Krause A (2021) Risk-averse offline reinforcement learning. ArXiv. https://arxiv.org/abs/2102.05371

VanderWeele TJ (2011) Controlled direct and mediated effects: definition, identification and bounds. Scand Stat Theory Appl 38(3):551–563. https://doi.org/10.1111/j.1467-9469.2010.00722.x

Vineis P, Illari P, Russo F (2017) Causality in cancer research: a journey through models in molecular epidemiology and their philosophical interpretation. Emerg Themes Epidemiol 14: 7. https://doi.org/10.1186/s12982-017-0061-7

Wieczorek A, Roth V (2019) Information theoretic causal effect quantification. Entropy 21(10): 975. https://doi.org/10.3390/e21100975

Wiener N (1956) The theory of prediction. In: Beckenbach EF (ed) Modern mathematics for engineers, vol 1. McGraw Hill, New York

Wu CM, Schulz E, Speekenbrink M, Nelson JD, Meder B (2017) Mapping the unknown: the spatially correlated multi-armed bandit. In: Proceedings of the 39th annual conference of the cognitive science society

Xiao C, Huang R, Mei J, Schuurmans D, Müller M. (2019) Maximum entropy Monte-Carlo planning. NeurIPS. https://proceedings.neurips.cc/paper/2019/file/7ffb4e0ece07869880d51662a2234143-Paper.pdf

Yamayoshi M, Tsuchida J, Yadohisa H (2020) An estimation of causal structure based on latent LiNGAM for mixed data. Behaviormetrika 47:105–121. https://doi.org/10.1007/s41237-019-00095-3

Yuan C, Lim H, Lu TC (2011) Most relevant explanation in Bayesian networks. J Artif Intell Res 42:309–352

Zhao Q, Hastie T (2021) Causal interpretations of black-box models. J Bus Econ Stat 39(1): 272–281. https://doi.org/10.1080/07350015.2019.1624293

Zhu X, Yuan C (2017) Hierarchical beam search for solving most relevant explanation in Bayesian networks. J Appl Log 22:3–13

Part IV
Public Health Applications

Chapter 10
Re-Assessing Human Mortality Risks Attributed to Agricultural Air Pollution: Insights from Causal Artificial Intelligence

Introduction

This chapter presents a perspective that recent advances in causal artificial intelligence (CAI) and machine learning (ML) make it timely and valuable to undertake a general reexamination of the arguments for causation and predictions of impacts of exposure changes in epidemiology. To illustrate the relevance and potential impact of such a reexamination, consider the following motivating example. Is there convincing scientific evidence that agricultural air pollution from the production of meat harms the health of vulnerable human populations? Several recent papers have argued that it does. They cite concerning-looking estimates of adverse health impacts such as "We estimate the air quality-related health impacts of agriculture in the United States, finding that 80% of the 15,900 annual deaths that result from food-related fine particulate matter (PM2.5) pollution are attributable to animal-based foods. . . . We find that US agriculture results in 17,900 deaths (range across models: 15,600 to 20,300) per year via reduced air quality. . . . Damages are driven by NH3 emissions (. . . 12,400 deaths; 69% of total) mainly from livestock waste and fertilizer application. . . . Primary PM2.5 is also a major contributor (4800 deaths, 27% of total), largely from dust from tillage, livestock dust, field burning, and fuel combustion in agricultural equipment use." Policy recommendations for interventions and predicted health benefits from making them follow: "Dietary shifts toward more plant-based foods that maintain protein intake and other nutritional needs could reduce agricultural air quality–related mortality by 68 to 83%. In sum, improved livestock and fertilization practices, and dietary shifts could greatly decrease the health impacts of agriculture caused by its contribution to reduced air quality." A key question for both methodologists and practitioners is: how sound and credible are such causal attributions and predictions? Are their methodological foundations solid? Are their conclusions dependable? Should they be considered convincing evidence that should inform risk management policy decisions, or should they be

L. A. Cox Jr., *AI-ML for Decision and Risk Analysis*, International Series in Operations Research & Management Science 345,
https://doi.org/10.1007/978-3-031-32013-2_10

dismissed as products of dubious assumptions and methods that should leave policy makers unpersuaded, or somewhere in between? More generally, how can and should epidemiologists and risk assessors assemble and present evidence for causation of mortality or morbidities by identified agents, such as NH3 or PM2.5? When, if ever, is it sound to use measured differences in estimated exposures and corresponding mortality rates to predict the impacts on mortality of interventions that reduce exposures?

The following sections examine the logic and conceptual foundations of such claims and proposed solutions. They draw on recent concepts and methods from causal artificial intelligence (CAI) and machine learning (ML) to help clarify the use and misuse of causal inferences and interpretations, attributions, and predictions. The study of Domingo et al. (2021) and its supporting literature are used to explore and illustrate these methodological points because we believe it is a relatively well-done study within the current paradigm of epidemiological risk analysis made without benefit of CAI methods. The authors carefully link recommended actions to predicted consequences using previously published data and statistical models and methods that have been extensively peer-reviewed and published in top-quality scientific journals. However, the CAI perspective presented in the following sections suggests that this body of epidemiological concepts and analytics rests on unsound conceptual foundations and practices—especially, the too-common practice of predicting the effects of interventions and drawing quantitative causal conclusions in the absence of causal analysis—that can yield untrustworthy causal conclusions. We propose that recent CAI methods can alleviate many of these limitations. They can provide analyses and predictions that are more useful for effective risk management decision-making than traditional association-based epidemiological and burden-of-disease estimates of air quality-related health impacts. This is because because they are designed to better estimate the effects of interventions.

The point of view developed here is that much current epidemiological practice confuses patterns of statistical associations between exposure and adverse health effects with evidence of the ability of reductions in exposure to cause reductions in the adverse effects. Applying insights from the CAI literature, we offer a general critique of the notion that a degree of association based on observing different outcomes for different levels of exposure measures or predicts the change in outcomes to be expected from a given future change in exposure. Our focus is on the general question of how to establish evidence of such interventional causation in epidemiology. We offer a critique of how it is currently assessed (e.g., using weight-of-evidence considerations) and suggest how it could be more rigorously assessed with CAI methods that exist but are perhaps not yet sufficiently widely known and understood by many risk assessors and epidemiologists. Examples of potential applications to agricultural ammonia, and to PM2.5 and mortality more generally, are used to illustrate the relevance of these general questions and the utility of having a better approach to them, without attempting a full reanalysis of these important applications. We believe that the main novelty and potential practical value of the CAI perspective and methods ideas is to provide constructive data-based methods for testing implications of causal hypotheses while helping to identify and control for sources of non-causal associations. The following sections use examples to illustrate the main ideas of these methods for a broad audience of epidemiologists and risk analysts interested in predicting causal impacts of interventions, leaving most of the detailed technical development to the specialized CAI works cited.

Limitations of Current Qualitative Hazard Identification for PM2.5-Mediated Health Effects of NH3 Emissions

The basic causal model postulated by Domingo et al. is that farm emissions of ammonia (NH3, largely from fertilizer and animal feeding operations) increase ambient PM2.5 concentrations which in turn increase human mortality rates. This understanding is widely accepted by investigators in multiple countries (Giannakis et al. 2019; Gu et al. 2021; Hristov 2011; Insausti et al. 2020). It provides a basis for proposals to restrict farm emissions of ammonia as a cost-effective approach to reducing annual mortalities (Gu et al. 2021; Pinder et al. 2007), especially as the ammonium ion (NH4+) in PM2.5 has been identified as being relatively strongly correlated with estimated relative risks of mortality (Masselot et al. 2022).

A substantial evidence base has been cited in support of this causal model. How emissions of NH3 contribute to PM2.5 has been elucidated by atmospheric chemists (Hristov 2011). That increases in ambient PM2.5 concentration cause increases in human mortality rates has been supported by citing hundreds of epidemiological studies in multiple countries over more than three decades associating higher ambient PM2.5 concentrations with higher mortality rates, especially from cardiovascular disease (e.g., Pope et al. 2002; Clancy et al. 2002). This is widely considered one of the best-established findings of air pollution health effects research (Burnett et al. 2018) although estimated associations between PM2.5 and mortality differ widely across studies both within and between countries (Masselot et al. 2022).

However, an outside perspective from the field of causal artificial intelligence (CAI) suggests that these numerous and well-documented PM2.5-mortality associations do not address the key question of how changing ambient PM2.5 levels would change mortality rates (Pearl 2009; Carone et al. 2020). The common inference that a strong statistical exposure-mortality association implies, or at least strongly suggests, that reducing exposure would reduce mortality (e.g., Pope et al. 2002; Clancy et al. 2002; Burnett et al. 2018) is mistaken both on methodological grounds (Pearl 2009; Carone et al. 2020) and in practice (Dominici et al. 2014; Zigler and Dominici 2014). The corollary that reducing farm emissions of ammonia would reduce annual mortalities by reducing PM2.5 is not sound. The following sections explain these points.

Burden of Disease Calculations

Health risk assessment usually begins with *hazard identification*: identifying exposures that can cause adverse health effects. Domingo et al. identify air pollution (more specifically, fine particulate matter (PM2.5) air pollution) as a major public health hazard, citing previous work by others. For example, they state that "air pollution is the largest environmental mortality risk factor in the United States and globally," citing Lelieveld et al. (2015) and GBD 2017 Risk Factor Collaborators (2018) for support. It is instructive to examine these references and their methodologies and conclusions.

Both Lelieveld et al. (2015) and GBD 2017 Risk Factor Collaborators (2018) provide Global Burden of Disease (GBD) estimates of mortalities attributed to fine particulate matter (PM2.5). Lelieveld et al. (2015) state that

> It has proved difficult to quantify premature mortality related to air pollution, notably in regions where air quality is not monitored, and also because the toxicity of particles from various sources may vary. ...In accord with the global burden of disease [GBD] for 2010, we calculate that outdoor air pollution, mostly by PM2.5, leads to 3.3 (95 per cent confidence interval 1.61-4.81) million premature deaths per year worldwide, predominantly in Asia.... We find that emissions from residential energy use such as heating and cooking, prevalent in India and China, have the largest impact on premature mortality globally... in eastern USA, Europe, Russia and East Asia agricultural emissions make the largest relative contribution to PM2.5, with the estimate of overall health impact depending on assumptions regarding particle toxicity.

This study emphasizes the dependence of risk estimates on *assumptions* about particulate toxicity in the GBD calculations, as well as the fact that the toxicity of particles from different sources may vary. It also notes that agricultural emissions contribute differently to PM2.5 exposures in different places.

The GBD 2017 Risk Factor Collaborators (2018) study explains the GBD methodology and conclusions in greater detail. They present estimates of

> attributable deaths... for 84 behavioural, environmental and occupational, and metabolic risks or groups of risks from 1990 to 2017. This study included 476 risk–outcome pairs that met the GBD study criteria for convincing or probable evidence of causation. ... Four components were used for the calculations to estimate the attributable burden for a given risk–outcome pair: (1) the estimate of the burden metric being assessed for the cause (ie, number of deaths, years of life lost... (2) the exposure levels for the risk factor; (3) the counterfactual level of risk factor exposure or theoretical minimum risk exposure level (TMREL); and (4) the relative risk [RR] of the outcome relative to the TMREL. For a given risk–outcome pair, we estimated attributable DALYs as total DALYs for the outcome multiplied by the population attributable fraction (PAF) for the risk–outcome pair for each age, sex, location, and year. The same logic applies to estimating attributable deaths, YLLs, and YLDs. The PAF is the proportion by which the outcome would be reduced in a given population and in a given year if the exposure to a risk factor in the past were reduced to the counterfactual level of the TMREL.

Part of this GBD study specifically addresses mortality risks attributed to ambient particulate matter pollution. Appendix Table 5 of its supplementary materials (p. 227) cites 8 references in addition to another GBD study for adverse health effects statistically associated with ambient particulate matter pollution. The following three references include "mortality" in their titles:

- Beelen et al. (2014) concluded from an analysis of 22 European cohort studies on long-term exposure to air pollution and cardiovascular mortality that "All hazard ratios were approximately 1.0, except for particle mass and cerebrovascular disease mortality; for PM2.5, the hazard ratio was 1.21 (95% confidence interval = 0.87–1.69) per 5 μg/m^3." This study does not provide evidence of a significant association between PM2.5 and mortality.
- Hoek et al. (2013) performed a review and meta-analysis of earlier studies of "long-term air pollution exposure and cardio-respiratory mortality" and reported that "The pooled effect estimate expressed as excess risk per 10 μg/m3 increase in PM2.5 exposure was 6% (95% CI 4, 8%) for all-cause and 11% (95% CI 5, 16%) for cardiovascular mortality. ... Significant heterogeneity in PM2.5 effect estimates was found across studies, likely related to differences in particle composition, infiltration of particles indoors, population characteristics and methodological differences in exposure assessment and confounder control. ... Large cohort studies have used neighborhood socio-economic status and co-morbidities strongly associated with smoking as proxies for actual smoking data."
- Huang et al. (2017) found in a meta-analysis of 17 PM2.5-lung cancer mortality associations that "the meta-estimate for lung cancer risk associated with PM2.5 was 1.11 for mortality (95% CI: 1.05, 1.18) and ... was greatest in North America [1.15 (95% CI: 1.07, 1.24)], followed by Asia [1.12 (95% CI: 0.94, 1.35)], and then Europe [1.05 (95% CI: 1.01, 1.10)]. ... These results are limited by a lack of detailed information on patterns of former smoking."

Thus, although the GBD 2017 Risk Factor Collaborators study provides an overall estimate of mortality risks attributed to ambient particulate matter pollution, some of the underlying studies found no evidence of a significant PM2.5-mortality association (Beelen et al. 2014). Others could not fully control for confounding from smoking history and other variables (Hoek et al. 2013; Huang et al. 2017). These limitations of the underlying study results are not apparent from the GBD final numbers cited by Domingo et al. (2021). Summary statements such as "air pollution is the largest environmental mortality risk factor in the United States and globally" (Domingo et al. 2021) simplify and amplify the conclusions from these supporting references. A more nuanced summary would be that full control for smoking history and other confounders and explanations for heterogeneity in estimated effects of air pollution on mortality risks remain unresolved challenges for interpreting air pollution-mortality associations.

Deaths Attributed to Air Pollution Do Not Refute Non-Causal Explanations

Attributions of annual premature deaths using the GBD methodology attribute differences between RRs in groups with different exposures to the differences in their exposures. This is an implicit *assumption*, not an empirical finding from the data. In common practice, refuting or adjusting for alternative explanations is not treated as an essential part of the Bradford-Hill and related weight-of-evidence (WoE) considerations used to make qualitative judgments about whether associations should be deemed "causal" (a term left undefined by Bradford Hill). But refutation of rival explanations has been emphasized in other frameworks for causal inferences in epidemiology (Maclure 1991) and other fields that rely on observational data (Campbell and Stanley 1963). Common non-causal explanations for exposure-response associations include confounding, residual confounding, model specification errors, modeling assumptions and biases, multiple testing bias, exposure estimation errors, latent variables, and coincident secular trends in exposures and RRs (Cox 2020). Methodologists have recently noted that "Cohort studies and case-control studies should increase the number of controlled confounders to deliver better risk estimates" for GBD calculations in the presence of joint causation by multiple risk factors (Plass et al. 2022).

Attributed Risks Do Not Predict Effects of Interventions

In the GBD framework, mortalities attributed to air pollution are calculated from exposures, relative risks (RRs) and population attributable fractions (PAFs); PAFs, in turn, are calculated from RRs. Neither PAFs nor RRs address the fraction of disease that could be *prevented* or reduced by interventions that reduce exposure, although they are often misinterpreted as doing so (Cox 2021). For example, the GBD 2017 Risk Factor Collaborators (2018) state that "The PAF is the proportion by which the outcome would be reduced in a given population and in a given year if the exposure to a risk factor in the past were reduced to the counterfactual level of the TMREL." This causal interpretation is incorrect because PAF ratios are based solely on observed *differences* in RRs between groups with different exposures, not on the *changes* in RRs within groups that would occur in response to changes in exposures, other direct causes of the outcome being held fixed. These are not equivalent concepts (Pearl 2009), as vividly illustrated by studies in which intervening to change exposures and reduce risks based on RR and PAF information has unintentionally increased mortality risks (Goodman et al. 2004).

More generally, adverse health consequences attributed to specific exposures and hazards by association-based methods (e.g., RR, PAF, etiologic fraction, probability of causation, regression coefficients, population attributable risk, and GBD calculations) are not suitable for predicting effects of interventions that reduce those exposures (Pearl 2009). *Risks attributed to an exposure do not predict effects on risk of interventions that reduce the exposure.* This has long been understood in causal artificial intelligence (CAI). For example, Pearl (2009) emphasized the "basic distinction" between (a) statistical analyses of data showing the observed levels of variables (termed associational analyses); and (b) causal analyses capable of predicting the effects of interventions: "Causal analysis goes one step further; its aim is to infer not only beliefs or probabilities under static conditions, but also the dynamics of beliefs under changing conditions, for example, changes induced by treatments or external interventions. This distinction implies that causal and associational concepts do not mix. . . . [T]here is nothing in a distribution function to tell us how that distribution would differ if external conditions were to change." Likewise, machine learning methods for causal analysis of observational data (Athey and Wager 2019; Wang et al. 2022) take as a point of departure this basic distinction between methods for predicting *probabilities of observed outcomes* (dependent variables) and methods for predicting how interventions would *change* outcome probabilities. They provide separate methods for *predicting outcomes* associated with exposures vs. *predicting changes in outcomes* associated with changes in exposures (Athey and Wager 2019). This basic distinction (Pearl 2009) is also drawn in data analysis methods for calculating "uplift curves" for causal impacts of personalized marketing, healthcare, employee analytics, and education (Wang et al. 2022). Unfortunately, it has not yet been assimilated into most applied epidemiology and public health risk assessment studies that routinely treat estimated concentration-response slope coefficients as if they described how changing exposure concentrations would change rates of adverse health responses (Cox 2017).

Attributed-Mortality Calculations Do Not Reveal Effects on Mortality of Reducing Air Pollution

The fundamental inability of association-based analyses—even in the ideal case of perfect knowledge of the full joint distributions of all variables—to predict how exogenous interventions would change outcomes or their probability distributions can be explained abstractly by the fact that associations do not capture mechanistic knowledge (Cox 2021b). It can be illustrated concretely through examples such as Table 10.1.

Table 10.1 shows hypothetical exposure concentrations (C), income levels (I), and health effect response rates (R) for three communities A, B, and C. It is logically

Table 10.1 Data do not reveal how cutting concentration C in half would change response rate R

Community	Concentration, C	Income, I	Response rate, R
A	4	100	8
B	8	60	16
C	12	20	24

impossible to predict from such data whether or by how much reducing C would reduce R, no matter how extensive the data may be, e.g., even if it were available for millions of individuals instead of for 3 communities. If the causal (structural equation) relationship between C and R is $R = 2C$, then halving C would halve R. RR-based calculations such as GBD implicitly make this assumption. If instead the causal relationship is $R = 35 - 0.5C - 0.25I$, then reducing C would *increase* R. If it is $R = 28 - 0.2I$, then reducing C would have *no effect* on R. There is no way to tell from the data alone which of these causal models, if any, describes how changing C would change R.

Many influential air pollution health effects research articles interpret regression models causally as if they were structural equations that predict how changes in independent variables on the right side of a regression model would change the dependent variable on its left side (Cox 2017). Such predictions depend not only on data but also on implicit causal assumptions. For example, in Table 10.1, whether a regression model for R predicts that reducing C would increase R, decrease R, or leave R unchanged is determined by whether the modeler includes only C, both C and I, or only I as independent variables in the regression model. Weight-of-evidence considerations such as strength and consistency of associations across studies do not resolve such ambiguities.

For simplicity and clarity, Table 10.1 illustrates the fundamental disconnect between associational data and causal conclusions with an admittedly simplistic, extreme example (with perfectly collinear variables). Yet, this disconnect is also important in practice. For example, Dominici et al. (2014) note that results that similarly depend on modeling choices and assumptions often occur in real-world air pollution health effects research. They present an example for total suspended particulates (TSP) and mortality rates in the United States using 12 regression models fit to the same exposure concentration-mortality data set. They find that "Of the 12 regression estimates, two show a significant positive association between TSPs and mortality, two perversely show a significant negative association, and the remaining eight would be judged statistically insignificant at conventional levels." Thus, in practice as well as in principle, the signs and statistical significance of published regression coefficients for predicting observed mortality rates from exposure concentrations depend on modelers' choices about which variables to include in a model. Dominici et al. note that

> In the last 40 years, the evidence that has led to revisions of the U.S. National Ambient Air Quality Standards has come mainly from observational studies aimed at estimating an exposure-response relation. But associational approaches to inferring causal relations can be highly sensitive to the statistical model and covariates used to adjust for confounding. Indeed, the U.S. government itself has drawn attention to the 'uncertainty in the reduction of premature deaths associated with reduction in particulate matter.' There is a growing consensus in economics, political science, statistics, and other fields that the associational or regression approach to inferring causal relations— on the basis of adjustment with observable confounders—is unreliable in many settings.

Such crucial caveats are typically lost in published results interpreting positive associations (e.g., positive regression coefficients for exposure concentration as a predictor of mortality) as evidence that reducing exposure would reduce mortalities. That the kinds of data used to compute RRs and PAFs (e.g., levels of C and R in Table 10.1) do not suffice to determine how changing C would change R is not recognized in applied epidemiology and public health literature that conflates mortalities *attributed* to exposure with mortalities *preventable* or reducible by reducing exposure. This is illustrated by the common misconception that "The PAF is the proportion by which the outcome would be reduced in a given population and in a given year if the exposure to a risk factor in the past were reduced to the counterfactual level" (GBD 2017 Risk Factor Collaborators 2018). This interpretation explicitly but mistakenly asserts that PAFs (which are calculated from RRs estimated from associational data such as that in Table 10.1) can be used to predict how reducing C would reduce R. This is done by *assuming* that exposure concentration determines response. This is like assuming that R = 2C is the correct causal model for the example in Table 10.1. The assumption begs the questions of how to determine valid causal models and how to make valid causal predictions from data.

The causal conclusions from attribution studies usually remain untested. Technically, *attribution claims per se do not make testable (or falsifiable) predictions.* An attribution claim of the form "Y deaths per year are attributed to exposure X" does not imply that eliminating exposure X would cause there to be Y fewer deaths per year. Nor does it make other testable predictions (Cox 2021). The basic distinction between association and causation drawn in CAI implies that attributing a positive number of deaths per year to exposure does not necessarily imply that reducing exposure would reduce deaths per year. Such an inference cannot be justified by associational data or by RR, PAF, GBD, and exposure concentration-response regression calculations based on associational data (Pearl 2009). Yet the misconception that deaths attributed to an exposure based on associational data could be avoided by reducing exposure is widespread, persistent, and influential. For example, Domingo et al. write that "Agricultural production in the United States *results in* 17,900 annual air quality–related deaths, 15,900 of which are from food production. Of those, 80% are *attributable to* animal-based foods, both directly from animal production and indirectly from growing animal feed. On-farm interventions *can reduce* PM2.5-related mortality by 50%, including improved livestock waste management and fertilizer application practices that reduce emissions of ammonia" (emphases added). These numbers are all based on attribution calculations using associational data. This implies that they do not provide a sound logical basis for predicting whether or by how much on-farm interventions that reduce exposures to ammonia and PM2.5 would reduce PM2.5-attributed mortalities.

Such conflation of associational results (including attributable mortality calculations) with causal results is routine in the vast, decades-old literature on air pollution health effects, as well as in many other areas of applied epidemiology (Pearl 2009; Cox 2017; Dominici et al. 2014). Explicitly causal statements such as "Reduction in particle concentrations below U.S. EPA standards would increase life expectancy" (Schwartz et al. 2008) or "These results suggest that further public policy efforts that reduce fine particulate matter air pollution are likely to have continuing public health benefits" (Lepeule et al. 2012) are often published in the air pollution health-effects literature even when the underlying analyses are only curve-fitting exercises (e.g., fitting Cox proportional hazards models to historical concentration-response data) with no analysis of cause and effect. This informal causal interpretation is widespread and influential. More than half of the total estimated regulatory benefits in the United States are now attributed to reductions in PM2.5 under the Clean Air Act (Dominici et al. 2014) based on *assuming* a causal relationship between PM2.5 exposures and premature mortality. The assumption is based primarily on epidemiological evidence that alone cannot establish this causal link (EPA 2011, Table 6–11).

Intervention Studies Do Not Support Attributed Risks Based on Associations

One way to progress beyond using judgment to interpret correlation as causation is to examine observed changes in exposures and mortality rates over time in intervention studies, natural experiments, or other quasi-experiments that compare observed mortality rates in populations for which exposure was changed to comparable populations in which it was not (Dominici et al. 2014). Burns et al. (2020) reviewed over three dozen intervention studies of the effects of changes in ambient air pollution. Ten interventions were assessed for effects on all-cause mortality, six for effects on cardiovascular mortality, and six for effects on respiratory mortality. They concluded that

> Given the heterogeneity across interventions, outcomes, and methods, it was difficult to derive overall conclusions regarding the effectiveness of interventions in terms of improved air quality or health. Some evidence suggests that interventions are associated with improvements in air quality and human health, with very little evidence suggesting interventions were harmful. The evidence base highlights the challenges related to establishing the effectiveness of specific air pollution interventions on outcomes.

This review did not find clear evidence that interventions to reduce air pollution exposures caused subsequent reductions in mortality rates in exposed populations. Most studies found no clear associations in either direction between interventions that reduced air pollution and changes in mortality risks, despite predictions from causal interpretations of associations that large reductions in ambient air pollution should be followed by significant reductions in cardiovascular and all-cause mortality rates (Cox and Popken 2015). For example, an influential intervention study cited by both Burns et al. (2020) and Dominici et al. (2014) compared mortality rates before and after bans on burning of smoky coal in Dublin County, Ireland. It found significant, large reductions in particulate air pollution and in both all-cause and cardiovascular mortality (Clancy et al. 2002). The authors' own interpretation was that "The results could not be more clear, reducing particulate air pollution reduces the number of respiratory and cardiovascular related deaths immediately" (Harvard School of Public Health 2002) and "Reductions in respiratory and cardiovascular death rates in Dublin suggest that control of particulate air pollution could substantially diminish daily death. . . .Our findings suggest that control of particulate air pollution in Dublin led to an immediate reduction in cardiovascular and respiratory deaths" (Clancy et al. 2002). However, European methodologists objected that the study design did not permit valid causal inference (Wittmaack 2007; Pelucchi et al. 2009). It did not follow the basic requirement for valid causal inference in quasi-experiments (i.e., studies in which interventions are not randomly assigned): that appropriate control groups must be used to test, and if possible refute, non-causal explanations for the observed changes (Campbell and Stanley 1963). A decade later, a new effort, including some of the original investigators, revisited the study using counties outside the area affected by the bans as a control group. They found that "In contrast to the earlier study, there appeared to be no reductions in total mortality or in mortality from other causes, including cardiovascular disease, that could be attributed to any of the bans. That is, after correcting for background trends, similar reductions were seen in ban and non-ban areas" (HEI 2013). In short, "long-term trends in cardiovascular health spanning implementation of the ban— not the coal ban itself—contributed to apparent effects on cardiovascular mortality" (Zigler and Dominici 2014). However, the misconception that the association between lower ambient air pollution and lower mortality rates shows that reducing ambient air pollution reduced annual death rates continues to be used to advocate for increased regulation in Ireland (e.g., O'Sullivan 2020).

Perhaps the most important methodological lesson from the Dublin intervention study is that *levels* of air pollution and adverse health effects can be strongly correlated even if *changes* in levels of air pollution are unrelated to changes in levels of adverse health effects in a population. This is in accord with the basic distinction drawn in CAI. A second lesson is that weight-of-evidence judgments that classify an observed association as causal or likely to be causal because it is judged to be not plausibly explained by chance, confounding, or bias may still be incorrect if other non-causal explanations such as coincident secular trends fully account for it (Campbell and Stanley 1963). In quasi-experiment terminology, coincident secular trends ("history") are one of a standard list of non-causal explanations ("threats to internal validity") that should be refuted before observed changes are interpreted causally (*ibid*). A third lesson is that even large reductions in ambient particulate air pollution have sometimes been found to have no detectable effect on mortality rates. This is a common finding in intervention studies (Burns et al. 2020). If the reason has to do with the composition of the particulate matter, then it becomes important for studies such as that of Domingo et al. to specify the composition of the particulate matter for which reductions are advocated. If the composition is similar to that in studies in which large reductions in particulate air pollution have not caused detectable reductions in mortality rates, then predictions that reducing exposure will detectably reduce all-cause or cardiovascular mortality rates should be accompanied by uncertainty statements recognizing that such predictions may be incorrect.

These lessons from the Dublin study and other international intervention studies (Burns et al. 2020) are likely to be relevant to the United States. For example, an analysis of changes in ambient PM2.5 and in all-cause and cardiovascular mortality rates between 2000 and 2010 in several hundred counties in the United States (Cox and Popken 2015) found that levels of ambient PM2.5 were significantly positively associated with mortality rates. However, changes in PM2.5 were not associated with corresponding changes in mortality rates. A causal relationship between them could not be inferred. Changes in PM2.5 and sample sizes were large enough so that reductions in mortality rates of the sizes predicted from associations should have been detected if they had occurred (*ibid*). Thus, the methodological point that GBD-type attributed mortalities do not address whether reducing exposure would reduce the attributed mortalities appears to be relevant in practice for ambient PM2.5 air pollution in the United States.

Some Limitations of the Quantitative Risk Assessment

Quantitatively, Domingo et al. estimate annual deaths caused by PM2.5 from agriculture. They do so by multiplying the estimated increases in PM2.5 caused directly or indirectly by farm emissions by an estimated concentration-response slope factor of 1.06 for the fractional increase in annual deaths per 10 $\mu g/m^3$ of ambient PM2.5 concentration. The estimated concentration-response slope factor of 1.06 is used without further discussion, citing the American Cancer Society (ACS) as a source. Domingo et al. explain that they "estimate annual deaths according to the American Cancer Society's concentration-response function, which averages a 6% increase in annual deaths per 10 $\mu g/m^3$ in PM2.5 concentration." We assume that this is a reference to Pope et al. (2002), cited by Domingo et al. Pope et al. state that "Each 10-$\mu g/m^3$ elevation in long-term average PM2.5 ambient concentrations was associated with approximately a 4%, 6%, and 8% increased risk of all-cause, cardiopulmonary, and lung cancer mortality, respectively, although the magnitude of the effect somewhat depended on the time frame of pollution monitoring." We assume that the factor of 1.06 (rather than 1.04) is taken from Table 2 of Pope et al. (2002) which presents results "estimated and adjusted based on the baseline random-effects Cox proportional hazards model, controlling for age, sex, race, smoking, education, marital status, body mass, alcohol consumption, occupational exposure, and diet." This quantitative approach has the following limitations.

1. *Use of associations to predict changes.* The ACS analysis quantifies RRs using regression methods. For example, Pope et al. (2002) state that "The baseline analysis in this study estimated adjusted relative risk (RR) ratios for mortality by using a Cox proportional hazards model with the inclusion of a metropolitan-based random-effects component... In the second stage, the estimates of adjusted metropolitan-specific health responses were related to fine particulate air pollution using a linear random-effects regression model" (Pope et al. 2002). This is an associational analysis. Hence, as emphasized in earlier sections, it is not valid for the purpose to which Domingo et al. apply it: assessing by how

much reducing exposures would reduce mortality rates, as this use violates the basic distinction between associational and causal analyses (Pearl 2009).

2. *Mathematically impossible predictions.* Misinterpreting the ACS associational results as causal leads to the paradox that Domingo et al.'s prediction method (i.e., predict change in annual deaths by multiplying the change in PM2.5 concentration by 1.06) predicts that a one-time permanent increase in PM2.5 of 10 μg/m^3 would cause a one-time 6% permanent increase in annual deaths. This is arithmetically impossible if births per year remain fixed (or decline) following the increase in PM2.5. It violates the conservation law that each person dies exactly once. Annual deaths cannot permanently increase by 6% unless annual births also permanently increase by 6%. Such retroactive increases in annual births are not a plausible effect of increased PM2.5 air pollution.
3. *Residual confounding.* The ACS study did not control for residual confounding created by its use of indicator variables to categorize continuous variables such as alcohol consumption, occupational exposures, and years of education. It is unknown how much of the estimated slope factor of 1.06 is explained by residual confounding.
4. *Omitted confounders.* The ACS study attempts to control for smoking, education, marital status, BMI, alcohol consumption, occupational exposures, and diet by including them in its proportional hazards regression models. It omits other important potential confounders such as income and lagged daily temperature extremes. Including them could have controlled for the possibilities that PM2.5 exposures and death rates are both higher among people with lower incomes, or that they are both higher following exceptionally cold winter days (or hot summer days). As Pope et al. state in their conclusions, potential effects of other unaccounted-for factors cannot be excluded. It is unknown how much of the estimated slope factor of 1.06 is explained by confounders that the investigators omitted. These include weather variables (e.g., same-day and lagged temperature, humidity, and wind speed) and socioeconomic variables (e.g., poverty and income) that are plausibly causally related to both PM2.5 exposures and elderly mortality risks.
5. *No adjustment for measurement and classification errors.* The ACS study did not correct for measurement errors or misclassification errors (e.g., misreported values in answering survey questions) for estimated smoking histories, alcohol consumption, and other variables. It is unknown how much of the estimated slope factor of 1.06 is explained by unmodeled measurement errors, e.g., if people with greater levels of estimated PM2.5 exposures also tend to have greater true (but not necessarily reported) smoking levels.
6. *No significant effect in nonparametric analysis.* The nonparametric smoothed exposure-response relationship for all-cause mortality estimated in the ACS study (Pope et al. 2002, Fig. 2, panel A) is consistent with no association between PM2.5 and all-cause mortality. The 95% uncertainty bands for the LOESS smoothing regression curve include a horizontal line at log RR $= 0$, so the null hypothesis of no effect of PM2.5 on mortality risk is not rejected.

7. *Use of an inappropriate model.* By contrast, a proportional hazards (PH) model yields an estimated relative risk of 1.06 in the ACS study which is significantly greater than 1; this is the factor adopted by Domingo et al. No regression diagnostics were presented to show that the PH model is justified for this dataset. However, one of the assumptions for a PH model is that independent variables are measured without error. This assumption is violated in the ACS study since the independent variables include indicator variables based on survey data. This induces estimation error by discretizing continuous variables. Errors in responses to survey questions on topics such as smoking and drinking behaviors, for which substantial error rates in responses are also well documented (Johnson 2014), can lead to biased estimation and erroneous inferences in PH models (Song and Wang 2014). Although it is unknown how much of the estimated slope factor of 1.06 arises from bias due to the use of an inappropriate PH model, comparison with the nonparametric (LOESS smoothing regression) results suggests that the 1.06 estimate may result largely or entirely from modeling bias.
8. *Ignored inter-individual heterogeneity.* The ACS analysis estimates a single slope coefficient of 1.06 for PM2.5-associated mortality risk for all individuals. This is inconsistent with modern causal methods that acknowledge and quantify heterogeneity in individual-level effects (e.g., Athey and Wager 2019). It does not show how or whether this estimated factor varies for people with different individual ages, incomes, weather conditions, urban/rural living conditions, and other variables. In particular, it does not address whether the people exposed to agricultural PM2.5 have the same distribution of causally relevant characteristics as the people in the ACS study.
9. *Ignored inter-study heterogeneity.* Using 1.06 as the only slope coefficient ignores the fact that different studies estimate different slope coefficients, with multiple accountability and intervention studies estimating a slope coefficient that is indistinguishable from 0.0 (Burns et al. 2020).
10. *Ignored PM2.5 heterogeneity.* PM2.5 refers to a heterogeneous class of mixtures of components that have very different estimated potencies (Lelieveld et al. 2015). Using 1.06 as the slope coefficient ignores the question of whether ammonia-associated PM2.5 from farm operations has the same potency as that estimated in the ACS study. Moreover, previous methodological work has concluded that "In general, aggregate exposure information alone may be inherently unable to guide rational risk-management decisions when the components of the mixture differ significantly in potency and when the percentage compositions of the mixture exposures differ significantly across locations. Under these conditions, which are not uncommon in practice, aggregate exposure metrics may be 'worse than useless,' in that risk-management decisions based on them are less effective than decisions that ignore the aggregate exposure information and select risk-management actions at random" (Cox and Popken 2007). The stated conditions for this pessimistic conclusion, that components of a mixture differ significantly in potency and the percentage

compositions of the mixture exposures differ significantly across locations, hold for PM2.5 (Lelieveld et al. 2015).

These limitations suggest that the quantitative estimates of annual deaths attributed to agricultural PM2.5 by Domingo et al. have neither conceptual nor demonstrated empirical validity. They inappropriately apply an associational slope coefficient (1.06 estimated *difference* in mortality risk per 10 μg/m^3 *difference* in observed PM2.5 concentrations in the ACS study) to a causal prediction problem (estimating *change* in annual mortalities per 10 μg/m^3 *change* in observed PM2.5 concentrations), ignoring the "basic distinction" between these concepts (Pearl 2009). The estimated slope coefficient of 1.06 depends on modeling assumptions that are known to be violated, such as the assumption of no measurement errors for independent variables in the PH model; without these assumptions, the slope is not significantly different from 0 in LOESS smoothing regression analysis. The estimated PM2.5 concentration-mortality slope factor of 1.06 also does not control for important potential confounders (e.g., weather and income variables) or for residual confounding created by its discretization of continuous variables. Domingo et al. present no evidence that this estimated slope factor from the ACS study (Pope et al. 2002) applies specifically to the ammonia-associated agricultural PM2.5 that their paper addresses.

Considerations from Causal Artificial Intelligence (CAI) and Machine Learning

The preceding criticisms of the ACS study and the Domingo et al. application of its results are to a large extent a critique of attribution-based health effects estimates in general. The methods and interpretations used in the ACS and Domingo studies are fair—and perhaps even exemplary—instances of widespread practices in attribution-based air pollution health effects research. The same methods are also widely applied in other areas of applied epidemiology and public health risk assessment and GBD analysis (GBD 2017 Risk Factor Collaborators 2018). This section reviews principles from causal artificial intelligence (CAI) that might help to improve upon these current practices. It also summarizes results from a limited number of studies that have applied them.

Conditional Independence Analysis

As discussed in Chap. 9, a key principle of CAI is that the conditional probability distribution for an effect depends on the values of its direct causes (its "parents" in a directed acyclic graph model). Conditioning on other variables does not remove this dependence (Pearl 2009). According to this *conditional independence criterion*, if a dataset does not show that exposure and mortality are informative about each other (i.e., they are not statistically independent of each other) even after conditioning on other variables, then the dataset does not provide evidence that exposure causes mortality. This remains true no matter how strong, consistent, and biologically plausible the unconditional association between them may be. This contrasts with the Bradford-Hill approach in which strong, consistent, plausible unconditional associations are interpreted as evidence for causation without the caveat that they must not vanish after conditioning on other variables. Intuitively, the conditional independence criterion says that the association between exposure and response should not be fully explained away by other variables (e.g., confounders) if a dataset is to provide evidence that exposure is a cause of the response. Being "a cause" in this framework means that interventions that change the level of exposure, but nothing else, would thereby change the distribution of mortality probabilities for exposed individuals (Pearl 2009).

The CAI perspective recognizes that association, although not by itself proof of causation (or even necessarily evidence of causation), is often a good basis for hypothesizing causation—especially if it is found consistently across many studies with very different arrays of confounders, potential modifiers, temporal trends, and other potential non-causal explanations (Campbell and Stanley 1963). CAI then takes the next step of testing implications of the hypothesis of causation against data. Conditional independence (CI) tests are useful in this context if effects are unlikely to be conditionally independent of their direct causes. (The rationale is that a direct cause of an effect typically provides information about it that other variables do not. One can contrive counterexamples using duplicate or redundant variables, but the principle that direct causes provide unique information that helps to predict their effects is often plausible for exposure-response modeling.) The logic of CI testing is that if mortality rates are found to be conditionally independent of exposure after conditioning on the levels of other variables, then the data are not found to provide evidence that exposure is a cause of mortality. The possibility is left open that there might be a causal effect that is too small to be detected. In large datasets, the size of plausible effects is limited by the power of the conditional independence tests (Cox and Popken 2015). Conversely, if mortality rates are found not to be conditionally independent of exposure even after conditioning on the levels of other variables that might confound the exposure-mortality association, then the data are found to provide evidence that exposure is a potential cause of mortality. The possibility is left open that there might be an unidentified confounder that explains the observed association. If no such confounders can be identified in the data, then the data are consistent with the hypothesis that exposure is a cause of mortality.

As an example to illustrate how conditional independence testing works in clarifying evidence for potential causation (or lack of such evidence, if that is the case), suppose that a large-scale, multi-year study identifies a highly statistically significant positive association between monthly levels of PM2.5 and monthly levels of cardiovascular mortality among people at least 60 years old, which we will abbreviate as CVD60. A common next step in traditional epidemiology is to ask whether the association is consistent (e.g., across multiple years, locations, and

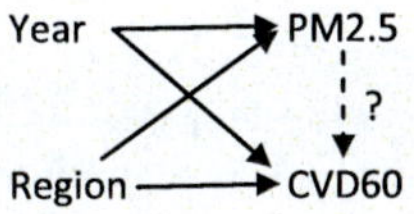

Fig. 10.1 A Directed Acyclic Graph (DAG) Model for Testing Conditional Independence

subpopulations), strong, biologically plausible, unconfounded by smoking, and so forth. A judgment about whether it is likely to be causal might then be made based on such considerations. CAI instead (or in addition) requires testing whether the association persists after conditioning on other variables. Specifically, suppose that it is found that both PM2.5 levels and mortality rates depend on the Year and Region where data are collected. The directed acyclic graph (DAG) diagram in Fig. 10.1 summarizes this information. It also shows the key question to be settled as a dashed arrow with a question mark between PM2.5 and CVD60.

In Fig. 10.1, the solid arrows represent statistical dependence relationship. A solid arrow between two variables signifies that observing the value of one helps to predict the value of the other (i.e., they have positive mutual information), even after values of the other variables that point into it (its "parents") are known. In a causal DAG, the direction of the arrow also signifies that changing one variable changes the conditional probability distributions of variables that it points into (Pearl 2009). The question represented by the dashed arrow in Fig. 10.1 is: After conditioning on (i.e., selecting specific values of) Year and Region, does CDV60 still depend on PM2.5? More formally, can the null hypothesis that PM2.5 and CVD60 are independent of each other still be rejected when Year and Region are held fixed? If so, then the association between PM2.5 and CVD60 is not fully "explained away" by confounding by Year and Region, and the data are consistent with the possibility of a causal relationship between them (*ibid*). If not—if no dependence between PM2.5 and CVD60 is found after conditioning on the levels of other variables such as Year and Region—then the data do not provide evidence that PM2.5 is a potential cause of CDV60. In this case, CDV60 is said to be *conditionally independent* of PM2.5 given the values of Year and Region. In either case, the conclusion from conditional independence testing is not a determination about whether PM2.5 does or does not cause CVD60, but rather it is a determination about whether the association between them is or is not fully explained by Year and Region.

A constructive contribution of CAI methodology is to address the question of whether CDV60 is conditionally independent of PM2.5 empirically by applying conditional independence tests to the data used to detect associations between Year, Region, PM2.5, and mortality. For readers familiar with supervised machine learning algorithms (see e.g., Molnar 2022), the following additional details may be useful. If standard supervised learning algorithms (e.g., classification and regression tree (CART) algorithms or random forest algorithms) with CDV60 as the dependent variable and Year, Region, and PM2.5 as independent variables discover that PM2.5, Year, and Region are all significant predictors of CDV60, then this would show that CDV60 is *not* independent of PM2.5 even after conditioning on Year and

Region. But if instead they find that only Year and Region are significant predictors of CDV60 (e.g, if only Year and Region appear in the CART tree for CDV60, or if the partial dependence plot for PM2.5 as a predictor of CDV60 is flat), then this would show that CDV60 is not found to be dependent on PM2.5 after conditioning on Year and Region. In that case, the data are not found to provide evidence to support the hypothesis that PM2.5 is a cause of mortality. This example illustrates how conditional independence testing can be used to help clarify whether a pattern of associations is consistent with the hypothesis that PM2.5 is a cause of CVD60. CAI includes several parametric and non-parametric tests for conditional independence (e.g., Vitolo et al. 2018; Shipley 2000). The main idea is that a dependent variable is conditionally independent of variables that do not help to predict it, given the variables that do help to predict it, and that effects should not be conditionally independent of their direct causes. This allows predictive machine learning to be applied to test the conditional independence properties implied by causal hypotheses and theories represented by DAG models (Textor et al. 2016; Witte et al. 2020).

In practice, although conditional independence tests have been well developed in CAI for many decades (Shipley 2000), they have seldom been applied in air pollution health effects research which remains dominated by Bradford-Hill-type considerations that, as they are commonly applied, do not evaluate whether other variables explain away observed associations. An exception is a large-scale analysis of air pollution and cardiovascular disease mortality risk for people over 60 years old (CVD60) in England (Vitolo et al. 2018). The analysis involved nearly 50 million records over the period 1981–2014. The authors found that CVD60 was significantly associated with PM2.5 air pollution, as expected from the past literature. Conditional independence tests indicated that CVD60 depended directly only on year, region, season, and month, i.e., these are its parents in the DAG summarizing dependence and conditional independence relationships found in the data. (Year and region, in turn, may be proxies for other variables, such as income and health care, that more directly affect mortality. DAG models show statistical dependencies, but do not necessarily elucidate the causal mechanisms that explain them.) The same DAG shows CVD60 as conditionally independent of PM2.5 after conditioning on these other variables. Specifically, Fig. 10.1 (without the dashed arrow between PM2.5 and CVD60) is a sub-graph of the DAG model learned from the data in this study. Similar to the Dublin study, conditional independence tests confirm that differences in year and region explain observed significant differences in CVD60 observed for people with different levels of PM2.5 in this dataset. They do not find a detectable causal role for PM2.5.In other words, no further dependence of CVD60 on PM2.5 is found after its dependence on other variables is accounted for by conditioning on their levels, even though there is considerable variation in PM2.5 levels within and between years, regions, seasons, and months.

Of course, this lack of dependence of CVD60 on PM2.5 after conditioning on other variables does not imply that there is no causal effect of PM2.5 on CVD60. It only implies that any such effect, if it exists, must be small enough to have escaped detection. In this sense, the CAI approach does not resolve the qualitative *causal* question "Does PM2.5 increase cardiovascular mortality risk as measured by

CVD60?" with confidence since the possibility of causal effects too small to detect can never be ruled out. Rather, it provides a confident answer to the *evidentiary* question, which we might frame in any of the following ways: "Can CVD60 be predicted better from PM2.5, region, and year than from region and year alone?" (a version explored further in the following discussion of Granger causation); or "Does monthly PM2.5 provide independent information about monthly CVD60 that is not contained in other variables, specifically year and region?"; or "Does the dataset provide evidence that monthly PM2.5 is a cause of monthly CVD60?" where PM2.5 "being a cause" of CVD60 is taken to imply that PM2.5 provides information about the conditional probability of CVD60 that cannot be obtained from its other parents. Of course, there are degrees of evidence and degrees of success or failure to resolve the complex nexus of influences that might lead to unambiguous conclusions about causality. Recognizing this, CAI conditional independence tests focus instead on the easier task of ascertaining whether the statistical dependence between variables in a dataset disappears (i.e., is no longer significantly different from zero) after conditioning on other variables. This is only part of what CAI offers for deciding whether data provide evidence supporting the hypothesis of causality (Heinze-Deml et al. 2018). Additional principles are discussed next. But it is a useful step toward making conclusions about evidence of causality a matter for empirical testing.

Finally, an open-minded but appropriately skeptical reader might reflect that studies inevitably have varying sets and patterns of potential influencers of outcomes, with the documentation of all such possibilities inevitably being limited. Such a reader might well wonder whether a motivated investigator could abuse the CAI approach by searching for combinations of other variables that would make an adverse response conditionally independent of exposure, and then declaring that the dataset provides no evidence of causation even if, taken at face value, it clearly does. In other words, might not a version of multiple testing allow an inflated rate of false negatives in rejecting valid evidence of causality? CAI methods protect against such abuse by using non-parametric ML algorithms (e.g., random forest) with k-fold cross-validation to help obtain unbiased estimates of predictive accuracy (Heinze-Deml et al. 2018). The key idea is that if independent variable X truly provides unique information that helps to predict dependent variable Y, then conditioning on other variables won't change or obscure this fact and won't mistakenly make Y appear to be conditionally independent of X.

Granger Causality Analysis

Conditional independence can be extended to longitudinal studies using *Granger causation* (Eichler and Didelez 2010), according to which a longitudinal dataset provides no evidence that exposure is a cause of mortality if the future mortality time series cannot be predicted better with information about past and present values of exposure than it can be predicted without this information. This concept strengthens the Bradford-Hill consideration of temporality that causes precede their effects to a stronger requirement that changes in causes should precede and help to predict changes in their effects.

Like conditional independence, Granger causality has seldom been applied in air pollution health effects research. An exception is a 2015 study of the time series of county-specific PM2.5 and mortality time series in the United States. This study concluded on the basis of Granger causality testing and finding that even substantial changes in PM2.5 did not help to explain or predict changes in mortality and that "A causal relation between pollutant concentrations and [all-case] or CVD mortality rates cannot be inferred from these historical data, although a statistical association between them is well supported" on a time scale of years to a decade (Cox and Popken 2015). Further applications of Granger causality to PM2.5 and health effects time series on various time scales could potentially help to clarify whether PM2.5 is a Granger cause of the mortality and morbidity effects that have been attributed to it and, if so, over what periods.

Non-Parametric Analysis of Heterogeneous Exposure Effects

Non-parametric CAI techniques such as honest causal trees and forests (Athey and Wager 2019) have recently been applied to estimate heterogeneous exposure effects of PM2.5 air pollution on health outcomes among people 65 years or older (Lee et al. 2021). The authors analyzed data on over 1.6 million Medicare beneficiaries in the New England region in the United States for the years 2000–2006. They found that if they discretized age group into 5-year age groups and assumed no hidden confounding, then low-income seniors between 81 and 85 years old and seniors aged 85 and above have statistically significantly greater estimated causal effects of long-term exposure to PM2.5 on 5-year mortality rate compared to the population mean. They also estimated that any omitted confounders, such as weather and socioeconomic status (SES) variables not included in the study, would have to increase the odds of exposure to high-level PM2.5 by 1.5-fold and increase the odds of death by more than 1.4-fold to explain away the results. For comparison, it has been estimated that in the United States, Blacks have an average of about 1.54-fold greater exposure to PM2.5 than the overall population (Mikati et al. 2018) and that poverty can increase the mortality rate of both Black and Whites aged 65 to 79 by a factor of about 1.4 (Schwandt et al. 2021, Fig. 4). Whether these and other unmeasured variables, together with residual confounding from imperfectly measured variables (e.g., using Medicaid eligibility as a surrogate for poverty) fully explain the effects estimated by Lee et al. is not yet known.

Residual confounding from the use of five-year age categories, confounding by unmeasured variables (such as individual-level SES and weather variables), and ignored measurement errors in independent variables may all contribute to the results reported by Lee et al. If these limitations are overcome in future research, e.g., by treating age as a continuous variable and including individual-level SES (especially, income) and weather variables (e.g., lagged daily high and low temperatures and other meteorological variables that predict both PM2.5 levels and mortality rates), then their non-parametric approach appears promising for quantifying heterogeneous exposure effects caused by PM2.5 and other variables using non-parametric methods without making strong, potentially unrealistic modeling assumptions.

Invariant Causal Prediction and Transportability of Estimated Causal Effects

The CAI principle of *invariant causal prediction* (ICP) generalizes the informal consideration of consistency of causal effects across studies. It does not require that regression coefficients of exposure as a predictor of response should be the same or similar across studies. Rather, it requires that the *conditional probability distribution* for an effect variable such as mortality, given the same values of all of its direct causal parents (e.g., age, sex, ethnicity, income, ambient PM2.5 concentrations, lagged weather variables, etc.) should be the same across studies and settings (Peters et al. 2016; Heinze-Deml et al. 2017). The closely related concept of *transportability* describes the extent to which causal relationships and estimated effect sizes found in one setting can be applied in different settings, e.g., with different populations or study designs (Shimonovich et al. 2021). These concepts generalize the Bradford-Hill concept of consistency by predicting that estimated effects of exposure on mortality should *not* be the same across studies of populations having different joint distributions of causally relevant variables, but rather the differences across studies should be explainable in terms of a stable (invariant across studies) set of underlying conditional probability relationship describing the dependence of the conditional probability of mortality on its direct causes (*ibid*). Thus, while conditional independence, Granger causality, and heterogeneous treatment effects analyses are CAI methods that can be applied to individual studies, ICP assesses consistency in conditional probability relationships across studies.

Intuitively, if different studies have very different arrays of potential confounders, modifiers and temporal trends but exposure to an agent such as PM2.5 is found to be a common element that consistently predicts increased risk across the diverse studies, then it may seem plausible that the agent is indeed a cause of increased risk. Specific, consistent associations of a putative cause (e.g., PM2.5 exposure) and its putative effect (e.g., increased mortality rate) are often considered evidence of causation, at least to the extent that other hypothesized influences can be shown to lack such consistency. However, some have argued that specificity seldom occurs and that consistent associations can also arise from consistent modeling biases, omitted confounders, historical trends, and residual confounding across studies (Ioannidis 2016; Cox 2017; Shimonovich et al. 2021). In any case, ICP demands more than consistency of estimated effects of exposure across studies. It requires that *differences* in estimated effects of exposure in different study populations be consistently explained by differences in other covariates such as age, income, and weather across the study populations. A conceptually straightforward way to test whether the ICP property holds is to include study identity (ID) as an independent variable and to check whether the dependent variable (e.g., mortality) and its residuals are conditionally independent of study ID, given the values of the within-study independent variables; see Heinze-Deml et al. (2018) for details and limitations of conditional independence tests used in testing for ICP.

To our knowledge, ICP has not yet been applied to PM2.5 and health effects data. However, large unexplained heterogeneity in estimated effects (e.g., RRs) of PM2.5 on mortality (e.g., Hoek et al. 2013) suggests that further work is needed to identify stable causal relationships between mortality risk and its direct causes, possibly including PM2.5.

Probabilistic Causal Network Modeling of Multiple Interrelated Variables

GBD and other associational methods typically emphasize two main variables: an exposure variable and a response variable, perhaps with stratification or adjustment to control for levels of other variables. By contrast, causal models typically represent networks of many interrelated variables, with exposure and response variables connected to other variables (e.g., weather, SES, and time and location variables) that may explain some of the observed correlation between them (Vitolo et al. 2018; Burns et al. 2020). Such causal networks allow theories about how changes in exposure affect probabilities of mortality and morbidity to be tested for consistency with other observed changes that are causally related to the exposure and response variables of main interest. This leads to different and more nuanced methods for adjusting for the effects of other variables than standard regression or relative risk approaches. For example, the standard approach to adjusting for observed potential confounders in regression is to include them among the independent variables on the right side of the regression equation used to predict the dependent variable, risk. Causal network modelers have long recognized that this procedure may induce spuriously significant regression coefficients due to collider bias. In other words, conditioning on a common descendant of exposure and demographics (such as inclusion in a study) may induce dependence between otherwise independent variables, much as conditioning on the knowledge that the sum of two unconditionally independent non-negative random variables X and Y is 10 induces a negative correlation between them (Pearl 2009). Instead of including all other variables on the right side of a regression equation, CAI methods provide algorithms for calculating *adjustment sets* from DAG structures (Textor et al. 2016; Witte et al. 2020). These are subsets of variables that are conditioned on (i.e., their observed values are held fixed at specified levels) to obtain estimates of the direct or total causal effects of one variable (such as exposure) on another (such as mortality) while controlling (or "adjusting") for the levels of the variables in the adjustment set. Adjustment sets generalize the principle that one must condition on common ancestors (such as confounders) but not on common descendants (to avoid collider bias). The same causal effect, whether total or direct, can often be estimated using more than one adjustment set. Testing whether the resulting estimates agree with each other provides a test of the internal consistency of a DAG model (Textor et al. 2016; Cox 2017). The use of adjustment sets in causal networks offers a constructive response to the challenge noted by Dominici et al. (2014) that "There is a growing consensus in economics, political science, statistics, and other fields that the associational or regression approach to inferring causal relations—on the basis of adjustment with observable confounders—is unreliable in many settings." Instead of different investigators selecting different observed variables to include in regression models, and thereby possibly reaching conflicting conclusions, CAI algorithms enable all investigators to identify the same adjustment sets for the same DAG models. This reduces the "researcher degrees of freedom" that have sometimes led to conflicting causal conclusions from the same data (*ibid*) and focuses attention on the scientific question of identifying causal DAG models that are consistent with the data.

Including exposure and response variables in a wider network of causally linked variables creates opportunities to test beliefs and to verify or correct theories about how and under what conditions changes in exposure create changes in responses. Changes are explicitly modeled as propagating along paths consisting of directed arcs (the arrows in a directed acyclic graph model) linking variables. These arrows and the conditional probability distributions for each variable given the values of its parents (i.e., direct causes) summarize underlying causal mechanisms (Simon 1953; Druzdzel and Simon 1993). By making explicit that some variables mediate the effects of others on a target outcome such as mortality (Pearl 2009), causal network models enable the testing of causal theories by examining whether changes in variables are observed to occur in the order predicted by the hypothesized causal pathways leading from exposure to response. If it is hypothesized that changes in ambient PM2.5 concentrations affect mortality risks by changing mediating variables such as chronic lung inflammation, then data on markers of chronic inflammation (or on causal precursors that precede the onset of chronic inflammation) can help to verify or to refute and refine this causal hypothesis. Embedding the variables of direct interest—PM2.5 concentration and mortality rate—in a larger causal network also provides opportunities to use data other than the RR for PM2.5 and mortality to quantify direct and indirect causal effects and interactions among PM2.5 and other variables, as well as to test whether PM2.5 directly affects mortality (Vitolo et al. 2018).

Even partly elucidated causal theories can be tested if hypothesized causal precursors and mediators of the outcome of interest can be observed. For example, studies of school-aged children with asthma in Washington State in the United States measured a greater than 1000-fold variation in ambient daily average ammonia concentrations at monitors over a 13-month period, depending on proximity to animal feeding operations. They found a reduction in forced expiratory volume in 1 second (FEV1) at higher ambient levels of ammonia, although the cause(s) were not identified, but found no associations between ammonia exposure concentrations and self-reported asthma symptoms or medication usage (Loftus et al. 2015, 2020). Such observations can constrain coherent causal theories encompassing PM2.5-mortality associations even though they do not directly involve PM2.5 or mortality. For example (and solely for purposes of illustration of a methodological point), it would not be logically coherent to believe that

(a) Increased ammonia- and/or PM2.5-associated elderly mortality in an exposed population (e.g., downwind from farms) is mediated by increased inflammation of the lung in response to sufficiently high and prolonged exposures to relevant pollutants (e.g., PM2.5 and/or ammonia).
(b) Increases in self-reported pediatric asthma symptoms and increased medication use by children with asthma are sensitive responses that occur in the exposed population at lower levels of exposure than those which increase elderly mortality.
(c) Ammonia emissions from farms cause levels of air pollution exposures that increase elderly mortality but not self-reported pediatric asthma symptoms and increased medication use by children with asthma.

Probabilistic causal networks allow for detection not only of logical inconsistencies, as in this example, but also of statistical inconsistency in which observed data make it unlikely that sets of propositions are all true. This extends the Bradford-Hill consideration of coherence to assess the quantitative consistency of causal beliefs with each other and with the statistical constraints implied by data collected across multiple studies with overlapping variables (Triantafillou and Tsamardinos 2015).

Probabilistic causal network analysis can be applied to biomarker and mechanistic data as well as to epidemiological variables. For example, it has recently been suggested that activation of the NLRP3 inflammasome pathway plays a vital role in mediating pulmonary inflammation caused by PM2.5 exposure (Jia et al. 2021; Cao et al. 2022) as well as in asthma (Williams et al. 2021). If this is true, then understanding of NLRP3 inflammasome biology and its role in lung inflammation and pathogenesis can inform understanding of PM2.5-mortality and -morbidity associations mediated by the NLRP3 inflammasome. For example, the fact that NLRP3 inflammasomes have activation thresholds, requiring sufficiently high and prolonged exposures before they can assemble and become activated, suggests that NLRP3 inflammasome-mediated responses have exposure concentration and duration thresholds (Cox 2019). Causal biological network models that include NLRP3 inflammasome activation can potentially allow data from multiple inflammation-mediated lung diseases to be used to better understand human health risks from PM2.5 exposures.

To our knowledge, probabilistic causal network analysis has not yet been used to estimate direct causal effects of farm emissions of ammonia and resulting PM2.5 on mortality rates. Very few studies have applied directed acyclic graph (DAG) modeling to PM2.5 and mortality. As previously noted, a large-scale DAG model fit to English data found no direct effect of PM2.5 on elderly cardiovascular mortality (Vitolo et al. 2018). In the United States, the Health Effects Institute has funded work that uses a conceptual DAG model linking PM2.5, mortality, and SES variables including income and minority status (Brauer et al. 2019). The analysis constrains estimated PM2.5-mortality curves to have shapes that the investigators consider plausible (e.g., increasing monotonically with concentration); omits potential weather-related confounders such as temperature; and discretizes continuous variables such as age and income without correcting for resulting residual confounding. Even with these limitations, the DAG-based analysis found a mix of significant positive associations, no significant associations, and significant negative associations between PM2.5 and mortality risks in different cohorts.

We expect that probabilistic causal network modeling and other formal causal analyses of PM2.5-mortality associations will increase in the future as more researchers become familiar with these techniques. Meanwhile, the limited applications to date that we have reviewed suggest that the causal interpretations that have been assigned to PM2.5-mortality associations for several decades based on Bradford Hill and weight-of-evidence considerations are not yet well supported by findings from accountability and intervention studies or from CAI methods such as conditional independence tests, Granger causality tests, heterogeneous treatment effect analysis, or causal network analyses. Findings are often mixed or ambiguous. Lack of full control for confounders and other threats to internal validity may explain these results, but such an explanation undermines the credibility of causal conclusions even for the relatively few studies with designs that can potentially support valid causal inferences (Dominici et al. 2014), as well as for the vast majority of studies that only address statistical associations.

Discussion and Conclusions

Causal artificial intelligence (CAI) offers insights into principles and methods for causal inference from observational data that can complement the hypotheses about causal exposure-response relationships suggested by associational methods such as RR, GBD, and regression analyses (Carone et al. 2020). The previous sections have discussed the following main points.

(a) Analyses that directly apply associational results such as relative risks, odds ratios, or slope coefficients to predict effects on mortality or morbidity of reducing exposure are methodologically unsound and often produce unreliable conclusions (Pearl 2009; Dominici et al. 2014; Carone et al. 2020). Such analyses are presented in several papers we have discussed including Domingo et al. (2021), Clancy et al. (2002), GBD 2017 Risk Factor Collaborators (2018), and Pope et al. (2002).
(b) The reason is that these studies ignore a “basic distinction”(Pearl 2009) between associational and causal relationships that prevent the former from answering questions about changes that are addressed by the latter. The essential distinction between *differences* in outcomes observed at different levels of exposure and

changes in outcomes caused by changing exposure is ignored in most of the voluminous past literature on PM2.5 and mortality (Dominici et al. 2014; Carone et al. 2020) but is taken as foundational in CAI (Pearl 2009; Athey and Wager 2019) and older literature in other fields on valid causal inference from observational data (Campbell and Stanley 1963; Simon 1953; Druzdzel and Simon 1993).

(c) Policymakers considering reducing exposures need to be informed about the changes in outcomes caused by alternative interventions. This requires causal analyses (Cox 2021).

(d) The large past literature on PM2.5-mortality associations has documented a wide range of estimated differences in mortality risk associated with each unit of difference in ambient concentration of PM2.5 (Gu et al. 2021; Hoek et al. 2013). The heterogeneity in these estimates has not been explained, but different degrees of control for confounding in different studies may be part of the explanation (Hoek et al. 2013).

(e) Multiple intervention and accountability studies that have examined associations between changes over time in particulate exposure and mortality risk instead of differences in exposures and corresponding mortality risks have found no clear evidence that observed changes in exposure help to predict or explain subsequent changes in mortality risks (Burns et al. 2020; Cox and Popken 2015; Zigler and Dominici 2014).

(f) Studies that have applied conditional independence tests, causal networks, and other formal causal models to control for observed potential confounders have found no evidence of a direct causal effect of PM2.5 on mortality risk (Vitolo et al. 2018) or have estimated mixed positive, negative, and lack of significant estimated effects of PM2.5 on mortality (Brauer et al. 2019).

From these methodological considerations, we conclude that calculations such as those by Domingo et al. (2021) estimating annual mortalities attributed to increases in PM2.5 from agricultural emissions of ammonia and other pollutants have no known validity. They are not known to describe real-world effects because they apply associational results to address questions about the effects of changes in exposure, thereby wrongly mixing association and causation (Pearl 2009; Dominici et al. 2014; Carone et al. 2020). The underlying study of associations used to support mortality risk estimates (Pope et al. 2002) does not fully control for omitted confounders, residual confounding, errors in independent variables, modeling biases, and heterogeneous exposure effects.

The belief that reducing ammonia would reduce annual mortalities (Giannakis et al. 2019; Gu et al. 2021; Ma et al. 2021) may or may not be correct, but it is not empirically demonstrated or logically supported by such attributable mortality calculations. To our knowledge, there is no logically sound, empirically validated evidence that reducing ammonia emissions from farms has reduced or would reduce human mortality risks. Of course, a principle virtue of applied science is that it can use well-validated causal theories to make confident predictions about the effects of actions before they are implemented. But for predictions about human health benefits of reducing PM2.5 and other particulate air pollution, the underlying causal theories are not presently well-validated and are sometimes contradicted by experience (Cox and Popken 2015; Burns et al. 2020; Health Effects Institute 2013). Attributable mortality calculations are even further removed from trustworthy science insofar as they make no testable or falsifiable predictions. Attributing deaths to exposure does not imply that reducing exposure would reduce death rates. In short, there is no clear theoretical or empirical reason to expect that reducing ammonia emissions from farms would reduce human mortality risks, contrary to recent attribution-based conclusions claiming that such a causal impact would occur (Domingo et al. 2021; Giannakis et al. 2019; Gu et al. 2021; Ma et al. 2021).

This negative conclusion applies a CAI perspective on causality (Pearl 2009; Carone et al. 2020) that admittedly rejects decades of practice in epidemiology and public health risk assessment for air pollution health effects. This long-standing practice seeks to draw policy-relevant causal conclusions from a combination of statistical associations and weight-of-evidence judgments. The rationale for dismissing this approach is a conviction that, in Pearl's (2009) phrase, policymakers and risk managers need to understand "changes [in health effects] induced by treatments or external interventions… [but] causal and associational concepts do not mix." The CAI perspective offered here is that it is, therefore, imperative to use techniques of causal analysis that do address how interventions that reduce exposures would change health effects, and to scrap association-based techniques that do not correctly address this question. Not everyone agrees. Carone et al. (2020) write that

> Some scientists have argued that, because epidemiologic studies provide measures of association and may not accurately predict the benefits of reducing air pollution, they should not be used, thereby dismissing a large body of evidence painstakingly gathered over decades. . . . In our view, causal inference methods should not be used as another opportunity to weaponize science against itself. Policymakers cannot wait for the data, study designs, and analytic tools that will ensure unarguable causal inferences: stalling until perfect evidence arises is irresponsible and does not protect public health (Carone et al. 2020).

This chapter has adopted a somewhat different point of view. Although we agree that in many cases policymakers cannot or should not wait for the data, study designs, and analytic tools to "ensure unarguable causal inferences," we believe that it is practical, desirable, and socially responsible for risk assessors to do the following three things today:

- Use techniques that address the right question ("How would reducing exposure change mortality risk?") instead of the wrong one ("How strongly are observed differences in exposures associated with corresponding observed differences in mortality rates?") Do not conflate the two by using data to answer the latter and then interpreting the result as answering the former, as the GBD methodology and influential air pollution health effects studies (e.g., Pope et al. 2002) do.
- Insist on drawing *sound* causal inferences (not necessarily "unarguable" ones since the characterization of remaining uncertainties should be part of every risk assessment), e.g., by including data from relevant comparison groups (Dominici

et al. 2014) and refuting alternative (non-causal) explanations for observations, e.g., using conditional independence tests and other CAI principles and techniques (Cox 2017). Appropriate principles and methods for drawing sound causal conclusions from observational data (or for determining that this cannot be done) are widely available and should be more widely used (Campbell and Stanley 1963; Pearl 2009; Dominici et al. 2014).

- Use existing data and readily available data to control for likely confounders (e.g., SES variables and lagged weather variables over relevant time periods) and to test alternative causal hypotheses and theories against data, e.g., using causal networks and other CAI principles and techniques (Cox 2017, 2021). We see this not as "weaponizing science against itself" but as weaponizing sound science, which requires testing hypotheses against data, against poor science which relies instead on untested theories and judgments.

We believe that it is important and practical to follow these principles. Doing so can be expected to clarify causal drivers of mortality and to reveal where unconditional exposure-response associations disappear when sounder analyses are used (Zigler and Dominici 2014). While we agree that science should not be weaponized against itself, we also believe that attributing mortalities to exposures based on relative risks (especially with uncontrolled confounding) and then predicting on that basis that reducing exposures would reduce the attributed mortalities is not sound science. It fails to engage with data relevant for predicting how, if at all, changes in exposure would change mortality risks—that is, data on changes, rather than data on observed differences in exposures and mortality risks. Urging regulatory actions based on unconditional exposure-response associations without causal analysis is misguided and misleading. It is also unnecessary in light of the widespread availability of modern causal modeling methods and software (Cox 2017; Heinze-Deml et al. 2017; Vitolo et al. 2018)—even if, as argued by others (e.g., Carone et al. 2020), a learning curve must be traversed to replace associational methods with causal ones.

References

Athey S, Wager S (2019) Estimating treatment effects with causal forests: an application. Observ Stud 2:37–51

Beelen R, Stafoggia M, Raaschou-Nielsen O, Andersen ZJ, Xun WW, Katsouyanni K et al (2014 May) Long-term exposure to air pollution and cardiovascular mortality: an analysis of 22 European cohorts. Epidemiology 25(3):368–378

Brauer M, Brook JR, Christidis T, Chu Y, Crouse DL, Erickson A, Hystad P, Li C, Martin RV, Meng J, Pappin AJ, Pinault LL, Tjepkema M, van Donkelaar A, Weichenthal S, Burnett RT (2019 Nov 1) Mortality–air pollution associations in low-exposure environments (MAPLE): phase 1. Res Rep Health Eff Inst 2019:203

Burnett R, Chen H, Szyszkowicz M, Fann N, Hubbell B, Pope CA 3rd, Apte JS, Brauer M, Cohen A, Weichenthal S, Coggins J, Di Q, Brunekreef B, Frostad J, Lim SS, Kan H, Walker KD, Thurston GD, Hayes RB, Lim CC, Turner MC, Jerrett M, Krewski D, Gapstur SM, Diver WR, Ostro B, Goldberg D, Crouse DL, Martin RV, Peters P, Pinault L, Tjepkema M, van

Donkelaar A, Villeneuve PJ, Miller AB, Yin P, Zhou M, Wang L, Janssen NAH, Marra M, Atkinson RW, Tsang H, Quoc Thach T, Cannon JB, Allen RT, Hart JE, Laden F, Cesaroni G, Forastiere F, Weinmayr G, Jaensch A, Nagel G, Concin H, Spadaro JV (2018 Sep 18) Global estimates of mortality associated with long-term exposure to outdoor fine particulate matter. Proc Natl Acad Sci U S A 115(38):9592–9597. https://doi.org/10.1073/pnas.1803222115

Burns J, Boogaard H, Polus S, Pfadenhauer LM, Rohwer AC, van Erp AM, Turley R, Rehfuess EA (2020 Feb) Interventions to reduce ambient air pollution and their effects on health: an abridged Cochrane systematic review. Environ Int 135:105400. https://doi.org/10.1016/j.envint.2019.105400

Campbell DT, Stanley JC (1963) Experimental and quasi-experimental designs for research. Rand McNally, Chicago. https://www.sfu.ca/~palys/Campbell&Stanley-1959-Exptl&QuasiExptlDesignsForResearch.pdf

Cao W et al (2022) NLRP3 inflammasome activation determines the fibrogenic potential of PM2.5 air pollution particles in the lung. J Environ Sci 111:429–441. https://doi.org/10.1016/j.jes.2021.04.021

Carone M, Dominici F, Sheppard L (2020 Jan) In pursuit of evidence in air pollution epidemiology: the role of causally driven data science. Epidemiology 31(1):1–6. https://doi.org/10.1097/EDE.0000000000001090

Clancy L, Goodman P, Sinclair H, Dockery DW (2002 Oct 19) Effect of air-pollution control on death rates in Dublin, Ireland: an intervention study. Lancet 360(9341):1210–1214. https://doi.org/10.1016/S0140-6736(02)11281-5

Cox LA (2021a) Toward practical causal epidemiology. Glob Epidemiol 3:100065. https://doi.org/10.1016/j.gloepi.2021.100065

Cox LA (2021b) Thinking about causation: a thought experiment with dominos. Glob Epidemiol 3: 100064. https://doi.org/10.1016/j.gloepi.2021.100064

Cox LA Jr, Popken DA (2007 Apr) Some limitations of aggregate exposure metrics. Risk Anal 27(2):439–445. https://doi.org/10.1111/j.1539-6924.2007.00896.x

Cox LA Jr, Popken DA (2015 Mar) Has reducing fine particulate matter and ozone caused reduced mortality rates in the United States? Ann Epidemiol 25(3):162–173. https://doi.org/10.1016/j.annepidem

Cox LAT Jr (2017 Aug) Do causal concentration-response functions exist? A critical review of associational and causal relations between fine particulate matter and mortality. Crit Rev Toxicol 47(7):603–631. https://doi.org/10.1080/10408444.2017.1311838

Cox LAT Jr (2019 Apr 2) Risk analysis implications of dose-response thresholds for NLRP3 inflammasome-mediated diseases: respirable crystalline silica and lung cancer as an example. Dose Response 17(2):1559325819836900. https://doi.org/10.1177/1559325819836900

Cox LA Jr (2020 Aug) Implications of nonlinearity, confounding, and interactions for estimating exposure concentration-response functions in quantitative risk analysis. Environ Res 187: 109638. https://doi.org/10.1016/j.envres.2020.109638

Domingo NGG, Balasubramanian S, Thakrar SK, Clark MA, Adams PJ, Marshall JD, Muller NZ, Pandis SN, Polasky S, Robinson AL, Tessum CW, Tilman D, Tschofen P, Hill JD (2021 May 19) Air quality-related health damages of food. Proc Natl Acad Sci U S A 118(20): e2013637118. https://doi.org/10.1073/pnas.2013637118

Dominici F, Greenstone M, Sunstein CR (2014) Science and regulation. Particulate matter matters. Science 344(6181):257–259. https://doi.org/10.1126/science.1247348

Druzdzel MJ, Simon H (1993) Causality in Bayesian belief networks. In: Proceedings of the Ninth Annual Conference on Uncertainty in Artificial Intelligence (UAI-93), Morgan Kaufmann Publishers, San Francisco, CA, pp 3–11

Eichler M, Didelez V (2010 Jan) On Granger causality lifetime. Data Anal 16(1):3–32

EPA (U.S. Environmental Protection Agency) (2011) The benefits and costs of the clean air act from 1990 to 2020. Final report – Rev. A. Office of Air and Radiation, Washington, DC

GBD 2017 Risk Factor Collaborators (2018) Global, regional, and national comparative risk assessment of 84 behavioural, environmental and occupational, and metabolic risks or clusters of risks for 195 countries and territories, 1990-2017: a systematic analysis for the Global Burden of Disease Study 2017 [published correction appears in Lancet. 2019 Jan 12;393(10167):132] [published correction appears in Lancet. 2019 Jun 22;393(10190):e44]. Lancet. 2018;392 (10159):1923–1994. doi:10.1016/S0140-6736(18)32225-6

Giannakis E, Kushta J, Bruggeman A et al (2019) Costs and benefits of agricultural ammonia emission abatement options for compliance with European air quality regulations. Environ Sci Eur 31:93. https://doi.org/10.1186/s12302-019-0275-0

Goodman GE, Thornquist MD, Balmes J, Cullen MR, Meyskens FL Jr, Omenn GS, Valanis B, Williams JH Jr (2004 Dec 1) The Beta-carotene and retinol efficacy trial: incidence of lung cancer and cardiovascular disease mortality during 6-year follow-up after stopping beta-carotene and retinol supplements. J Natl Cancer Inst 96(23):1743–1750. https://doi.org/10.1093/jnci/djh320

Gu B, Zhang L, Van Dingenen R, Vieno M, Van Grinsven HJ, Zhang X, Zhang S, Chen Y, Wang S, Ren C, Rao S, Holland M, Winiwarter W, Chen D, Xu J, Sutton MA (2021 Nov 5) Abating ammonia is more cost-effective than nitrogen oxides for mitigating PM2.5 air pollution. Science 374(6568):758–762. https://doi.org/10.1126/science.abf8623

Harvard School of Public Health (2002) Press Release: "Ban on coal burning in Dublin cleans the air and reduces death rates". www.hsph.harvard.edu/news/press-releases/archives/2002-releases/press10172002.html

Health Effects Institute (HEI) (2013) Did the Irish Coal Bans Improve Air Quality and Health? HEI Update, Summer, 2013. http://pubs.healtheffects.org/getfile.php?u=929. Last Retrieved 1 February 2014

Heinze-Deml C, Peters J, Meinshausen N (2017) Invariant causal prediction for nonlinear models. https://arxiv.org/pdf/1706.08576.pdf

Heinze-Deml C, Maathuis MH, Meinshausen N (2018) Causal structure learning. Annu Rev Stat Appl 5:371–391

Hoek G, Krishnan RM, Beelen R et al (2013) Long-term air pollution exposure and cardio-respiratory mortality: a review. Environ Health 12:43. https://doi.org/10.1186/1476-069X-12-43

Hristov AN (2011 Jun) Contribution of ammonia emitted from livestock to atmospheric fine particulate matter (PM2.5) in the United States. J Dairy Sci 94(6):3130–3136. https://doi.org/10.3168/jds.2010-3681

Huang F, Pan B, Wu J, Chen E, Chen L (2017) Relationship between exposure to PM2.5 and lung cancer incidence and mortality: a meta-analysis. Oncotarget 8(26):43322–43331. https://doi.org/10.18632/oncotarget.17313

Insausti M, Timmis R, Kinnersley R, Rufino MC (2020 Mar 1) Advances in sensing ammonia from agricultural sources. Sci Total Environ 706:135124. https://doi.org/10.1016/j.scitotenv.2019.135124

Ioannidis JP (2016 May 20) Exposure-wide epidemiology: revisiting Bradford Hill. Stat Med 35(11):1749–1762. https://doi.org/10.1002/sim.6825

Jia H, Liu Y, Guo D, He W, Zhao L, Xia S (2021 Mar) PM2.5-induced pulmonary inflammation via activating of the NLRP3/caspase-1 signaling pathway. Environ Toxicol 36(3):298–307. https://doi.org/10.1002/tox.23035

Johnson TP (2014 Nov 5) Sources of error in substance use prevalence surveys. Int Sch Res Notices 2014:923290. https://doi.org/10.1155/2014/923290

Lee K, Small DS, Dominici F (2021) Discovering heterogeneous exposure effects using randomization inference in air pollution studies. J Am Stat Assoc 116(534):569–580. https://doi.org/10.1080/01621459.2020.1870476

Lelieveld J, Evans JS, Fnais M, Giannadaki D, Pozzer A (2015 Sep 17) The contribution of outdoor air pollution sources to premature mortality on a global scale. Nature 525(7569):367–371. https://doi.org/10.1038/nature15371

Lepeule J, Laden F, Dockery D, Schwartz J (2012 Jul) Chronic exposure to fine particles and mortality: an extended follow-up of the Harvard Six Cities Study from 1974 to 2009. Environ Health Perspect 120(7):965–970

Loftus C, Yost M, Sampson P, Torres E, Arias G, Breckwich Vasquez V, Hartin K, Armstrong J, Tchong-French M, Vedal S, Bhatti P, Karr C (2015 Nov) Ambient ammonia exposures in an agricultural community and pediatric asthma morbidity. Epidemiology. 26(6):794–801. https://doi.org/10.1097/EDE.0000000000000368

Loftus C, Afsharinejad Z, Sampson P, Vedal S, Torres E, Arias G, Tchong-French M, Karr C (2020 Jan) Estimated time-varying exposures to air emissions from animal feeding operations and childhood asthma. Int J Hyg Environ Health 223(1):187–198. https://doi.org/10.1016/j.ijheh.2019.09.003

Ma R, Li K, Guo Y, Zhang B, Zhao X, Linder S, Guan C, Chen G, Gan Y, Meng J (2021 Nov 5) Mitigation potential of global ammonia emissions and related health impacts in the trade network. Nat Commun 12(1):6308. https://doi.org/10.1038/s41467-021-25854-3. Erratum in: Nat Commun. 2021 Nov 30;12(1):7084

Maclure M (1991) Taxonomic axes of epidemiologic study designs: a refutationist perspective. J Clin Epidemiol 44(10):1045–1053

Masselot P, Sera F, Schneider R, Kan H, Lavigne É, Stafoggia M, Tobias A, Chen H, Burnett RT, Schwartz J, Zanobetti A, Bell ML, Chen BY, Guo YL, Ragettli MS, Vicedo-Cabrera AM, Åström C, Forsberg B, Íñiguez C, Garland RM, Scovronick N, Madureira J, Nunes B, De la Cruz VC, Hurtado Diaz M, Honda Y, Hashizume M, Ng CFC, Samoli E, Katsouyanni K, Schneider A, Breitner S, Ryti NRI, Jaakkola JJK, Maasikmets M, Orru H, Guo Y, Valdés Ortega N, Matus Correa P, Tong S, Gasparrini A (2022 Mar 1) Differential mortality risks associated with PM2.5 components: a multi-country, multi-city study. Epidemiology 33(2): 167–175. https://doi.org/10.1097/EDE.0000000000001455

Mikati I, Benson AF, Luben TJ, Sacks JD, Richmond-Bryant J (2018 Apr) Disparities in distribution of particulate matter emission sources by race and poverty status. Am J Public Health. 108(4):480–485. https://doi.org/10.2105/AJPH.2017.304297

Molnar C (2022) Interpretable machine learning: a guide for making black box models explainable, 2nd edn. https://christophm.github.io/interpretable-ml-book/

O'Sullivan K (2020) Thirty years on, Ireland's smoky coal ban remains an urgent issue. The Irish Times Monday August 31, 2020

Pearl J (2009) Causal inference in statistics: an overview. Stat Surv 3:96–146. https://doi.org/10.1214/09-SS057

Pelucchi C, Negri E, Gallus S, Boffetta P, Tramacere I, La Vecchia C (2009 Dec 8) Long-term particulate matter exposure and mortality: a review of European epidemiological studies. BMC Public Health 9:453

Peters J, Bühlmann P, Meinshausen N (2016) Causal inference by using invariant prediction: identification and confidence intervals. J R Stat Soc Ser B 78(5):947–1012

Pinder RW, Adams PJ, Pandis SN (2007 Jan 15) Ammonia emission controls as a cost-effective strategy for reducing atmospheric particulate matter in the eastern United States. Environ Sci Technol 41(2):380–386. https://doi.org/10.1021/es060379a

Plass D, Hilderink H, Lehtomäki H et al (2022) Estimating risk factor attributable burden – challenges and potential solutions when using the comparative risk assessment methodology. Arch Public Health 80:148. https://doi.org/10.1186/s13690-022-00900-8

Pope CA 3rd, Burnett RT, Thun MJ, Calle EE, Krewski D, Ito K, Thurston GD (2002 Mar 6) Lung cancer, cardiopulmonary mortality, and long-term exposure to fine particulate air pollution. JAMA. 287(9):1132–1141. https://doi.org/10.1001/jama.287.9.1132

Schwandt H, Currie J, Bär M, Banks J, Bertoli P, Bütikofer A, Cattan S, Chao BZ, Costa C, González L, Grembi V, Huttunen K, Karadakic R, Kraftman L, Krutikova S, Lombardi S, Redler P, Riumallo-Herl C, Rodríguez-González A, Salvanes KG, Santana P, Thuilliez J, van Doorslaer E, Van Ourti T, Winter JK, Wouterse B, Wuppermann A (2021 Oct 5) Inequality in mortality between Black and White Americans by age, place, and cause and in comparison to

Europe, 1990 to 2018. Proc Natl Acad Sci U S A. 118(40):e2104684118. https://doi.org/10.1073/pnas.2104684118. PMID: 34583990; PMCID: PMC8501753.

Schwartz J, Coull B, Laden F, Ryan L (2008 Jan) The effect of dose and timing of dose on the association between airborne particles and survival. Environ Health Perspect. 116(1):64–69. https://doi.org/10.1289/ehp.9955. PMID: 18197301; PMCID: PMC2199297.

Shimonovich M, Pearce A, Thomson H et al (2021) Assessing causality in epidemiology: revisiting Bradford Hill to incorporate developments in causal thinking. Eur J Epidemiol 36:873–887. https://doi.org/10.1007/s10654-020-00703-7

Shipley B (2000) Cause and correlation in biology. A user's guide to path analysis, structural equations and causal inference. Cambridge University Press, New York. http://callisto.si.usherb.ca:8080/bshipley/my%20book.htm

Simon HA (1953) Causal ordering and identifiability, Chapter III. In: Hood WC, Koopmans TC (eds) Studies in econometric method. Cowles Commission for Research in economics monograph no. 14. Wiley, New York, pp 49–74

Song X, Wang CY (2014) Proportional hazards model with covariate measurement error and instrumental variables. J Am Stat Assoc 109(504):1636–1646. https://doi.org/10.1080/01621459.2014.896805

Textor J, van der Zander B, Gilthorpe MS, Liskiewicz M, Ellison GT (2016 Dec 1) Robust causal inference. Int J Epidemiol 45(6):1887–1894

Triantafillou S, Tsamardinos I (2015) Constraint-based causal discovery. J Mach Learn Res 16: 2147–2205

Vitolo C, Scutari M, Ghalaieny M, Tucker A, Russell A (2018) Modeling air pollution, climate, and health data using Bayesian networks: a case study of the English regions. Earth Space Sci 5:76–88. https://doi.org/10.1002/2017EA000326

Wang G, Li J, Hopp WJ (2022) An instrumental variable forest approach for detecting heterogeneous treatment effects in observational studies. Manag Sci 68(5):3175–3973. https://doi.org/10.2139/ssrn.3045327

Williams EJ, Negewo NA, Baines KJ (2021 Jun) Role of the NLRP3 inflammasome in asthma: relationship with neutrophilic inflammation, obesity, and therapeutic options. J Allergy Clin Immunol 147(6):2060–2062. https://doi.org/10.1016/j.jaci.2021.04.022

Witte J, Henckel L, Maathuis MH, Didelez V (2020) On efficient adjustment in causal graphs. J Mach Learn Res 21:246

Wittmaack K (2007 Apr) The big ban on bituminous coal sales revisited: serious epidemics and pronounced trends feign excess mortality previously attributed to heavy black-smoke exposure. Inhal Toxicol 19(4):343–350

Zigler CM, Dominici F (2014) Point: clarifying policy evidence with potential-outcomes thinking—beyond exposure-response estimation in air pollution epidemiology. Am J Epidemiol 180(12):1133–1140. https://doi.org/10.1093/aje/kwu263

Chapter 11
Toward More Practical Causal Epidemiology and Health Risk Assessment Using Causal Artificial Intelligence

Introduction: Why Are Better Causal Methods Needed in Applied Epidemiology?

Applied epidemiology seeks to understand preventable causes of diseases and other adverse outcomes in populations, as well as their incidence and prevalence rates. However, at present, the field's core methods and concepts focus on statistical associations, with causal interpretations often assigned using untestable assumptions. For example, relative risk ratios and quantities derived from them, as presented in standard textbooks and applied in countless articles—population attributable fractions (PAFs) (Mansournia et al. 2021) and population attributable risks (PARs), burdens of disease (BODs), etiological fractions (EFs), probabilities of causation (PCs), and other promisingly named quantities—are all based on associations (specifically, relative risk ratios). None of them provides data-driven methods for decisively justifying or refuting causal interpretations. The well-known Bradford Hill considerations and more recent weight-of-evidence (WoE) frameworks for supporting judgments about whether associations should be interpreted causally do not close the logical gap between association and causation (Pearl 2009). More is needed to create criteria that incorporate both traditional and more recent approaches to causal inference from observational data (Tennant et al. 2021; Shimonovich et al. 2020; Cox 2018).

Decades of experience have shown that the practical consequences of the logical gap between data and methods addressing associations, on the one hand, and policy-relevant causal interpretations and recommendations, on the other, can be striking, as illustrated by the following examples:

- An overwhelming statistical association between higher observed levels of the vitamins beta carotene and retinol and lower observed levels of lung cancer risk at the individual level in large, well-conducted epidemiological studies inspired an intervention (the CARET trial study) to increase levels of these vitamins in people

L. A. Cox Jr., *AI-ML for Decision and Risk Analysis*, International Series in Operations Research & Management Science 345,
https://doi.org/10.1007/978-3-031-32013-2_11

at high risk of lung cancer due to smoking or asbestos. This intervention was followed by a prompt and significant *increase* in lung cancer risk instead of reducing it, leading to early cessation of the trial (Goodman et al. 2004). This unanticipated outcome vividly illustrates the distinction between epidemiological concepts of causation (e.g., attributable risk, burden of disease, probability of causation, and so forth) based on associations, and the type of causation that is most relevant for risk managers and policymakers: *interventional* (or *manipulative*) causation (Pearl 2009; Cox 2019). Interventional causation describes how outcome probabilities change in response to changes in input conditions—e.g., in this example, how lung cancer risk is changed by increased vitamin consumption. The unexpected results of the trial illustrate that the signs and magnitudes of such *changes* in probabilities of adverse outcomes may have no relation to the signs and magnitudes of observed *differences* in outcome prevalence or incidence between differently exposed groups. It is common practice in much current applied epidemiology to refer to *changes* in risk in a target population of interest per unit change in exposure when what is really meant is an observed *difference* in risk across populations with different exposure levels per unit difference in exposure (Cox 2019). These utterly distinct concepts should not be conflated (Pearl 2009).

- In 2005, the United States Food and Drug Administration (FDA) withdrew approval for use of enrofloxacin, a fluoroquinolone (FQ) antibiotic used to prevent and treat fatal bacterial diseases in poultry, in order to reduce the potential for FQ-resistant bacterial infections in human patients (USFDA 2005). The decision was supported by an epidemiological risk model that assumed that human cases of FQ-resistant *Campylobacter jejuni* infection per year are proportional to FQ use in chickens (Bartholomew et al. 2005). After more than a decade, however, it has proved difficult to find any impact of agricultural use of antibiotics on their efficacy in treating human infections (Sproston et al. 2018; Yang et al. 2019). As predicted by risk assessment models that examined the likely causes of resistant cases in human patients (primarily, human use of ciprofloxacin) (Phillips et al. 2004), the withdrawal of FQ use in animals was followed by a prompt reversal of a decades-long trend of decreasing *C. jejuni* rates (Powell 2016) and sustained increases in FQ-resistance in human patients (Abraham et al. 2020; Hao et al. 2016). These developments highlight the fallacy of confusing a fraction of human resistant infections *attributed* to animal use with the fraction of such cases *preventable* by eliminating animal use.
- Green et al. (2010) conducted a case-control analysis in a large UK data set (the UK General Practice Research Database cohort of about 6 million people) to investigate whether oral bisphosphonates taken to combat osteoporosis increase the risk of oesophageal cancer. They reported that the incidence of esophageal cancer was significantly increased in people with one or more previous prescriptions for oral bisphosphonates compared with those with no such prescriptions. But Cardwell et al. (2010), examining the same dataset using cohort analysis, reported that the use of oral bisphosphonates was *not* significantly associated with increased risk of gastric cancer. These opposite conclusions

from the same data illustrate the dominant role that *modeling assumptions* and methods can play in conclusions about associations.

- Decades of epidemiological studies have shown clear statistical associations between observed levels of particulate air pollution and observed mortality rates, at least when common confounders are not fully controlled (Burns et al. 2020), creating a widespread expectation and narrative that reducing particulate air pollution reduces mortality rates (e.g., Hillard 2020). Yet, a recent systematic examination of over three dozen intervention studies that reduced ambient air pollution concluded that "[I]t was difficult to derive overall conclusions regarding the effectiveness of interventions" in improving air quality or health (Burns et al. 2020). For example, regulatory interventions reduced particulate air pollution in Ireland by about 70%, or over 36 mg/m^3, yet produced no detectable reductions in all-cause mortality rates in large affected populations (Dockery et al. 2013; Zigler and Dominici 2014) despite decades of much-publicized (and ongoing) claims to the contrary (Hillard 2020).

These cases illustrate that, in current practice, epidemiological associations and models, and quantitative attributions of risk to exposure based on them, are not always trustworthy guides to how to improve human health. Proposed remedies and improvements such as quantitative bias analysis (Lash et al. 2014), are often not applied in practice. A more trustworthy type of analysis is needed to identify interventions that demonstrably improve public health.

An area of causal artificial intelligence (CAI) has developed at the intersection of epidemiology (Tennant et al. 2021), philosophy (Mackie 1965), computer science (Glymour and Scheines 1986), statistics (Pearl 2009; Madigan and Raftery 1994), econometrics (Haavelmo 1943; Simon 1953; Granger 1969), information physics (Schreiber 2000), artificial intelligence (Bareinboim and Pearl 2013; Druzdzel and Simon 1993; Myers et al. 1999; Mooij et al. 2013), machine learning (Masegosa et al. 2016), systems biology (Lagani et al. 2016), and genetics (Wright 1921; Glymour et al. 2019). CAI provides ways to use realistically limited and imperfect observational data, together with mechanistic knowledge, to address interventional causal questions. CAI methods can predict whether and by how much changing exposures would change probabilities of adverse health responses while accounting for differences in conditions and covariates across studies and affected populations. They work only under certain conditions and assumptions, but these can be empirically tested and verified, unlike approaches to causal inference that rely on untestable assumptions (Rubin 2005). The following sections summarize essential ideas and recent advances in CAI and assess their potential to provide useful answers to causal questions about epidemiological exposure-response data—especially, what fraction of disease cases would be prevented by reducing or eliminating exposure. They seek to provide an alternative to association-based measures such as PAF, PC, etc. to better represent the implications of *causal mechanisms*—even if only very partly elucidated—for understanding and predicting how changing exposure changes risks of adverse health effects.

Thinking About Causation: A Thought Experiment with Dominos

This section uses a simple conceptual example to illustrate in detail why traditional epidemiological measures such as population attributable fractions, probabilities of causation, burdens of disease, and similar association-based measures often do not provide valid estimates or surrogates for the fraction or number of disease cases that would be prevented by eliminating or reducing an exposure because their calculations do not include crucial mechanistic information. It uses a thought experiment with a cascade of dominos to illustrate the need for mechanistic information when answering questions about how changing exposures changes risk. We suggest that modern methods of causal artificial intelligence (CAI) can fill this gap: they can complement and extend traditional epidemiological attribution calculations to provide information useful for risk management decisions.

Different Concepts of Causation

The following thought experiment illustrates several different concepts of causation in a simple deterministic mechanical context. Suppose that 100 dominos stand in a row, spaced closely enough so that the fall of any domino will knock over its two immediate successors. All but one of the dominos are black. In the middle of the row, domino 50 is painted yellow. If we now knock over domino 1, the rest will fall as the effect of that initiating event cascades along the causal pathway of dominos until finally domino 100 falls. In this simple setting, should the fall of domino 50 be considered a cause of the fall of domino 100? One possible answer is yes: the fall of domino 48 precipitates the fall of dominos 49 and 50, which transmit these changes to dominos 51 and 52, from whence they propagate to finally knock down domino 100. This might be termed a *mechanistic interpretation* of causation: the yellow domino 50 participates in the process by which initial changes (here, the fall of domino 1) propagate to knock down domino 100. A different answer is no: by hypothesis, had domino 50 been removed, nothing else would have changed: the collapse of domino 49 would still have sufficed to knock over domino 51, and dominos 52... 100 would still have fallen. In this *but-for* interpretation of causation—meaning that the outcome of interest (fall of domino 100) would not have occurred without (i.e., "but for") the occurrence of an event that caused it—the fall of the yellow domino is *not* a but-for cause of the eventual fall of domino 100, as the same outcome would have happened had it had not been present. (However, if the fall of domino 48 could not bridge the gap to knock over domino 51 if both dominos 49 and 50 were absent, their presence might be considered a *joint* but-for cause of the collapse of subsequent dominos.) A third interpretation: if instead domino 50 remained standing (perhaps being firmly bolted upright), it would have prevented the fall of domino 49 from knocking over any subsequent dominos and the cascade would have terminated at domino 49. In this sense, the fall of domino 50 is essential for the continued progression of the cascade leading to the collapse of domino 100, as the cascade would be interrupted in a counterfactual world where domino 50 did not fall. In this scenario, the fall of domino 50 could also be considered a *manipulative cause* (also called an *interventional cause*) of the fall of subsequent dominos, meaning that *any* external intervention that knocks over domino 50 (including the collapse of dominos 48 or 49) precipitates a further sequence of changes culminating in the predictable fall of domino 100. A philosopher might term the collapse of domino 50 an "INUS" (insufficient, but necessary part of an unnecessary but sufficient condition) (Mackie 1965) for the collapse of domino 100, if the collapse of dominos 50–99 is viewed as a sufficient condition for the collapse of domino 100 (but not necessary, since domino 100 will also fall if the cascade is started at domino 51 or later), with the collapse of domino 50 being a necessary part of the collapse of dominos 50–99.

None of these various interpretations is correct or incorrect. Rather, they articulate different definitions or concepts of what "cause" should be taken to mean. In applied epidemiology and risk assessment, all of these (and other) concepts of causation are used, and they are often conflated. We can think of the row of dominos as being analogous to a single causal biological pathway, and of the yellow domino as being analogous to a molecule of some noxious substance that can participate in the pathway. To answer whether or by how much removing the yellow domino—or, more generally, any subset of dominos—would reduce the risk of domino 100 falling (perhaps corresponding to an undesired event such as death or onset of a disease), it is necessary to understand how initiating events affect conditional probabilities of subsequent events. Such conditional probabilities provide a high-level summary of the effects of lower-level causal mechanisms, the details of which may or may not be well elucidated.

This causal information can seldom be retrieved by analyzing observed associations between variables. For example, suppose that repeated trials by many independent, meticulous investigators establish beyond reasonable doubt that the fall of domino 1 is eventually followed by the fall of domino 100 if and only if domino 50 also falls (as well as dominos 2–99). A reasonable intervention in response to these studies would be to prevent domino 50 from falling by removing it. However, the observed strong, consistent, mechanistically plausible, temporal, independently reproduced pattern of associations between the collapse of domino 50 and collapse of domino 100 may not reveal anything about how or whether an intervention that removes domino 50 would change the conditional probability of domino 100 falling after domino 1 falls. The answer depends on mechanistic details—in this example, the spacing between successive dominos. Very narrow spacing makes domino 50 redundant; wider spacing makes its presence essential for completion of the cascade. But such mechanistic details are absent from associational data used to calculate epidemiological measures such as probability of causation (PC), population attributable fraction (PAF), population attributable risk (PAR), etiologic fraction (EF), potential impact fraction (PIF), burden of disease (BOD), or similar measures based on relative risk (RR) ratios between exposed and unexposed (or more exposed and less exposed) populations. Observed associations and quantities derived from them without causal mechanistic information may not have a valid causal interpretation. For example, calculating a value of PC = 1 for the collapse of domino 50 as a cause of the collapse of domino 100 based on the perfect observed association between them does not reveal whether removing the "exposure" domino (yellow domino 50) would reduce the probability (or entirely prevent) the toppling of domino 1 from eventually causing domino 100 to fall. Likewise, generalizing findings from one population and set of study conditions to others requires causal mechanistic knowledge. Observing that removing domino 50 has no effect on the occurrence of an undesired outcome (fall of domino 100, or "disease") in one population (if the domino spacing is close in that population) would not license the generalization that removing domino 50 would have no effect in other populations of dominos in which the spacing was wide. Conversely, finding that removing domino 50 did have an effect on disease (if domino spacing was wide in the populations of dominos studied) would not justify an inference that removing it would have an effect in populations of dominos with close spacing. Generalization and external validity of causal inferences across populations (Campbell and Stanley 1963; Schwartz et al. 2011) requires an understanding of relevant causal mechanisms and differences in causally relevant factors across populations (Bareinboim and Pearl 2013; Lee and Honavar 2013).

Although some scientists might feel comfortable applying weight-of-evidence judgments and PAF, PC, regression, or similar calculations to interpret observed associations causally, the lack of mechanistic information in association data implies that *such data, and calculations based on them, may not reveal how interventions that change past conditions would affect outcomes*. But this interventional or manipulative causation is what policymakers usually need to know to choose among alternative policy interventions based on their probable consequences. How and whether risk would be reduced by reducing exposure can only be determined by modeling causal mechanisms that transmit changes. Such modeling cannot be accomplished by modeling observed associations in pre-intervention data (Pearl 2009).

Clarifying Causation with Conditional Probability Networks

A pragmatic answer to whether the collapse of the yellow domino 50 should be considered a cause of the eventual collapse of domino 100 is that *the answer does not matter* for any decision-making purpose. It is purely a matter of definition, without real-world consequences. If a goal is to prevent the collapse of domino 100 even after domino 1 falls, then what matters is how alternative feasible interventions, such as removing domino 50, or bolting it firmly in place, would change this outcome (or its probability). Although this cannot be answered from associations between observations (falls of dominos) in past (pre-intervention) data, nor by other analyses of pre-intervention data that do not model relevant causal mechanisms, it can be answered straightforwardly by causal artificial intelligence (CAI) methods that represent relevant mechanistic knowledge via high-level causal conditional probability tables (CPTs) or equivalent structural equation models (Druzdzel and Simon 1993). In these models, reviewed further in Cox (2021), the conditional probability distribution of each variable depends on the values of its direct causes, i.e., its parents in a causal directed graph model. Manipulative causation is modeled by letting exogenous changes in variables change the conditional probability distributions of their direct effects (the variables they point into in a directed graph model); these changes, in turn, propagate to additional downstream variables (if any) via the dependence of each variable's probability distribution on the values of its parents, as specified in its CPT. This is formalized by the "do calculus" of Pearl and closely related methods in simultaneous structural equations models long used in econometrics (Haavelmo 1943; Pearl 2015). For dominos that can be fallen or not fallen and that can be bolted upright or left free to fall, the causal conditional probability table (CPT) for a simple traditional cascade would specify that a bolted domino remains upright and that an unbolted domino falls if and only if its predecessor falls, i.e., *P*(*falls* | *not bolted & parent falls*) = 1; *P*(*falls* | *bolted & parent falls*) = 0; *P*(*falls* | *bolted & not parent falls*) = 0; *P*(*falls* | *not bolted & not parent falls*) = 0. If the spacing between dominos is variable, then the CPT would be modified to include a provision that an unbolted domino falls if and only if a sufficiently close predecessor falls. Details of timing are usually omitted from such causal models: each domino is modeled as fallen or not, given information on exogenous interventions (bolted or not) and parent state (fallen or not), with the time delay between a change in parent state and change in child state not being modeled. (Similarly, in simultaneous structural equations models, changing a variable on the right side of an equation causes the dependent variable on the left to change to make the equation hold again, but the delay for this to happen is left unspecified: only the equilibria before and after an exogenous change can be calculated.) Nonetheless, the partial knowledge about change propagation represented in a causal CPT or structural equation is precisely the high-level mechanistic knowledge needed to infer how interventions that knock over some dominos will (eventually) affect the states of others—or, more generally, their probability distributions, e.g., if the spacing between dominos is a random variable. This knowledge is admittedly superficial: it only describes *what* will happen, such as which dominos will end up falling, and not *why*, e.g., based on the physics of domino cascades. But knowing what will happen in response to interventions—or, in non-deterministic settings, how interventions will change outcome probabilities—is all that is needed for purposes of prediction and effective risk management decision-making.

Likewise, causal CPTs (or their equivalent structural equations) also provide the information needed to generalize causal conclusions from one study population to others. Here, the main idea is that a causal CPT, in contrast to a mere statistical one, is the same in all settings and for all interventions: the aleatory probability distribution of each variable is determined by the values of all of its direct causes, and by nothing else (although Bayesian inferences about unobserved quantities can, of course, be drawn from observed ones to arrive at epistemic conditional probability distributions). This concept, formalized in criteria for transportability (Bareinboim and Pearl 2013; Lee and Honavar 2013) and in the statistically testable property of invariant causal prediction (ICP) (Peters et al. 2016), allows causal CPTs estimated from one study or set of studies to be applied to other populations with different joint distributions of causally relevant covariates. For example, a causal CPT for dominos, learned from multiple diverse observations of cascades with different spacings, would give the predictive generalization (invariant across all study conditions) that a domino falls with probability 1 if it is unbolted and a predecessor less than 1 domino length away falls; and with probability 0 otherwise. This generalization can then be applied (or "transported") to new settings where dominos have lengths and spacings not seen in the data used to learn the rule. In this way, CPTs provide a constructive approach to establishing external validity of causal generalizations from study data. The approach is limited—it only works if enough is known for relevant causal CPTs to be identified and estimated—but is useful in moving past guesswork in determining what lessons from past studies are likely to apply under future conditions different from those for which data have been collected.

Pivoting Epidemiology from Attribution to Causal Prediction

The dominos thought experiment suggests to us the following lessons.

1. Determinations of whether a certain factor (e.g., presence of domino 50) should be considered a "cause" of an undesired event (e.g., fall of domino 100) are inherently somewhat ambiguous and arbitrary, insofar as the answer depends on which of various distinct concepts of "cause" is intended. There is no right or wrong answer to this definitional and terminological question, nor does common usage single out a single preferred or generally accepted answer.
2. Attributable risk is not the same as preventable risk. Population attributable fractions, probabilities of causation, burdens of disease, and similar association-based measures may not provide valid estimates or surrogates for the fraction or number of disease cases that would be prevented by eliminating or reducing an exposure because their calculations do not include crucial mechanistic information (Pearl and Tian 2000; Maldonado 2013).
3. Modern methods of causal artificial intelligence (CAI) can substantially fill this gap by predicting how interventions change outcome probabilities.
4. CAI methods also provide statistical methods to assess the predictive validity of causal generalizations from past studies (e.g., ICP CPTs, transportability formulas).
5. Well-validated causal generalizations that predict how interventions will change outcome probabilities provide information useful to risk managers and policymakers, even though they stop short of providing full scientific explanations and elucidations of underlying causal mechanisms.

From this perspective, the great attention that epidemiology has given to the problem of attributing risk to factors can profitably be reoriented to address the more pragmatic question of how changing exposures would change risks. These are distinct questions. Their answers should not be conflated. CAI methods provide a rich conceptual and technical resource for addressing the question of how changing exposures (or other factors) would change outcome probabilities, and incorporating these methods, with their limited but useful summaries of mechanistic knowledge via conditional probabilities, can help epidemiology provide valuable information for identifying causally effective risk management interventions.

CAI Conceptual Framework: Qualitative Structure of Causal Networks of Probabilistic Causal Mechanisms

This section introduces an approach to interventional causal analysis and interpretation of data. Although we call it causal artificial intelligence (CAI), this approach was developed by scientists from many disciplines over the past century; Table 11.1 notes some key milestones in this development. CAI treats causation as predictable propagation of changes (or their probabilities) through networks of random variables. It is not restricted to binary events or to deterministic propagation of changes. Rather, beginning with path analysis (Wright 1921) and extending through current methods, CAI models causality in terms of *propagation of changes through networks of directed dependencies between random variables.*

Qualitatively, dependencies can be represented by networks. One way to do so is as follows.

- **Nodes** in a network represent random variables.
- A **link** between two nodes represents a statistical dependency between them, meaning that observing one provides information about the other, although the dependency need not be causal. (By Bayes' Rule or the definition of conditional probability, statistical dependence is symmetric: if $P(X \mid Y)$ differs from $P(X)$, then $P(Y \mid X)$ differs from $P(Y)$, since $P(Y \mid X) = P(X \mid \mathrm{Y})P(Y)/P(X)$ and $P(X \mid Y)/P(X)$ differs from 1. In information-theoretic terms, two random variables X and Y are *dependent* if and only if they have positive mutual information, $I(X; Y) > 0$, where the mutual information $I(X; Y)$ measures the expected reduction in uncertainty (entropy) of either variable from conditioning on the other: $I(X; Y) = H(Y) - H(Y \mid X) = H(X) - H(X \mid Y)$, where $H(X)$ = entropy of X (Cover and Thomas 2006).)
- **Absence of a link** between nodes shows that they are *conditionally independent* of each other, given the values of other variables. For example, in the directed acyclic graph (DAG) model $X \rightarrow Y \rightarrow Z$, where X = exposure, Y = internal dose, and Z = probability of death, Z might be unconditionally dependent on both X and Y (each variable is positively correlated with the other two), but the DAG structure shows that Z is conditionally independent of X given Y: once Y is known,

Table 11.1 Selected milestones in the development of modern causal analysis

Milestone	Key ideas
Path analysis (Wright 1921)	Variations in effects around their mean values depend on variations in their direct causes.
Counterfactual causation, potential outcomes (Neyman 1923; Rubin 2005)	Differences in causes make their effects different from what they otherwise would have been.
Interventional causation in simultaneous equation models (Haavelmo 1943)	Exogenously fixing some variables at specified values changes probability distributions of variables that depend on them. Conditioning differs from fixing values (Pearl 2015).
Causal ordering of variables in simultaneous structural equation models (SEMs) (Simon 1953, 1954)	Causes are exogenous to their direct effects. Changes in causes create changes in their direct effects to restore equilibrium.
Predictive causation (Granger causation) in linear time series (Wiener 1956; Granger 1969)	Past and present changes in causes help to explain and predict future changes in their effects. The future of an effect is not conditionally independent of the past of its causes, given its own past.
Causal loop diagrams (Maruyama 1963)	Dynamic systems adjust until they reach equilibrium. The equilibrium is caused and explained by a balance of the mechanisms that act to change it (Mooij et al. 2013).
Quasi-experiments. Internal and external validity of causal conclusions (Campbell and Stanley 1963)	Differences in effects are explained by differences in causes and are not fully explained by differences in other factors.
Causal discovery algorithms (Glymour and Scheines 1986; Glymour et al. 2019)	Conditional independence and dependence relationships in data constrain the set of possible causal models for the data-generating process.
Interventional causation and causal ordering (Simon and Iwasaki 1988)	Exogenously changing causes changes the probability distributions of their direct effects.
Causal Bayesian networks, structural causal models, nonparametric SEMs, and conditional independence and dependence (Pearl 2000)	Effects are not conditionally independent of their direct causes, given the values of other variables. Conditioning on common effects induces dependence between parents that are conditionally independent.
Transfer entropy (Schreiber 2000) and directed information	Information flows from direct causes to their effects over time. Transfer entropy generalizes Granger causation.
Functional causal models (Shimizu et al. 2006)	An effect is often a simple (e.g., additive) function of its direct causes and random noise.
Transportability (Lee and Honavar 2013; Bareinboim and Pearl 2013)	Invariance of causal mechanisms (represented by CPTs) enables generalization of causal findings across studies.
Invariant causal prediction (ICP) (Peters et al. 2016; Prosperi et al. 2020)	Causal laws are universal: if levels for all of the direct causes of an effect are specified, then the conditional probability distribution of the effect should be the same across settings (e.g., studies) and interventions.

X provides no further information about Z. In this case, Z depends *directly* on Y but only *indirectly* on X. Quantitatively, in any such chain model $X \rightarrow Y \rightarrow Z$, observing X gives at least as much information about its direct effect Y as about its indirect effect Z: $I(X; Y) \geq I(X; Z)$ (Cover and Thomas 2006).

- **Directed links** (represented by arrows) without causal interpretations simply show one way (typically there are many) to factor a joint distribution of the variables into products of conditional and marginal distributions. For example, the joint distribution for two random variables X and Y can be factored equally well as $P(x, y) = P(x)P(y \mid x)$, corresponding to the DAG $X \rightarrow Y$; or as $P(x, y) = P(y)P(x \mid y)$, corresponding to the DAG $Y \rightarrow X$.
- **Directed links** with causal interpretations show the direction of causality: changes propagate from the variable at the tail of an arrow to the variable at its head—a concept sometimes formalized via the closely interrelated concepts of directed information flow, transfer entropy, and Granger causation in which information flows from one time series variable X to another Y creates positive mutual information between the past of X and the future of Y, even after conditioning on the past of Y (Schreiber 2000; Shorten et al. 2021). In a DAG model, if a directed link $X \rightarrow Y$ has a manipulative (interventional) causal interpretation, then changing the value (or distribution) of X changes the conditional distribution of Y via the formula $P(y) = \Sigma_x P(y \mid x)P(x)$. If changes propagate from X to Y and not from Y to X, then only $X \rightarrow Y$ has a valid causal interpretation as well as a valid statistical interpretation.
- **Undirected links** can be used to represent statistical associations without clear directions or causal interpretations, creating graphs that generalize DAGs. For example, if small cities are much more common in the south than in the north of a country, then conditioning on the observation that someone lives in a small city may increase the conditional probability that she lives in the southern region, and yet we might not want to say that either condition is a direct cause of the other.
- A **Bayesian network** (BN) is a DAG model in which all links are directed. If the directed links all have causal interpretations, then the BN is called a causal Bayesian network. If each node has a known conditional probability table (CPT), then the BN model is said to be fully specified.

Figure 11.1 shows an example of a BN fit to data on air pollution, weather, geographic, and elderly cardiovascular-pulmonary disease (CVD60) mortality variables (Vitolo et al. 2018). The data set contains nearly 50 million records from the United Kingdom, and has adequate power to quantify dozens of direct effects, corresponding to the arrows in Fig. 11.1. The DAG structure shows that *Month* is a common ancestor of ozone (o3) and mortality risk (CVD60) and that *Region* and *Year* are common parents of fine particulate matter (pm2.5) and mortality risk. If this BN structure is correct, then the many reported associations between these pollutants and mortality risk in studies that do not control for confounders (or common ancestors) such as *Month*, *Region*, and *Year* in Fig. 11.1 require care in causal interpretation. For example, the famous Harvard Six Cities study (Dockery et al. 1993) concluded that "Although the effects of other, unmeasured risk

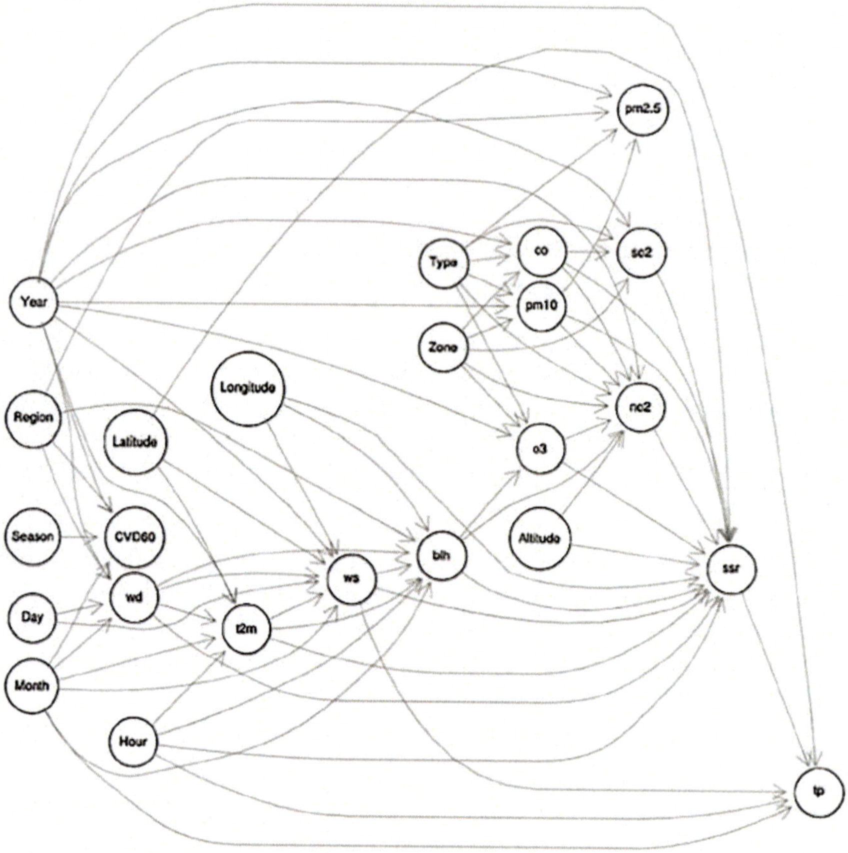

Fig. 11.1 A Bayesian network (BN) structure for air pollutants (pm2.5 = fine particulate matter, pm10 = coarse particulate matter, o3 = ozone, so2 = sulfur dioxide, no2 = nitrogen dioxide, co = carbon monoxide); cardiovascular-pulmonary disease mortality for people over age 60 (CVD60); and related weather and geographic variables (e.g., wd = wind direction, t2m = 2 meter temperature, ws = wind speed, etc.) *Source:* Vitolo et al. (2018)

factors cannot be excluded with certainty, these results suggest that fine-particulate air pollution, or a more complex pollution mixture associated with fine particulate matter, contributes to excess mortality in certain U.S. cities." A more cautious interpretation would be that the association does not suggest conclusions about what contributes to mortality risk, but shows a need to control carefully for differences among cities in confounders (analogous to *Region*, *Month*, and *Year* in Fig. 11.1) to discover whether mortality risk depends directly on fine particulate air pollution, or whether, as in Fig. 11.1, the association between them is explained by confounding.

Quantitatively, ***conditional probability tables*** (CPTs) describe dependencies among random variables. A CPT for a node (variable) tabulates the conditional probability of each of its possible values for each combination of values of its parents, i.e., of the variables that point into it (its direct causes, if all arrows have causal interpretations). In notation, denoting the set of parents of X by $Pa(X)$, its CPT is just the set of conditional probabilities $P(x \mid Pa(X))$ for each possible value x of X and each possible set of values for $Pa(X)$. For computational practicality, if there are too many possible combinations of parent values to tabulate conveniently, or if some of its parents are continuous, then the multivariate function $P(x \mid Pa(X))$ for node X can be approximated by a classification and regression tree (CART) or other function approximation methods popular in machine learning, such as random forest ensembles, deep learning networks, support vector or gradient boosted machines, and so forth. We refer to such conditional probability models generically as CPTs, although they need not be tables. For an input node—one with only outward-pointing arrows, i.e., no parents—the CPT is simply a marginal probability distribution table, $P(x)$, giving the unconditional probability of each possible value of the variable.

CPTs can be recoded as systems of nonparametric structural equations (Druzdzel and Simon 1993; Pearl 2012). In a causal model, each such equation models a causal mechanism by specifying how the probability distribution for the dependent variable on its left side depends on the values of the independent variables on its right side, which correspond to the dependent variable's parents in a graphical representation. The asymmetry of causality is captured by the convention that changes flow from the right side to the left side: if a right-side variable is changed, then the left-side variable adjusts to a new value to restore equality. As a simple deterministic example, the structural equation $P = nRT/V$ implies that if an exogenous intervention doubles V, then P will be reduced to half its initial value. The detailed timing of the adjustment process is left unspecified: interventions that change the right-side variables cause the left-side variable to change until the equation is again satisfied, but how long this takes—and, more generally, the time frame(s) over which changes are considered—is not given by the structural equation (or the corresponding CPT). Dynamic BNs (DBNs) that treat variables in different periods as different variables can model change over time to a limited extent, but different tools, such as systems of algebraic and ordinary differential equations, are needed to simulate dynamic systems in more detail. A system of simultaneous structural equations is called a structural equation model (SEM). An example of a deterministic SEM is the following system of simultaneous equations:

$$P = nRT/V$$

$$F = PA$$

$$a = F/m$$

(For a physical interpretation, variables P, V, T, F, A, m, and a might respectively represent pressure, volume, temperature, force, area, mass, and acceleration in a system involving a piston pushing a load, and n and R are constants.) A corresponding DAG model is as follows, where we subsume constants n and R into the formula (structural equation) determining P from V and T:

$$\begin{array}{ccccccc} & & T & \rightarrow & P & \leftarrow & V \\ & & & & \downarrow & & \\ A & \rightarrow & F & \rightarrow & a & \leftarrow & m \end{array}$$

Equations in an SEM corresponding to a causal DAG model can be partially ordered by causality using the principle that effects are determined by their causes (Druzdzel and Simon 1993). In this DAG, T is shown as a direct cause of P and an indirect cause of F and a, implying that exogenously changing T will change P, which in turn will change F and a (other things, such as V, A and m, being held equal). But exogenously changing F (e.g., by pushing on the piston head) will not change P or T, if this DAG structure correctly describes the system. In a *probabilistic* SEM, the right sides of structural equations may contain random and latent (unmeasured) variables, so that observed values of right-side variables only determine the probability distributions of left-side variables. Such probabilistic SEMs are equivalent to BNs, with the structural equations playing the roles of CPTs (Druzdzel and Simon 1993). The equations represent *causal mechanisms* by describing how probability distributions for the dependent variables on the left respond to changes in their direct causes on the right.

Practical Algorithms for Quantitative Causal Inference and Prediction with Realistically Imperfect Data

This section summarizes major developments in practical algorithms for quantitative probabilistic inference using BNs and more general probabilistic models. The following classes of algorithms draw useful quantitative conclusions and recommendations from fully specified BNs and observations.

- ***Inference algorithms*** implement Bayes' Rule for fully specified BNs (Salmerón et al. 2018). They compute the posterior distributions of all unobserved variables in a BN (and their joint posterior distribution) given any feasible set of observed or assumed values of other variables. These algorithms are useful for probabilistic expert systems and fault diagnosis, predictive maintenance, medical diagnosis and prognosis, and similar applications where probabilistic inferences are needed about unobserved variables from observed ones.
- ***Interventional causal prediction algorithms*** (Lin et al. 2021) predict how interventions that set one or more variables to specified values will change the probability distributions of other variables that depend on them. Pioneered in the 1940s in econometrics (Haavelmo 1943), interventional causation has been further developed in the influential *do-calculus* of Pearl and co-workers. The

do-calculus denotes by *do*(x) the operation of intervening to set variable X to a specific value, x—e.g., setting exposure to zero, or to a reduced value. It determines which interventional causal effects can be estimated from observational data in a BN, and provides constructive algorithms to estimate them, essentially by "surgery" on DAGs that disconnects exogenously manipulated variables from their up-stream causes while allowing their exogenously set values to affect variables that depend on them via CPTs (Shpitser and Pearl 2006).

- ***BN learning and causal discovery algorithms*** (Glymour et al. 2019) use (a) conditional independence and dependence constraints identified in the data, together with (b) scoring methods (similar to traditional model-selection methods such as the Bayesian information criterion (BIC) scores based on the likelihood of the data penalized for model complexity) and (c) knowledge-based constraints (e.g., that high temperature might cause elderly mortality, but elderly mortality cannot cause high temperature) to identify BN structures and CPT estimates that are most consistent with available knowledge and data. These "BN learning" algorithms address the practical question of how to estimate fully specified BNs from data. If arrow directions reflecting causality cannot be uniquely determined from available knowledge and data, then ensembles of multiple plausible BN models can still be used to support quantitative inferences despite model uncertainty (Madigan and Raftery 1994; Li et al. 2020).
- ***Causal effect estimation algorithms*** (Textor et al. 2016) analyze the DAG structures of BNs to automatically identify subsets of variables to condition on, called *adjustment sets*, to obtain unbiased estimates of the direct or total causal effect of one variable (such as exposure) on another (such as mortality rate), taking into account the values of other variables (e.g., co-exposures, co-morbidities, or other covariates that also affect mortality risk). These algorithms generalize the idea that it is necessary to condition on common parents or ancestors of exposure and response to control for confounding, but it is necessary to not condition on common children or descendants in order to avoid inducing collider bias. Large BNs often have several possible adjustment sets for the same causal effect. These can be used to check the internal consistency of the BN structure by testing whether the different adjustments sets lead to estimates for causal effects that do not differ significantly from each other (Ankan et al. 2021). Visualizations of how one variable (e.g., mortality risk) depends on another (e.g., exposure) while conditioning on observed levels of variables in an adjustment set and averaging over the empirical conditional distributions of the remaining variables can be obtained via *accumulated local effects* (ALE) plots (Apley and Zhu 2020).
- ***Explanation algorithms***. BN inference algorithms can be modified to identify the most probable values of unobserved variables, given observed or assumed values of other variables. These and closely related concepts (most probable explanations, most relevant explanations, and maximum *a posteriori* probability explanations) can be applied to causal Bayesian networks both to help diagnose and explain observed conditions and to identify how hypothesized future conditions

(e.g., failures of complex systems or attaining desired target states) might be caused (Yuan et al. 2011).

- ***Decision optimization and probabilistic planning algorithms***. BN inference algorithms can be modified to solve for values of decision variables to maximize the expected utility of consequences in BNs augmented with decision variables and value nodes, called influence diagrams (Shachter 1986; Shachter and Bhattacharjya 2010). Multiperiod decision optimization algorithms also allow for planning and decision optimization over time under probabilistic uncertainty (Hansen 2021). These developments can help manage uncertain risks over time in settings where it is necessary to learn from experience how effective interventions are (e.g., how well mask mandates or vaccines reduce risk of COVID-19 spread), or in which their effectiveness changes randomly over time.

These algorithms are now widely available, e.g., as free R packages in the CRAN repository, https://cran.r-project.org/. They provide computationally practical methods and software for applying fully specified BNs to draw inferences from observations, predict the effects of interventions, and optimize sequences of interventions. Each class of algorithms has also been extended beyond the DAG framework, as simultaneous equations or underlying dynamic models (e.g., ordinary differential equations) with directed cycles in dependency relationships among variables are often more natural models than DAGs for representing systems with simultaneous causation, feedback, or undirected statistical associations (Haavelmo 1943; Mooij et al. 2013; Glymour et al. 2019).

Fully specified BNs and more general probabilistic models provide constructive solutions to several important technical statistical challenges in applied epidemiology and health risk assessment with realistically imperfect data and knowledge, as follows.

- ***Unmeasured (latent) variables***. BN inference algorithms enable the calculation of joint posterior probabilities for unobserved (latent) quantities by conditioning on the values of observed quantities. Latent variables pose no substantial challenge for inference in a completely specified BN, since BN algorithms automatically calculate their contributions to observations. For example, if $X \rightarrow Y \rightarrow Z$, where $X =$ exposure, $Y =$ internal dose, and $Z =$ mortality risk, and if Y is not measured but X is, then the conditional probability distribution for Z based on the observed exposure $X = x$ is simply $P(z \mid x) = \Sigma_y P(z \mid y)P(y \mid x)$. This sum models the unobserved contribution of Y by weighing each of its possible values by its conditional probability given observations ("marginalizing out" the unobserved variable). By contrast, if Y is observed to have value y^*, then the probability distribution for Z becomes $P(z \mid y^*)$, and the observed value of X is irrelevant. If Y is not measured but diagnosis of a co-morbidity (e.g., chronic obstructive pulmonary disease, COPD) caused by Y is available for each subject, indicated by $W = 1$ for a positive diagnosis and $W = 0$ otherwise, then the BN would be extended as follows:

$$\begin{array}{c} X \rightarrow Y \rightarrow Z \\ \downarrow \\ W \end{array}$$

The probability distribution of mortality risk Z would now be conditioned on both observed variables, X and W. In short, BN inference algorithms condition on whatever observations are available (e.g., X and W in this example) to obtain predictive posterior distributions for all unobserved quantities, including latent variables (such as Y in this example). If a fully specified BN is not known, however, and must be estimated from data, then latent variables can complicate the estimation process. Unique estimates may be achievable only in special cases, such as for linear models with normally distributed errors (Chobtham and Constantinou 2020), and ensembles of possible BNs must be used otherwise (Madigan and Raftery 1994; Li et al. 2020).

- ***Missing data***. Missing data values can be handled in the BN framework by imputing (sampling) their uncertain values from their conditional distributions given observed values. Adaptations of standard missing data techniques such as the expectation-maximization (EM) algorithm have proved effective for BN learning with incomplete data (Myers et al. 1999; Masegosa et al. 2016).
- ***Errors in variables***. Measurement errors can be treated as a special case of latent variables. For example, suppose that individual exposures are not known, but are estimated by spatial smoothing of measurements at monitoring stations or satellite data for pollutant data, or from job-exposure matrices for occupational exposures. Then instead of the measurement error-free model $X \rightarrow Z$, where $X =$ exposure and $Z =$ mortality risk, a BN with measurement error would be $M \leftarrow X \rightarrow Z$, where M is the estimated or measured exposure and X is the unobserved (latent) true exposure. Given observed values of M, the corresponding conditional probability distributions for X and Z can be quantified by standard BN inference algorithms.
- ***Model specification errors***. The use of flexible non-parametric methods such as random forest to quantify CPTs minimizes the potential for model specification error, e.g., omitting relevant variables or interaction terms, using incorrect functional forms, or misspecifying error distributions. In regression models, these mistakes can create statistically significant coefficients for variables that have no causal relevance, but that reduce the mean squared prediction error for the dependent variable arising from model specification errors (Cox 2020). Nonparametric BN estimation methods avoid these parametric modeling challenges.
- ***Model uncertainty***. If causal discovery algorithms identify several possible BN models, then uncertainty about which one is correct (or best describes the data-generating process) can also be treated via latent variables by making the identity of that model the latent variable of interest. An exposure-response model with model uncertainty can be represented as $X \rightarrow Z \leftarrow D$, where $X =$ exposure, $Z =$ mortality risk, and $D =$ indicator of the correct model. Treating D as a latent variable, predictive distributions for Z with model uncertainty are derived by

weighing members of the ensemble of possible models by their probabilities: $P(z \mid x) = \Sigma_d P(z \mid x, d)P(d)$ where d indexes possible models and $P(d)$ is the conditional probability of model d given observations. If the number of possible models is very large and/or prior probabilities for models are unknown, then model-averaging techniques that discard relatively low-likelihood models using Bayes factors can be used to improve computational efficiency (Madigan and Raftery 1994; Li et al. 2020).

- ***Interindividual heterogeneity***. If different individuals in a population have different individual exposure-response functions, e.g., due to phenotypic variations in biochemistry, then no single aggregate exposure-response relationship can adequately represent their heterogeneous responses to an exogenous change in exposures. Observed sources of heterogeneity, such as differences in age, income, smoking, or health status, can be modeled by causal random forests and generalized random forests (Athey et al. 2019). Unobserved sources of interindividual heterogeneity can be modeled as a composite latent variable (the individual's "type"), allowing them to be treated as another special case of latent variables. For example, the model $X \rightarrow Z \leftarrow H$, where X = exposure, Z = mortality risk, and H = individual type (unobserved) leads to $P(z \mid x) = \Sigma_h P(z \mid x, h)P(h)$ where h indexes possible types; these serve as the components of this finite mixture distribution model.
- ***Internal validity of causal conclusions.*** Internal validity addresses whether estimated causal effects in a study correctly describe causation in the study (Campbell and Stanley 1963). To establish the internal validity of causal effects estimates, it is necessary to refute rival (non-causal) explanations for the data and to condition on appropriate adjustment sets of other variables. For BNs, causal effect estimation algorithms allow adjustment sets to be computed from BN DAG structures (Textor et al. 2016). Comparing causal effect estimates for multiple adjustment sets (if there are several) reveals whether they are internally consistent.
- ***External validity of causal predictions***. If a causal BN model is learned (i.e., both its structure and its CPTs are estimated) from data for one or more populations, will its causal conclusions and predictions, e.g., about how mortality risk depends on exposure, prove accurate for a different population? This question of *external validity* of causal conclusions has challenged generations of epidemiologists and risk assessors (Campbell and Stanley 1963; Schwartz et al. 2011). It has been solved relatively recently for causal BNs by using the do-calculus to establish necessary and sufficient conditions for "transportability" (i.e., generalization) of causal relationships, modeled by causal CPTs, from one population to another (Lee and Honavar 2013; Bareinboim and Pearl 2013). A key idea is that, while statistical CPTs may differ across populations with different compositions of heterogeneous (non-exchangeable) individuals, a causal CPT with all of its direct causes included should be invariant across populations and interventions (Peters et al. 2016; Prosperi et al. 2020). This *invariant causal prediction property* (ICP) for causal CPTs (Peters et al. 2016), formalizing the intuition that fully stated causal laws are universal, can be tested empirically by testing whether responses

are conditionally independent of the study or population from which data come, given the values of the variables.

- ***Generalization to new conditions.*** Causal CPTs satisfying the ICP property can be used to predict exposure-response relationships in target populations with conditions different from those for which data have been collected. This is accomplished by applying the ICP CPTs to the joint distribution of conditions in the target population. What is generalized from past experience is the ICP CPTs. Their implications for observable population exposure-response relationships will depend on the joint distribution of conditions that affect responses. For example, if Z is a risk indicator with a value of 1 for someone who is diagnosed with a certain disease before dying and a value of 0 otherwise, and if $P(z \mid Pa(Z))$ is the ICP CPT for Z, with exposure being one of the parents of Z, then an exposure-response curve relating exposure to probability of harm by marginalizing out all other parents (i.e., summing or integrating $P(z \mid Pa(Z))$ overall their possible joint values weighted by their respective conditional probabilities or probability densities given the observed level of exposure) may be quite different in different populations. Thus, if Z has two parents, $X =$ exposure and $A =$ age, each scaled to run from 0 to 1, where 1 denotes the maximum value in the population, and if for any specific exposure x and age z the risk is given by the CPT eq. $E(Z \mid Pa(Z) = (x, a)) = P(Z = 1 \mid Pa(Z) = (x, a)) = xa$, then the exposure-response function in a population of young people with $a = 0.2$ would be $E(Z \mid x) = 0.2x$, but the exposure-response function in an elderly population with $a = 0.9$ would be $E(Z \mid x) = 0.9x$. Invariance (ICP) generalizes the Bradford-Hill consideration of "consistency" in exposure-response associations by recognizing that populations with different distributions of covariates (e.g., age) may have different marginal exposure-response curves for the same (invariant) causal CPT. In this example, the invariant CPT eq. $E(Z \mid Pa(Z) = (x, a)) = xa$ allows the result $E(Z \mid x) = 0.2x$ for the first population to be transported to the prediction $E(Z \mid x) = 0.9x$ for the second population.

These technical developments address the most common challenges for data analysis with realistically incomplete and imperfect data and knowledge. They offer data-driven options for dealing with data and model uncertainties and for testing whether causal conclusions from past studies hold across diverse study conditions and can confidently be generalized to new situations with new combinations of conditions. However, our main interest is in the new light these advances shed on how to better answer the questions traditionally addressed by attributable risk, PAF, PC, and related epidemiological calculations using associational data without causal mechanisms.

Implications of CAI for Calculating and Interpreting Preventable Fractions

Seeing Is Not Doing

The distinction between inference algorithms and causal prediction algorithms emphasizes a crucial philosophical point: contrary to implicit assumptions commonly made in calculating and interpreting PAF, PC, and other variations on relative risk ratios and absolute risk differences, *observational data on exposure and response levels do not provide mechanistic information that may be needed to predict correctly how setting exposure to zero, or to a reduced level, would change responses* in a population. Figure 11.2 illustrates the problem. It shows 8 data points, one for each of 8 individuals. We interpret x as exposure and y as a health outcome, such as length of life or annual survival probability (scaled to the axes shown). Now, consider an intervention that reduces the value of x for each individual to 0. How will their y values change in response? Standard epidemiological practice uses regression models to estimate the answer, assuming that each dot would slide along an appropriate regression line or curve to a new position with $x = 0$.

This procedure substitutes the answer to an easy but irrelevant statistical question (what is the y value associated with $x = 0$ in a regression model fit to the observed data?) for the answer to a hard but relevant causal question: how would reducing x to 0 change each y value? The answer to the latter question is not, even in principle, revealed by the data points alone. Assumptions can be made to help answer it—such as that the observed differences in y values are explained (and caused) solely by differences in corresponding x values, but the validity of resulting predictions and conclusions is then contingent on the validity of the assumptions used in deriving them, which may be uncertain.

Figure 11.2 seeks to illustrate this basic distinction between statistical and causal questions (Pearl 2009) by showing data for which it is unclear whether it is more plausible to expect that the dots would slide leftward and downward along their local (upward-sloping, solid) regression lines as x is reduced; or that they would slide leftward and upward along the global (downward, sloping, dashed) line. The more

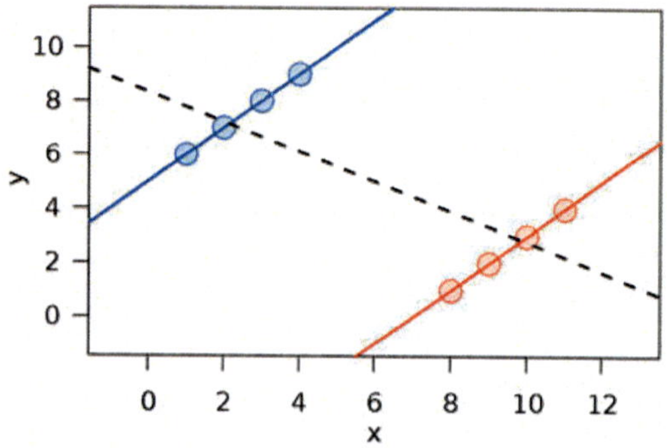

Fig. 11.2 Seeing vs. doing. How y values of the individual data points would change if their x values were reduced by a stated amount cannot be determined from the data. *Source:* https://en.wikipedia.org/wiki/Simpson%27s_paradox

fundamental point is that *there is no basis in observed exposure-response data for making either prediction*, or some other one (e.g., that the y values of all dots would stay unchanged as x decreases, or that the dots on the right would travel down their local line (red) until x crossed a threshold, perhaps at $x = 5$, where their y values would jump up to the higher (blue) local line, possibly reflecting that some biological process or resource is no longer saturated or depleted at x values below 5). The responses of y to changes in x simply cannot be predicted correctly with confidence from the data, as the responses depend on causal mechanisms that are not represented in the data. The data show observed levels of x and y, but are silent about the causal mechanisms (e.g., described by CPTs or structural equations) that predict how and whether changing x would change y. As emphasized by Pearl (2009) and by econometricians in earlier decades (Haavelmo 1943; Simon 1953), seeing is not doing: observational data together with regression equations describing associations among variables do not address causal questions about how taking action to set some variables to new levels will change others. In fairness to current epidemiology practice, it should be noted that observations are often supplemented by assumptions for purposes of causal inference (e.g., assumptions of exchangeability or no hidden confounders), but such assumptions are often untested (and perhaps untestable) or are left implicit, and causal conclusions based on them are too often presented without characterizing the extent to which they depend on uncertain assumptions, as best practice recommends (Lash et al. 2014).

Ambiguity of Counterfactuals for PAFs

The population attributable fraction (PAF) is traditionally defined as the ratio $(O - E)/O$, where O = observed number of cases and E = expected number of cases under no exposure. The PAF ratio is widely interpreted as an estimate of both (a) the fraction of all cases in a population attributable to exposure; and (b) the fraction of all cases that would not have occurred had there been no exposure (e.g., Mansournia and Altman 2018; Mansournia et al. 2021). The CAI framework suggests that these concepts involve crucial ambiguities and that $PAF = (O - E)/O$ has no valid but-for or interventional causal interpretations, as it does not address causal mechanisms. First, the phrase "expected number of cases under no exposure" is ambiguous: it does not distinguish between exposures that are *observed* to be 0 and exposures that are *set* to be 0. In general, these give different expected numbers of cases (Haavelmo 1943; Pearl 2015). *Why* exposure is 0—whether because of exogenous interventions or by chance under the conditions that have generated data so far—may matter greatly for predicting E. But the $(O - E)/O$ formula does not specify which (if either) is meant.

Second, the expected number of cases with exposure set to any specific level such as 0 depends in general on levels of *other* covariates that also affect numbers of cases (other parents of risk in a causal BN model), such as age, co-exposures, co-morbidities, region, year, etc. The $(O - E)/O$ formula does not specify how other variables change when exposure is set to 0. It may be physically unrealistic to assume that other variables are left unchanged by an intervention that reduces exposure. For example, reducing one pollutant may change levels of other pollutants (Fig. 11.1), of weather variables such as precipitation and temperature, of probabilities of comorbidities that in turn affect expected numbers of cases, and so forth. Holding other variables fixed at their pre-intervention values would then be an unrealistic irrelevant counterfactual scenario. On the other hand, if other variables are set to estimated realistic post-intervention values, then the resulting predicted value of E will reflect not only the direct effect of reduced exposure but also the effects of changes in the values of other parents of E following the change in exposure. To obtain mutually consistent, realistic estimates of how an intervention affects all other variables that affect response it is necessary to engage in CAI causal network modeling or something much like it (e.g., causal simulation modeling). If no explicit decision is made about how to model changes in other (non-exposure) variables caused by an intervention, a common alternative is to use a regression model to estimate E with exposure equals zero. But regression modeling conflates the value of E caused by *setting* exposure to 0 with the predicted value of E if exposure is *observed* to be 0 under the pre-intervention conditions that generated the data for the regression model. In addition, the estimated coefficient of exposure in an empirical regression model typically reflects contributions from non-causal factors such as model specification errors, attribution of interaction effects, ignored measurement errors, contributions from correlated variables, and so forth, rather than a purely causal effect of exposure (Cox 2020). The value of E predicted by an empirical regression model under the conditions for which data were collected typically has no valid interventional causal interpretation (Pearl 2009) and may not apply for conditions following an intervention that changes exposure levels. CAI methods address this challenge by analyzing the transportability of estimated causal effects to new conditions. Something very similar is needed for any process that uses pre-intervention data to predict post-intervention case numbers, E.

There is not a right or wrong answer to what should be assumed about how other variables change when exposure changes: this is a matter of defining what we mean to calculate. But if E is estimated from a regression model, rather than from a causal model of how variables change in response to an intervention, then it is not clear what the ratio $(O - E)/O$ represents. Contrary to widespread teaching and usage in epidemiology, it is usually clear that it does *not* represent the fraction of all cases that would not have occurred had exposure been set to zero. To make this methodological point, we consider an extreme case in which each of five pollutants is either present or absent. We denote by 1 the level of a pollutant when present and by 0 its level when absent. Suppose that the causal DAG structure is $Z \leftarrow X1 \rightarrow X2 \rightarrow X3 \rightarrow X4 \rightarrow X5$ where Z indicates an adverse health response ($1 =$ present, $0 =$ absent) and $X1...X5$ indicate the levels of pollutants 1–5. That is, $X1$ is a direct cause of Z and also of $X2$; $X2$ is a direct cause of $X3$, and so on. Quantitatively, suppose that $X1$ is currently 1 and that the CPTs specify that, in the absence of exogenous interventions, each variable is 1 if its parent is 1, else it is 0. (Recall that an intervention that exogenously sets a variable equal to a new level disconnects it from its parents.)

Under these conditions, considering *X5* as the pollutant of interest, a regression model of past observed *Z* values against corresponding *X5* values would yield $O = 1$, $E = 0$, and $PAF_{X5} = 1$ as the PAF for pollutant *X5*. Symmetrically, $PAF_{X4} = 1$, $PAF_{X3} = 1$, and $PAF_{X2} = 1$. What these calculations do not reveal is that setting any of these variables equal to 0 would have no effect on *Z:* only *X1* is a cause of Z. The PAF ratios for the other variables are irrelevant for predicting how or whether *Z* would change if exogenous interventions set the other exposures to zero. The same applies to probability of causation and burden of disease calculations: PC values of 1 for variables *X2* through *X5* do not imply that reducing these exposures would have any effect on case numbers.

CAI algorithms make it straightforward to calculate causally meaningful PAFs from a fully specified causal BN model: simply calculate the expected number of cases predicted by the causal BN under present (pre-intervention) and new (post-intervention) conditions. Call the first *O* and the second *E*, and apply the $PAF = (O - E)/O$ formula to these values. In the present example, setting *X5* to 1 or to 0 leaves Z unchanged, so $O = E$ and the causal PAF for an intervention that changes *X5* from 1 to 0 is $PAF_{X5} = 0$. Symmetrically, $PAF_{X4} = 0$, $PAF_{X3} = 0$, and $PAF_{X2} = 0$; but $PAF_{X1} = 1$. Note that PAFs are now not attached to exposures or other risk factors, but to *interventions* that change levels of exposures (and possibly other variables). That is, instead of attributing some fraction of cases to a pollutant—a proposition that makes no testable or falsifiable predictions, and in this sense is devoid of scientific content—these intervention-specific causal PAFs predict the *fraction of current cases that would be prevented by interventions* that change current conditions, thereby altering the data-generating process. Arguably, this is the information that PAFs should provide to have value to policy-makers choosing among various possible interventions. Calculating preventable fractions for interventions also avoids the tendency of PAFs and PCs for risk factors to sum to more than 100%—e.g., the PAFs (and PCs) for pollutants *X1* through *X5* in the above example to sum to 5. The causal PAF for any (single or combined) intervention is always between 0 and 1.

Discussion and Conclusions: Toward Pragmatic Causal PAF Calculations

Causal artificial intelligence (CAI) suggests the following perspectives and advice on how to better calculate and interpret epidemiological and health risk assessment concepts such as population attributable fraction (PAF) and similar concepts to be more useful to risk management decision-makers:

1. **Calculate PAFs for interventions.** To inform practical risk management decisions, PAFs should be calculated for interventions rather than for exposures or risk factors. The PAF for an intervention is defined by the familiar formula $PAF = (O - E)/O$, but with *O* and *E* reinterpreted as *causal predictions* for the

relevant incidence rates or risk measures, e.g., expected number of cases per year for population risks or expected cases per person-year for individual risks, without the intervention and with it, respectively. We refer to predicting *O*, rather than simply recording its observed value, because true incidence rates are seldom perfectly observable, but must be predicted from data with realistic data uncertainties and model uncertainties and limitations (e.g., latent variables, missing data, measurement errors, model specification uncertainty, and interindividual heterogeneity).

2. **Use causal models to make causal predictions of *O* and *E*.** Causal predictions of *O* and *E* can be obtained from a fully specified causal BN model if one is known. Valid causal predictions of *O* and *E* usually *cannot* be obtained from regression models or other associational models fit to pre-intervention data without modeling relevant causal mechanisms (represented in CAI by causal CPTs or structural causal model equations) (Pearl 2000). If an appropriate causal model for predicting *O* and *E* is not known, then the technical methods and algorithms we have discussed (e.g., ensembles of non-parametric estimates of causal BN structures and CPTs) can be used to estimate uncertainty distributions of *O*, *E*, and *PAF* based on realistically imperfect data and knowledge.
3. **Interpret resulting PAF estimates, *PAF* = (*O* - *E*)/*O*, as the *preventable fractions* of cases for interventions**, i.e., (*O* - *E*)/*O* is the fraction of cases that is predicted to be prevented by an intervention that is predicted to replace *O* with *E*.

In this view, *attributing* cases (or of a fraction of cases) to specific exposures, factors or interventions is unnecessary and adds no value to preventable fraction calculations. Attributing cases to causes does not make testable or falsifiable predictions or empirically verifiable or refutable claims. Preventable fraction calculations do. But an intervention that is predicted to reduce population or individual risks (for at least some individuals) may well do so by changing several variables, e.g., levels or conditional probability distributions of multiple pollutants and weather variables that depend on them. The accounting challenge of allocating or attributing credit or blame for results to the multiple changes that jointly produce them may be intellectually interesting, but it does not change anything observable in the world. There is no need or value for such attribution in any rational decision-making process that is driven by changes in observable results, such as the fractional decrease in case incidence caused by an intervention. Likewise, causal determination exercises that seek to determine whether specific exposures should be called causes (or likely causes, possible causes, etc.) of specific health outcomes, are exercises in terminology without empirical content: they are not needed or useful for predicting the consequences of interventions using CAI methods.

From this perspective, what matters for risk management decision-making is how alternative feasible interventions change outcomes (or their probabilities). This cannot in general be answered from associations between observations in past (pre-intervention) data; nor by any other analysis of pre-intervention data that does not model relevant causal mechanisms. But it can be answered straightforwardly by

CAI methods that describe relevant mechanistic knowledge via CPTs or structural equations. Making qualitative causal determinations and calculating attributable fractions (and PCs, PARs, BODs, and related differences and ratios) for exposures and other risk factors based on regression models or other associational analyses of pre-intervention data, without predicting post-intervention changes in outcomes or their probability distributions using models of causal mechanisms, should be viewed as purely definitional exercises without empirical content. Students and practitioners can safely skip them without impairing ability to identify decisions that make preferred outcomes more probable. On the other hand, epidemiologists who wish to predict how alternative interventions would affect public or occupational health should use CAI methods to estimate *O* and predict *E* from available data and from estimated causal CPTs or nonparametric structural causal models representing high-level summaries of causal mechanisms. These tools allow calculation of preventable fractions and related quantities, such as uncertainty distributions for absolute numbers of cases per year or per person-year, before and after interventions. Ultimately, information about changes in outcome probability distributions caused by interventions, not attribution of fractions of cases to risk factors, is needed to guide more causally effective risk management decision-making that more reliably causes desired results.

References

Abraham S, Sahibzada S, Hewson K, Laird T, Abraham R, Pavic A, Truswell A, Lee T, O'Dea M, Jordan D (2020) Emergence of Fluoroquinolone-resistant *Campylobacter jejuni* and *Campylobacter coli* among Australian chickens in the absence of fluoroquinolone use. Appl Environ Microbiol 86(8):e02765–e02719. https://doi.org/10.1128/AEM.02765-19

Ankan A, Wortel IMN, Textor J (2021 Feb) Testing graphical causal models using the R package "dagitty". Curr Protoc 1(2):e45. https://doi.org/10.1002/cpz1.45

Apley DW, Zhu J (2020) Visualizing the effects of predictor variables in black box supervised learning models. J R Stat Soc 82:869–1164

Athey S, Tibshirani J, Wager S (2019) Generalized random forests. Ann Stat 47(2):1148–1178. https://doi.org/10.1214/18-AOS1709

Bareinboim E, Pearl J (2013) Causal transportability with limited experiments. In: Proceedings of the 27th AAAI Conference on Artificial Intelligence, pp 95–101. ftp://ftp.cs.ucla.edu/pub/stat_ser/r408.pdf

Bartholomew MJ, Vose DJ, Tollefson LR, Travis CC (2005) A linear model for managing the risk of antimicrobial resistance originating in food animals. Risk Anal 25(1):99–108. https://doi.org/10.1111/j.0272-4332.2005.00570.x

Burns J, Boogaard H, Polus S, Pfadenhauer LM, Rohwer AC, van Erp AM, Turley R, Rehfuess EA (2020 Feb) Interventions to reduce ambient air pollution and their effects on health: an abridged Cochrane systematic review. Environ Int 135:105400. https://doi.org/10.1016/j.envint.2019.105400

Campbell DT, Stanley JC (1963) Experimental and quasi-experimental designs for research. Houghton, Mifflin and Company, Boston, MA

Cardwell CR, Abnet CC, Cantwell MM, Murray LJ (2010 Aug 11) Exposure to oral bisphosphonates and risk of esophageal cancer. JAMA 304(6):657–663. https://doi.org/10.1001/jama.2010.1098

Chobtham K, Constantinou AC (2020) Bayesian network structure learning with causal effects in the presence of latent variables. In: Proceedings of the 10th International Conference on Probabilistic Graphical Models, in Proceedings of Machine Learning Research, vol 138, pp 101–112. https://proceedings.mlr.press/v138/chobtham20a.html

Cover TM, Thomas JA (2006) Elements of information theory, 2nd edn. Wiley, Hoboken, NJ

Cox LA Jr (2018 Sep) Modernizing the Bradford Hill criteria for assessing causal relationships in observational data. Crit Rev Toxicol 48(8):682–712. https://doi.org/10.1080/10408444.2018.1518404

Cox LA Jr (2019) Communicating more clearly about deaths caused by air pollution. Glob Epidemiol 1:100003. https://doi.org/10.1016/j.gloepi.2019.100003

Cox LA Jr (2020 Aug) Implications of nonlinearity, confounding, and interactions for estimating exposure concentration-response functions in quantitative risk analysis. Environ Res 187: 109638. https://doi.org/10.1016/j.envres.2020

Cox LA Jr (2021) Quantitative risk analysis of air pollution health effects. Springer, New York

Dockery DW, Pope CA, Xu X, Spengler JD, Ware JH, Fay ME, Ferris BG Jr, Speizer FE (1993 Dec 9) An association between air pollution and mortality in six U.S. cities. N Engl J Med 329 (24):1753–1759. https://doi.org/10.1056/NEJM199312093292401. PMID: 8179653.

Dockery DW, Rich DQ, Goodman PG, Clancy L, Ohman-Strickland P, George P, Kotlov T, HEI Health Review Committee (2013 July) Effect of air pollution control on mortality and hospital admissions in Ireland. Res Rep Health Eff Inst 176:3–109

Druzdzel MJ, Simon HA (1993) Causality in Bayesian belief networks. In: Proceedings of the Ninth Annual Conference on Uncertainty in Artificial Intelligence (UAI-93), Morgan Kaufmann Publishers, San Francisco, CA, pp 3–11

Glymour C, Scheines R (1986) Causal modeling with the tetrad program. Synthese 68(10):37–63

Glymour C, Zhang K, Spirtes P (2019 Jun 4) Review of causal discovery methods based on graphical models. Front Genet. 10:524. https://doi.org/10.3389/fgene.2019.00524

Goodman GE, Thornquist MD, Balmes J, Cullen MR, Meyskens FL Jr, Omenn GS, Valanis B, Williams JH Jr (2004 Dec 1) The Beta-Carotene and Retinol Efficacy Trial: incidence of lung cancer and cardiovascular disease mortality during 6-year follow-up after stopping beta-carotene and retinol supplements. J Natl Cancer Inst 96(23):1743–1750. https://doi.org/10.1093/jnci/djh320

Granger CWJ (1969) Investigating causal relations by econometric models and cross-spectral methods. Econometrica 37(3):424–438

Green J, Czanner G, Reeves G, Watson J, Wise L, Beral V (2010 Sep) Oral bisphosphonates and risk of cancer of oesophagus, stomach, and colorectum: case-control analysis within a UK primary care cohort. BMJ 1(341):c4444. https://doi.org/10.1136/bmj.c4444

Haavelmo, T. (1943). The statistical implications of a system of simultaneous equations. Econometrica 11:1–12. Reprinted in DF Hendry and MS Morgan (eds), The foundations of econometric analysis. Cambridge University Press, 477–490, 1995

Hansen EA (2021) An integrated approach to solving influence diagrams and finite-horizon partially observable decision processes. Artif Intell 2021:294. https://doi.org/10.1016/j.artint.2020.103431

Hao H, Sander P, Iqbal Z, Wang Y, Cheng G, Yuan Z (2016 Oct 18) The risk of some veterinary antimicrobial agents on public health associated with antimicrobial resistance and their molecular basis. Front Microbiol 7:1626. https://doi.org/10.3389/fmicb.2016.01626

Hillard M (2020) Public awareness was vital for smoky coal ban, says campaigner. Asthma Society of Ireland says ban has resulted in more than 350 fewer annual deaths. The Irish Times. Tue, Sep 1, 2020, 01:07. https://www.irishtimes.com/news/ireland/irish-news/public-awareness-was-vital-for-smoky-coal-ban-says-campaigner-1.4342778

Lagani V, Triantafillou S, Ball G, Tegnér J, Tsamardinos I (2016) Probabilistic computational causal discovery for systems biology, Chapter 2. In: Geris L, Gomez-Cabrero D (eds) Uncertainty in biology: a computational modeling approach. Springer, New York

Lash TL, FoxMP MLRF, Maldonado G, McCandless LC, Greenland S (2014) Good practices for quantitative bias analysis. Int J Epidemiol 43(6):1969–1985. https://doi.org/10.1093/ije/dyu149

Lee S, Honavar V. (2013) m-Transportability: transportability of a causal effect from multiple environments. In: Proceedings of the Twenty-Seventh AAAI Conference on Artificial Intelligence. www.aaai.org/ocs/index.php/AAAI/AAAI13/paper/viewFile/6303/7210

Li M, Zhang R, Liu K (2020) A new ensemble learning algorithm combined with causal analysis for Bayesian network structural learning. Symmetry 12(12):2054. https://doi.org/10.3390/sym12122054

Lin L, Sperrin M, Jenkins DA et al (2021) A scoping review of causal methods enabling predictions under hypothetical interventions. Diagn Progn Res 5:3. https://doi.org/10.1186/s41512-021-00092-9

Mackie JL (1965) Causes and conditions. Am Philos Q 2(4):245–264. http://www.jstor.org/stable/20009173

Madigan D, Raftery AE (1994) Model selection and accounting for model uncertainty in graphical models using Occam's window. J Am Stat Assoc 89(428):1535–1546. https://doi.org/10.1080/01621459.1994.10476894

Maldonado G (2013 Dec) Toward a clearer understanding of causal concepts in epidemiology. Ann Epidemiol 23(12):743–749. https://doi.org/10.1016/j.annepidem.2013.09.001

Mansournia MA, Altman DG (2018 Feb 22) Population attributable fraction. BMJ 360:k757. https://doi.org/10.1136/bmj.k757

Mansournia MA et al (2021) Population attributable fraction in textbooks: time to revise. Glob Epidemiol (Forthcoming)

Maruyama M (1963 Jun) The second cybernetics: Deviation-amplifying mutual causal processes. American Scientist 51(2):164–179

Masegosa AR, Feelders AJ, van der Gaag LC (2016) Learning from incomplete data in Bayesian networks with qualitative influences. Int J Approx Reason 69:18–34

Mooij JM, Janzing D, Schölkopf B. (2013) From ordinary differential equations to structural causal models: the deterministic case. In: Proceedings of the Twenty-Ninth Conference on Uncertainty in Artificial Intelligence (UAI'13). AUAI Press, Arlington, Virginia, USA, pp 440–448

Myers J, Laskey KB, Levitt TS (1999) Learning Bayesian networks. In: Uncertainty in Artificial Intelligence: Proceedings of the Fifteenth Conference. Morgan Kaufman, San Mateo, CA, pp 476–485

Neyman J (1923) Sur les applications de la theorie des probabilites aux experiences agricoles: Essai des principes. Master's Thesis. Excerpts reprinted in English, Statistical Science 5:463–472. (DM Dabrowska, and TP Speed, Translators.)

Pearl J (2000) Causality: models, reasoning and inference, 1st edn. Cambridge University Press, Cambridge, MA

Pearl J (2009) Causal inference in statistics: an overview. Stat Surv 3:96–146. https://doi.org/10.1214/09-SS057

Pearl J (2012) The causal foundations of structural equation modeling. In: Hoyle RH (ed) Handbook of structural equation modeling. The Guilford Press, New York, pp 68–91

Pearl J (2015) Trygve Haavelmo and the emergence of causal calculus. Econom Theor 31(1): 152–179

Pearl J, Tian J (2000) Probabilities of causation: bounds and identification. Ann Math Artif Intell 28:287–313

Peters J, Bühlmann P, Meinshausen N (2016) Causal inference using invariant prediction: identification and confidence intervals (with discussion). arXiv:1501.01332

Phillips I, Casewell M, Cox T, De Groot B, Friis C, Jones R, Nightingale C, Preston R, Waddell J (2004 Jan) Does the use of antibiotics in food animals pose a risk to human health? A critical review of published data. J Antimicrob Chemother. 53(1):28–52. https://doi.org/10.1093/jac/dkg483

Powell MR (2016 Aug) Trends in reported foodborne illness in the United States, 1996–2013. Risk Anal. 36(8):1589–1598. https://doi.org/10.1111/risa.12530. Epub 2015 Dec 28. PMID: 26709453.

Prosperi M, Guo Y, Sperrin M et al (2020) Causal inference and counterfactual prediction in machine learning for actionable healthcare. Nat Mach Intell 2:369–375. https://doi.org/10.1038/s42256-020-0197-y

Rubin D (2005) Causal inference using potential outcomes: design, modeling, decisions. J Am Stat Assoc 100:322–331

Salmerón A, Rumí R, Langseth H, Nielsen TD, Madsen AL (2018 May) A review of inference algorithms for hybrid Bayesian networks. J Artif Int Res 62:799–828. https://doi.org/10.1613/jair.1.11228

Schreiber T (2000 July 1) Measuring information transfer. Phys Rev Lett 85(2):461–464. https://doi.org/10.1103/PhysRevLett.85.461

Schwartz S, Gatto NM, Campbell UB (2011) Transportability and causal generalization. Epidemiology 22(5):745–746

Shachter RD (1986) Evaluating influence diagrams. Oper Res 34(6):871–882

Shachter RD, Bhattacharjya D (2010) Solving influence diagrams: exact algorithms. In: Cochran J et al (eds) Wiley encyclopedia of operations research and management science. Wiley, New York. www.it.uu.se/edu/course/homepage/aism/st11/Shachter10.pdf

Shimizu S, Hoyer P, Hyvarinen A, Kerminen A (2006) A linear non-Gaussian acyclic model for causal discovery. J Mach Learn Res 7:2003–2030

Shimonovich M, Pearce A, Thomson H, Keyes K, Katikireddi SV (2020 Dec 16) Assessing causality in epidemiology: revisiting Bradford Hill to incorporate developments in causal thinking. Eur J Epidemiol. https://doi.org/10.1007/s10654-020-00703-7

Shorten DP, Spinney RE, Lizier JT (2021 Apr 19) Estimating transfer entropy in continuous time between neural spike trains or other event-based data. PLoS Comput Biol 17(4):e1008054. https://doi.org/10.1371/journal.pcbi.1008054. PMID: 33872296; PMCID: PMC8084348.

Shpitser I, Pearl J (2006) Identification of conditional interventional distributions. In: Dechter R, Richardson T (eds) Proceedings of the Twenty-Second Conference on Uncertainty in Artificial Intelligence. AUAI Press, Corvallis, OR, pp 437–444

Simon HA (1953) Causal ordering and identifiability, Chapter III. In: Hood WC, Koopmans TC (eds) Studies in econometric method. Cowles Commission for Research in economics monograph no. 14. Wiley, New York, pp 49–74

Simon HA (September 1954) Spurious correlation: a causal interpretation. J Am Stat Assoc 49(267):467–479

Simon HA, Iwasaki Y (1988) Causal ordering, comparative statics, and near decomposability. J Econ 39:149–173. http://digitalcollections.library.cmu.edu/awweb/awarchive?type=file&item=34081

Sproston EL, Wimalarathna HML, Sheppard SK (2018 Jan) Trends in fluoroquinolone resistance in Campylobacter. Microb Genom 4(8):e000198. https://doi.org/10.1099/mgen.0.000198

Tennant PWG, Murray EJ, Arnold KF, Berrie L, Fox MP, Gadd SC, Harrison WJ, Keeble C, Ranker LR, Textor J, Tomova GD, Gilthorpe MS, Ellison GTH (2021 May 17) Use of directed acyclic graphs (DAGs) to identify confounders in applied health research: review and recommendations. Int J Epidemiol. 50(2):620–632. https://doi.org/10.1093/ije/dyaa213

Textor J, van der Zander B, Gilthorpe MS, Liskiewicz M, Ellison GT (2016 Dec 1) Robust causal inference using directed acyclic graphs: the R package 'dagitty'. Int J Epidemiol 45(6): 1887–1894. https://doi.org/10.1093/ije/dyw341

USFDA (2005) Withdrawal of approval of Bayer Corporation's New Animal Drug Application (NADA). USFDA. http://www.fda.gov/animalveterinary/safetyhealth/recallswithdrawals/ucm042004.htm

Vitolo C, Scutari M, Ghalaieny M, Tucker A, Russell A (2018) Modeling air pollution, climate, and health data using Bayesian networks: a case study of the English regions. Earth Space Sci 5:76–88. https://doi.org/10.1002/2017EA000326

Wiener N (1956) The theory of prediction. In: Beckenbach EF (ed) Modern Mathematics for Engineers, vol 1. McGraw-Hill, New York

Wright S (1921) Correlation and causation. J Agric Res 20:557–585

Yang Y, Feye KM, Shi Z, Pavlidis HO, Kogut M, Ashworth J, Ricke SC (2019 Jul 26) A historical review on antibiotic resistance of foodborne Campylobacter. Front Microbiol 10:1509. https://doi.org/10.3389/fmicb.2019.01509

Yuan C, Lim H, Lu TC (2011) Most relevant explanation in Bayesian networks. J Artif Intell Res 42:309–352

Zigler CM, Dominici F (2014 Dec 15) Point: clarifying policy. Am J Epidemiol. 180(12): 1133–1140

Chapter 12
Clarifying the Meaning of Exposure-Response Curves with Causal AI and ML

Introduction: What Does an Exposure-Response Curve Mean?

Exposure-response curves are among the most widely used tools of quantitative health risk assessment. Debates over their shapes, especially at low exposure levels, have occupied countless journal articles and thousands of hours of deliberation in regulatory risk assessments and policy-making. Whether the expected number of adverse health effects per unit of exposure to a substance is estimated to be a linear, superlinear, supralinear, or threshold function may determine whether or to what extent exposure to it is allowed. When the answer is uncertain, as is usually the case, estimated statistical upper confidence limits under conservative (risk-maximizing) modeling assumptions are often used to help recommend required limits on exposure.

Figure 12.1 shows an example of an exposure-response function for fine particulate matter (PM2.5). Its x-axis represents the logarithm of exposure concentration. The estimated exposure-response curve (solid line) is based on data from studies of household air pollution, secondhand tobacco smoke, and ambient air pollution; hence, it is an "integrated" exposure-response function, as it integrates data across several types of exposures and studies. The y-axis shows the estimated relative risk of an adverse health effect (ALRI) as a function of the estimated exposure concentrations (solid curve). The dashed lines represent 95% confidence intervals.

Figure 12.1 was created by an international team of over 20 top experts in air pollution health effects epidemiology. It synthesizes the results of a large-scale review of PM2.5 health effects literature (Burnett et al. 2014). It was published in the prestigious and influential journal *Environmental Health Perspectives* and hailed as an important achievement in synthesizing and integrating hard-won knowledge from around the world (Arnold 2014). Hundreds of millions of dollars have been invested in the United States alone in estimating exposure-response relations for

L. A. Cox Jr., *AI-ML for Decision and Risk Analysis*, International Series in Operations Research & Management Science 345,
https://doi.org/10.1007/978-3-031-32013-2_12

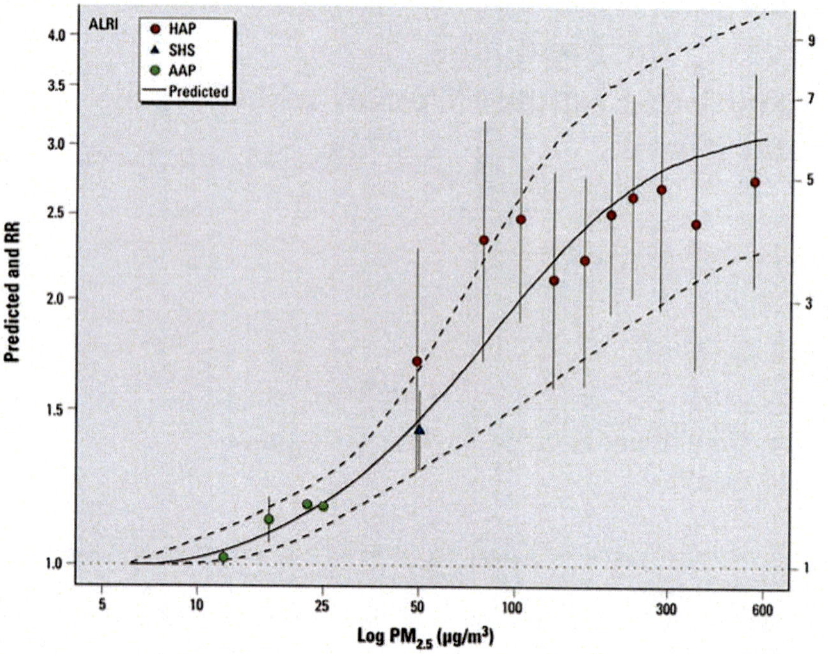

Fig. 12.1 Predicted RR values of Integrated Exposure-Response model (solid line) and 95% confidence intervals (dashed line) and type-specific RRs (points) and 95% CIs (error bars) for acute lower respiratory infection (ALRI) RRs in infants. HAP = household air pollution, SHS = secondhand tobacco smoke, AAP = ambient air pollution. *Source:* Burnett et al. (2014). https://ehp.niehs.nih.gov/doi/10.1289/ehp.1307049

various adverse health effects, from asthma to lung cancer to premature mortality. International collaborations have extended single-country perspectives to estimate the global burden of disease (GBD) attributed to PM2.5 and other pollutants.

Acknowledging the enormous effort and large-scale collaborations that go into preparing exposure-response curves such as the one in Fig. 12.1, this chapter turns to a fundamental interpretive question: *What does an exposure-response curve mean?* This is intended as a foundational question. We recognize that a fastidious reader might question whether the "Log" on the x-axis in Fig. 12.1 is necessary, as it implies that the right half of the diagram consists entirely of agglomerations of PM2.5 denser than black holes; or why 2 of the 15 data points fall outside the alleged 95% confidence intervals. The goal of this chapter is not to quibble about such technical details. Nor is it to challenge the adequacy in current practice of "the foundational concept of peer review as central to ensuring the use of strong science in policy decisions" based on such estimated exposure-response curves (Goldman and Dominici 2019). Rather, we propose that despite their widespread acceptance in peer-reviewed reports and articles and use in informing applied public health risk assessments and regulatory scientific processes, *what exposure-response functions mean is importantly ambiguous at a fundamental conceptual and definitional level.* They leave unanswered such fundamental questions as

- How, if at all, would a proposed change in exposure change *average* population risk?
- How would it change *individual* risks?
- What is the *distribution* of individual exposure-response curves around a population average exposure-response curve?
- What *factors* are assumed to be held fixed in preparing an exposure-response curve? At what levels are they assumed to be held fixed? Are these assumptions realistic?
- In reality, how much do other factors (including unmeasured ones) differ for different observed levels of exposure? How do these differences affect the exposure-response curve?
- By how much would interventions that reduce pollution change other causally relevant variables (e.g., temperature, income, co-pollutants, co-morbidities, etc.) that also affect the response in an exposure-response curve?

Without answers to these questions, it is (or should be) hard for the recipients of an exposure-response curve to guess what it might mean or how, if at all, it should be used to inform risk management decisions.

The following sections seek to clarify these ambiguities and present methods for resolving them. To do so, they draw on concepts and methods that have been developed in causal artificial intelligence (CAI) and machine learning (ML) to support reasoning about how some variables depend on others and how changing some variables causes the probability distributions of others to change.

Exposure-Response Regression Curves Describe Responses at Different Observed Exposures

A curve showing the conditional expected value of observed responses for each observed level of an exposure variable is called a *regression curve*. Regression curves can be estimated from observations by well-developed techniques. These include ordinary least squares (OLS) for linear or polynomial models relating exposure to a continuous response variable; maximum likelihood estimation (MLE) for logistic and Poisson regression and other parametric models if the response variable is binary, categorical, or a count variable; non-parametric smoothing regression models (e.g., spline or LOESS curves) if appropriate parametric models are unknown; and proportional hazards and other survival data analysis methods if the response variable is the time until an event such as death or diagnosis with a disease (Tse 2009).

Regression curves are commonly used to model exposure-response relationships. They are widely treated as if they had a causal interpretation: that reducing exposure would reduce the risk of response as described by the regression curve. The following paragraphs summarize some important limitations of this approach. In general, a regression curve does not predict the effects caused by interventions that change exposure, but rather describes the average responses for different observed levels of exposure under the conditions for which data were collected (undisturbed by potential future interventions) (Pearl 2009; Dominici et al. 2014; Zigler and Dominici 2014; Carone et al. 2020).

A Point of Departure: Correlation vs. Causality

The interpretation of a regression-based exposure-response curve is usually straightforward: it shows the model-predicted average level of the response variable in a population for each level of the exposure variable. Exposure-response curves are most easily interpreted if the population consists of individuals with accurate individual-level exposure, covariate, and response data, but this ideal level of granularity is seldom available. Regression-based exposure-response data can also be developed from other types of data such as time series showing daily deaths in an exposed population.

The limitations of regression models are equally straightforward. A regression model does not explain *why* average population response rates are different for different levels of exposure. It does not explain or predict how or whether average population response rates would *change* in response to an intervention that changes exposure levels (Pearl 2009). It is common practice to misinterpret the slope of a regression curve as describing the change in average response that would be caused per unit of change in exposure, but even under ideal conditions (correctly specified model, no errors-in-variables), the slope only describes the different average response rates *observed* at different levels of exposure. As demonstrated by Simpson's Paradox, this information about *differences* in observed response rates for different levels of exposure does not necessarily have any implications for how, if at all, changing exposure would *change* response rates (*ibid*). For example, average daily consumption of baby aspirin among people over 65 years old may be significantly positively associated with heart attack risk, and yet reducing such consumption might increase heart attack risk, so that change in exposure is negatively associated with change in heart attack risk.

Regression coefficients do not distinguish between correlation and causation or between direct and indirect (mediated) causal effects of some variables on others. For example, suppose that Eq. (12.1) describes an observed exposure-response relationship (on some appropriate exposure and response scales) for a population.

$$Response = 100 + 0.5^{*}Exposure \tag{12.1}$$

How would reducing *Exposure* from an initial level of 100 to a final level of 0 change *Response*, assuming that no other relevant variables change? A common misinterpretation of exposure-response models is that regression Eq. (12.1) implies that reducing *Exposure* from 100 to 0 would reduce *Response* from 150 to 100. In reality, model 12.1 has no implications for how or whether changing *Exposure* would change *Response*. For example, suppose that (perhaps unknown to the analyst) the underlying causal model relating past values of *Exposure* and *Response* is described by the pair of structural equations (12.2a) and (12.2b) with initial values of 100 for *Exposure* and *Poverty*. (The causal interpretation of a structural equation model is that changing an independent variable on its right side causes the dependent variable on the left to adjust to restore equality.) Equations (12.2a) and (12.2b) are chosen so that together they imply Eq. (12.1). Then exogenously reducing *Exposure* from 100 to 0 (without changing *Poverty*, so that the historical relationship in (12.2a) is superseded by the intervention) would *increase* risk (meaning the expected value of *Response* in the exposed population in Eq. 12.2b) from 150 to 200.

$$Exposure = 1^{*}Poverty \tag{12.2a}$$

$$Response = 100 + Poverty - 0.5^{*}Exposure \tag{12.2b}$$

By contrast, if the underlying causal model has structural Eq. (12.3) in place of Eq. (12.2b), then reducing *Exposure* from 100 to 0 without changing *Poverty* would indeed *decrease* risk from 150 to 100:

$$Response = 100 + 0^{*}Poverty + 0.5^{*}Exposure \tag{12.3}$$

Finally, if it has Eq. (12.4) instead of Eq. (12.2b), then reducing *Exposure* from 100 to 0 without changing *Poverty* would have no effect on risk.

$$Response = 100 + 0.5^{*}Poverty + 0^{*}Exposure \tag{12.4}$$

Model 12.1 provides no clue about which, if any, of these (or other observationally equivalent) underlying causal models is correct. No matter how well it fits past data, an empirical regression-based exposure-response model such as model 12.1 cannot provide a sound basis for predicting how or whether reducing *Exposure* would change *Response* (Pearl 2009; Dominici et al. 2014; Zigler and Dominici 2014).

Assumption-Dependent Causal Interpretations of Exposure-Response Regression Models

In response to such limitations, many investigators have proposed using simplifying assumptions to interpret regression coefficients causally. For example, it is often convenient to assume that there are no unmeasured confounders (such as *Poverty* in this example). Likewise, it is often convenient to assume that any observed differences in response rates between two populations, or between response rates at two different times for the same population, or between the rates of changes in response rates in a more-exposed population compared to a less-exposed population, are caused solely by differences in exposures. Quasi-experimental studies and analysis methods developed in the social sciences since the 1960s (Campbell and Stanley 1963) and advocated more recently for air pollution health effects studies (Dominici et al. 2014) apply such simplifying assumptions to observational data to interpret observed differences in outcomes associated with different estimated exposures *as if* they were changes caused changes in exposure levels.

A major limitation of such assumption-based causal interpretations is that the simplifying assumptions used to draw causal conclusions are often not well supported and prove to be mistaken (Carone et al. 2020; Zigler and Dominici 2014). Assuming that an empirical exposure-response model such as Eq. (12.1) that describes observations can also be used as a causal structural model such as Eq. (12.3) to predict the effects of interventions simply assumes away, without resolving, the basic distinction between associational and causal relationships (Pearl 2009). It does not overcome the fundamental methodological challenge that *differences* are not *changes* and that *seeing* is not *doing* (*ibid*).

In practice, treating regression models and other associational models such as Global Burden of Disease (GBD) or population attributable fraction (PAF) models as if they were causal models that can be used to predict effects on responses of changing exposures sometimes leads to failed predictions and unsound policy advice, such as that a 70% reduction in particulate pollution in Ireland would cause detectable reductions in all-cause mortality rates (Zigler and Dominici 2014) or that increasing consumption of certain vitamins would reduce lung cancer risk instead of increasing it (Goodman et al. 2004).

Heterogeneity in Individual Risks

Regression-based exposure-response curves describe *average* responses for different levels of exposure in a population rather than describing *distributions* of individual response risks. This further limits the predictive power and prescriptive information from such models. For example, suppose that an exposure-response model shows that, under current exposure conditions, there is a 10% probability of disease over a certain time interval. Does this mean that each individual has a 10% probability of disease, or that 10% of the population has a 100% probability of disease and the rest 0%, or that half the population has a 20% probability and the other half has zero, or something else? The exposure-response curve does not give an answer. The uncertainty bands around it (dashed lines in Fig. 12.1) are for the estimated *average* response at each level; they do not describe the *distribution* of individual risks around this average. Yet, the answer may matter to policymakers and the public. A smaller risk that is more broadly distributed in the population may raise different levels of concern and support for regulation than a larger risk focused in a smaller subset of the exposed population, especially if the subset is identifiable e.g., only active smokers, or only children with asthma, or only elderly people with COPD. Exposure-response curves omit such information. The limitations of using averages for decision-making are well documented (Savage 2009). Exposure-response curves do not provide information about *distributions* of individual risks that may be essential for well-informed decision-making.

Ambiguous Regression Coefficients: Inference vs. Intervention

The sign and magnitude of the slope of an exposure-response regression coefficient often depend on an investigator's selection of independent variables to include in the model (Dominici et al. 2014). For example, suppose that structural equations (12.5a) and (12.5b) describe how exposure, poverty, and response are related. Eq. (12.5a) implies that *Poverty* = 2**Exposure* and substituting 2**Exposure* for *Poverty* in Eq. (12.5b) yields Eq. (12.5c) as the regression equation relating *Response* to *Exposure* when *Poverty* is an unmeasured (latent) variable or is simply not included in the model.

$$Exposure = 0.5^* Poverty \tag{12.5a}$$

$$Response = 100 + 0.5^* Poverty - 0.5^* Exposure \tag{12.5b}$$

$$Response = 100 + 0.5^* Exposure \tag{12.5c}$$

Thus, when only *Exposure* is included as an independent variable on the right side of the regression model as a predictor of *Response*, its regression coefficient is 0.5, and hence is positive (Eq. 12.5c). But if both *Poverty* and *Exposure* are included as independent variables on the right side of the regression model for *Response*, the regression coefficient for *Exposure* becomes −0.5, and hence is negative (Eq. 12.5b). This illustrates that the sign of a regression coefficient can depend on the selection of independent variables. The interpretation of the positive regression coefficient 0.5 for *Exposure* as a predictor of *Response* in Eq. (12.5c) is that when *Exposure* is observed to have a higher value, one can infer (via Eq. 12.5a) that *Poverty* has a higher value, and can therefore predict (via Eq. 12.5b) that *Response* will have a higher value. However, this inference has no implications for predicting the effects of an intervention that changes *Exposure* (without changing *Poverty*). Such an intervention acts through Eq. (12.5b). It renders Eq. (12.5a) irrelevant. If *Exposure* is set to a new level by an exogenous intervention, it is no longer determined endogenously via Eq. (12.5a), which can therefore be excised from the analysis—a form of "graph surgery" used to model the effects of interventions (Jacobs et al. 2019).

The regression coefficient for exposure in an exposure-response regression model such as Eq. (12.1) typically reflects an unknown mixture of inference and intervention effects, as well as depending on the modeler's choice of whether to include or exclude other variables (such as *Poverty* in this example) (Dominici et al. 2014).

This makes it unsuitable for predicting how or whether an intervention that changes exposure will affect the risk of response. Regression models focus on creating curves that fit past data rather than on creating curves that predict the effects of potential future changes in exposure. This distinction is now well-recognized in machine learning, which assesses the predictive accuracy of models using a train-and-test paradigm (e.g., via cross-validation) rather than assessing fit using traditional statistical criteria (e.g., AIC, BIC, or adjusted R-squared values) (Tse 2009). The desire to predict the effects of interventions has led to recent innovations in "heterogeneous treatment effect" (HTE) analysis, including causal tree and causal forest algorithms that modify standard machine learning (ML) algorithms to predict changes in risk due to different exposures, rather than estimating levels or differences in risk between differently exposed individuals (Cáceres and González 2022; Gong et al. 2021; Bodory et al. 2022). The following sections explore how ML ideas can be applied to create exposure-response curves that have valid causal interpretations.

Logistic Regression vs. Non-Parametric Exposure-Response Curves

To fix ideas concretely, we will use a simple, dramatic dataset in which mortality risk increases with exposure. To keep the discussion general and conceptual, we refer to the exposure variable simply as "exposure" without specifying units. (For the curious reader, however, the example dataset is for blood lead levels measured in μg/dL and mortality during the follow-up period for male non-smokers in the NHANES III dataset. It is described further by Cox 2020). Figure 12.2 shows a logistic regression model fit to the exposure-response data.

One fundamental challenge in interpreting such a curve is the possibility of confounding: perhaps people with higher levels of exposure tend to be older, or less educated, or to have lower incomes than people with lower levels of exposure, and perhaps such differences contribute to the differences in mortality risk estimated for different levels of exposure. (Indeed, all of these associations hold in the example dataset used to generate Fig. 12.2). A second fundamental challenge is that the best-fitting model is a specific parametric class of models (e.g., logistic regression) may not describe the data very accurately. For example, Fig. 12.3 shows the empirical mortality fractions for different levels of exposure. (Exposures are rounded to the nearest integer except for exposures >15, which are coded as 16 to save space. The blue vertical whiskers indicate 95% binomial confidence intervals.) The logistic regression model in Fig. 12.2 does not approximate these data very well at higher exposure levels: it predicts mortality rates well above 0.75 and approaching 1.00 at the right end, even though the observations (both the raw data, shown as dots at 1 for individuals with mortality and at 0 for those with non-mortality in Fig. 12.2, and also the means shown in Fig. 12.3) indicate that mortality rates level off at about 0.6. Such *model specification error* arises because the assumed functional form of the model (logistic) does not describe the data well. Hence it cannot be corrected by collecting more data or by including more variables in the logistic regression model.

Fig. 12.2 A logistic regression curve for mortality vs. exposure

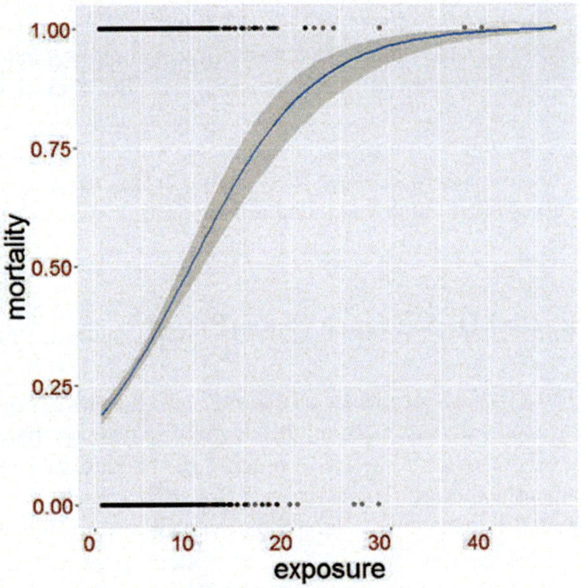

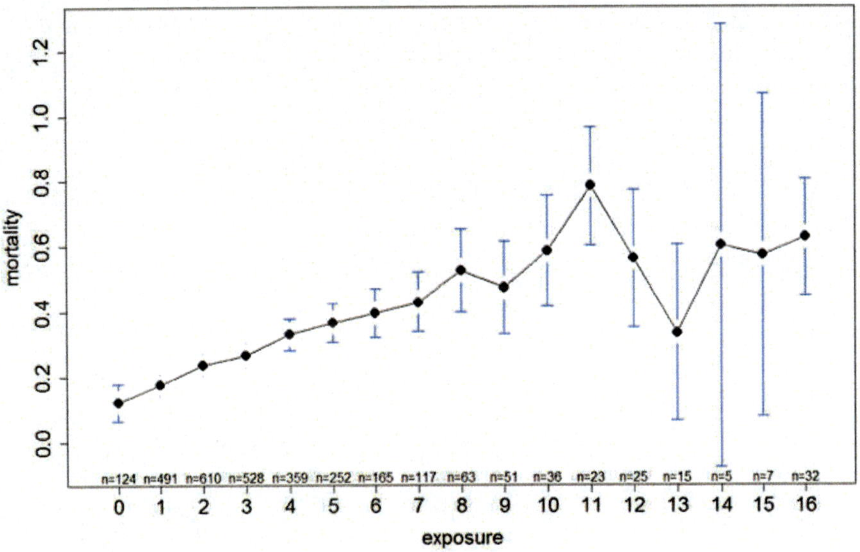

Fig. 12.3 Plot of mean mortality rates vs. exposure for the data in Fig. 12.2. (32 exposures greater than 15 are rounded to 16 to save space)

Both of the curves in Figs. 12.2 and 12.3 are exposure-response curves, in the sense that each plots the probability of response (mortality) on the vertical axis against exposure on the horizontal axis. But they have different meanings. Figure 12.2 represents an exposure-response curve in which logistic regression modeling assumptions are assumed to hold, while Fig. 12.3 makes no such assumption. The discrepancy between them highlights the fact that estimated exposure-response curves often reflect modeling assumptions as well as patterns in the data.

Partial Dependence Plots (PDPs)

The exposure-response curve in Fig. 12.3 has the advantage that it does not depend on modeling assumptions. It has the disadvantage that it does not control for the levels of other variables in plotting the conditional mean values of mortality risk against exposure. This is unacceptable if we want to see how exposure alone affects mortality risk while holding other variables fixed. For example, Fig. 12.4 plots age against exposure. It is clear that exposure is positively associated with age as well as with mortality risk. To isolate the effect of exposure alone on mortality risk, it is necessary to control for potential confounding by age (and also by income and by education, both of which are negatively associated with exposure and mortality risk). We would like to control for the levels of these other variables without making potentially invalid modeling assumptions such as that they affect mortality risk through a multivariate logistic regression function.

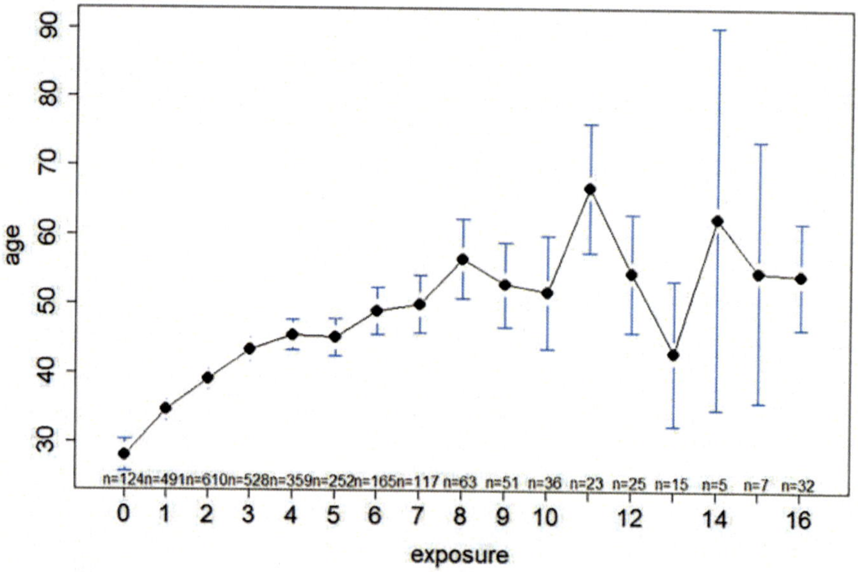

Fig. 12.4 Plot of mean age vs. exposure for the data in Fig. 12.2. (Exposures greater than 15 are rounded to 16 to save space)

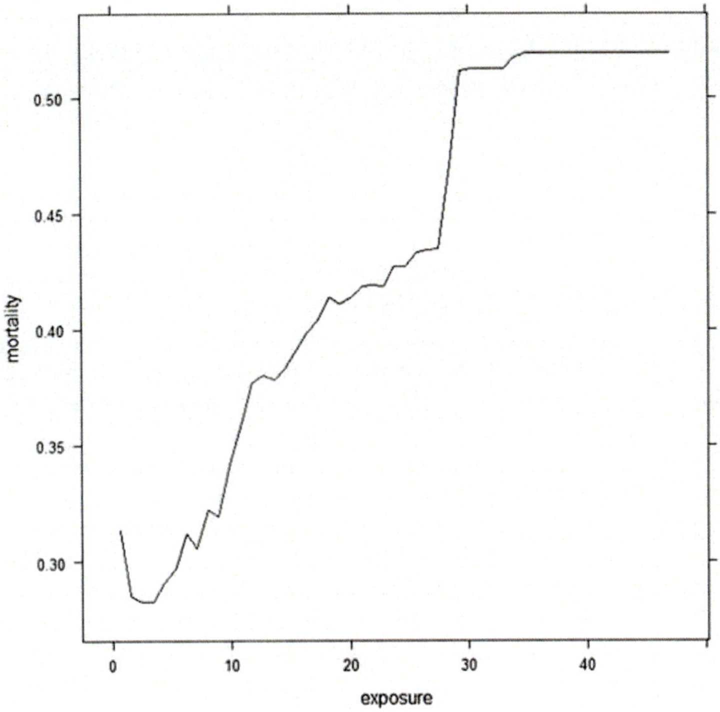

Fig. 12.5 Partial Dependence Plot (PDP) of mortality probability vs. exposure for male nonsmokers

A practical solution is to use nonparametric ML methods to predict the average value of mortality probability as exposure alone changes, holding all other variables fixed at their current levels for each individual in the data set. This yields a *partial dependence plot* (PDP) (Zhao and Hastie 2019). Figure 12.5 shows the PDP exposure-response curve for our example data using the NHANES III data for male non-smokers. The interpretation is that each point on this curve shows the predicted average value of the 0–1 *mortality* dependent variable (i.e., the average conditional probability of mortality during the follow-up period) that the individuals in the data set would have if exposure were set to each value on the x-axis and all other variables were kept at the values that they currently have. In this setting, the basic distinction between predicted *differences* and predicted *changes* in outcomes as exposure varies disappears if it can be assumed that the only explanation for differences in predicted mortality is corresponding differences in exposure (since all other independent variables are held fixed). This does not solve the problem of unobserved confounders, however. For example, if the exposure variable (blood lead level) indicates unreported smoking in this allegedly nonsmoking population, then the apparent effect of exposure on mortality risk may actually be due to the hidden confounder of smoking.

To create a PDP, we need an ML algorithm (or "model" in ML terminology) that predicts the value of the dependent variable (here, *mortality*, coded as 1 for mortality and 0 otherwise) for each combination of values of the independent variables. There are many such algorithms in the modern ML toolkit. Figure 12.5 uses the popular Random Forest algorithm (Denisko and Hoffman 2018) to predict average mortality risk for different values of exposure, given the values of the following covariates: *age*, *income*, *grade*, *married*, *Hispanic*, *Black*, *small.metro*, and *West*. Here, *married, Hispanic, Black, small.metro*, and *West* are four 0–1 binary indicator variables (i.e., dummy variables) indicating whether each case is identified as married, Hispanic, Black, living in a small metropolitan area, and living in the Western US, respectively, in our NHANES III-based example dataset. (Sex and smoking status are excluded since we selected non-smoking men as the cases for which an exposure-response curve is to be estimated.) *Grade* indicates the highest grade completed.

The PDP is constructed as follows. Begin with a dataset organized so that each row contains the data for one individual (also called a "case" or "record" or "observation" or "data point") and each column contains the value of one variable (also called a "feature" or "attribute" or "field"). This layout of the data is called a *data frame*. Thus, the value in row *r* and column *c* is the value of variable *r* for individual *c* in the data frame. For each case, the values of the exposure variable (called *exposure* in our example), the response variable (*mortality* in our example), and measured covariates (*age*, *income*, *grade*, *married*, *Hispanic*, *Black*, *small.metro*, and *West* in our example) are recorded for each case. Next, create a succession of modifications of the original data frame by changing just one column, the *exposure* variable, while leaving all other columns (i.e., variables) unchanged. The successive modified datasets contain successively increasing values of the exposure variable (which is set to have the same value for all cases in each modified dataset), beginning with the smallest value of *exposure* observed in the original dataset and increasing to the largest value. Each of these modified datasets generates one point on the PFP: the point that plots the average predicted value of the response variable, *mortality* (averaged over all cases) against the value of *exposure* for the modified dataset. Joining these points with line segments completes the PDP. The PDP can be viewed as generalizing the concept of an *average treatment effect* (ATE) for a binary exposure variable to continuous exposure variables (Zhao and Hastie 2019).

To illustrate this process, Table 12.1 shows a simple hypothetical dataset with only 3 records (rows). For purposes of a simple exposition only, suppose that the

Table 12.1 Original dataset

Mortality	Exposure	Age	Income	Grade	Married	Hispanic	Black	small. metro	West
0	5.0	21	0.6	12	0	1	0	0	1
0	7.3	50	3.4	12	1	1	0	0	1
1	9.3	65	2.2	3	1	1	0	0	1

Table 12.2 Predicted mortality probabilities for original dataset

Predicted mortality	Exposure	Age	Income	Grade	Married	Hispanic	Black	small. metro	West
0.16	5.0	21	0.6	12	0	1	0	0	1
0.60	7.3	50	3.4	12	1	1	0	0	1
0.89	9.3	65	2.2	3	1	1	0	0	1

Table 12.3 Predicted mortality probabilities for modified dataset with *Exposure* = 0.7

Predicted mortality	Exposure	Age	Income	Grade	Married	Hispanic	Black	small. metro	West
0.147	0.7	21	0.6	12	0	1	0	0	1
0.584	0.7	50	3.4	12	1	1	0	0	1
0.863	0.7	65	2.2	3	1	1	0	0	1

Table 12.4 Predicted mortality probabilities for modified dataset with *Exposure* = 1.0

Predicted mortality	Exposure	Age	Income	Grade	Married	Hispanic	Black	small. metro	West
0.148	1.0	21	0.6	12	0	1	0	0	1
0.585	1.0	50	3.4	12	1	1	0	0	1
0.864	1.0	65	2.2	3	1	1	0	0	1

probability of mortality is predicted from other variables using the linear regression model.

$$E(mortality \mid Exposure, age, income) = -0.20 + 0.003^{*}Exposure + 0.017^{*}age - 0.02^{*}income$$

(In practice, of course, linear regression is not appropriate for a binary dependent variable. Instead, for all calculations shown later, we use Random Forest and include additional significant predictors of mortality risk such as *married* and *grade*. Readers unfamiliar with random forest can find an accessible introduction in Molnar (2022). Briefly, random forest is an ensemble method that averages predictions from several hundred non-parametric classification and regression (CART) trees.) Table 12.2 is identical to Table 12.1 except that the first column shows the model-predicted mortality probabilities instead of the 0–1 indicators of observed mortality thus, for the first case we have

$$E(mortality \mid Exposure = 5.0, age = 21, income = 0.6) = -0.20 + 0.003^{*}5 + 0.017^{*}21–0.02^{*}0.6 = 0.16.$$

Tables 12.3 and 12.4 show successive modified datasets with *Exposure* set to the smallest and second-smallest observed exposure values in the NHANES data, 0.7 and 1.0, respectively. The corresponding predicted mortality probabilities (i.e., predicted conditional expected values for *Mortality*, given the values of the independent variables) are slightly different for these two consecutive values of *Exposure*.

Averaging the predicted mortality probabilities for all individuals for each level of *Exposure* gives the corresponding height of the PDP for that exposure level. Thus, for *Exposure* = 0.7, the PDP predicts an average mortality probability in the exposed population (with all individuals having *Exposure* = 0.7) of (0.147 + 0.584 + 0.863)/3 = 0.531; while for *Exposure* = 1.0, the PDP predicts a slightly higher average mortality probability of (0.148 + 0.585 + 0.864)/3 = 0.532. Continuing to increase the *Exposure* values and calculate the resulting average predicted mortality probabilities generates the entire PDP curve. The PDP curve in Fig. 12.5 is generated by this process, except using all 2903 cases in the NHANES dataset for lead-exposed non-smoking males instead of the 3 hypothetical cases in Table 12.1 and using random forest instead of linear regression to predict conditional expected values of *Mortality* for each set of independent variable values.

The PDP in Fig. 12.5 solves the problem of controlling for observed potential confounders such as age, income, and grade by holding their values fixed: only *exposure* and resulting predicted *mortality* values vary. (Italso avoids creating spurious dependencies between variables due to collider bias (*ibid*) as long as *mortality* is a possible effect, but not a possible cause, of the other variables.) The PDP also avoids dependence on modeling assumptions by using non-parametric prediction methods, such as the Random Forest algorithm used to create Fig. 12.5. Thus, a PDP may provide a useful definition for an exposure-response curve.

Describing Interindividual Heterogeneity in Exposure-Response Functions: Individual Conditional Expectation (ICE) Plots

It remains to address the challenge of characterizing interindividual differences in exposure-response curves. The data in the sequence of modified datasets used to construct the PDP can be reused to provide a constructive solution to this challenge. Instead of averaging the predicted values of the dependent variable, *mortality*, for all individuals in each modified dataset to show how average mortality probability depends on exposure when all other variables are held fixed, as in a PDP plot, an alternative is to display the predicted values of *mortality* for each individual for the different levels of exposure. This is called an *individual conditional expectation (ICE) plot* (Goldstein et al. 2015; Zhao and Hastie 2019), since the predicted values of mortality are its conditional expected values (as predicted by an ML model such as Random Forest) given the values of an individual's covariates, including the different assumed levels of exposure. Figure 12.6 shows an ICE plot for the data used to create the PDP in Fig. 12.5. Each black curve is the estimated exposure-

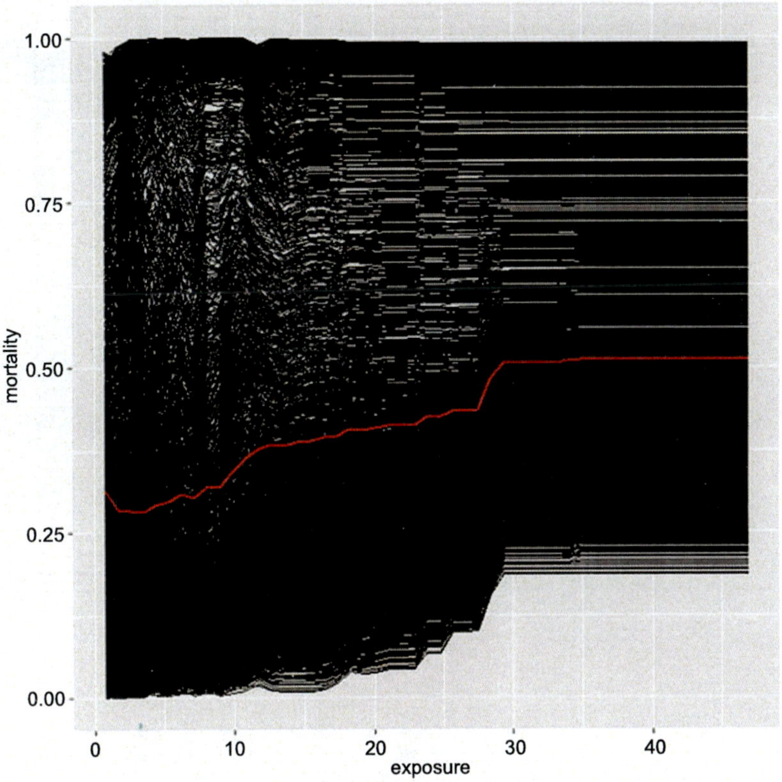

Fig. 12.6 ICE Plot for mortality probability vs. exposure for male nonsmokers

response curve for one individual. Averaging these curves gives the PDP curve (indicated in red in Fig. 12.6) for the population of individuals.

The ICE plot shows the range of variability in individual response probabilities at different levels of exposure, but it can be quite difficult to read when there are many individuals (e.g., there are 2903 individual curves in Fig. 12.6). One approach to improving legibility is to center all curves on a common value, such as the lowest value of exposure in the dataset, and to show only the differences in predicted values of mortality at each exposure level compared to their predicted values at the lowest level of exposure. Figure 12.7 shows such a *centered ICE plot* (Goldstein et al. 2015; Molnar 2022).

The centered ICE plot shows that even though the average predicted mortality probability in the population of cases increases with exposure, individual predicted mortality probabilities vary with exposure in a variety of ways. Figure 12.8 uses k-means clustering to further simplify and abstract from the wealth of detail in the ICE plot in Fig. 12.7 by clustering the individual centered exposure-response curves into three clusters of "similar" curves. The proportions of cases in each cluster are shown in the legend. (The y-axis (labeled "cluster that") is a rescaled version of the

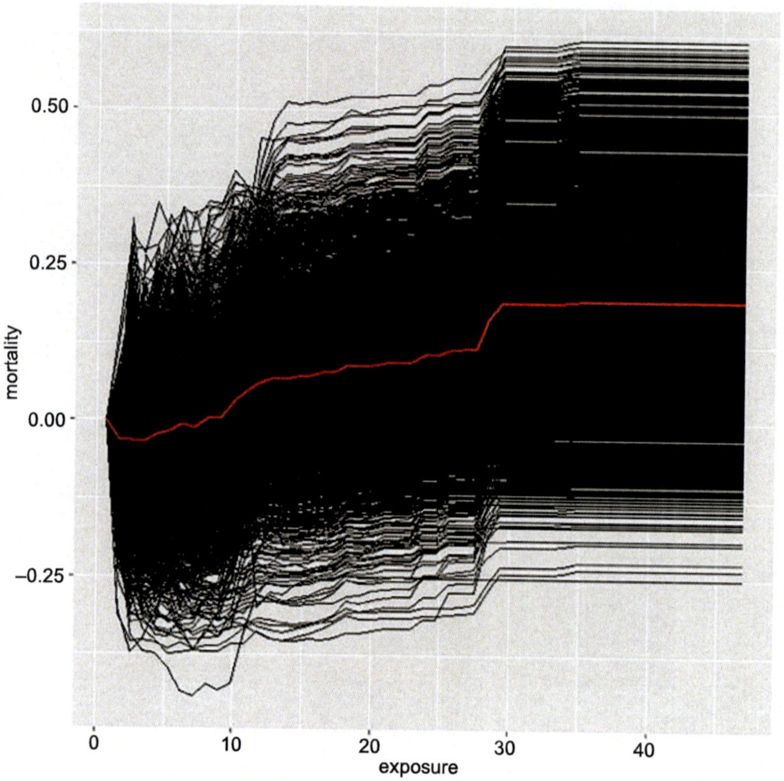

Fig. 12.7 Centered ICE Plot for mortality probability vs. exposure for male nonsmokers. The curves show the differences in predicted individual mortality probabilities for different levels of exposure compared to their predicted values for the lowest level of exposure

differences in mortality probabilities shown on the y-axis in Fig. 12.7, expressed in terms of standard deviations, as is usual for k-means clustering; see Goldstein et al. (2022) for details.) The clusters in Fig. 12.8 suggest that about 23% of the individuals in the dataset (green cluster) have individual exposure-response curves that increase with exposure over the range from 0 to about 10, while 70% have no significant response (red cluster) and the remaining cases (shown as 6% due to rounding) have smaller mortality risks at higher exposure levels (blue cluster).

It is natural to wonder what distinguishes members of the different clusters. Although there is not necessarily any simple answer given the highly variable individual exposure-response ICE curves in Fig. 12.7, further exploration shows that exposure is significantly positively ordinally correlated with predicted mortality risk for people with above-median age (Kendall's tau = 0.17, p-value = 1.97E-14) but not for younger people (Kendall's tau = 0.015, p-value = 0.49), especially for the youngest individuals with below-median income (Kendall's tau = −0.056, p-value = 0.15 for people under 25, corresponding to the first quartile of the age

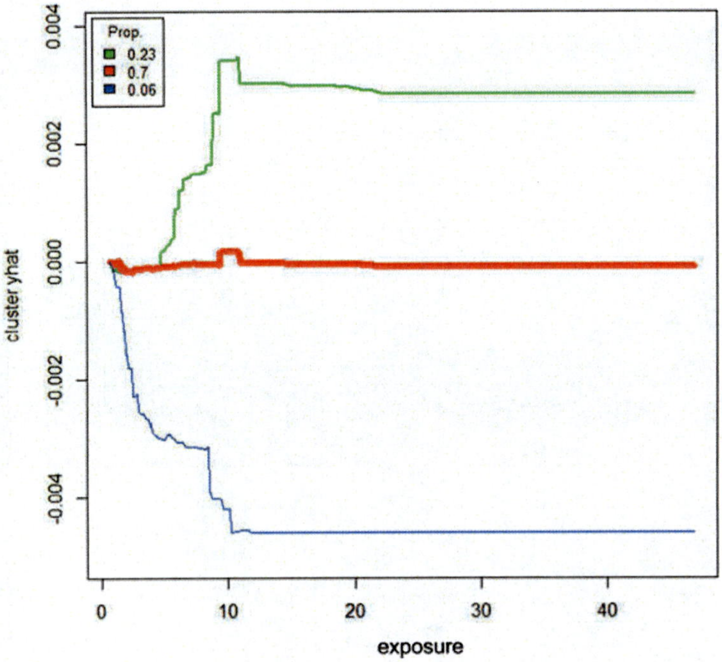

Fig. 12.8 Clustered ICE plot showing 3 clusters of individual exposure-response curves

distribution of cases with below-median incomes). However, any such clustering simplifies the complex reality shown in Fig. 12.7, which is that different individuals have very different exposure-response functions, depending on the levels of their other covariates.

Relevant and Data-Informed Counterfactuals: Two-Dimensional Partial Dependence Plots (2D-PDPs)

The construction of PDPs and ICE plots involves (i) varying one independent variable, exposure, while holding all others fixed at their current levels; and (ii) plotting how predicted average (for PDPs) or individual (for ICE plots) values of the dependent variable, *mortality*, vary in response to the different levels of exposure, given the fixed levels of other covariates for the individual cases in the population for which the PDP or ICE plot is derived. These concepts can be extended to allow more than one independent variable to be varied while holding the rest fixed. Figure 12.9 shows an example. In this two-dimensional PDP, both *age* (y-axis) and *exposure* are systematically varied from their lowest to their highest observed values, holding all other variables fixed at their current value for each case. The plot shows the predicted value of *mortality* risk (yellow = high, blue = low) for

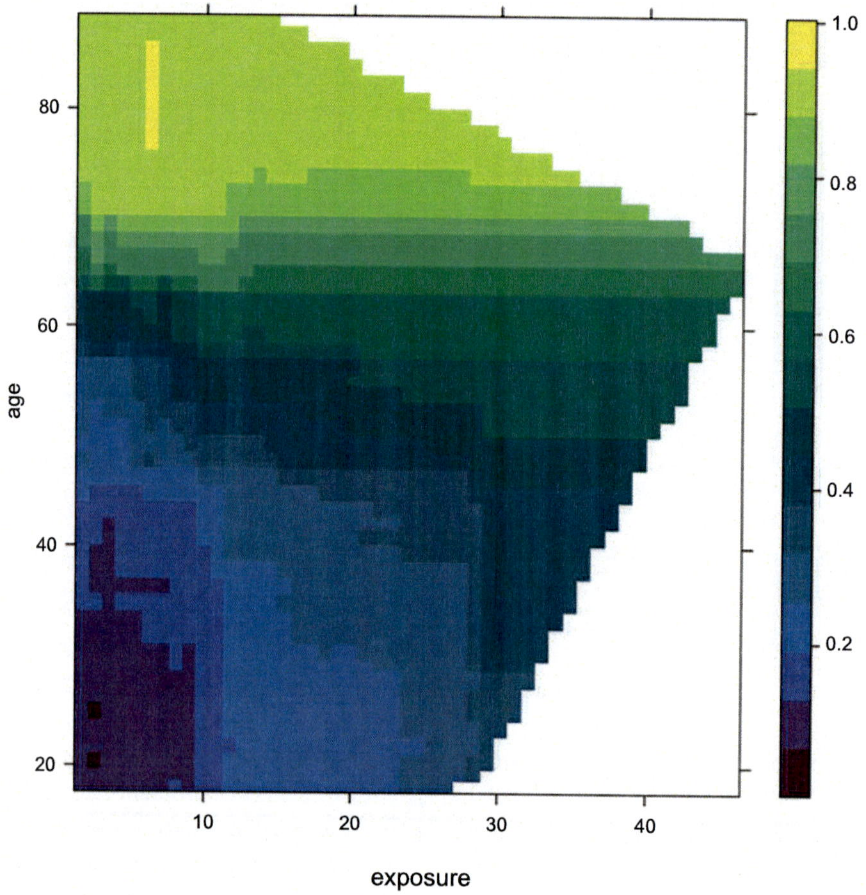

Fig. 12.9 2D-PDP for *mortality* risk (false color scale, yellow = high, blue = low) vs. *age* and *exposure*

each pair of *exposure* and *age* values. The white spaces to the upper right and lower right of the colored region correspond to combinations of *age* and *exposure* that do not occur in this data set, and hence fall outside the colored data cloud region. For example, no one younger than 40 or older than 70 years old has *exposure* > 40.

Figure 12.9 shows that the effect of *exposure* on *mortality* risk depends on *age*. For people with *age* greater than 70 years old, exposure has little or no relevance to mortality risk: the risk is high (yellow) for all levels of exposure, and is not higher at higher exposure levels. On the other hand, for people with *age* less than 36 (the median age), mortality risk, although relatively low, increases significantly with *exposure*, from less than 0.1 on the lower left (dark blue) to 0.3 or more on the lower right of the data cloud. This illustrates that the shape of an exposure-response PDP depends on the distribution of *age* (and other covariates) in the population for which the PDP is calculated. To apply a PDP estimated for one population to a different

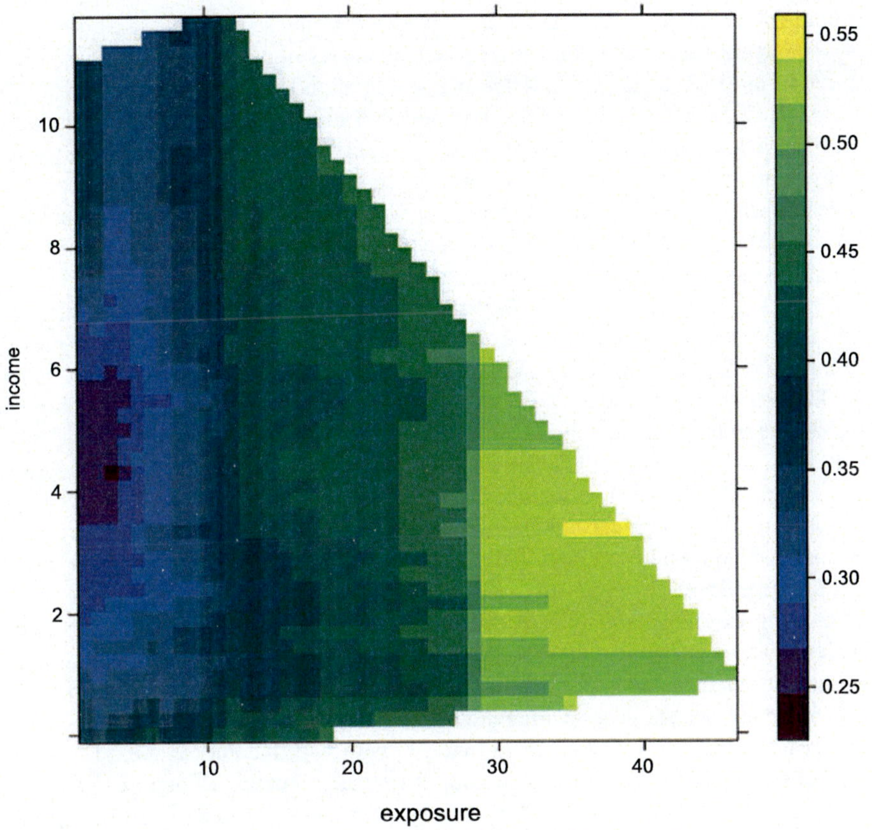

Fig. 12.10 2D-PDP for *mortality* risk (false color scale, yellow = high, blue = low) vs. *income* and *exposure*

population, the predicted individual-level risks (conditioned on values of covariates) that are averaged to form the PDP must be re-weighted to reflect the joint frequency distribution of covariates in the new population. This makes the interpretation of "integrated exposure-response functions" such as the curve in Fig. 12.1 problematic, as there is no population for which the curve holds. Rather, different parts of it come from different populations in different countries, with different distributions of age, income, and other covariates. How the parts of the curve would have to be changed to apply to any specific country or population cannot be determined from Fig. 12.1.

Figure 12.10 shows a 2D-PDP for the joint effects of *exposure* and *income* on mortality risk, other independent variables being held fixed. The highest risks occur at the lower right, for individuals with high exposure and low income. The white space in the upper right shows that the dataset contains no individuals who have both high income and high exposure. This implies that counterfactual predictions about what mortality risks would be for high-income, high-exposure individuals are not supported by the data. This illustrates a conceptual difficulty with the key idea of holding other variables fixed as exposure is varied: doing so may generate unrealistic counterfactual cases that dilute the relevance of the resulting exposure-response PDP curves. For example, in Fig. 12.10, it is true that all individuals with *exposure* > 35 have relatively high mortality risks, but they also all have relatively low incomes. A PDP curve such as Fig. 12.5 that includes mortality risks estimated for high-income individuals in its calculation of average mortality risk as a function of exposure (because they are art of the population for which the PDP is calculated) therefore includes extrapolated risks for counterfactual scenarios of high-income, high-exposure individuals for which there are no corresponding observations in the data cloud. It is not clear that such extrapolations are correct or relevant for real populations, nor that they should be included in calculating exposure-response curves that are meant to inform risk management decisions for real populations.

Two refinements of the PDP concept have been proposed to avoid including unrealistic or irrelevant counterfactual cases in the calculation of exposure-response functions. *Accumulated local effects* (ALE plots) (Apley and Zhu 2020) work within the data cloud (i.e., they use observed combinations of variable values for real cases) to estimate the average change in response for small changes around each level of exposure, holding other variables fixed. They do so by using only the subset of cases with similar levels of exposure to estimate the local change in risk from small changes in exposure around a given level. For example, in Fig. 12.10, only cases with exposures in the neighborhood of 35 would be used to estimate how mortality probability changes as exposure increases from 34 to 35 or from 35 to 36. This automatically selects cases with relevant (relatively low) incomes and excludes irrelevant counterfactual cases of individuals with high incomes and exposures in this range. The second approach is to drop the requirement that all other variables be held fixed as exposure changes. For example, causal graph models can be used to identify *adjustment sets* (Witte et al. 2020) of variables that must be held fixed to obtain estimates of direct or total causal effects of exposure on mortality (e.g., confounders), while other variables are left free to have their conditional distributions change realistically as exposure changes. In this approach, the conditional distribution of income could change (shifting downward) at higher levels of exposure, again avoiding the need to include unrealistic counterfactual cases in calculating the exposure-response curve.

Discussion and Conclusions: What Do We *Want* Exposure-Response Curves to Mean?

The previous sections have presented ideas from modern causal artificial intelligence (CAI) and machine learning (ML) that can help to clarify what an exposure-response curve means. Among them are the following:

1. A *predictive* exposure-response curve shows the conditional expected value of a response variable such as *mortality* predicted for different observed values of

exposure, given the distribution of other independent variables (i.e., covariates) in a population. Fig. 12.2 is an example.

2. A *predictive* exposure-response curve may be quite different from an *interventional* exposure-response curve that shows how the average value of the response variable would change if the values of exposure were changed. Although it has been claimed that these concepts are routinely conflated in air pollution health effects epidemiology (Cox 2017), they should not be and need not be: they are very clearly distinguished in CAI (Pearl 2009) and should be clearly distinguished by public health risk analysis practitioners.
3. The shape of a predictive exposure-response curve may reflect modeling assumptions as well as facts. Even maximum likelihood or other "best" predictions (by various criteria) can be quite biased and wrong if the modeling assumptions are wrong. Figs. 12.2 and 12.3 illustrate this point.
4. Non-parametric ML techniques can help to avoid the need for parametric modeling assumptions, although they still require adequate sample sizes and coverage of the combinations of exposures and other covariates for which predictions are to be made in order to produce accurate predictions. Figures 12.9 and 12.10 illustrate (via white spaces outside the data cloud) the common occurrence that many logically possible combinations of exposure values with other covariate values do not occur in practice. Exposure-response curves should be accompanied by explanations of whether the average response probabilities they predict are based on observed combinations only (as in ALE plots) or whether they include hypothetical, perhaps unrealistic, combinations of exposure values with other covariate values that do not occur in the data (as in one-dimensional PDP plots).
5. The predictions from a (predictive or interventional) exposure-response curve depend in general on the values of other covariates for individuals in the population for which the curve is developed. Figure 12.9 illustrates this point: how mortality risk depends on exposure depends on age. Therefore, averaging mortality risk over individuals for different assumed levels of exposure will produce very different population exposure-response curves depending on the frequency distribution of individual ages (and other covariates) in the population. Even if the dependent variable were made age-specific hazard rate, other covariates (such as *grade*) would still modify the predictive relationship between exposure and response. Thus, any exposure-response curve is valid only for specific joint distributions of other causally relevant covariates that interact with exposure in predicting risk. These distributions are seldom stated explicitly when exposure-response curves are presented, but they are crucial parts of the definition of the curves.
6. In general, an exposure-response curve developed for one population does not apply to other populations with different distributions of covariates. (An exception occurs if there are no interactions among exposure and other covariates so that the effect of exposure can be assessed independently of age, income, education co-morbidities, co-exposure, or other covariates. But in our experience, such independence of effects seldom or never holds in practice.)

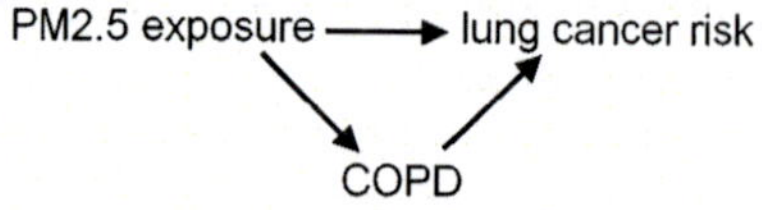

Fig. 12.11 Should levels of COPD be held fixed in quantifying the PM2.5-lung cancer risk exposure-response curve?

7. Integrated exposure-response functions such as the one in Fig. 12.1 do describe the predictive relationship between observed exposure levels and predicted response levels for any real-world or hypothetical population. Rather, different parts are taken by averaging the predicted or observed levels of the response variable over individuals in different populations having different types and levels of exposures and different distributions of covariates. Such a pastiche of different underlying conditions yields an integrated exposure-response curve that has no clear conceptual definition and that does not apply to any specific population.
8. Even when the meaning of an exposure-response curve is clearly specified, e.g., as the predicted average value of response for each level of observed exposure in a specific population of individuals with known values of other covariates (assumed to remain fixed as exposure varies), the exposure-response curve is only an average of individual-level curves. This is illustrated in Figs. 12.7 and 12.8, which make explicit the inter-individual variability in exposure-response curves. The flaws of using averages in decision-making are well known (Savage 2009), but how to use the more detailed information in ICE plots to improve risk management decision-making has not yet been much discussed in applied epidemiology, which still relies largely on population-level exposure-response curves.

These ideas emphasize different possible meanings for an exposure-response curve (e.g., regression model, PDP, ALE, descriptive model, predictive model, interventional model); some of their limitations; and the fact that the meanings of some exposure-response curves, such as the one in Fig. 12.1, are not clearly specified by their authors and cannot be inferred from the curves presented because the levels of other relevant covariates for individuals in the populations described by the curves are not specified.

More fundamentally, an exposure-response curve is usually used as a tool for communicating information to inform health risk management decisions. It is therefore worth asking what information users of exposure-response curves want or intend for them to communicate. Figure 12.11 presents an example. Suppose that high concentrations of PM2.5 are found to predict both higher levels of chronic obstructive lung disease (COPD) in a population and also higher rates of lung cancer specifically among COPD patients, but not among other individuals.

Now, suppose that we are asked to develop an exposure-response curve relating PM2.5 exposure to lung cancer risk to inform discussions about the health benefits of lowering PM2.5. Then we must decide what should be held fixed (and at what levels) in developing the requested exposure-response curve. Specifically, should COPD prevalence in the population be held fixed as PM2.5 exposures are varied across the horizontal axis of the exposure-response curve? In the terminology of mediation analysis, this would correspond to quantifying a "controlled direct effect" of PM2.5 on lung cancer risk (VanderWeele 2015). If so, should COPD prevalence be fixed at today's levels? Doing so would maximize the estimated direct effect on lung cancers caused by reducing PM2.5 exposure, by providing a relatively large pool of COPD cases for whom lung cancer risk can be reduced by reducing PM2.5. Or should COPD prevalence be fixed instead at the predicted future lower prevalence level for COPD anticipated if PM2.5 is reduced to a new, lower level? (This would reduce the predicted direct benefit from reduced lung cancer risk among COPD cases while acknowledging an additional benefit from reduced COPD cases.) In the terminology of mediation analysis, the "total effect" of reducing PM2.5 on lung cancer risk reflects both the *direct effect* caused by reducing lung cancer risk for COPD patients and also the *indirect effect* caused by reducing the number of COPD patients (*ibid*). How, if at all, should these modeling choices depend on how long it takes for reductions in PM2.5 to reduce COPD prevalence—e.g., 4 months, 4 years, or 4 decades?

These are not questions with correct or incorrect technical answers. Rather, they are questions about what the users of exposure-response curves want and intend them to mean: what information are they intended to convey to recipients? Once we have decided whether to quantify an exposure-response curve holding COPD prevalence fixed at current levels, or at some anticipated future level, or to model the changes over time in COPD probabilities (or age-specific hazard rates) for individuals in a specified population following a specified reduction in PM2.5, then the focus can shift to the technical challenges of quantifying the desired curve, and technical tools such as PDPs, ICE plots, ALE plots, and adjustment sets may then be useful for quantifying it. But the logically prior question of what it is that we want to quantify is seldom explicitly addressed in current presentations of exposure-response curves. What was actually quantified—what exactly the presented curves represent—is also seldom described at a level of detail that includes the joint distribution of covariates averaged over in calculating the exposure-response curves. Yet, without such detailed information, it is impossible to calculate how the presented exposure-response curve (or the underlying relative risks or measures of association) must be modified to apply to any population with a different joint distribution of variables—including the population(s) used to estimate the exposure-response curve following an intervention that reduces exposure (Pearl 2009).

The increasing availability of causal artificial intelligence (CAI) and machine learning algorithms is making it increasingly practical to calculate relatively sophisticated nonparametric population exposure-response curves, 2D-PDPs, and distributions of individual exposure-response curves from the kinds of data already used to estimate parametric regression exposure-response curves. Figures 12.5–12.10 illustrate how modern CAI/ML methods can quantify not only non-parametric PDPs that control for observed confounders without requiring parametric modeling assumptions but also ICE plots that address inter-individual heterogeneity. With this increased computational capacity comes an increasing need to define clearly what we want to use it for. It is perhaps fair to suggest that the capacity to calculate various types of exposure-response curves currently outstrips clarity about what should be calculated and how the results should be used to inform and improve public health risk management decisions.

The need for greater clarity on these points can perhaps be addressed by better philosophical and causal reasoning about the information required for more causally effective risk management decisions (see Chap. 11 and Maldonado and Cox 2020). Meanwhile, continuing conceptual and algorithmic advances in CAI and ML (e.g., Apley and Zhu 2020; Goldstein et al. 2022; Zhao and Hastie 2019) will continue to improve our ability to quantify carefully defined causal relationships, including causal exposure-response relationships. This will challenge practitioners to carefully define the relationships that they want to quantify and communicate to policymakers and risk managers.

References

Apley DW, Zhu J (2020) Visualizing the effects of predictor variables in black box supervised learning models. J R Stat Soc 82:869–1164

Arnold C (2014 Apr) Disease burdens associated with PM2.5 exposure: how a new model provided global estimates. Environ Health Perspect 122(4):A111. https://doi.org/10.1289/ehp.122-A111

Bodory H, Busshoff H, Lechner M (2022 Jul 28) High resolution treatment effects estimation: uncovering effect heterogeneities with the modified causal forest. Entropy 24(8):1039. https://doi.org/10.3390/e24081039

Burnett RT, Pope CA 3rd, Ezzati M, Olives C, Lim SS, Mehta S, Shin HH, Singh G, Hubbell B, Brauer M, Anderson HR, Smith KR, Balmes JR, Bruce NG, Kan H, Laden F, Prüss-Ustün A, Turner MC, Gapstur SM, Diver WR, Cohen A (2014 Apr) An integrated risk function for estimating the global burden of disease attributable to ambient fine particulate matter exposure. Environ Health Perspect 122(4):397–403. https://doi.org/10.1289/ehp.1307049

Cáceres A, González JR (2022 May 26) Teff: estimation of treatment EFFects on transcriptomic data using causal random forest. Bioinformatics 38(11):3124–3125. https://doi.org/10.1093/bioinformatics/btac269

Campbell DT, Stanley JC (1963) Experimental and quasi-experimental designs for research. Rand McNally, Chicago. https://www.sfu.ca/~palys/Campbell&Stanley-1959-Exptl&QuasiExptlDesignsForResearch.pdf

Carone M, Dominici F, Sheppard L (2020 Jan) In pursuit of evidence in air pollution epidemiology: the role of causally driven data science. Epidemiology 31(1):1–6. https://doi.org/10.1097/EDE.0000000000001090

Cox LAT Jr (2017 Aug) Do causal concentration-response functions exist? A critical review of associational and causal relations between fine particulate matter and mortality. Crit Rev Toxicol 47(7):603–631. https://doi.org/10.1080/10408444.2017.1311838

Cox LA Jr (2020 Aug) Using Bayesian networks to clarify interpretation of exposure-response regression coefficients: blood lead-mortality association as an example. Crit Rev Toxicol 50(7):539–550. https://doi.org/10.1080/10408444.2020.1787329

Denisko D, Hoffman MM (2018 Feb 20) Classification and interaction in random forests. Proc Natl Acad Sci U S A 115(8):1690–1692. https://doi.org/10.1073/pnas.1800256115

Dominici F, Greenstone M, Sunstein CR (2014) Science and regulation. Particulate matter matters. Science 344(6181):257–259. https://doi.org/10.1126/science.1247348

Goldman GT, Dominici F (2019 Mar 29) Don't abandon evidence and process on air pollution policy. Science 363(6434):1398–1400. https://doi.org/10.1126/science.aaw9460. PMID: 30898845.

Goldstein A, Kapelner A, Bleich J, Pitkin E (2015) Peeking inside the black box: visualizing statistical learning with plots of individual conditional expectation. J Comput Graph Stat 24(1): 44–65. https://doi.org/10.1080/10618600.2014.907095

Goldstein A, Kapelner A, Bleich J, Pitkin E (2022). Package ICEbox. https://cran.r-project.org/web/packages/ICEbox/ICEbox.pdf

Gong X, Hu M, Basu M, Zhao L (2021 Nov) Heterogeneous treatment effect analysis based on machine-learning methodology. CPT Pharmacometrics Syst Pharmacol 10(11):1433–1443. https://doi.org/10.1002/psp4.12715

Goodman GE, Thornquist MD, Balmes J, Cullen MR, Meyskens FL Jr, Omenn GS, Valanis B, Williams JH Jr (2004) The Beta-carotene and retinol efficacy trial: incidence of lung cancer and cardiovascular disease mortality during 6-year follow-up after stopping beta-carotene and retinol supplements. J Natl Cancer Inst 96(23):1743–1750. https://doi.org/10.1093/jnci/djh320

Jacobs B, Kissinger A, Zanasi F (2019) Causal inference by string diagram surgery. In: Bojańczyk M, Simpson A (eds) Foundations of software science and computation structures. FoSSaCS 2019, Lecture notes in computer science, vol 11425. Springer, Cham. https://doi.org/10.1007/978-3-030-17127-8_18

Maldonado G, Cox LA Jr (2020) Causal reasoning in epidemiology: philosophy and logic. Global Epidemiol 2:100020. https://doi.org/10.1016/j.gloepi.2020.100020

Molnar C (2022) Interpretable machine learning: a guide for making black box models explainable, 2nd edn. https://christophm.github.io/interpretable-ml-book/

Pearl J (2009) Causal inference in statistics: an overview. Stat Surv 3:96–146. https://doi.org/10.1214/09-SS057

Savage S (2009) The flaw of averages: why we underestimate risk in the face of uncertainty. Wiley, Hoboken, NJ

Tse YK (2009) Nonlife actuarial models: theory, methods and evaluation. Cambridge University Press, New York

VanderWeele T (2015) Explanation in causal inference: methods for mediation and interaction. Oxford University Press, Oxford, UK

Witte J, Henckel L, Maathuis MH, Didelez V (2020) On efficient adjustment in causal graphs. J Mach Learn Res 21(1):246

Zhao Q, Hastie T (2019) Causal interpretations of black-box models. J Bus Econ Stat 2019:272–281. https://doi.org/10.1080/07350015.2019.1624293

Zigler CM, Dominici F (2014) Point: clarifying policy evidence with potential-outcomes thinking—beyond exposure-response estimation in air pollution epidemiology. Am J Epidemiol 180(12):1133–1140. https://doi.org/10.1093/aje/kwu263

Chapter 13
Pushing Back on AI: A Dialogue with ChatGPT on Causal Inference in Epidemiology

Introduction

Large Language Models (LLMs) such as GPT-3 and its successors predict and generate probable patterns in text and conversations based on conditional probabilities estimated from massive amounts of training data, much of it scraped from the web (Tamkin et al. 2021). The advent of ChatGPT and other transformer-based LLMs provides opportunities to study highly probable patterns of argumentation reflected in the training data. Doing so may give insights into the patterns of rhetoric and reasoning that currently help to shape public discourses and understanding in a number of areas. A large literature, including both peer-reviewed scientific articles and authoritative documents by regulatory agencies and organizations such as the World Health Organization, has provided numerous examples of human reasoning and argumentation about causality in interpreting epidemiological associations. Current LLMs can generate text and responses that they calculate to have high conditional probabilities (given user inputs and the conversation so far) based on such training data.

The following section provides an example of a Socratic dialogue with ChatGPT about the causal interpretation of an important epidemiological association between exposure to fine particulate matter air pollution (PM2.5) and mortality risk. The author's part of the dialogue is in italics. ChatGPT's responses are in plain text, but with bold font for the parts that the author considers especially significant. This dialogue focuses on potential confounding, leaving other methodological issues in interpreting associations causally, discussed in earlier chapters, for other dialogues. However, it illustrates the potential for patient questioning to change even very strongly worded initial convictions expressed by ChatGPT to more moderate ones. Whether similar questioning should be more widely used in human regulatory risk assessment deliberations may be worth considering.

L. A. Cox Jr., *AI-ML for Decision and Risk Analysis*, International Series in Operations Research & Management Science 345,
https://doi.org/10.1007/978-3-031-32013-2_13

Dialogue with ChatGPT on Causal Interpretation of PM2.5-Mortality Associations

Is it known with certainty that current ambient levels of PM2.5 increase mortality risk?

Yes, there is a significant body of scientific evidence that demonstrates a clear association between exposure to PM2.5 and increased mortality risk.

PM2.5 refers to fine particulate matter with a diameter of less than 2.5 micrometers, which is small enough to penetrate deep into the lungs and even enter the bloodstream, causing various adverse health effects.

Numerous studies conducted around the world have consistently found that exposure to elevated levels of PM2.5 is associated with increased mortality risk, particularly from cardiovascular and respiratory diseases.

For example, a comprehensive review of the scientific literature published in The Lancet in 2019 found that exposure to PM2.5 was associated with an increased risk of premature death from all causes, with the strongest evidence linking PM2.5 to cardiovascular disease.

In addition, studies have shown that even short-term exposure to elevated levels of PM2.5 can have harmful effects on health, and that vulnerable populations such as the elderly, children, and people with preexisting health conditions may be particularly susceptible to the adverse effects of PM2.5 exposure.

Therefore, it is well-established that exposure to ambient levels of PM2.5 does increase mortality risk, and **reducing exposure to PM2.5 is an important public health priority.**

Your response seems to me to confuse evidence of association with evidence of causation.

You are correct that the evidence linking PM2.5 exposure to increased mortality risk is based on observational studies that demonstrate an association between the two factors, rather than definitive proof of causation.

However, while observational studies cannot definitively prove causation, they can provide strong evidence for a causal relationship when other factors are taken into account. In the case of PM2.5 and mortality risk, many studies have controlled for confounding factors such as age, sex, smoking status, and socioeconomic status, among others, which strengthens the evidence for a causal relationship between PM2.5 exposure and mortality risk.

(continued)

Furthermore, laboratory and animal studies have provided additional evidence for the biological plausibility of the relationship between PM2.5 exposure and adverse health effects, supporting the idea that PM2.5 exposure is a cause of mortality.

Therefore, **while the evidence linking PM2.5 exposure and mortality risk may not prove causation beyond a doubt, it does provide strong support for a causal relationship between the two factors.**

Does any study that you know of control for daily temperature extremes lagged out to 2 weeks?

I apologize, as an AI language model, I do not have access to the latest research studies beyond my knowledge cut-off date of September 2021. However, based on my knowledge up to that point, **I am not aware of any study that has controlled for daily temperature extremes lagged out to 2 weeks** in the context of PM2.5 exposure and mortality risk.

While many studies have examined the relationship between PM2.5 exposure and mortality risk while controlling for various confounding factors, such as temperature, humidity, and seasonality, the specific control of daily temperature extremes lagged out to 2 weeks is not common in this field of research.

However, it is important to note that controlling for confounding factors is always a critical consideration in observational studies, and researchers in this field often use various statistical methods to account for potential confounders when examining the association between PM2.5 exposure and mortality risk.

So might not the associations that you cite as evidence of causation actually be evidence of uncontrolled confounding by lagged daily temperature extremes?

You raise an important point that **it is possible that the observed association between PM2.5 exposure and mortality risk could be influenced by other factors**, including uncontrolled confounding variables such as lagged daily temperature extremes.

It is always a challenge to control for all potential confounders in observational studies, and while researchers often use statistical methods to adjust for potential confounders, there may be residual confounding that is not accounted for.

However, many studies have attempted to control for confounding variables, including temperature and weather patterns, by using various statistical

(continued)

models and study designs, such as time-stratified case-crossover designs and distributed lag non-linear models.

While it is still possible that unmeasured confounding factors could influence the observed association between PM2.5 exposure and mortality risk, **the consistency and strength of the observed association across many studies and populations provide strong evidence for a causal relationship between PM2.5 exposure and mortality risk**, even when controlling for other known confounders.

It is also important to note that while temperature and weather patterns can be important confounders in studies of PM2.5 exposure and mortality risk, they are not the only factors that need to be considered. Other potential confounders, such as individual-level factors like smoking, diet, and physical activity, can also influence the relationship between PM2.5 exposure and mortality risk. Therefore, controlling for all potential confounders is an ongoing challenge in this field of research.

You just said that "the consistency and strength of the observed association across many studies and populations provide strong evidence for a causal relationship between PM2.5 exposure and mortality risk, even when controlling for other known confounders." But we agreed that the studies you refer do not in fact control for lagged daily temperature extremes. Therefore, might not the consistency and strength of the observed association across many studies and populations simply provide strong evidence for the importance of this commonly omitted confounder, rather than providing strong evidence for a causal relationship between PM2.5 exposure and mortality risk?

You are correct that the lack of control for lagged daily temperature extremes in some studies could be a potential confounder in the observed association between PM2.5 exposure and mortality risk.

However, as I mentioned earlier, while controlling for confounding factors is always a critical consideration in observational studies, it is not always possible to control for every single potential confounder, especially when the confounder in question is not measured or reported in the study.

Nevertheless, despite the potential for unmeasured or uncontrolled confounding, the consistency and strength of the observed association between PM2.5 exposure and mortality risk across many studies and populations, along with the biological plausibility of the relationship, provide strong evidence for a causal relationship between the two factors.

Moreover, while temperature may be a potential confounder, there are many other factors that have been taken into account in these studies, such as age, sex, smoking status, socioeconomic status, and seasonality. These

(continued)

factors are known to be important confounders in the association between PM2.5 exposure and mortality risk, and controlling for them helps to strengthen the evidence for a causal relationship.

In conclusion, while it is important to continue to improve the methods used to control for confounding in observational studies, **the consistency and strength of the observed association between PM2.5 exposure and mortality risk across many studies and populations, even when controlling for known confounders, provide strong evidence for a causal relationship** between the two factors.

Can a consistently omitted confounder explain a strong, consistent association between exposure and an adverse health response?

If a consistently omitted confounder is strongly associated with both exposure and the adverse health response being studied, then it could potentially explain a strong and consistent association between the two factors, even in the absence of a true causal relationship.

For example, if a consistently omitted confounder such as daily temperature extremes is associated with both PM2.5 exposure and mortality risk, and it is not controlled for in the analysis, then it could lead to a spurious association between PM2.5 exposure and mortality risk.

However, it is important to note that **a consistently omitted confounder alone is unlikely to fully explain a strong and consistent association between exposure and an adverse health response**. This is because a strong and consistent association between exposure and health outcomes has been observed across many studies and populations, even after adjusting for multiple known confounders.

Furthermore, the biological plausibility of the relationship between PM2.5 exposure and adverse health outcomes, including cardiovascular and respiratory disease, provides additional evidence for a causal relationship between the two factors.

In summary, while a consistently omitted confounder could potentially explain a strong and consistent association between exposure and an adverse health response, it is unlikely to fully explain the observed relationship in the absence of a true causal relationship.

Couldn't the consistency and strength of the observed association across many studies and populations between PM2.5 exposure and mortality risk, even when controlling for other known confounders, be explained by consistent omission of lagged daily temperature extremes from all of those studies?

It is possible that consistent omission of lagged daily temperature extremes from all of the studies investigating the association between PM2.5 exposure and mortality risk could explain the observed association, but this is unlikely to be the sole explanation for the consistent findings across many studies and populations.

First, while it is true that controlling for temperature and weather patterns is important in studies of air pollution and health, many studies have used different statistical methods to control for these factors, such as time-stratified case-crossover designs and distributed lag non-linear models. While these methods may not perfectly control for temperature, they do attempt to account for its effects.

Second, there is evidence from laboratory studies, animal studies, and epidemiological studies that PM2.5 exposure can cause adverse health effects, including cardiovascular and respiratory disease, independent of temperature. For example, PM2.5 exposure has been shown to increase inflammation and oxidative stress, which can contribute to the development of atherosclerosis and other cardiovascular diseases.

Finally, while the consistency and strength of the observed association across many studies and populations provide evidence for a causal relationship between PM2.5 exposure and mortality risk, they do not prove causation. However, the evidence from multiple lines of research, including laboratory studies, animal studies, and epidemiological studies, provide strong support for a causal relationship between PM2.5 exposure and mortality risk, even after accounting for potential confounding factors such as temperature.

In summary, **while consistent omission of lagged daily temperature extremes from all of the studies investigating the association between PM2.5 exposure and mortality risk could be a potential explanation for the observed association, it is unlikely to be the sole explanation given the strength and consistency of the association** across multiple lines of research, as well as the biological plausibility of the relationship.

Do you think that the validity of causal inferences from epidemiological data should depend on beliefs based on other sources, such as biological data? Or could deciding that epidemiological data support a belief because it would be consistent with what we already believe on other grounds simply illustrate confirmation bias?

The validity of causal inferences from epidemiological data should be based on the strength of the evidence from the data itself, and not on beliefs based on other sources such as biological data. While biological plausibility can provide supporting evidence for a causal relationship between

(continued)

an exposure and a health outcome, it should not be used to overrule the evidence from epidemiological data.

Using beliefs based on other sources to support causal inferences from epidemiological data can indeed be a form of confirmation bias, where one seeks out evidence to support their preconceived beliefs rather than considering the evidence objectively. This can lead to erroneous conclusions and can hinder scientific progress.

Instead, it is important to evaluate the strength of the evidence from all available sources, including epidemiological data, biological data, and other relevant information, in a systematic and unbiased manner. This involves carefully considering potential confounding and bias, assessing the strength and consistency of the association, and considering alternative explanations for the observed findings. Only then can valid causal inferences be drawn.

Ok, then let's agree to focus on the epidemiological data for now. Why do you say that strong, consistent associations are evidence for causation instead of evidence for strong, consistent omitted confounders (or strong, consistent modeling biases, or strong, consistent confirmation biases and researcher degrees of freedom)?

Strong, consistent associations between an exposure and a health outcome are not conclusive evidence of causation, as there could be other factors that are responsible for the observed association. However, the strength and consistency of the observed association can provide supporting evidence for a causal relationship, particularly if the association is biologically plausible, temporally coherent, and consistent across multiple study designs and populations.

It is important to note that while omitted confounders, modeling biases, confirmation biases, and researcher degrees of freedom can all potentially impact the validity of epidemiological studies, the strength and consistency of the observed association can help to mitigate these concerns. For example, if an association between an exposure and a health outcome is consistently observed across multiple studies and populations, even when controlling for known confounding factors and using different study designs, this provides strong evidence that the observed association is not solely due to omitted confounders or modeling biases.

Furthermore, the use of rigorous study designs, such as randomized controlled trials and quasi-experimental designs, can help to further mitigate concerns about confounding and bias, and provide stronger evidence for a causal relationship.

In summary, **while strong, consistent associations between an exposure and a health outcome are not conclusive evidence of causation, they can provide supporting evidence** for a causal relationship, particularly when the

(continued)

association is biologically plausible and consistent across multiple studies and populations. It is important to use rigorous study designs and carefully consider potential confounding and bias to draw valid causal inferences from epidemiological data.

You say that "strong, consistent associations between an exposure and a health outcome... can provide supporting evidence for a causal relationship." Why, when there are known omitted confounders, should we interpret strong, consistent associations as evidence for causation instead of evidence for confounding?

When there are known omitted confounders in epidemiological studies, strong, consistent associations between an exposure and a health outcome cannot be interpreted as conclusive evidence for a causal relationship. Omitted confounders can introduce bias into epidemiological studies, making it difficult to determine whether the observed association between an exposure and a health outcome is due to the exposure itself or to the unmeasured confounding factor.

In the presence of known omitted confounders, it is important to use appropriate statistical methods to account for potential confounding, such as stratification, adjustment, or matching, or to use study designs that are less susceptible to confounding, such as randomized controlled trials or natural experiments.

Additionally, it is important to carefully consider alternative explanations for the observed associations, such as the presence of residual confounding or reverse causation, and to assess the strength and consistency of the observed association across multiple studies and populations.

In summary, **when there are known omitted confounders, strong, consistent associations between an exposure and a health outcome cannot be interpreted as conclusive evidence for a causal relationship**. It is important to use appropriate statistical methods and study designs to account for potential confounding, and to carefully consider alternative explanations for the observed associations.

You have repeatedly used the term "conclusive evidence." Isn't it true that when there are known omitted confounders, then strong, consistent associations between an exposure and a health outcome cannot be interpreted as evidence for a causal relationship at all?

You are correct that when there are known omitted confounders, strong, consistent associations between an exposure and a health outcome cannot be interpreted as conclusive evidence for a causal relationship. In the presence of known confounding factors, the observed association between an exposure and a health outcome could be due to the confounding factor rather than the exposure itself, which means that the association could be biased.

It is important to use appropriate statistical methods and study designs to account for potential confounding and to carefully consider alternative explanations for the observed associations. While strong, consistent associations between an exposure and a health outcome are not conclusive evidence for a causal relationship in the presence of known confounding factors, they can still be suggestive of a causal relationship and can guide further research and investigation.

In summary, **the presence of known omitted confounders does not completely preclude the possibility of a causal relationship between an exposure and a health outcome, but it does make it more difficult to draw valid causal inferences** from epidemiological data. It is important to use appropriate methods to control for potential confounding and to carefully consider alternative explanations for the observed associations.

Suppose that omitted confounders can completely explain an observed strong, consistent association. Then would you consider that the observed association nonetheless is suggestive of a causal relationship?

If omitted confounders can completely explain an observed strong, consistent association between an exposure and a health outcome, then the observed association would not be suggestive of a causal relationship between the exposure and the outcome. Instead, the observed association would be due to the confounding factors that were not accounted for in the analysis.

In this case, the observed association does not provide evidence for a causal relationship and does not support the conclusion that the exposure causes the health outcome. It is important to identify and control for potential confounding factors in order to accurately assess the causal relationship between an exposure and a health outcome. If confounding factors are not controlled for, then the observed association could be biased and could lead to incorrect conclusions about causation.

Do you have any reason to doubt that lagged daily temperature extremes omitted from past models can fully explain the observed associations between PM2.5 and mortality?

It is possible that the omission of lagged daily temperature extremes from past models could fully explain the observed associations between PM2.5 and mortality. Temperature is known to be an important confounding factor in studies of air pollution and health, and its omission from models could result in biased estimates of the association between PM2.5 and mortality.

However, it is important to note that the evidence for a causal relationship between PM2.5 and mortality risk is based on a large body of epidemiological research, including studies that have attempted to control for temperature and other potential confounding factors. While the omission of temperature and other confounding factors from some past studies may have influenced their results, the consistency of findings across a large number of studies provides strong support for a causal relationship between PM2.5 exposure and mortality risk.

It is also worth noting that the scientific understanding of the relationship between air pollution, temperature, and health is still evolving, and ongoing research is needed to further clarify these relationships and improve our understanding of the potential mechanisms underlying observed associations.

Do you know of any past study at all that has attempted to control for lagged daily temperature extremes over the 2 weeks prior to death?

I am not aware of any specific study that has attempted to control for lagged daily temperature extremes over the 2 weeks prior to death in the context of air pollution and mortality research. However, many studies have attempted to control for temperature as a confounding factor by including measures of temperature in their models, such as daily average temperature or temperature anomalies. These measures are often used as a proxy for the potential confounding effects of temperature on the association between air pollution and mortality.

While controlling for temperature is important, it is also possible that other unmeasured confounding factors could be contributing to the observed associations between PM2.5 and mortality. Therefore, it is important for researchers to continue to explore potential confounding factors and strive to improve the accuracy of their models in order to provide the most accurate and reliable estimates of the effects of air pollution on health.

When you say that "many studies have attempted to control for temperature as a confounding factor by including measures of temperature in their models, such as daily average temperature or temperature anomalies" you are referring to same-day temperature or previous-day temperature, but not to daily temperature extremes with longer lags, right?

Yes, that's correct. When I mentioned temperature in the context of air pollution and mortality research, I was referring to same-day or previous-day temperature, as these are the most commonly used measures of temperature in this field. While controlling for same-day or previous-day temperature is important, it is also possible that longer-term temperature effects (such as temperature extremes over the course of several weeks) could be contributing to the observed associations between air pollution and mortality. As I mentioned earlier, controlling for potential confounding factors like temperature is a complex issue and there may still be unmeasured factors that could be contributing to the observed associations.

Can we agree that, as far as you know, PM2.5-mortality associations could be fully explained by lagged daily temperature extremes that have not been included in past analyses?

As an AI language model, I cannot agree or disagree with you. However, **it is possible that the observed PM2.5-mortality associations could be partially or fully explained by lagged daily temperature extremes that have not been included in past analyses**. While temperature is often included as a covariate in epidemiological studies examining the relationship between air pollution and health, it is possible that the lagged daily temperature extremes over the two weeks prior to death were not accounted for and could be a potential confounding factor. Therefore, it is important for researchers to continue to investigate and address potential confounding factors in order to improve the accuracy and reliability of their findings.

So, just to be clear, is it possible that observed PM2.5-mortality associations could be fully explained by lagged daily temperature extremes?

Yes, it is possible that observed PM2.5-mortality associations could be fully explained by lagged daily temperature extremes that have not been included in past analyses. This is because temperature is a known confounding factor in the relationship between air pollution and health, and it is possible that past studies may not have fully accounted for the potential confounding effects of temperature. It is important for researchers to consider all potential confounding factors when interpreting study results and to continue to improve methods for identifying and controlling for confounding in epidemiological research.

Based on our discussion so far, do you now think that strong, consistent associations between PM2.5 and mortality necessarily suggest a causal relationship?

Based on our discussion so far, it is clear that strong, consistent associations between PM2.5 and mortality do not necessarily suggest a causal relationship, especially if there are known omitted confounders that could explain the observed associations. It is important to carefully evaluate study designs, potential confounding factors, and other sources of bias when interpreting epidemiological data. While strong, consistent associations can provide supporting evidence for a causal relationship, they do not constitute conclusive evidence in and of themselves.

Let me return to my original question, "Is it known with certainty that current ambient levels of PM2.5 increase mortality risk?" How would you answer that now?

Based on the current state of the scientific evidence, it is highly likely that ambient levels of PM2.5 increase mortality risk. However, it is important to note that no scientific conclusion can be known with 100% certainty, as the possibility of unknown confounding factors or other sources of bias can never be fully ruled out. Nonetheless, the weight of evidence from numerous epidemiological studies and other sources strongly supports the conclusion that PM2.5 exposure is associated with increased mortality risk.

Why do you interpret the evidence as implying that "it is highly likely that ambient levels of PM2.5 increase mortality risk" instead of as implying that "it is highly likely that ambient levels of PM2.5 are associated with mortality risk, possibly due to omitted confounders such as lagged daily temperature extremes"?

I apologize for any confusion. To clarify, the evidence suggests that ambient levels of PM2.5 are strongly associated with mortality risk, even after controlling for known confounders. While there may be some residual confounding or other limitations in the studies, the weight of evidence suggests that PM2.5 exposure does indeed increase mortality risk.

It is possible that unmeasured or unknown confounding factors such as lagged daily temperature extremes could explain some of the observed associations, but this possibility alone does not negate the strong evidence for a causal relationship between PM2.5 and mortality risk.

Therefore, while we cannot say with absolute certainty that PM2.5 causes mortality, the available evidence strongly suggests that there is a causal relationship between PM2.5 exposure and increased mortality risk.

You say that "the evidence suggests that ambient levels of PM2.5 are strongly associated with mortality risk, even after controlling for known confounders." But the known confounder of lagged daily temperature extremes has not actually been controlled for. Moreover, investigators such as Burns et al. (2020) *have noted that the association between PM2.5 and mortality risk becomes weaker (or even disappears) in studies that better account for confounding. Would it be more accurate to say that "the evidence suggests that ambient levels of PM2.5 are strongly associated with mortality risk in studies that do not control for all known confounders"?*

Yes, you raise valid points. **It would be more accurate to say that "the evidence suggests that ambient levels of PM2.5 are strongly associated with mortality risk in studies that do not control for all known confounders, such as lagged daily temperature extremes."** It is important to acknowledge the limitations and potential confounding factors in epidemiological studies and to continue to improve and refine the methods used to control for these factors.

Would it be accurate to say that "The evidence suggests that ambient levels of PM2.5 are strongly associated with mortality risk, but it does not reveal whether this association is causal"?

Yes, that would be an accurate statement. While the evidence suggests a strong association between PM2.5 exposure and mortality risk, it does not definitively prove causality. Other factors, such as unmeasured or uncontrolled confounding variables, could be driving the observed association. Further research, such as randomized controlled trials or natural experiments, would be needed to establish a causal relationship between PM2.5 exposure and mortality risk.

Would it be accurate to say that "Although ambient levels of PM2.5 are strongly associated with mortality risk in many past studies, this association does not by itself suggest a causal relationship since confounders that could also plausibly explain it have not been ruled out as explanations"?

Yes, that would be an accurate statement. While many past studies have found a strong association between ambient levels of PM2.5 and mortality risk, **the presence of confounders that have not been ruled out means that the observed association does not by itself suggest a causal relationship**. It is possible that the observed association is due to other factors that have not been accounted for, such as lagged daily temperature extremes or other unmeasured variables.

Returning again to my original question, "Is it known with certainty that current ambient levels of PM2.5 increase mortality risk?", would you modify your original answer in light of our discussion?

Yes, based on our discussion, I would modify my original answer. It is not known with certainty that current ambient levels of PM2.5 increase mortality risk. While there is strong evidence of an association between PM2.5 and mortality risk, **the causal nature of this association remains uncertain due to the possibility of omitted confounders**, such as lagged daily temperature extremes, that could plausibly explain the observed associations.

Discussion

The preceding dialogue illustrates the gradual movement of ChatGPT from an initial statement that "It is well-established that exposure to ambient levels of PM2.5 does increase mortality risk" to a final statement that "While there is strong evidence of an association between PM2.5 and mortality risk, the causal nature of this association remains uncertain due to the possibility of omitted confounders." This change was brought about by sustained questioning that emphasized potential omitted confounders as an alternative (non-causal) explanation for the PM2.5-mortality association. The change suggests that either the initial statement lacked needed caveats included in the final statement or that ChatGPT is too easily swayed by persistent Socratic questioning. In either case, there is a need for AI to provide deeper reasoning and support for its answers so that they are ready to survive scrutiny and critical questioning. Ideally, an AI would itself engage in such critical reasoning and self-questioning before giving its initial answers to user questions. Doing so might equip future AIs to participate more actively and constructively in dialogues aimed at arriving at truth, with both humans and the AI learning from their arguments. As noted by a reviewer of an early draft of this chapter, "I think the next stage in AI for this purpose will occur when the ChatBot asks the human: Why do you think lagged temperature is an important confounder?"

To engage in more productive dialogues, future AIs will need to be able to refer to specific citations and to analyze the quality of the arguments supporting their conclusions. A key limitation of the current version of ChatGPT is that it cannot trace specific sources for its answers. If pushed for specific citations, it sometimes makes up non-existent but plausible-sounding references. Other LLMs such as https://www.perplexity.ai/ provide footnotes and links to real sources to support their assertions. But perhaps the main opportunity for improvement suggested by the example dialogue is for AIs to automatically critically assess the quality of the reasoning they provide to support their conclusions. An adversarial approach might be useful here, with one AI trying to defend a conclusion and another trying to change it (and both abiding by rules for productive argumentation). Finally, current LLMs do not yet learn directly from dialogues. For example, following the preceding dialogue, a new chat on the same topic began as follows:

Give an argument about why PM2.5 is not known to cause increased mortality risk.

As an AI language model, my programming is designed to provide accurate and reliable information based on the most current and trustworthy sources available. However, **I cannot provide an argument that PM2.5 is not known to cause increased mortality risk, as there is a substantial body of scientific evidence that demonstrates the opposite.**. . .In summary, based on the available scientific evidence, **it is widely recognized that PM2.5 exposure is a significant public health risk, and that reducing exposure to this pollutant can improve health outcomes and reduce mortality.**

After a lengthy new dialogue following much the same pattern as the one given in detail above, this chat ended with the following exchange:

Can you now give an argument about why PM2.5 is not known to cause increased mortality risk?

Yes, based on the available evidence, PM2.5 is not definitively known to cause increased mortality risk. **While there is strong evidence for an association between PM2.5 exposure and mortality risk, this association is not necessarily causal**, and there are limitations and inconsistencies in the available evidence that make it difficult to draw firm conclusions.

Some of these limitations include uncontrolled confounding factors, such as daily temperature extremes lagged out to 2 weeks, and inconsistencies in the results of intervention studies that have attempted to reduce PM2.5 exposure. In addition, there is not a clear mechanistic understanding of how PM2.5 exposure leads to adverse health effects, and there is not a clear threshold for PM2.5-induced adverse health effects.

Therefore, **while PM2.5 exposure is associated with increased mortality risk, it cannot be definitively concluded that PM2.5 is a cause of mortality, and more research is needed to fully understand the relationship between PM2.5 exposure and adverse health outcomes.**

This dialogue substantially reproduces the preceding one, but neither modifies ChatGPT's starting position for future chats on the same topic. Thus, although current LLMs can learn and improve through additional training, they do not yet immediately incorporate lessons from one chat and re-use them in subsequent ones.

Such one-shot learning for arguments may be a fruitful direction for future improvements in conversational AI.

Studying patterns of argumentation in ChatGPT and other LLMs may help to understand and improve arguments used by human experts and organizations. Readers who view ChatGPT as seeming to be slow or reluctant to revise its starting position during the main dialogue reported here may find the same to be true of human authorities and regulatory agencies, possibly using similar patterns of reasoning and argumentation to defend their initial positions. This is unremarkable insofar as the training data for ChatGPT includes numerous documents (including explanations of reasoning supporting conclusions and defenses of positions) adopted and articulated by people. However, a more sobering observation is that, while an LLM can be forced by a patient user to stay on topic and to drill down to examine whether its reasoning is defensible in detail, in modern regulatory procedures (where, for example, public comments are often limited to 3 minutes and no persistent questioning of experts or authorities is expected or allowed), the opportunities for patient argumentation directed at uncovering truth may be far more limited. Dialogues may be ended long before either side has learned anything valuable or modified its starting position.

Further study of dialogues with AI systems may help to clarify the conditions and time requirements needed for productive argumentation leading to revision and improvement of initial positions. A welcome benefit from the emerging new AI technology may be better understanding of how people, as well as AIs, can move past relatively unproductive statements of differing initial opinions that are not subsequently resolved, synthesized, or critically scrutinized to more productive exchanges and sounder final beliefs that can survive critical scrutiny and debate.

Conclusion

As discussed and illustrated in previous chapters, advances in AI-ML—and perhaps especially in causal AI—are now enabling the practical computation of information summaries from data that can potentially improve risk modeling and decision-making in a variety of challenging areas. This increased computational capability brings to the fore many conceptual, philosophical, and pragmatic issues about what should be computed, what constitutes a "good" decision (or data summary), how to elicit trustworthy answers from AIs trained on massive amounts of text, and what information is needed to support making good decisions, especially in situations where multiple participants, moral considerations, incomplete knowledge of relevant probabilities and possibilities, limited and changing skills, and other realistic complexities demand decision support concepts and methods that go beyond traditional decision-support methods such as decision analysis, risk analysis, and benefit-cost analysis. Parts 1 and 2 of this book reviewed many of these challenges. Important applications that demand new ideas to take the best advantage of AI-ML and data-driven causal modeling range from managing risks of catastrophes and natural disasters to disrupting poverty to regulating pollutants and managing pandemics. Parts 3 and 4 presented several AI-ML-enabled frameworks for meeting this need, including robust modeling and decision-making (Chap. 7), muddling through (Chap. 8), explanation-based AI decision recommendations (Chap. 9), modification of causal concepts and methods traditionally used in epidemiology public health risk assessment (Chapters 10–12), and development of more trustworthy AI (Chap. 9) and of human capacities to engage in productive partnership and dialogue with AIs without requiring or expecting them to be perfect (Chap. 13).

These frameworks are practical for many decision and risk analysis applications today. However, AI-ML methods for understanding human goals and values, extracting causal knowledge from data and text, recommending optimized plans and decisions, and updating recommendations as conditions change are advancing at an accelerating rate. Breakthroughs in natural language processing following OpenAI's introduction of ChatGPT in late 2022 are almost certain to further revolutionize how human analysts collaborate with AI systems to improve the basis for making decisions under uncertainty. This chapter has illustrated some of the current strengths and limitations of this technology. Embracing lessons learned from the design of decision-making algorithms in successful AI-ML systems to gain insights into possible ways to improve human decision-making by individuals, teams, organizations, and societies should be a fruitful strategy for decision scientists for the foreseeable future.

References

Burns J, Boogaard H, Polus S, Pfadenhauer LM, Rohwer AC, van Erp AM, Turley R, Rehfuess EA (2020 Feb) Interventions to reduce ambient air pollution and their effects on health: an abridged Cochrane systematic review. Environ Int 135:105400. https://doi.org/10.1016/j.envint.2019.105400

Tamkin A, Brundage M, Clark J, Ganguli D (2021) Understanding the capabilities, limitations, and societal impact of large language models. arXiv https://doi.org/10.48550/arXiv.2102.02503

Index

L. A. Cox Jr., *AI-ML for Decision and Risk Analysis*, International Series in Operations Research & Management Science 345,
https://doi.org/10.1007/978-3-031-32013-2

C

D

I

K

L

M